Thomas Bedürftig and Roman Murawski
Philosophy of Mathematics

Thomas Bedürftig and Roman Murawski

Philosophy of Mathematics

DE GRUYTER

Mathematics Subject Classification 2010
00A30, 03A05, 00A35, 01A05; secondary 03A10, 01Axx, 01A70, 01A72, 97E20

Authors
Prof. Dr. Thomas Bedürftig
Leibniz Universität Hannover
Institut für Didaktik der Mathematik und Physik
Welfengarten 1
30167 Hannover
Germany
beduerftig@idmp.uni-hannover.de

Prof. Dr. Roman Murawski
Adam Mickiewicz Universität
Fakultät für Mathematik und Informatik
ul. Umultowska 87
61614 Poznań
Poland
rmur@amu.edu.pl

ISBN 978-3-11-046830-4
e-ISBN (PDF) 978-3-11-046833-5
e-ISBN (EPUB) 978-3-11-047077-2

Library of Congress Control Number: 2018947914

Bibliographic information published by the Deutsche Nationalbibliothek
The Deutsche Nationalbibliothek lists this publication in the Deutsche Nationalbibliografie; detailed bibliographic data are available on the Internet at http://dnb.dnb.de.

Cover image: Tridland/iStock/Getty Images
Typesetting: Dimler & Albroscheit, Müncheberg
Printing and binding: CPI books GmbH, Leck

www.degruyter.com

With thanks to our wives Michaela and Hania
for patience and sympathy
in the time of our work on this book.

Contents

Preface

A human being must philosophize
whether he wants it to or not
– he is not free here.[1]

Ludwig Feuerbach

The present book is an introduction to the philosophy of mathematics. It represents a special approach to it – one that does not look philosophically at mathematics from the outside but rather attempts to ask philosophical questions from the standpoint of a mathematical (research and teaching) practice concerning fundamental concepts, constructions and methods. It looks for answers both in mathematics and in the philosophy (of mathematics) from their beginnings till today. We think that without the historical component the philosophy of mathematics would be empty. Moreover, the historical approach is important for we should consider (even current) mathematics as being in the process of permanent development.

Philosophy of mathematics is a field between philosophy and mathematics. This is in fact a difficult position. Those being outside mathematics will ask: what at all has mathematics – the firmly stated and infallible system of numbers, formulas and methods – in common with philosophy. On the other hand philosophy – to which the philosophy of mathematics belongs – and philosophers themselves have and had always problems with mathematics, the discipline forming in a sense a paradigm of the scientific thinking and developed to such a degree that it sometimes appears to be completely unmanageable and unclear. Finally, the subject of the philosophy of mathematics is just mathematics – and mathematicians are generally not inclined to consider it in a philosophical way. This is in some sense a rational attitude because all that mathematicians are doing seems to be far away from any philosophy.

We will try to show that the situation is different. Mathematics is a vivid scientific discipline improving itself and altering permanently inside a changeful history. Its basic concepts, methods and principles are traditionally important subjects of philosophy. Furthermore, if we look into the foundations of mathematics we realize that mathematical practice, teaching and learning of mathematics are close to philosophy.

The fundamentals on which mathematics is built permit and even need to be reflected. To do this mathematically is the task of *the foundations of mathematics*, set theory and logic. However, no one is dispensed by that from further reflection because the foundations of mathematics extend to the everyday base of numbers, to mathe-

1 Der Mensch muß philosophieren, ob er mag wollen oder nicht, hierin ist er nicht frei.

https://doi.org/10.1515/9783110468335-203

matical methods, to teaching and learning mathematics as well as to mathematical speaking and thinking. Many questions asked in the foundations of mathematics have philosophical origin and – vice versa – various answers given and results obtained there have philosophical meaning. In this new context new philosophical questions arise. The reflection on questions coming from mathematical practice, from the foundations of mathematics and from the philosophy leads to a deeper consciousness in mathematical research practice as well as in teaching and learning mathematics.

The authors of the present book are mathematicians interested in the philosophy and with a certain philosophical background. Our aim is to introduce mathematicians and teachers of mathematics as well as students of mathematics into the philosophy of mathematics. Our introduction is suitable also for professional philosophers as well as for students of philosophy, just because it approaches philosophy from the side of mathematics. The knowledge of mathematics needed to understand the book is elementary. Parts of the book where this is not sufficient are printed in small size. The same is also done when we speak about very special advanced results of mathematics and mathematical logic or about special items in philosophy. Those fragments of the book can be omitted by first reading without any serious loss. The text printed in normal size (almost all of the book!) can be in principle understood by any layman with school mathematics background.

The starting point as well as a reference point of our text are all the time real numbers. In Chapter 1 the way to them is sketched and some mathematical and philosophical problems and questions connected with them are indicated. Those questions lead to the foundations of mathematics and to the philosophy of mathematics. Chapter 2 provides a survey of various views, positions and tendencies that were formulated and developed from the very beginning of the philosophical reflection on mathematics till today. They form a sort of a background for considerations in Chapter 3 where questions and problems formulated in Chapter 1 are discussed. Chapter 2 can be also treated as an independent compendium of views and trends in the philosophy of mathematics. One can read the Introduction, Chapter 1 and Chapter 3 together and from time to time – when there appears such a demand – consult Chapter 2.

Chapter 4 is devoted to the universally accepted and applied background of the mathematical way of writing and speaking, i.e., to set theory. The usage of set-theoretical language influences the mathematical thinking. The aim of this chapter is to draw interest to that usually unconscious fundament of mathematical speaking and thinking. We shall consider two different approaches to sets and indicate some problems connected with them. Chapter 5 is devoted to the axiomatic method and to the second fundament of mathematics, namely to logic. A short survey of fundamental logical concepts as well as some information on the development of logic are provided. Mathematical logic has adopted many previously philosophical questions about mathematics and obtained results of philosophical meaning and significance.

In Chapter 6 we go out in search of infinitesimals, the infinitely small quantities – we spoke about them already in Section 3.3. They can be potentially found in school mathematics – it seems that thinking in terms of infinitesimals is not alien for pupils, however, it disappeared from mathematics. In further parts of Chapter 6 we consider elements of nonstandard analysis and discuss philosophical and mathematical problems and consequences connected with this way of thinking. In Chapter 7 we look back at what has been said in previous chapters. We attempt to characterize shortly the philosophy of mathematics introduced in the book and to state what is it for. We conclude thoughtfully. The Appendix contains short biographies of selected philosophers and mathematicians. At the end of the book one finds indices of symbols, names and concepts.

For simplicity we use in the book the masculine form (he/him/his). Whenever it appears it should be understood inclusively (and not exclusively) as referring both to her and him.

We would like to thank the editors of De Gruyter Verlag Berlin, in particular Nadja Schedensack and Apostolos Damialis for their always patient and helpful cooperation.

Hanover and Poznań
in April 2018

Thomas Bedürftig
Roman Murawski

Introduction

I am not a religious man,
but it's almost like being in touch with god
when you're thinking about mathematics.

Paul Halmos

The mathematical path of a human being begins early. First mathematics arises from contention and accord with the reality. Numbers – connected with counting and becoming natural numbers in mathematics – receive from that point their meaning. The same holds in a similar way also for negative numbers, fractions and rational numbers that arise from the contact with the everyday magnitudes by abstraction.

The situation with reals – considered in Chapter 1 at the very beginning of our book – is different. Here one has to do with an essentially new situation. To obtain reals starting from natural and rational numbers one uses in mathematics a completely new way. It is detached from ties with concrete applications. The old, simple abstraction from everyday and physical magnitudes does not work here any longer. It is just the conflict with magnitudes that causes the *theoretical* construction of the reals or the postulation of their desired properties in an axiomatic way. Some geometrical and theoretical needs challenge mathematics and demand appropriate concepts and methods that not so long ago were very new and revolutionary. Here arose and arise mathematical as well as philosophical questions. These questions point to various directions in the philosophy and history of mathematics. We indicate some of them in this introduction and mention some problems.

The reals are – and this is the common opinion today – a safe mathematical wealth. One has forgotten the heavy discussions – that still a hundred years ago affected the heads and hearts even consciences of mathematicians – or they are now treated as closed. But the problems did not disappear. One suspends them in the foundations of mathematics, omits problems and pragmatically runs over to the order of business that begins with $\mathbb{R}$. In teaching mathematics – where the main aim is to introduce as quickly as possible concepts and methods – one starts from the reals, possibly avoiding the conceptual and methodological questions and throwing away at this crucial point the possibility presenting an interesting material and deeper insight into the foundational elements of mathematics.

In Chapter 1 we consider a way to the reals that uncovers very concretely and in an elementary way problems that still today occupy the foundations as well as the philosophy of mathematics. After having presented in Chapter 2 various conceptions developed in the philosophy of mathematics, in Chapter 3 we will analyze the problems

https://doi.org/10.1515/9783110468335-001

that lead to a better understanding of the conflicts that caused quite a deep-rooted excitement at the time of Kronecker, Frege, Cantor and Dedekind.

Those conflicts are partially or completely unnoticed in teaching at universities and at schools. In the middle of teaching mathematics at German colleges the reals are usually introduced in such a way that hides the meaning of this step. Even for a well-disposed reader this step possibly had been unnoticed during his own time at school. Before considering the problems in details in Chapter 1, now we would like to indicate shortly some important points in teaching mathematics both at schools and universities. They concern some simple issues that usually are overlooked.[2]

Teaching mathematics at universities begins usually with real numbers. When it is done properly they are introduced in an axiomatic way, i.e., by listing their properties without questioning them. This can proceed as follows.

In a handbook for teachers about number domains [258, p. 159], in the part dealing with the introduction of reals, one finds the following: At the beginning of the chapter about the reals it is stated – after a remark on the diagonal of the unit square – a remark on the number line (see below) and an indirect proof of the irrationality:

> "Hence the number $\sqrt{2}$ well known to us does not belong to the set of rational numbers."

This reflects the usual practice of introducing reals in the process of teaching mathematics and is a bit surprising in a handbook. Where from and why is the number $\sqrt{2}$ known to us? Just before that the rational numbers were introduced. Where does the number $\sqrt{2}$ belong? It is suggested that it belongs to a special type of numbers, namely to reals that are already here in a way. Where – this will be seen soon.

Such an attitude in "introducing" reals is in some cases in fact the background of teaching mathematics. How should teachers who themselves were trained and taught in such a way be able to convey pupils what in fact takes place here, at what conceptual threshold they stand? A chance to understand – at least partially – the problem and the peculiarity of solving it will be squandered. It will remain unexplained what $\sqrt{2}$ as a *number* actually *is* and why one can *calculate* with the term $\sqrt{2}$ that has not appeared so far.

An important point appears afterwards in teaching: the procedure of approximating, e.g., to $\sqrt{2}$ – whatever $\sqrt{2}$ is. Then follows a phrase like:

> "To the finite and periodic decimal fractions one adds the infinite non-periodic decimal fractions. All together are the reals." ([258, p. 189].)

2 Notice that we refer here only to the standard theory of reals. Other interpretations are considered for example in the book [91] where also historical explanations can be found – we recommend it for further study of reals.

What are infinite non-periodic decimal fractions? This is in fact only a negative description, the denial of a great problem. It concerns the *handling of the infinite*. The issue the mathematicians of the 19th century struggled heavily and in fact undecidedly for in university and school teaching will be a precondition in a subordinate clause. Still today the infinite – being a base for various further questions – is an important problem of the foundations of mathematics. The complex of problems connected with the infinite will appear throughout the whole book.

Behind such a procedure there is the so-called number line – it is not distinguished whether one speaks here about points or numbers. It is in fact a customary and justified exercise in the process of teaching mathematics to *illustrate* numbers as points on a number line. But is it justified to *identify* numbers and points? The chapter about "introduction to the reals" of the above-mentioned handbook begins with the following explanation:

> "Hence the reals will be at the beginning explained by the totality of **all** points of the number line and seen as given." [Bold in the original.]

This is useful – and murderous. The whole complex of problems of the continuum in all facings is settled at one blow. But what is in fact *the* number line? What is "the totality of all points"? How is it constructed? Does one set the points or can one take them? What type of object is a point? Is the continuum of a line exhausted by points, hence a set or "the totality" of points? If this is the case: points are initially not numbers. Can points be declared to be numbers? What numbers are they? How does one calculate with points?

The problem of the completeness of the set of the real numbers $\mathbb{R}$ will be removed by this explanation. As a set of all points of a line, the real numbers seemingly have no gaps. In fact, today one proceeds reversely. The real numbers $\mathbb{R}$ are constructed or introduced axiomatically and then one declares copies of $\mathbb{R}$ to be lines. Hence points do not become numbers, but vice versa – numbers become points.

The aim of set-theoretical constructions of reals, whatever they are, is always the completeness. Here arises ultimately and clearly the question about the connection of the constructed domain of numbers and the geometrical line. Is there a *difference* between the *set* of reals and possible conditions in the *continuum* of the geometrical line? This problem is not recognizable any more when $\mathbb{R}$ is represented as the set of all points of a line or lines are treated as copies of $\mathbb{R}$. The quite possible difference refers to the so-called nonstandard analysis. In the background appears the old idea of *infinitely small* quantities. We come to this in Chapter 3 and in Chapter 6.

Which questions referring to philosophy do appear here? We have indicated them. They are philosophical fundamental questions: it is the old question about the infinity that concerns the construction of infinite sets as well as the acceptance of infinitely small quantities, the infinitesimals. It is the question about mathematical concepts such as number or magnitude. It is the problem of the classical continuum and of the contemporary conception of the continuum as a set of elements and the identification of points and numbers. It is the question about axioms and the axiomatic method. And it is generally the question about the relation of mathematics and its concepts with the reality. What is the status of mathematical concepts? What are numbers? What is their source?

Such questions – that are relevant for the teaching of mathematics – will be made more precise in Chapter 1 on the way to $\mathbb{R}$. In Chapter 2 we shall face them again and again in the extensive outline of the history of philosophy of mathematics. There we present numerous mathematicians and philosophers as well as positions in the philosophy of mathematics from which answers can be expected.

In Chapter 3 we consider the fundamental questions in the background of the history of philosophy and mathematics. It is the question what numbers actually are. It is the question about the concept of the infinity which appears in various positions presented in Chapter 2, about the concept of a magnitude, the continuum and the infinite small. Magnitudes and the infinitesimals disappeared from pure mathematics when the classical continuum has been replaced by $\mathbb{R}$. Only names of magnitudes remained here and bare denotations of the infinitesimals.

Set theory and logic form the mathematical domain of the foundations of mathematics. They are the source of the today's way of speaking and explaining notions of a proof, a consequence, a theory. A new method of securing and presenting the mathematical knowledge provides the renewed axiomatic method. Set theory, which is in fact a theory of the infinite, together with logic as well as axiomatics will be presented in Chapters 4 and 5 – their development will be described and their problems and results will be shown and interpreted. New foundational problems naturally arise where new foundations are.

Above we have emphasized some *problems* which are hidden behind the reals. One should necessarily notice which mathematical *possibilities* they opened and what progress they produced. The step from the classical continuum into the continuum of reals made in the second half of the 19th century was in fact revolutionary. Only the reals and the set theory in their background made possible to grasp mathematically properties of the continuum hidden in the visual continuum. To them belong the fundamental concepts of completeness, continuity or dimension. Eventually, the concepts of limit, differential and integral – being used for a long time but unsecured – could be made precise. All this was possible only thanks to abstracting from some problems mentioned above – all the time against a massive resistance, e.g., of intuitionism. Today we are in another position. The foundations being then new have been for a long time proved of value. We can maintain distance and realize today on what ground we stay – without disputes concerning the foundations and problems which are still here. We can admire the mathematical achievements and attempts to overcome the problems and at the same time enter upon the mathematical venture that lies behind them.

The indicated aspects will be considered in the following chapters. But first in Chapter 1 we come to some problems connected with the foundations of $\mathbb{R}$ that in the everyday practice seem to be forgotten a bit. Chapter 1 is short, concrete, elementary and sketchy. The indicated steps on the way to the reals are known of course. However, the abstracting from any previous knowledge, the special attention paid to every single step as well as the inexorable distinguishing between geometrical and arithmetical level and the clear naming of problems can seem to be a bit unfamiliar in the beginning.

1 On the way to the reals

Between the intuitional idea and the mathematical formulation which should describe the scientifically substantial elements of our intuition in precise terms, there will always remain a gap.[1]

Richard Courant

In this chapter we will try to trace the mathematical base of the reals and to present briefly their setting and mathematical foundations, in order to formulate problems of a philosophical, methodological and mathematical nature. We state, once again, as already indicated in the Introduction, the ignorance of those problems as well as the pragmatism in putting reals as the universal base of mathematics. One pragmatically carries on doing mathematics. The reals seem to be always there. They became quasi "natural" numbers for mathematicians. In this chapter, first of all, we want to call attention to questions lying in the background of the reals by observing carefully the way they are introduced and indicating details and problems appearing there. For the sake of clarity we choose a short and pointed formulation of the problems.

The way to the real numbers $\mathbb{R}$ begins – as almost everything in mathematics does – with (genuine) natural numbers $\mathbb{N}$. We choose in this chapter a subsequent starting point: the rational numbers $\mathbb{Q}$. It is the starting point of someone learning mathematics who does not know anything about the reals. So we put ourselves consciously in the position of a student learning mathematics – it is similar to a position of the Pythagoreans 2500 years ago. *All we have and know are rational numbers and nothing else.* It is a requirement in this chapter to abandon really completely all our previous knowledge. About the way from natural to rational numbers we write briefly in Chapter 3.

1.1 Irrationality

What is irrationality? Consider the standard example. One looks for a number whose square is 2. It is called $\sqrt{2}$. The following result is stated everywhere.

Theorem. $\sqrt{2}$ *is irrational.*

1 Zwischen der intuitiven Idee und der mathematischen Formulierung, welche die wissenschaftlich wichtigen Elemente unserer Intuition in präzisen Ausdrücken beschreiben soll, wird immer eine Lücke bleiben.

https://doi.org/10.1515/9783110468335-002

A standard indirect proof of this is then given. For completeness we also shall give a proof here, and we do this by the oldest way of arguing, which one finds in *Elements* by Euclid [109, Book X, Section 115a].

Proof. Suppose that $\sqrt{2}$ is rational, say $\sqrt{2} = \frac{m}{n}$, where m and n have no common divisor (i.e., they are relatively prime). Hence $2 = \frac{m^2}{n^2}$ and $2 \cdot n^2 = m^2$. So m^2 and consequently also m are even, e.g., $m = 2 \cdot a$. So n should be odd because by assumption m and n have no common divisor. On the other hand since $4 \cdot a^2 = m^2 = 2 \cdot n^2$, one gets $n^2 = 2 \cdot a^2$, so n^2 is even, and hence n is also even. But this is a contradiction! □

What does this theorem actually say? The suggestion, made even to experts, is: $\sqrt{2}$ is a number of another type, just an irrational number.

So what is our situation? There is nothing other than rational numbers. Hence "irrational" can mean only

$\sqrt{2}$ is not rational.

But "not rational" simply means, in the absence of other numbers, the following result.

Theorem. *$\sqrt{2}$ is not a number.*

This means that there is no number whose square is 2. So what is $\sqrt{2}$?

$\sqrt{2}$ is a term without any meaning.

One can however write this term to dramatize the question how it can have a sense. But at first it has no sense. So one tries to give $\sqrt{2}$ a different sense. Can it not be seen in the next standard example?

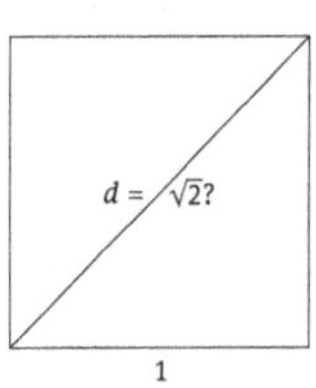

We argue thus: from the theorem of Pythagoras it follows that the square with the side being a diagonal of the unit square has a surface area equal to 2, hence the diagonal d has length $\sqrt{2}$. If we put d on the *number line* then $\sqrt{2}$ can be seen there as the length of d:

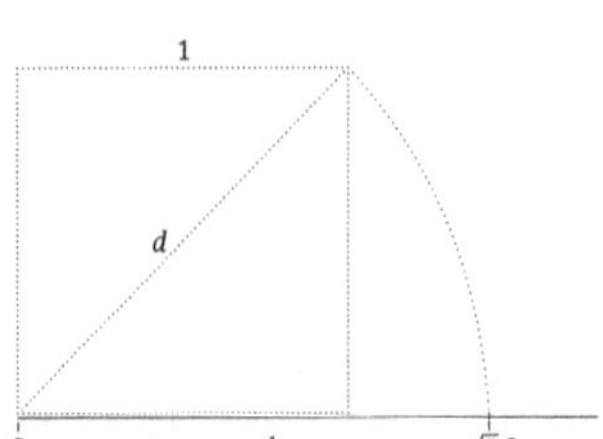

This is the next *suggestion* which misleads an insider and lecturer and seduces a learner to accept $\sqrt{2}$ as a number without further ado.

What error have we committed? We naively assume that every point on a line on which we have visualized rational numbers as points represents a number. For numbers lie there densely. But what numbers do we have? Rational numbers! And just to such numbers belongs a point on a number line. Since $\sqrt{2}$ is not rational, we have found no *number* $\sqrt{2}$ on the number line but only a *point* to which no number corresponds. This means that there is no number corresponding to the length of the diagonal d. So:

$\sqrt{2}$ is not a number. There is no measure of the length of d.

Hence, one infers the following result.

Theorem. *The diagonal d in the unit square is not measurable.*

This is an astonishing situation. The diagonal d has a length and our experience says that there is no problem about measuring lengths. So it must be accepted that our experiences deceive us. There are magnitudes to which – by a given unit – we cannot assign a number. We are unable to measure them. In fact, the unit and a magnitude can be *incommensurable*. The domain of magnitudes cannot be grasped by our numbers.

Let us turn back to the beginning where something goes wrong. One cannot start out on the way to the reals by stating "$\sqrt{2}$ is irrational". In fact, this pretends that $\sqrt{2}$ is a number. One should start in the following way.

Theorem. *There is no number whose square is 2. There is no measure for the diagonal of the unit square.*

To operate from the very beginning with the symbol $\sqrt{2}$ is essentially problematic because this suggests "number" and pretends that the solution of our problems is available. The profound difficulty of the problems and the fact that there is a long way to go to get to the reals is usually hidden. Teachers and lecturers are shy to admit complete failure: the arithmetic developed so far does not work. The fascinating and productive question is overlooked: what can be done mathematically?

Behind such seemingly only methodological and didactic problems there are fundamental philosophical questions.

1.2 Incommensurability

Of what type the problems are can be seen in the best way when one looks back in the history of mathematics. The discovery made by us was also made by the Pythagoreans probably about 450 BC – in another, most probably more direct way. Records are unclear at this point and sources are poor. One of the reconstructions of historians in connection with this says that the phenomenon of incommensurability was discovered on the example of a regular pentagon, the emblem of Pythagoreans as well as their symbol of the cosmos.[2]

The task was to determine the proportion of the side and the diagonal in a regular pentagon. The Pythagoreans developed a procedure to *determine the proportion* of segments that could be applied everywhere – their “reciprocal subtraction” – and that became the well-known *Euclidean algorithm* in the domain of natural numbers. This was an important procedure in the history of mathematics – it will play a role again in Chapters 2 and 3. Since it is rarely applied in a geometrical form we present it briefly and apply it to the regular pentagon.

Let a and b be two segments. The Greeks did not have standardized units of measure for segments with the help of which a and b could be measured, corresponding to fixed numbers, and in such a way their proportion determined. They proceeded as follows:

Remove from a the smaller segment b as many times as possible. One gets the remainder r_1. Next remove from b twice the remainder r_1. The remainder r_2 remains, etc.

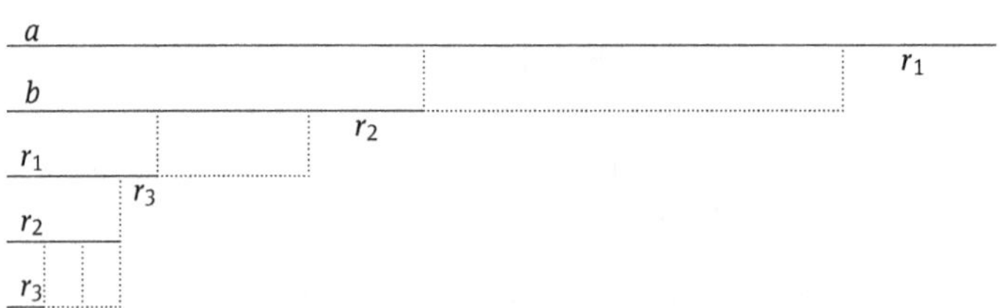

$$a = 2 \cdot b + r_1$$
$$b = 2 \cdot r_1 + r_2$$
$$r_1 = r_2 + r_3$$
$$r_2 = 3 \cdot r_3$$

In our example r_3 is contained in r_2 three times. Hence $r_1 = 4 \cdot r_3$, $b = 11 \cdot r_3$, $a = 26 \cdot r_3$ and the proportion of a to b is $26 : 11$. Now r_3 is a common measure for a and b. This was the aim of the reciprocal substraction: to determine the common measure for two given segments.

Let us transfer this procedure to the case of a regular pentagon shown in the picture below. We would like to determine the proportion of the diagonal d to the segment a. By symmetries in the regular pentagon the reciprocal substraction proceeds as follows:

$$\begin{array}{ccccc} d & = & a & + & d_1 \\ d_1 & = & a_1 & + & d_2 \\ d_2 & = & a_2 & + & d_3 \\ \vdots & \vdots & \vdots & \vdots & \vdots \end{array} \qquad \begin{array}{ccccc} a & = & d_1 & + & a_1 \\ a_1 & = & d_2 & + & a_2 \\ a_2 & = & d_3 & + & a_3 \\ \vdots & \vdots & \vdots & \vdots & \vdots \end{array}$$

What can one see? What did the ancient Greeks see?

2 Cf. the paper by H.-G. Bigalke [36]. Other authors – e.g., H. Boehme – reject the idea of this version (see [37, 39]) and establish an arithmetic-algebraic and indirect reasoning similar to the one indicated above.

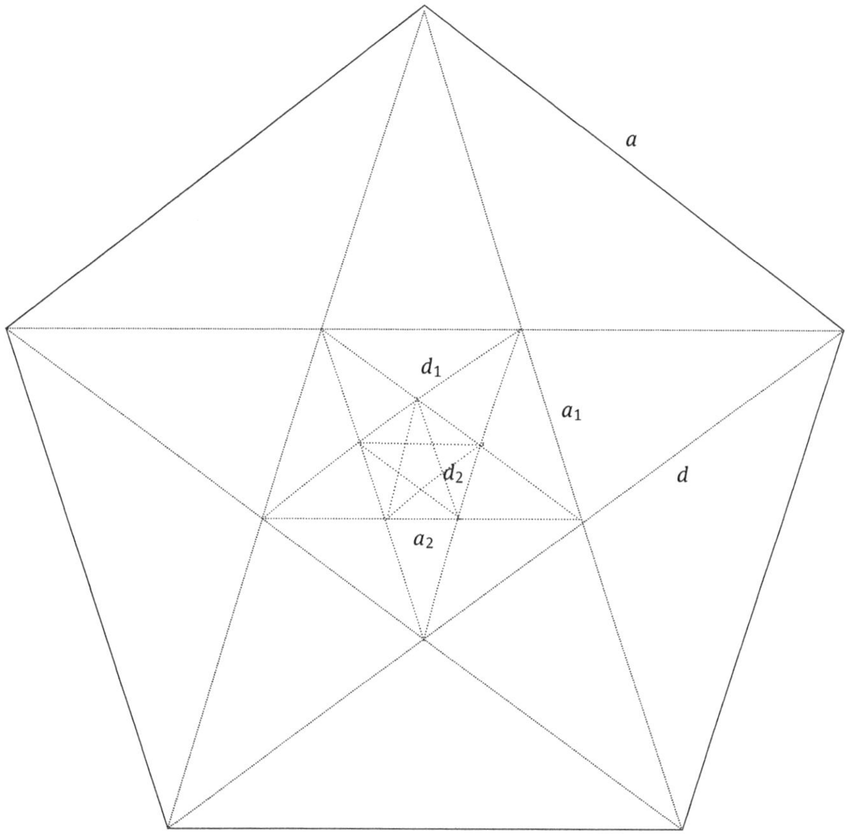

Fig. 1.1. The beginning of an infinite sequence of regular pentagons.

The procedure does not stop. It continues in a sequence of pentagons that will become smaller and smaller and that disappear at infinity. That is:

> There is no common measure for the diagonal and the segment of a regular pentagon.

The corollary is the following.

Theorem. *Diagonals and sides in a regular pentagon are incommensurable. There is no number expressing the proportion of the diagonal and a side.*

A “visible” and direct proof of the incommensurability such as the one just given is something other than an indirect reasoning for the “irrationality” of $\sqrt{2}$ given above. It provides a different experience for a learner that makes the incommensurability understandable and – much later – justifies the concept of irrationality. Also in the case of $\sqrt{2}$ there is possibly an in*sight* – aesthetically not so convincing – into the incommensurability of the diagonal and the side of a square.

In the above picture one can additionally see that, since $d_1 = d - a$, the inner and outer proportions of d and a coincide: $d : a = a : (d - a)$. Hence the proportion of d and a is the *golden ratio*. So:

There is no number expressing the proportion of the golden ratio.

The discovery of the incommensurability ca. 2450 years ago shocked the ancient Greeks. Why?

Numbers provided the foundation of mathematics and – this is noteworthy – the *metaphysics* of the Pythagoreans. For them numbers had real power that affected the material world from the higher world of numbers and formed the real things as well as their relations according to relations between numbers. This was a deep philosophical conviction of the Pythagoreans: *all is number*. And now under their eyes this conviction and their philosophy broke down. They had discovered line-segments whose relation could not be represented by natural numbers and in this way the central principle of their philosophy had been negated. Their philosophy broke in mathematics and metaphysics which before were a oneness.

Legends entwine the discovery of the incommensurability that in the fraternity of Pythagoreans was guarded as a secret and revealed by Hippasos of Metapont – in fact with dramatic consequences for the betrayer. He was cursed by Pythagoras. During his escape on the sea he was engulfed by waves stirred up by a storm.

Plato declared a hundred years later in clear words that it was a shame to know nothing about incommensurability:

> "And it seemed to me as if it were impossible for humans but rather for pigs, and I felt ashamed not only for myself but also for all Hellenes." ([276, Nomoi, Band 7, 819 d–e], translation by the authors.)

Similarly, for people beginning in mathematics today, the world breaks apart, just as it did for the Pythagoreans and for Plato. His mathematics, which had held good up to that point, now fails. But this experience can be made by someone learning mathematics only if he is given such an opportunity. Here belongs an arithmetical admission of complete failure as above. It cannot be made when one gives him $\sqrt{2}$ and the irrational numbers and proceeds as if they have always been there and everything could be measured. He needs this experience to be able to recognize the big step into a new and theoretical mathematics and to experience how mathematics invents "theoretical numbers" and *theoretically* solves and overcomes the problem of incommensurability.

Tutors who are lecturing are probably not always sensitive enough to all this. Recall the handbook mentioned in the Introduction [258]. In fact, one willingly forgets what we undertook above, namely really to take the point of view of a learner and to abstract from any previous knowledge. The reals are so close and so comfortable. Everything is measurable by them. And it is in fact hard to follow every step on the way to the reals. But the insight into the "abyss" of incommensurability and the next steps to $\mathbb{R}$ are important. One thing is clear: at the moment when the reals are there the incommensurability disappears. The irrationality remains.

Today – as it was back then – in connection with this certain philosophical questions arise. Numbers and number theory that for Pythagoreans were the *base* of philosophy detached themselves from the philosophy and now faced it. They became *subjects* of philosophical reflection. In Plato's philosophy ideas appeared in the place of numbers. What was then the status of numbers? Along with the mathematical problem there arose also a philosophical question concerning numbers:

- What are numbers?

First of all the simplest numbers, natural numbers are meant here – to them all other numbers including the later reals can be traced back. Just this question along with the question about reals led in the 19th century to the rise of the new mathematical discipline – the foundations of mathematics. It should explain *mathematically* what numbers are. We shall write about such attempts and about philosophical positions relating to them in Chapters 2 and 3.

The next question will be:

- What are reals?

This question is connected not only with mathematical problems but – as will be seen soon – also with philosophical ones. What in fact are those objects we have to do with every day? General questions from the philosophy of mathematics can arise:

- What is the status of mathematical concepts?
- What is the relation of mathematics and its concepts to reality?

The philosophical program of the Pythagoreans did not fail then in practice. The limits of numbers were indicated by geometrical magnitudes. The mathematical answer of the Greeks was plausible: they developed a theory about magnitudes that removed the theory about numbers and that so to say got priority now. This theory about magnitudes was, similarly to Greek geometry, axiomatic, i.e., it was founded on – not always explicitly given – basic statements describing the relations and the uses of magnitudes.

In today's axiomatic approach one excludes the question what in fact mathematical objects are. What in a mathematical and philosophical sense therefore remains is the question:

- What are, or better, what *were* magnitudes?

Today the old magnitudes are expatriated from pure mathematics. They have been replaced by numbers, by reals.

1.3 Calculating with $\sqrt{2}$?

Above the following question was formulated: what can be mathematically done with arithmetically empty hands? For example: how should one calculate with $\sqrt{2}$ when we do not know what $\sqrt{2}$ is and what its relation to the rational numbers is? The terms

$\sqrt{2}$, $\sqrt[3]{2}$, $\sqrt{3}$, $\sqrt{5}$ etc. are all without meaning. Then how should we understand

$$3 \cdot \sqrt{2}, \quad 2 + 3 \cdot \sqrt{2}, \quad \sqrt{2} \cdot \sqrt{3}$$

etc.? Even $\sqrt{2} \cdot \sqrt{2}$ is a term without meaning.

Nevertheless, one calculates without any doubt as if the reals were given: one adjoins, as is usually said, $\sqrt{2}$ to $\mathbb{Q}$ and proceeds in such a way as if the set of all terms $a + b \cdot \sqrt{2}$, that is,

$$\mathbb{Q}(\sqrt{2}) = \{a + b \cdot \sqrt{2} : a, b \in \mathbb{Q}\},$$

were a subset of $\mathbb{R}$ – where $\mathbb{R}$ and calculations on it are tacitly assumed. But this is not correct as it stands. At least some words about calculating with $\sqrt{2}$ and other terms as above should be necessarily said.

One should be clear in learning and teaching and at least the following should be said: *We proceed* as if one could calculate with $\sqrt{2}$, and calculate formally with $\sqrt{2}$ in such a way *as if* its square were equal 2. We grasp, e.g., $3 \cdot \sqrt{2}$ and $2 + 3 \cdot \sqrt{2}$ as *formal* terms but calculate with them as usual, i.e., as if the calculation rules working in $\mathbb{Q}$ held also for those formal terms.

It should be made clear that something new and *theoretical* takes place here, that one deals here with a *formal expansion, formal* adjunction of $\sqrt{2}$ to $\mathbb{Q}$. The elementary algebraic procedure of formal adjunction should be presented early in academic lecturing – and this can be done without mentioning $\sqrt{2}$.

1.4 Procedure of approximating, nesting of intervals and completeness

To repeat once again: we have *nothing except the rational numbers* $\mathbb{Q}$. We cannot speak about the *number* $\sqrt{2}$ even when we calculate with $\sqrt{2}$.

Even when one cannot measure d, the diagonal in a unit square, one can approximate its length by rational numbers. Many procedures of approximating have been invented for this case as well as for many others. We choose an old procedure that was known already in the antiquity and which is named after Heron of Alexandria (about 100 AD).

We want to approximate a square over d having a surface area 2 by rectangles whose surface area is 2 and which have rational side lengths. We begin quite simply and very imprecisely with the rectangle R_1 whose sides have lengths $a_1 = 2$ and $b_1 = 1$. As a next approximation we take the rectangle R_2 whose side a_2 is the arithmetical mean of a_1 and b_1, that is, $a_2 = \frac{a_1+b_1}{2}$. Then there should be $b_2 = \frac{2}{a_2}$. We proceed with a_3 and b_3, respectively, etc. The following picture shows the sequence of the rectangles.

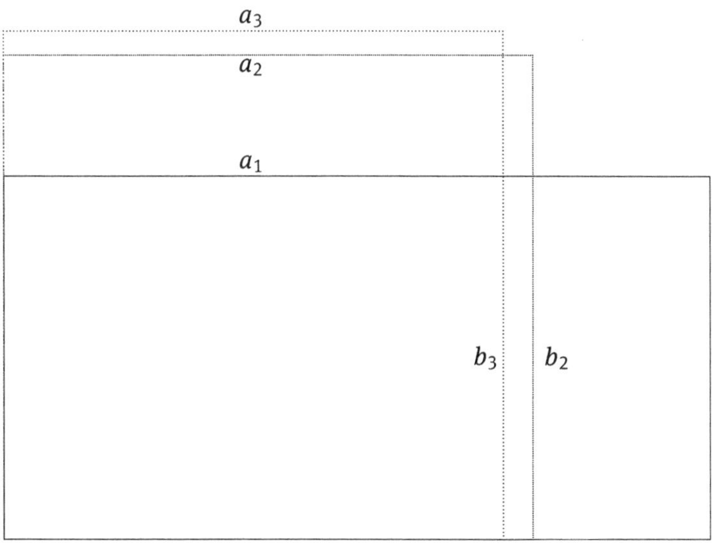

The sides in the rectangle R_n have the lengths

$$a_n = \frac{a_{n-1} + b_{n-1}}{2} \quad \text{and} \quad b_n = \frac{2}{a_n}.$$

All measures for the sides a_i, b_i of the rectangle R_i are rational numbers. The values a_i decrease and the values b_i increase, the differences $a_i - b_i$ approach 0.

Assume now that the sequence of the sides a_i, b_i is put on a line g. The sides a_i, b_i begin with the point 0, hence are represented as *points* on g denoted by strokes. We stress that we are talking about sides as *geometrical* intervals and their lengths. The sequence of the sides looks like this:

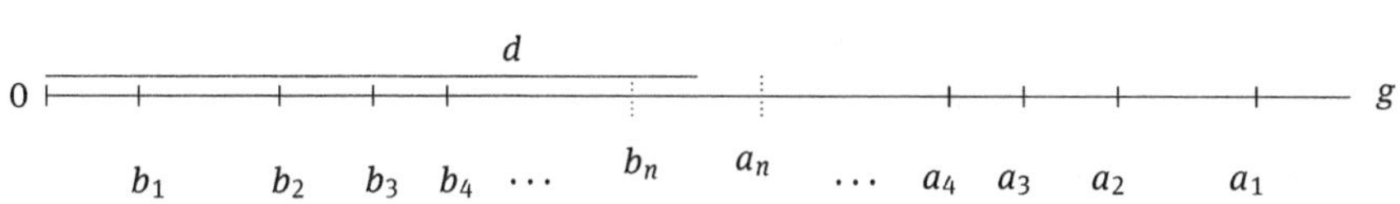

One sees a nesting of intervals that are always bounded by end points b_i and a_i. This is shown more clearly in the next picture:

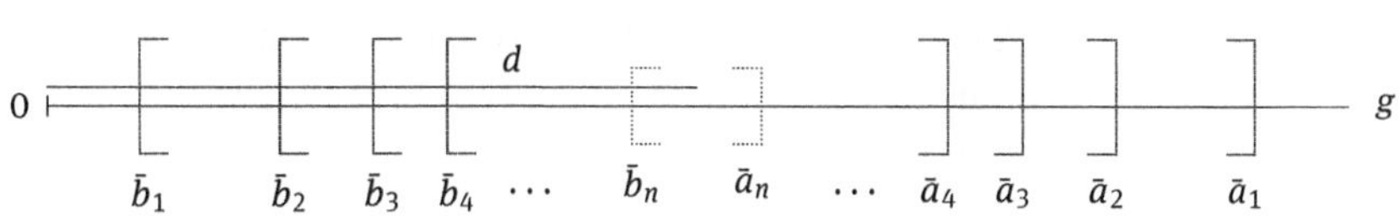

Here a_i and b_i are *end points of sides of rectangles* with rational lengths. The dots "..." indicate that the procedure goes on and always leads to new sides a_i, b_i that with every step move closer to the point for d. The interval $[b_n, a_n]$ is a position with probably large n in the never-ending sequence of intervals $[b_i, a_i]$ all of them including the end point d. This is in fact a *geometrical setting.*

Besides – and this should be seen completely separately – there is the arithmetical setting in the domain $\mathbb{Q}$ of rational numbers. We imagine all rational numbers presented as points on a line in the following way: in the next picture we want to see *only* "rational points", i.e., the "number strokes" of rational numbers and the intervals of rational numbers that give the lengths a_i, b_i of sides of the rectangles. The analogy "point – number" is so commonly used that the rational measures of the sides a_i, b_i are simply marked by a_i, b_i. We should be strict and distinguish between numbers and points. So we denote the measures by $\overline{a}_i$, $\overline{b}_i$. Note that there is no equivalent for d since there is no rational number for the length of d.

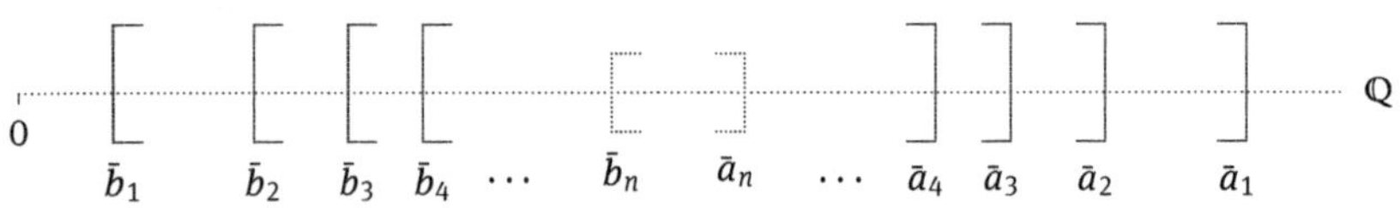

Though the arithmetical positions seem to be similar to the geometrical ones, they are in fact different. Above in the geometrical representation the end point of d is located in all intervals $[b_i, a_i]$. Here in $\mathbb{Q}$ there is no number corresponding to d:

– The intersection of all intervals $[b_i, a_i]$ is empty in $\mathbb{Q}$.

What is to be done? One *postulates* an appropriate number.

– *Claim:* There is exactly one number belonging to all intervals $[b_i, a_i]$.

It is called $\sqrt{2}$. It fills the gap in $\mathbb{Q}$ and measures d.

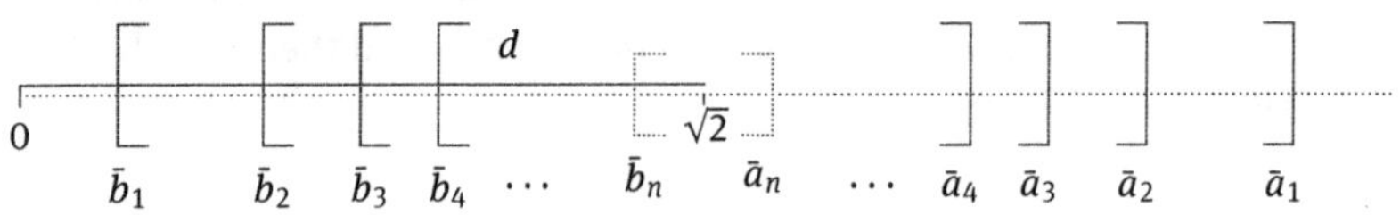

Another question is: What is this entity $\sqrt{2}$ that has turned up here. We will come to this in a moment. The procedure described above exemplifies the decisive step to the reals. We have closed a *gap* in $\mathbb{Q}$ by $\sqrt{2}$ just by a claim.

Now we claim to get the whole set of reals that should be everywhere *without gaps* and *complete.*

Claim: Let $\mathbb{R}$ be a set. One calculates with elements of $\mathbb{R}$ in a way similar to the way in which one calculates with the rational numbers. Also the order of elements should equal the order of elements of rational numbers with the following additional property of completeness.

Axiom (Axiom of completeness). Let $[r_n, s_n]$ be a nesting of intervals in $\mathbb{R}$. Then there is exactly one number $x \in \mathbb{R}$ that belongs to all intervals $[r_i, s_i]$.

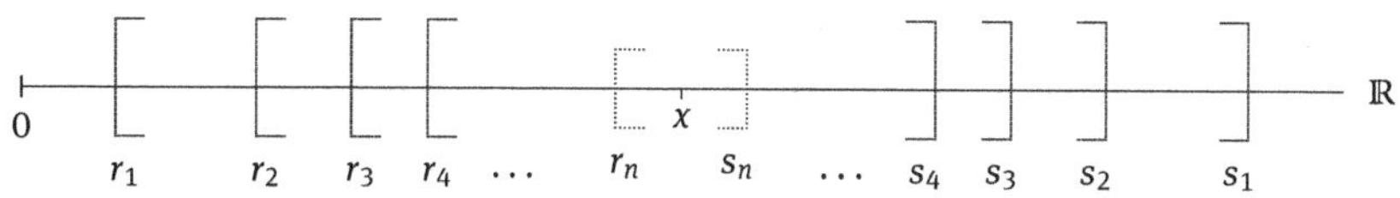

"Axiom" is the Greek word for "claim", a claim that is treated as "fair", "suitable" or plausible and the statement of which should be "obvious". The conflict with intuitively clear geometrical magnitudes has provoked the axiomatic claim of completeness. It claims that from now on geometrical magnitudes can be measured and represented by reals. Just this was the intention of the claim.

Here however most essential questions about the axiomatic method arises.

- Where do axioms come from, how are they chosen, how are they justified?

And quite fundamentally:

- Is the axiomatic method suitable for grasping real or perceptual phenomena?

It is said that axioms are obvious statements. Note that the obviousness of the claim of completeness was borrowed from the perceptual completeness of the geometrical line and did *not* come out of the domain of the rational numbers itself. It is just a geometrical obviousness that is added to arithmetic – one can even say "forced into arithmetic". In the case of classical arithmetic the fact that the range up to $\mathbb{Q}$ is incomplete is simply obvious. By the axiomatic setting of $\mathbb{R}$ together with the completeness axiom a new arithmetical level, an axiomatic one, i.e., a *theoretical* one, is attained.

There still lacks a construction of the reals that would say what the claimed real numbers really are. We come to this in the next section.

Now the situation has been crucially changed: to every possible geometrical length marked by a point on a line there corresponds – this is assured by the completeness axiom – a real number. And vice versa: all those real numbers can be again *illustrated* on a geometric line as "number strokes", as points. This is a big step – thanks to the mutual correspondence of points and numbers one can identify numbers and lengths. Or better: lengths and magnitudes in general can be *replaced* by real numbers.

The consequence of this is: the number *line* becomes a line of *numbers*. A geometric line will become a *copy* of $\mathbb{R}$. We have briefly seen what gradually happened in 19th century mathematics. A linear geometrical continuum has been replaced by an arithmetical continuum. Since that time $\mathbb{R}$ *is* the continuum. The continuum became a set. Translated back to geometry: the linear continuum became a set of points.

Now when one looks at that outcome the following *question* can be asked: is this reversible correspondence between real numbers and points of the geometric line really suitable with respect to the perceptible line? Is it at all possible to understand and to grasp the geometrical "continuum" of a line by points located on it? Or is it a matter of quite alien things in the case of points and lines between which there only exists an *external relation*, the relation of incidence: points are located on lines, lines range through points?

- Can the phenomenon of continuity be grasped by points?
- Is the geometric continuum a set of points?

If these questions are answered affirmatively then one should for example be able to build a line as a set of points. However, if one wants to build a set of points, one should know what a point is. Nobody wishes to know this anymore since one has the new axiomatic approach. Everybody smiles at the "definition" by old Euclid and writes in quotation marks: "A point is that which has no part". We will anew ask this question, and we must do this:

What is a point?

The situation yielded by the construction of the reals given in the next section seems to clarify and solve the problem of building a set of points. We can put a system of coordinates on a line as real numbers when we have them, i.e., to depict reals on a line. But is "every point" caught in this way? What are "all points" of a line? The answer supposes yet again the assumption that a line is a completely built and given set of points – and corresponds point by point to $\mathbb{R}$.

Just this is assumed. And just this makes a line into a copy of $\mathbb{R}$, turning it into the number line. But the general philosophical question remains:

Can the set of reals $\mathbb{R}$ *be* the linear continuum?

Mathematically the decision has been made, but the question has not been answered.

Once again this question arises, in another way. It is associated with the obviousness of the axiom of completeness in which the central mathematical assumption is to be found. Is the axiom of completeness really geometrically obvious?

If one considers the nesting of intervals *geometrically* then segments correspond to intervals. Consider the perceptual intersection over all segments of a nesting of intervals. Then, by the axiom of completeness, this intersection is geometrically exactly one point. Why *exactly* one point? Can there be more points? Their distance would then be infinitely small, infinitesimal. This is possible when one does not make an assumption expressed by "exactly one" in the formulation of the axiom of completeness given above. This assumption is the following.

Axiom (Axiom of Archimedes). If ε is a real then there exists a natural number n such that $\frac{1}{n} < \varepsilon$.

This requirement completely excludes infinitesimals and consequently also infinitesimal distances between points on a number line. But beside real numbers there also are non-archimedean "hyperreal" domains of numbers with infinitesimal numbers that are all smaller than all $\frac{1}{n}$. In Section 3.3 and especially in Chapter 6 we shall give an example.

One more observation should be made. Above, from a continuum by an operation of intersection over all intervals arises a point: hence something being not continuous. Can it be so? Is it then not the case that it is possible, or even not the case that it is necessary that the result of the intersection should be an interval, a *continuum*? It should be infinitely small, that is "infinitesimal". Hence we may ask: could it be as follows:

Let $([r_n, s_n])$ be a nesting of intervals with real bounds on a line g. Then there exists a *continuum* I included in all intervals $([r_n, s_n])$. This continuum is infinitesimal.

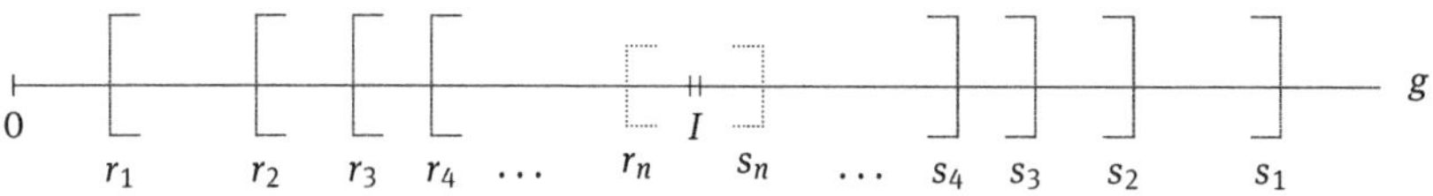

The idea of something being infinitely small is a challenge. Infinite smallness is today no mathematical problem; it is a philosophical problem – like the problem of the infinity that we face soon.

How is infinite smallness thinkable? What is its relation with numbers?

The idea of infinitesimal magnitudes occurred early in mathematics, and led in the 17th century to the calculus of infinitesimals, blossomed in the 18th century in spite of the problematic nature of infinitesimals and was rejected in the 19th century. In the middle of the 20th century mathematically infinitesimal magnitudes were rehabilitated – with rather little influence on mathematical thinking. We write about this in the third section of Chapter 3 as well as in Chapter 6.

Note that the intersections over intervals that we considered above and that led to the question about the infinitely small contain also an element of the *infinitely large*: in the case of an intersection one has to do with infinitely many intervals.

1.5 On the construction of the reals

Now we come to the question what kind of an object the number x can be claimed in the axiom of completeness. This question is about the construction of those numbers and the construction of the domain of numbers $\mathbb{R}$.

One is doing here something that is very mathematical and that possibly surprises a naive reader or perhaps appears strange. Imagine a nesting of intervals as above that – as it is said – converges to the claimed number $\sqrt{2}$. As one cannot say anything about what this $x = \sqrt{2}$ is, one raises the nesting of intervals given above into the mathematical identity and says – in a first attempt: this nesting of intervals is $\sqrt{2}$. Concisely but illegally said:

$\sqrt{2}$ is the nesting of intervals converging to $\sqrt{2}$.

In such a formulation we have made the very mistake we wanted to avoid: at the end of the sentence we proceed as if $\sqrt{2}$ already there were. In this way the formulation becomes circular. In other words, however it turns laborious and vague: *This* nesting of intervals is $\sqrt{2}$. The idea is to explain the formalterm $\sqrt{2}$ as a given, concrete nesting of intervals. Briefly and less clearly:

$\sqrt{2}$ is a nesting of intervals.

This formulation is also to some extent curious from the psychological and philosophical point of view. The problem is to construct a number postulated by the axiom of completeness. We have at our disposal as an instrument a nesting of intervals, a process. However, the constructing process provides no number. It leads to no result because it is infinite. But since the only thing one has at disposal is the process, it is *declared* to be its own result. So:

The process itself is put in place of its own result.

This seems to be paradoxical. Since the problem of constructing a number is represented by a nesting of intervals, it becomes almost absurd:

The problem is a solution.

When a problem is as concrete as in the example of $\sqrt{2}$ then it is mathematically legitimate. It is a typical theoretical procedure that is however very unusual and difficult for a learner to accept. A nesting of intervals is in fact no number.

What is done here is also noteworthy from the point of view of epistemology. Even Richard Dedekind had problems with this sort of procedure. It was he who in the 19th century participated in a vital way in the construction of the real numbers. In his efforts he used something that was other than the nestings of intervals, but ultimately comparable to them. He would have defended himself against identifying a number *imagined* within all intervals with a nesting itself. In thinking – according to him – something is added to the infinite process. Dedekind spoke about $\sqrt{2}$ as a "mental creation" generated by man on the base of the imagination of the nesting of intervals. It is however a philosophical way of speaking and mathematically not precise. In our example only the nesting of intervals is a mathematical entity.

This mathematical procedure – the declaration that the nesting of intervals is a number and then calculations with such nestings of intervals – illustrates very clearly the mathematical step towards the theoretical. For an unprepared beginner this step is a special challenge. The construction of real numbers is a construction in a new theoretical sense. It is usually – with reason – not mentioned at school. Also in university teaching it is deliberately avoided.

In the provisionally described construction arises an additional problem that makes the first attempt to define $\sqrt{2}$ more complicated though in principle not different. There are various nestings of intervals that can represent $\sqrt{2}$. It suffices in the example to change only the initial conditions of Heron's procedure, given above, and one gets another nesting of intervals that can represent $\sqrt{2}$. Since there is a free choice and there are various procedures, there are unboundedly many nestings of intervals that can be $\sqrt{2}$? What is then to be done? One takes all such "equivalent" nestings of intervals and declares them together to be $\sqrt{2}$. Particular nestings of intervals – such as those described above – *represent* then $\sqrt{2}$ in this new sense. This is a *second and final attempt* to say what $\sqrt{2}$ is.

- $\sqrt{2}$ is a set of nestings of intervals.
- $\mathbb{R}$ is the set of all sets of equivalent nestings of intervals.

There are different constructions of the reals that proceed in a similar way. And all constructions indicate – as the axiom of completeness did – that in the domain of numbers something new and special happens:

By the *requirement* of reals expressed especially in the axiom of completeness the reference to reality is set aside. The *constructions* provide representatives for the reals that are not single objects but infinite processes.

The axiom of completeness is – as remarked above – *not* obvious from the point of view of numbers.

The axiom of completeness is for the domain of numbers something abstract, borrowed from the geometry of the line.

The aim was the correspondence between points on a line and numbers in order to represent quantities and points on a line by numbers and then to be able to replace one with the other. This aim forced the unexplained assumption that a line is a completed set of points. Only in this way can one speak about *the* points of a line. This way of speaking became a custom so that we do not notice it any more. A consequence and the explicit intention was the arithmetization of analysis and of mathematics.

Constructions of the reals – we described one of them – are no abstractions from the physical or perceptual reality. In the next subsection we shall say something about their degree and problems of formality.

Axiomatization and construction of reals that accept infinite processes as mathematical entities and capture a line as a set of points makes mathematics ascend to a new level of abstraction and being theoretical.

1.6 On the handling of the infinite

In the construction of reals described above nestings of intervals were considered as clearly determined objects. Nowadays this has become so common that one does not notice what in fact is happening here any more.

A nesting of intervals is an infinite sequence of intervals

$$[b_1, a_1], [b_2, a_2], [b_3, a_3], [b_4, a_4], \dots,$$

that goes with the sequence of natural numbers $1, 2, 3, 4, \dots$. Such a sequence is an open and never closed process. This will be especially clear when we think for example about the construction of the nesting of intervals given above and converging to $\sqrt{2}$. One estimates interval after interval and never comes to an end. The bounds of succeeding intervals cannot be realized in a similar way as in the sequence of natural numbers. Dots "..." in the expression $[b_1, a_1], [b_2, a_2], [b_3, a_3], [b_4, a_4], \ldots$ send a signal that the process always continues.

At first glance one considers such sequences *potentially infinite*. Here sequences are not clearly defined terminated objects like other mathematical objects. They can be scarcely mathematically grasped. Sequences themselves are so unfinished; they are not concrete and cannot be grasped as concrete individual objects. But the latter is in fact necessary when one wants to say what $\sqrt{2}$ is.

What is done in this situation? One acts – and this is again an *assumption* – as if the infinite process of members of a sequence – that proceeds like counting – were closed. In such an approach the sequences of intervals are treated as "actually infinite", as clearly determined wholes. Instead of $a_1, a_2, a_3, \ldots$ and $b_1, b_2, b_3, \ldots$ one writes (a_n) and (b_n), and $([b_n, a_n])$ instead of the corresponding sequence of intervals, and in this way symbolizes the sequence as a completely given object.

What is happening here will become more clear in the case of natural numbers. The term "$1, 2, 3, \ldots$" becomes here "$\{1, 2, 3, \ldots\}$". In this case the assumption is particularly clear: the dots "..." symbolize a never-ending process and by putting the curly bracket "}" the end of this process is symbolized. This is connected with the set-theoretical Axiom of Infinity claiming that one can think about infinite sets as *actually infinite*, i.e., as being given in a way similar to other mathematical objects and usually treated in a similar way as finite sets.

Today such an approach is mathematical daily routine. It is usually practiced without any comment both in lecturing and teaching. A hundred years ago mathematicians discussed whether this is legitimate, whether infinite sets can be thought and applied. We shall see that the decision to accept the actual infinity first promoted by Georg Cantor (1845–1918) involves philosophical as well as mathematical problems. The problem of the infinite will accompany us in all the following chapters.

The question that arises here is the question about the infinity:

Is it legitimate to treat and grasp infinite sequences as ordinary entities?

In Chapter 2 various historical and current positions with respect to this question will be presented. In Section 3.2, "Infinities", we discuss in detail the question of the infinity as well as various answers to it. In Section 4.3 the Axiom of Infinity is the decisive one.

Notice that despite the problems connected with the actual infinity, actual infinite sets – as well as the reals based on them – are reliable instruments in mathematics today. There are indications of a further foundational question. The reals are theoretical constructions or they are claimed in an axiomatic way. We know how well they function in applications. But why is this the case?

- Why do the highly abstract real numbers solve concrete problems?

Why does the concept of an actual infinity – that probably has nothing to do with the real world – appear to be so effective in applications?

- Why is infinite mathematics applicable?

We shall look for answers to this philosophical question about applicability in Section 3.4.

1.7 Infinite non-periodic decimal fractions

The calculation of bounds of intervals to determine $\sqrt{2}$ for example by Heron's algorithm leads to finite sequences of decimal fractions whose difference is still smaller and smaller.

When it is said that the decimal fractions approach root 2 more and more closely, what it means is first of all that the length of the diagonals of the unity square will be approximated even more closely:

$$\sqrt{2} \approx 1,41421356.$$

If one wants to give $\sqrt{2}$ "exactly" then it is written

$$\sqrt{2} = 1,41421356\ldots.$$

The dots "..." say "and so on" and this suggests that one knows how it goes on – as in the case when one writes $1, 2, 3, \ldots$. At each step it is open what the next position will look like. Nevertheless, it is assumed – on the basis of an actual infinite sequence of all possible calculations – that all infinitely many positions of $\sqrt{2}$ are on hand. This indicates again the power of the assumption contained in the *Axiom of Infinity*. Such ways of writing and thinking make great demands on someone learning mathematics for the first time.

It is still more problematic when one speaks about *arbitrary* infinite non-periodic decimal fractions in general that should be a complement of the rational numbers with respect to the reals. In the Handbook [258] indicated in the Introduction one finds the following sentence:

> "It is right to expect that the set of real numbers (i.e., by our definition the totality of all points of a number line) *completely corresponds to the set of all possible decimal fractions* (finite, periodic or non-periodic)."

How should such an arbitrary infinite non-periodic fraction be specified? For example in such a way: 3,33526788 ...? What can "..." mean here? We are not given a procedure that would provide values for position after position that interprets "..." as "and so on". The term "..." cannot be understood. "non-periodic" is only a negative expression that cancels the meaning of "..." and "and so on". Using terms like "infinite non-periodic decimal fractions" is problematic. This comprises an intangible non-denumerable dimension. It is alarming when such problems are omitted in teaching and lecturing.

Already fifty years ago Paul Lorenzen (1915–1994) commented on the "leger-demain" of infinite non-periodic decimal fractions as follows.

> "To speak about infinitely many digits following each other is – supposing that it is not a complete nonsense – at least a big risk. In teaching however is usually no word about this."[3] ([234, p. 5], quoted from [340, p. 327].)

Till today nothing has changed here.

We would like to give two small examples to what troubles and dilemmas "infinite non-periodic decimal fractions" can lead, even if a calculating procedure is given. The last example comes from the intuitionist L. E. J. Brouwer (1881–1966) who at the beginning of the 20th century vigorously declaimed against the actual infinite. One finds this example, e.g., in [339] depicted in a slightly different form. It indicates how severe the problem was.

We construct on the base of an infinite decimal expansion of π a new number ψ as follows:

ψ_1 begins with 0 or a point. Digits after the point are determined as follows: the n-th position of ψ_1 is 1 if the n-th position of the decimal expansion of π is 0 followed by the series 1, 2, 3, 4, 5, 6, 7, 8, 9. Otherwise the nth position is equal to 0.

Is

$$\psi_1 = 0 \quad \text{or} \quad \psi_1 \neq 0?$$

Can this be decided? 50 years ago an answer to this question was utopian. Today however this can be decided thanks to computers we have at our disposal – and in which we trust.

The answer:

The 17387594880-th position of ψ_1 is 1.

This means – even if ψ_1 is "in practice" 0 – mathematically it holds that

$$\psi_1 \neq 0.$$

Now we can construct, in a similar way, another number ψ_2 on the base of π:

ψ_2 begins again with 0 and a point.

(i) If the first position after the point in a decimal expansion of π is 7 then put 1 in the 1-st position after the point in ψ_2, otherwise one puts 0.
(ii) If then in the decimal expansion of π there follow two digits 7 then put 1 in the next, 2-nd position of ψ_2, otherwise put 0.

3 "Von einer Aufeinanderfolge unendlich vieler Ziffern zu reden, ist also – wenn es überhaupt nicht Unsinn ist – zumindest ein großes Wagnis. Hierüber wird im mathematischen Unterricht zur Zeit aber meist kein Wort verloren."

(iii) If then in the decimal expansion of π there follow three digits 7 then put 1 in the next, 3-rd position of ψ_2, otherwise put 0.
(iv) If then in the decimal expansion of π there follow four digits 7 then put 1 in the next, 4-th position of ψ_2, otherwise put 0.
(v) Etc.

Is $\psi_2 = 0$ or $\psi_2 \neq 0$? Can it be decided?

This leads us to a curious situation. One would bet that

$$\psi_2 = 0.$$

However, no computer – neither today nor in the future – will be able to settle this bet. Is the finite calculating ability of all available computers the reason for this? No! If we want to win the bet then we would with the help of our giant but finite computers running in a finite time through infinitely many positions of the decimal expansion of π. But this is *in principle* impossible.

Or does the following hold:

$$\psi_2 \neq 0?$$

Is it really possible to exclude the fact that eventually, beyond all imagined accessibility, somewhere among the infinitely many series of numbers the required series of sevens does appear?

The situation is as follows: We are even not able to decide whether the undecidability of the disjunction "$\psi_2 = 0$ or $\psi_2 \neq 0$" is of principled or practical nature.

It is assumed that the infinite expansion of π is on hand. Then the infinite decimal expansion of ψ_2 will be also on hand. What does the dilemma of ψ_2 about our "infinite non-periodic decimal fractions" say to us? We know as much or as little about the decimal positions of ψ_2 as we know about positions in the decimal expansion of π. Is our belief that the *infinite* decimal expansion of π is on hand – when we consider the above quotation of Lorenzen – maybe after all absurd rather than merely risky?

2 On the history of the philosophy of mathematics

Which kind of philosophy one chooses depends on which kind of human one is.[1]

Johann Gottlieb Fichte

Chapter 1 was an elementary introduction to problems of the philosophy of mathematics. It itself belongs already to the philosophy of mathematics. The latter begins where mathematics and mathematical activity are considered. We asked questions and detected problems we would like to answer and to solve. However, we are still unable to present answers and solutions. Hence we are looking for hints and support in the history of mathematics and philosophy since antiquity till modern times. Therefore we shall follow the historical development, describe it and present various views and positions of famous and important philosophers and mathematicians concerning mathematics and its objects. We will try to describe them in a neutral way and to provide a wide objective basis for possible answers to our questions we will propose in Chapter 3 as well as for possible solutions of problems that will appear in further chapters. The reader will be able to build – according to the motto – his own conception using the variety of presented views. However, there will be no true, right position implying unique true answers.

Our survey is chronological and necessarily sketchy. At first we follow great names of famous philosophers who considered mathematics and its objects. Since the 19th century among them were more and more mathematicians themselves who began to consider their own discipline and to formulate conceptions concerning it. Hence there will appear names of great mathematicians whose conceptions will then orientate our survey. Finally, there arose schools and tendencies represented by various mathematicians. At the turn of the 19th and 20th centuries mathematicians took charge of the foundations of mathematics and consequently also a part of the philosophy of mathematics connected with foundational problems. There are still philosophers dealing with the phenomenon of mathematics but also they are doing this primarily taking into account the foundations.

Positions in the philosophy of mathematics we will consider are influenced by general philosophical systems and conceptions that can be in principle classified as belonging to the neighbourhood of one of three *basic philosophical positions*. Intuitively they can be presented as points in the following triangle whose distance from the vertices indicates how close they are to the basic positions.

1 Was für eine Philosophie man wähle, hängt davon ab, was für ein Mensch man ist.

https://doi.org/10.1515/9783110468335-003

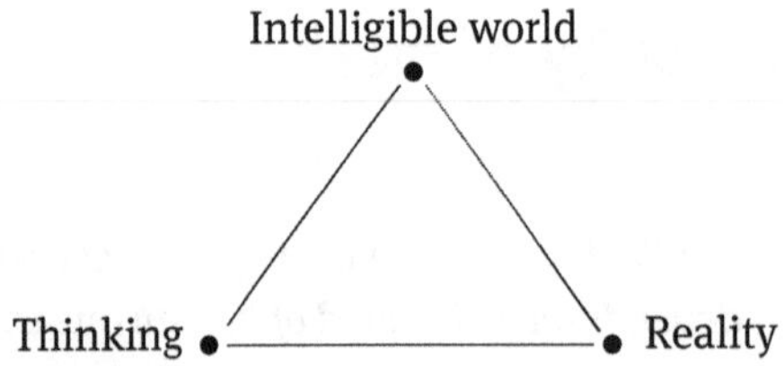

The fundamental question is: Where are the sources of the knowledge and its concepts? Positions represented by the vertices are the classical basic positions:
- The idealistic position claiming that the source of knowledge lies in the spiritual, intelligible world.
- The empirical position according to which the reality through experience gives us knowledge.
- The rationalistic position seeing the foundations of our knowledge in structures of thinking.

A famous representative of the idealistic position is Plato. Positivists, e.g., represent the empirical position. Kant represents what we call – a bit against common philosophical and linguistic usage – "rationalistic". Those positions will be described more exactly when they come in. Another new basic position will be presented when we will speak about new tendencies in the philosophy of mathematics.

Conceptions concerning the sources of knowledge are always connected with ontological questions, i.e., questions on the nature and existence of objects. Investigations of them belong to philosophical doctrines of being, to the so-called ontology where according to positions different answers are given.

As guiding thread we follow the views on the first and simplest mathematical objects, i.e., natural numbers, through the development of the philosophy of mathematics. Questions on their essence and mode of existence are the oldest questions asked in the philosophy of mathematics. We want to characterize briefly and concisely particular conceptions concerning natural numbers and to emphasize

characterizations in italic and as distinguished text.

This will be done in all cases in which representatives of concrete conceptions did formulate their views or we are able to reconstruct them authentically.

2.1 Pythagoras and Pythagoreans

In the prehistory of human beings an elementary mathematics has been developed in order to manage everyday problems. It served to solve economic and practical geometrical problems of measuring the areas of soil and of the early astronomy. Pythagoras and the Pythagoreans seem to be the first who not only applied and developed this elementary mathematics but also reflected on it.[2]

Pythagoras (ca. 570–ca. 500 BC) brought to Greece from his travels and stays in ancient Egypt and Babylon the arithmetical, astronomical and geometrical knowledge of priests. At the end of sixth century B.C, about 530 BC, he founded an ethical-religion secret society of Pythagoreans in Croton (Lower Italy) in west Greek colonies. Croton became soon the center of science at this time and Pythagoreans were the leading philosophers in ancient Greece. They were scientific pioneers in particular in the domain of mathematics and natural sciences. They were interested first of all in mathematics, music and astronomy. Various mathematical results in arithmetic, number theory and geometry are ascribed to them. Characteristic for them was that they connected religious mysticism with scientific principles and exact research methods.

Only few Pythagoreans are known by their names. The reason seems to be the Pythagorean world-view according to which it was forbidden to stress one's personal achievements. Also scientific discoveries have been apparently kept a secret. The blossom of Pythagorean mathematics took place at the turn of the fifth and fourth centuries BC Members of the school were then Archytas of Tarentum, Timaeus, Eudoxus of Cnidus, Philolaus of Tarentum and Eurytas. It is worth noting that – according to Diogenes Laërtius – Pythagoreans were the authors of the first definitions in mathematics. From them probably come the majority of definitions from the first book of Euclid's *Elements*.

At the center of mathematics of the early Pythagorean school around Pythagoras were numbers, arithmetic and elementary number theory. Together with mathematics Pythagoras brought also the mysticism of numbers cultivated by priests in Babylon and Egypt. Numbers were simultaneously measure numbers used to describe astronomical relations and as symbols having mystical meaning used in the astrology. This mysticism of numbers transformed Pythagoras into an early philosophical world-view. "All is number" was the philosophical motto of Pythagoreans. Only what is formed can be known – said Philolaus – and form is based on measure and number. Numbers formed for Pythagoreans a distinct, higher spiritual world according to which the earthly world was built.

This conception is noteworthy at least for two reasons:

- It explains the origin of a *theory*.
  Pythagoreans investigated the higher world of numbers in order to understand the material world. It was excluded to justify laws of this higher world of numbers by experience in the material world. Justification of them *should* be done inside the world of numbers. There possibly appeared the first real theory: number theory.

2 Texts on mathematical contributions of the Pythagoreans as well as their reconstructions are uncertain; cf. for example H. Boehme [37, 39] and W. Burkert [59]. We follow the common tradition; cf. for example van der Waerden [349].

– This theory was mathematics and philosophy in one.

Numbers in their philosophical role received by the Pythagoreans beside practical and theoretical meaning also metaphorical meaning that partially indicates their mystical and symbolic sources. Even numbers were "feminine", odd ones – "masculine". Five, the sum of the first even and odd number symbolized marriage. Ten being the sum of the first four numbers – *τετϱάκτυς* (tetractys) – was a "divine" number. There were – as still are in the contemporary number theory – *amicable* and *perfect* numbers. The later are numbers that are equal to the sum of their divisors. A task of number theory was to find principles of obtaining perfect numbers. Number one represented geometrically "point" – the geometrical "unit", two – a line, three – a surface, four – a solid body and three dimensions of the space. Numbers were for Pythagoreans no abstractions. They were *real powers* acting on the nature and in the nature. "Number" was the principle of all being. The structure of the real world was for them an image of the higher world of numbers. Number theory was metaphysics.[3]

Then there were – as Iamblichus reported about 300 BC – attempts to determine philosophically what numbers are. The following phrase comes probably from Pythagoras himself: numbers are an "unfolding" of "the generating principle lying in the unit". This phrase explains why the Pythagoreans did not treat one as an ordinary number but as a principle: as the principle of unit and simultaneously as source and origin of "common numbers".

Characterizing the philosophical position of the Pythagoreans concerning numbers, one can summarize it by saying the following:

> *Numbers are elements of a higher world – generated by the unit.*
> *They are spiritual powers of form* – over/beyond *things.*

The number theory of the Pythagoreans collapsed as the foundation of their world-view and this happened apparently in a dramatic way (cf. [36]). It happened in this way: For the Pythagoreans any segment was a number – relatively to a given unit of measure – or a proportion of numbers, and geometry was a subdiscipline of number theory. The absolute domination of numbers was finished by discovering the incommensurability about 450 BC There were no numbers characterizing the proportion of the side and the diagonal of the regular pentagon (cf. Section 1.2). This discovery led to a genuine crisis of the philosophical world-view. Perhaps it was the birth of mathematics that – after having lost its philosophical function – became now a proper discipline. How dramatic the whole situation was back then is indicated by the legend about the traitor who revealed the secret of the incommensurability – we have told it in Section 1.2.

The discovery of the incommensurability was not only a philosophical challenge. It dared mathematics to look for new mathematical foundations that would replace

3 Metaphysics is the fundamental philosophical discipline investigating origins and conditions of being.

number theory and would solve mathematical problems. Additionally there appeared a new challenge – the paradoxes of Zeno of Elea that indicated further troubles with the infinite.

The best known paradox is probably one of the paradoxes of motion, namely Achilles and the tortoise. Everybody knows that Achilles overtakes the tortoise even if he allows the tortoise a head start. On the other hand it is difficult to disprove the contrary claim that Achilles *cannot* overtake the tortoise. Whenever Achilles reaches somewhere the tortoise has been, he still has farther to go. For that time the tortoise will have advanced farther etc. (cf. [1, Book 6;9]).

New foundations have been then found in the so-called geometrical algebra. It consisted in replacing numbers and operations on them by geometrical magnitudes and operations on them. A definite solution of the crises was provided only by the so-called method of exhaustion coming from Eudoxus of Cnidus – it made possible to omit problems connected with the infinity. A further big step was the theory of magnitudes and the theory of proportions from the first half of the fourth century BC – it can be treated as a predecessor of today's theory of real numbers. All this can be found in Euclid's *Elements* (about 300 BC).

In Euclid's *Elements* geometry stays at the first place. Eudoxus' theory of magnitudes was presented by Euclid in Book V of the *Elements* before the number theory. The latter is since then in principle a part of the theory of magnitudes and their proportions even if it was in the *Elements* strongly separated and developed in an independent way and later on practiced in Greek mathematics.

Speaking about the early – before Plato – philosophy of mathematics, one should necessarily mention still one philosopher: *Socrates* (469–399 BC). He himself did not rather deal with mathematics, however his influence on the further development of mathematics was very great. In his ideas and his methods one can suppose the sources of Plato's theories, of the methodology of Aristotle as well as of the systematic deduction of Euclid. We tell about that in the next sections.

2.2 Plato

Plato (427–347 BC), the founder of the famous Academy (385 BC – 529 AD), a school of philosophers, is one of the most famous and important philosophers in the history. For him the most fundamental, perhaps basic philosophical problem was the problem of distinguishing between the seeming and the real. This distinction was for him not only a theoretical problem that is essential for philosophers and scientists, but it had also great practical and ethical meaning – for example for politics in ancient Athens. The result of this distinction and the background of Plato's philosophy was his theory of ideas.

By Plato ideas built – similarly to numbers in the perception of the Pythagoreans – a distinct higher world that penetrates the actual world of being, the material world. Ideas have their own existence and they determine the existence of real things.

Plato's theory of ideas distinguishes two modes of being forming two separate worlds: the unchanging, constant world of ideas that are clearly determined real entities existing beyond time, space and independently of human cognition. Ideas are unique: there is only one idea of unit, of beauty or of circularity.

On the other hand there are variable things of the material world being experienced by human beings with the help of their senses, possessing a lower level reality that appear as unstable shadows of ideas. Ideas do exist really, physical objects have their source in them. They receive their existence from ideas. Physical objects are for example circular when they participate in the idea of circularity. There are many circular objects. Things are like images of ideas, ideas are prototypes of things.

The relation between ideas and physical things can be seen like the following proportion:

$$\frac{\text{ideas}}{\text{things}} = \frac{\text{things}}{\text{shadows}}.$$

This means that the relation of ideas and things resembles the relation of things and shadows (of them). This has been vividly described by Plato in his famous allegory of the Cave (cf. [275, VII, 1–3]). It can be said that ideas are like forms for earthly things. The order of the material world is a reflection of the order governing the world of ideas. Material things are apprehended by senses and ideas – by concepts.

This ontology, this theory of existence and essence of things, is decisive (normative) when Plato deals with mathematics. Mathematical concepts are like ideas – immaterial and *real* and are characterized like ideas – as being unchangeable, clear and necessary. Ideas exist independently of space and time. The same holds also for mathematical concepts that are close to ideas and – in the interpretation of Aristotle – lie between the world of ideas and the world of physical things. There are the ideas of circularity, of proportion, of number and many mathematical realizations of them: there are many circles, many proportions, many numbers in the mathematical world.

At the end of Book IV of *Republic* Plato discusses the following classes of objects:

Tab. 2.1. Classes of objects by Plato.

immaterial	ideas	unit, beauty, goodness, justice, circularity ...
	mathematical objects	circle, number, relation, proportion ...
material	physical objects	wheel, table, house, tree, ...
	images of physical objects	image of a wheel, image of a tree ...

A consequence of Plato's conception is that a mathematician does not create or invent mathematical objects. He discovers and describes them. Therefore for Plato is clear (decided) that the fundament of mathematical knowledge is reason and the appropriate method for mathematics is the axiomatic method: axioms describe the basic properties of mathematical basic concepts. Plato was probably the first who introduced this method and propagated it. Mathematics was very close to Plato's ideal of knowledge because it disregards changeable phenomena and concentrates on unchangeable, timeless, mind-independent and definite objects and relations between them.

Plato admitted that a mathematician uses in his research practice observations and drawings or performs constructions. However, they serve only the process of recollection of the proper mathematical objects (ideas) and not the creation of them. The mathematician has in fact in his mind inborn knowledge about concepts and ideas and recalls them – they were seen by his soul in its life in the world of ideas. A human being attains in his earthly life the knowledge about ideas and mathematical concepts by recollection and not by constructing or active learning. This is Plato's theory of recollection, of *anamnesis* (cf. [274, XV–XXI]). Cognition consists of recollection by the soul of ideas that have been seen by it before its earthly existence.

Numbers are mentioned by Plato in the same breath together with ideas. They seem to have a special position among mathematical objects. Beside the unit (τὸ ἕν) from which already by the Pythagoreans originated numbers, there is the duality, the proportion, the λόγος, by whose incommensurable appearance collapsed the numbers by the Pythagoreans. Using both of them Plato explains – again according to Aristotle – in a bit unclear way the origin of numbers as well as of ideas (cf. [346, pp. 60 ff.]).

The character of numbers by Plato can be described as follows:

Numbers are immaterial intermediaries – between *the ideas and the material reality. Numbers form – like the geometry – the gate to the world of ideas.*

The theory of ideas made it possible for Plato to explain the problem of relations between pure and applied mathematics. According to him theorems of pure mathematics are suitable with respect to the material world of things because the latter are material images of immaterial ideas and concepts. A simple example: "One apple and one apple are two apples" – this is a theorem of applied mathematics since in the immaterial world of numbers holds "$1 + 1 = 2$" – this is a theorem of pure mathematics. It is so because *an apple* participates in the ideas of *unit*. Pure mathematics investigates and describes – according to Plato – the world of mathematical ideas and concepts and applied mathematics – the world of empirical things insofar as the latter participate in the former. Since physical things are only fuzzy images of ideas, mathematical theorems cannot be abstractions from perceptions of things.

Plato's conception can be described as radical realism of ideas and mathematical concepts. According to this conception – as we explained – a mathematician does not build mathematical objects but discovers them and their relations. Hence mathematics is a description of a world that is independent of time, space and human cognition. Even if there were no mathematician on the earth, there would be the world of numbers, of geometrical figures and other mathematical objects as well as relations between them. It is irrelevant for this world whether it is expressed as mathematics, as a system of definitions and theorems. A mathematician stands in front of a given reality that is eternal, independent and unchangeable, in front of the world of mathematical objects. His duty is to describe this reality.

2.3 Aristotle

Aristotle (384–322 BC) is the second great philosopher of the antiquity. His philosophy influenced – beside the Platonic one and maybe more effectively – the next 2 000 years of philosophy and science. Aristotle was a member of Plato's Academy; however, he never was his proper student as will be seen soon. He was the founder of the School of Peripatetics who, as is said, during tuition were walking and discussing along colonnades (*peripatoi*) in Lykeion, the place where the school was located.

No particular work by Aristotle devoted to mathematics is known. Diogenes Laërtius (about 250 AD) mentions in fact the work *On mathematics* but its content is unknown. However, in Aristotle's works devoted to logic and methodology one finds many fragments and remarks concerning mathematics.

The philosophy of Aristotle has been developed partly in opposition to that of Plato and partly independently of it. First of all he rejected Plato's theory of ideas. According to Aristotle mathematics is not a theory of independent ideal objects of a primary world that are prototypes of real things in a secondary material world. It is rather a theory of objects – called by Aristotle "mathematical objects" – that are obtained from real things by the process of abstracting and idealizing.

Hence mathematics is a theory of abstractions and idealizations that are results of a mental process. Those abstracts do not exist beyond things. They belong inseparably to the essence of things. Plato's ideas become forms by Aristotle – they are *in things* and are recognized in thinking. In various works of Aristotle one can find supporting fragments for this conception. In *Physics* [7, 193 b 22] he writes

> "[...] Obviously physical bodies contain surfaces and volumes, lines and points, and these are the subject-matter of mathematics. [...] Now the mathematician, though he too treats of these things, nevertheless does not treat of them as the limits of a physical body; nor does he consider the attributes indicated as the attributes of such bodies. That is why he separates them; for in thought they are separable from motion, and it makes no difference, nor does any falsity result, if they are separated."

In Metaphysics ([9], 1061 a 28) one finds the following words:

> "As the mathematician investigates abstractions before beginning his investigation he strips off all the sensible qualities, e.g., weight and lightness, hardness and its contrary, and also heat and cold and the other sensible contrarieties, and leaves only the quantitative and continuous, sometimes in one, sometimes in two, sometimes in three dimensions, and the attributes of these qua quantitative and continuous, and does not consider them in any other respect, and examines the relative positions of some and the attributes of these, and the commensurabilities and incommensurabilities of others, and the ratios of others; but yet we posit one and the same science of all these things – geometry – the same is true with regard to being."

Other relevant fragments are in *Posterior Analytics* [8, 83 a 32] and in *On the Heavens* [4, 299 a 15].

Let us try to summarize Aristotle's conception of mathematical objects on the example of numbers. For him for example "three" was not an immaterial object existing *per se* but like a property tied with material objects: a table is, e.g., three feet wide, a period last three years, etc. However, numbers are not detached from things like properties – as abstractions are understood today. They have not lost the power of forming that we saw by the Pythagoreans and by Plato: they form – together with other form powers – the material, the matter into things. They are inseparably connected with things. Briefly:

> *Numbers are (mental) form powers –* in *things. They are recognized in thinking in a kind of abstraction.*

Human being apprehends in numbers and forms the essence of things and turns towards the spiritual world a part of which he is – by his mental perception soul. Aristotle's position lies in the reality. However, in the triangle of basic positions (see the beginning of this chapter) he can be represented not as a point but rather by a vector from the reality above to the spiritual world.

Such ontology of numbers and of mathematical objects made it possible for Aristotle to explain in a simple way relations between pure and applied mathematics. Statements of pure mathematics are suitable to description of empiric things because mathematical objects are in real things and come from real things. Forms in things are representatives, the real things themselves are approximations of mathematical objects. Mathematics can be so much the better applied, so much the better the real things are in accord with mathematical objects.

Aristotle claimed that necessity and clearness of mathematics cannot be found in single statements but in logical connections between statements. Those connections are expressed by hypothetical statements, i.e., by implications. Hence only statements of the form "If p then q" can be true and not single mathematical statements claiming something (cf. [7, 200 a 15–19] and [9, 1051 a 24–26]).

Aristotle was more interested in the structure of theories than in single statements. Mathematics is only an example of a science in his methodology. According to him the base of any knowledge is formed by general notions which do not need to be defined, and by general propositions which do not need to be proved. All other notions should be defined and all other statements should be proved on the base of the undefined notions and unproved statements.

Aristotle distinguished in a theory the following four components (cf. *Posterior Analytics* [8]):
- Principles which describe fundamental notions and consequently are common to all theories. Aristotle called them axioms. – They correspond to logical axioms and axioms of identity in today's terminology.
- Specific principles describing specific properties of objects investigated in a given theory. Aristotle called them postulates. – They correspond to nonlogical axioms in nowadays terminology of formal logic.
- Definitions of specific notions. – Aristotle did not assume that what is defined exists.
- Existential hypotheses assuming that what has been defined exists independently of our perception and thought. – Aristotle claimed that such hypotheses seem not to be required for pure mathematics.

An example of usage of this methodology are the *Elements* of Euclid of Alexandria (ca. 300 BC).

One of the most important contribution of Aristotle to the philosophy of mathematics is the clear formulation of the problem of infinity. This has been done in his works *Physics* and *Metaphysics*. First he introduced the distinction of two types of infinity in mathematics: potential infinity and actual infinity. This distinction is connected with the distinction between actual and potential existence that is fundamental for Aristotle's ontology. Potentially infinite is a process that is open, i.e., boundless and at any stage can be prolonged. Simplest example can be here the process of counting "$1, 2, 3, \ldots$" by using numbers or the process of bisecting a segment. On the other hand the actual infinity is the final result of such an unlimited process that are commonly accepted in modern mathematics. The actual infinity is secured today by Axiom of Infinity (cf. Section 3.2.1 and Section 4.3). An example can be the set $\mathbb{N}$ of natural numbers. It is important that according to Aristotle this set does *not* exist. He allowed in mathematics only potential infinity (cf. [9, K 10]):

> "[...] a thing may be infinite in respect of addition or of subtraction, or both."

Aristotle declared on the contrary that the actual infinity is impossible:

> "The infinite cannot be a separate, independent thing." ([9, K 10].)

He claimed that the actual infinity is in mathematics unnecessary, moreover it is superfluous. Hence he allows the open sequence of natural numbers but not the set of them. Just the possibility of prolonging a series contradicts for him the existence of the totality of objects generated in this process. This decision of Aristotle *against the actual infinite* dominated in mathematical and philosophical attitudes during the next 2200 years.

The reserve of Aristotle towards the actual infinity was connected with troubles of Greeks in dealing with the infinite (compare with the paradoxes of Zeno of Elea). One did not find a solution to those problems arising here. The latter were left open, transcribed or eliminated from science.

It should be stressed that Aristotle's distinction between potential and actual infinity proved to be very useful and is accepted till today. It has been adopted for example by Cantor in his foundations of set theory (cf. Section 2.14).

Let us say at the end something about the aesthetic elements and beauty that Aristotle saw in mathematics. He claimed in *Metaphysics* that mathematics speaks, though not explicitly, about the beauty and reveals some of its elements. Just the beauty and aspiration for it are driving forces of mathematics. One reads in *Metaphysics* [9, Book 13, 1078 a 52–1078 b 4]:

> "Now since the good and the beautiful are different (for the former always implies conduct as its subject, while the beautiful is found also in motionless things), those who assert that the mathematical sciences say nothing of the beautiful or the good are in error. For these sciences say and prove a great deal about them; if they do not expressly mention them, but prove attributes which are their results or their definitions, it is not true to say that they tell us nothing about them. The chief forms of beauty are order and symmetry and definiteness, which the mathematical sciences demonstrate in a special degree. And since these (e.g., order and definiteness) are obviously causes of many things, evidently these sciences must treat this sort of causative principle also (i.e., the beautiful) as in some sense a cause. But we shall speak more plainly elsewhere about these matters."

It is worth noting that the neoplatonic philosopher Proclus Diadochus who lived in the fifth century AD wrote in his *Commentary to the First Book of Euclid's "Elements"* in a similar way about the beauty of mathematics. Similar statements can be again and again found in the history of philosophy, for example in the year 1908 in the work *Science et méthode* [280] by the famous French mathematician of the late 19th century Henri Poincaré. We write below (Section 2.16) about him.

2.4 Euclid

Euclid of Alexandria (ca. 365–ca. 300 BC) made no direct contribution to the philosophy of mathematics. However, his work *Elements* were so important for the methodology of mathematics that one should say some words about him here. The *Elements* were the presentation of Greek mathematics of the preceding 300 years and they gave a firm basis for the future development of mathematics. Influence of Plato and Aristotle can be spotted by Euclid. The influence of Plato can be seen in implicit philosophical presuppositions. For Euclid geometry is static. He tries to catch in definitions what is constant and unchangeable. Hence a straight line arising by movement of a point is inconceivable for him. Further traces of platonism can be seen in his attitude towards applications of geometry. Euclid did not accept any approximations or rough solutions typical for the practice of measuring – for, as stated by Plato, science has nothing to do with practice. Plato stressed in *Republic* that the geometry is science because it takes into account the necessary and unchangeable ideas and not unstable phenomena of things that appear and disappear. And last but not least one should notice that the very axiomatic-deductive method used by Euclid in the *Elements* comes from Plato who proposed it as a method for mathematics.

The *Elements* consists of 13 books (today one would say: chapters). Books I–IV and Book VI are devoted to plane geometry, Book V to Eudoxus' theory of magnitudes and proportions in its purely geometrical form, Books VII–IX to arithmetic and number theory, Book X to incommensurable magnitudes and geometrical algebra and Books XI–XIII to solid geometry. The method of Euclid consists of deducing theorems from definitions, axioms and postulates. Every new book began by definitions of new notions and by a list of axioms and postulates (the influence of Aristotle can be easily seen here). For example in Book I there are 35 definitions, 9 axioms and 5 postulates. Among the latter is the famous fifth postulate – the parallel postulate (for the following translations see [110]):

> "If a line segment intersects two straight lines forming two interior angles on the same side that sum to less than two right angles then the two lines, if extended indefinitely, meet on that side on which the angles sum to less than two right angles."

What cannot be omitted here as an example is the definition of the concept of a number given in Book VII of the *Elements*. The following formulation of Euclid summarizing his conception of number sounds in comparison with earlier philosophical attempts soberly:

1. An unit is that by virtue of which each of the things that exist is called one.
2. A number is a multitude composed of units.

One can easily see in the structure of the *Elements* the influence of Aristotle and his methodology. Aristotle recommended in *Posterior Analytics* to begin any scientific discipline by a list of definitions of its *specific* notions and by listing its axioms and postulates. And exactly in this way Euclid proceeds in his *Elements*. Axioms and postulates are distinguished just as Aristotle forced. For example postulates of the first Book are statements about specifically geometrical constructions. On the other hand axioms are statements of the general character describing fundamental properties of magnitudes, in particular of numbers, segments, surfaces and peripheries. Among postulates one finds such statements: “let the following be postulated: To draw a straight line from any point to any point” (Postulate 1), “to produce a finite straight line continuously in a straight line” (Postulate 2), “to describe a circle with any center and distance” (Postulate 3). Axioms are expressions of the following type: “things that are equal to the same thing are also equal to one another” (Axiom 1), “if equal be subtracted from equals, the remainders are equal” (Axiom 3), “the whole is greater than the the part” (Axiom 8).

As typical examples of definitions of the first Book let us quote the following ones: “a point is that which has no part” (Definition 1), “a plane surface is a surface which lies evenly with the straight lines on itself” (Definition 7), or “a circle is a plane figure contained by one line such that all the straight lines falling upon it from one point among those lying within the figure are equal to one another” (Definition 15). It can be easily seen that Euclid’s definitions are rather explanations of notions – often of the philosophical type – than definitions in the strict logical sense. It seems that there are by Euclid no primitive notions that would be consciously undefined.

Using Aristotle’s methodology Euclid established in the *Elements* the first deductive system in the history of mathematics. The exact analysis from today’s point of view indicates that there are certain gaps in Euclid’s system. However, this does not decrease in any way the meaning and importance of it for the development of the methodology of mathematics. The geometry of the *Elements* became for all scientific disciplines a pattern of a strict scientific method – the phrase *more geometrico* was used to describe it. The *Elements* have been for more than two millennia till the 19th century a canon and a pattern for what one understood by “mathematical”. On the base of them mathematics has been developed as an axiomatic (from today’s standard rather quasi-axiomatic) system. In fact, everywhere in the historical mathematics there were gaps in lists of axioms, postulates and definitions, they were not complete. One freely used in proofs various “obvious” truths or referred to the intuition. There were no attempts to investigate, to describe and to determine the language and symbols of mathematical theories. However, it should be stressed: thanks to Euclid and since his *Elements* mathematics has been developed as a system organized *by itself*.

Books of the *Elements* were used till the beginning of the 20th century – hence more than twenty three centuries – at schools and universities as mathematical handbooks. In the antiquity and in the Middle Ages they were frequently copied, later printed and translated into many languages. Since the discovery of print there have been more than thousand editions – only the *Bible* has more editions.

2.5 Proclus Diadochus

Proclus Diadochus (410–485) was a famous representative of the Athens school in the neoplatonism – beside this school there were also the Rome–Alexandria school of Plotinus and the Syrian school (with Iamblichus). Proclus is known in the history of the philosophy of mathematics by his work *Commentary to the First Book of Euclid's "Elements"* [285, 286]. There he presented and commented achievements of his predecessors and in this way gave a survey of positions in the academic (Platonic) and peripatetic (aristotelian) tradition. He developed them and derived his own conception of mathematics.

In the ontology of mathematics Proclus ascribes to mathematical objects a certain center position in the hierarchy of beings. He puts them between the highest beings and the lowest ones, i.e., material things. The highest objects are "undivided, simple, not composed" [286, 11, 1]. On the contrary the material objects are "divided and complex" [286, 11, 1]. The source of mathematical objects is by Proclus the soul in which their archetypes and their essence are located. This determines the method of mathematics that should be adjusted to the nature of investigated objects. Proclus claims that for mathematics the characteristic is not the intuitive but the discursive way of thinking, hence thinking in which conclusions are deduced from generally accepted or hypothetical assumptions.

Proclus adopted methodological principles formulated by Aristotle and applied by Euclid in the *Elements*. In particular he distinguishes in any theory definitions, axioms and postulates. In the *Commentary* [286, pp. 178 ff.] he considers the nature and the character of axioms and postulates and discusses the problem what the differences and similarities between them are. He comes to the conclusion that their common feature is that they need neither justification nor proof and that they are self-evident and certain. The difference between them is according to Proclus similar to the difference between theorems and problems. He writes:

> "In the case of theorems it is required to apprehend and to recognize that a conclusion follows from assumptions and in the case of problems one should find or do something. In axioms one assumes that what is immediately obvious and what does not make no troubles to our mind, while in postulates one attempts to find what can be easily found and what can be easily established without special effort and without using any sophisticated procedures and constructions." ([286, pp. 178 ff.].)

The *Commentary* contains also an interesting discussion of the parallel postulate of Euclid's *Elements* [286, pp. 191,16–193,9]. Proclus indicates there that already Geminus of Rhodes – a Greek philosopher, astronomer and mathematician, who flourished in the first century BC – noticed that there are certain lines which going into infinity are still closer and closer to each other but never intersect (for example a hyperbola or a conchoid and their asymptotes). One can discover in Proclus' *Commentary* possible indications on later non-Euclidean hyperbolic geometry of Gauß, Bolyai and Lobachevsky.

Let us consider for example the bundle of lines parallel to a given straight line in Klein's projective disk model of the hyperbolic geometry. There are two specific straight lines in this bundle of parallel lines: parallel lines that do not intersect the given straight line and separate other parallel lines from the straight lines that intersect the given straight line. Hence there are straight lines which on the one hand are parallel to the given straight line but intersect the given line when their rakes are changed. Proclus writes in the following way: "[...] it could be claimed that since two right angles can be made smaller and smaller without any bounds hence till certain grade of reduction the straight lines remain asymptotes and by a greater reduction – they intersect each other." (*Commentary*, 193)

Proclus discussed in his *Commentary* also the problem of the infinite. He noticed a paradoxical – as he claimed – property of infinite sets consisting in the fact that an infinite set can have exactly the same number of elements as its proper subset. This made him accept only the potential infinity and to reject the possibility of an actual infinity. He argued as follows:

> "A diameter divides a circle into two equal parts. When however by *one* diameter *two* semi-circles arise and by the center there can go infinitely many diameters then it will imply that there will be twice so many semi-circles than infinitely many diameters. Some people see in this a paradox of the infinite division of magnitudes. However, we claim that magnitudes are divided into infinity but not into infinitely many parts (*ἐπ' ἄπειρον, οὐκ εἰς ἄπειρα δέ, ad infinitum, sed non in infinita*). The latter would imply that there would be actually infinitely many parts, the former that only potentially; the latter grants to infinity the substantial existence, the former only the status of becoming. One diameter generates two semi-circles but there will be never (actually) infinitely many diameters even if there would be indefinitely many ones. Hence there will never be twice so many than infinitely many semi-circles even if actually arising semi-circles will be twice so many as always finite number of diameters. The number of "given" (i.e., actually constructed) diameters is always bounded." ([286, 158, 1–20].)

It should be noticed that such approach of Proclus is typical for the Greek way of thinking that goes back to Aristotle. One tries to avoid troubles with the infinite rather than to take it seriously and to attempt to solve them. Such attempts appeared only in the 19th century when Bolzano stated that "paradoxical" properties of infinite sets – i.e., the property that they are equipollent to their subsets – are characteristic for them, when Richard Dedekind used this property as a definition of infinite sets and Georg Cantor established set theory as the theory of the infinite.

2.6 Nicholas of Cusa

Nicholas of Cusa, usually in the Latin form Nicolaus Cusanus (1401–1464) was one of the last representatives of scholastic philosophy for which theology was the fundament of philosophy and of any science. He was a theologian, philosopher and mathematician and is seen as a thinker who prepared the philosophy of the modern era. Mathematical and theological ideas have been closely connected by him and influenced each other in his works. His philosophical position can be located – as will be seen – in the middle of the triangle of basic positions idealism, realism and empiricism.

The spiritual world is represented by Nicholas of Cusa by God. God created the world according to mathematical laws; numbers and ideas come from him. This conception can be seen again by Gottfried Wilhelm Leibniz (1646–1716) (see below). By Nicholas of Cusa numbers play an important role. Therefore we want to present his position concerning mathematical objects just by giving an example of them.

In Chapter 6 of the work *Liber de mente* [354] Nicholas studies the concept of number. He distinguished numbers being the object of mathematics and numbers coming from God. The former are coming from human intellect and mind, the latter have their origin in God's mind. The "human" and mathematical numbers are images (*ymago*) of "divine" numbers. Everything that has numerical character in things and in connection with things is for Nicholas an implementation of numbers of divine mind. He writes:

> "[...] and [when you see] how the essence of all things manifests so that the number from divine mind ought to be."[4] ([354, c6, 121v].)

And further:

> "You also see that number and a thing that is counted are not different; therefore the number is not something in the middle between the divine mind and the things what would have its own existence. Number is the matter of thing."[5]

And again in [354, c6, 121r]:

> "[...] only the mind counts; without the mind there exists no independent number."[6]

4 [...] ac quod quidditas rerum omnium exorta est, ut sit numerus divine mente.

5 Conspicis eciam, quomodo non est aliud numerus quam res numerate, ex quo habes, inter mentem divinam et res non mediare numerum, qui habeat actuale esse, sed numerus rerum res sunt.

6 nam sola mens numerat, sublata mente numerus discretus non est.

One can say that the real things have been created from divine numbers and so numbers in things. But how does a human being create his *mathematical* numbers? Cusanus indicates here the specific ability of human thinking consisting of simulating in its own constructions the divine creation processes and to build in this way mathematical numbers via counting. The simulation is based on an abstraction, on "comparative differentiation" of things.

Beside the idealistic position of Cusanus – the source of numbers is in God – one can recognize here a *rationalistic* and an *empirical* side of his thinking. Empirical is that "the number [...] is the matter of thing". In *Liber de mente* one finds: "[...] there is nothing in mind that earlier would not be in senses" ("[...] ut nihil sit in ratione, quod prius non fuit in sensu"). Rationalistic is the creative reconstruction of numbers in the mind of the human being in "comparative differentiation" of things.

One can summarize Nicholas of Cusa's conception of numbers in the following way:

> *Numbers of human are the rational reconstructions of divine-spiritual numbers. They are obtained by the process of comparison and differentiation from real things in which the divine numbers are implemented.*

In a similar way Cusanus describes also his conception of geometrical objects. According to him geometrical objects like points, lines, surfaces, circles etc. are creations of the human mind. He writes in [354, Chapter 11] that "the mind creates the point" and that "[the mind] creates the line by imagining the length without the width".[7] Or: "You know how we accomplish the mathematical figures on the base of spiritual power"[8] [354, c3, 117v]. Mathematical objects are – like mathematical numbers – images of entities existing in the mind of God. In [354, c9, 125r] the following words can be found:

> "Therefore the measure and boundary of any thing comes from [human being] mind. Trees and stones have measure and boundaries also outside of our minds, however they do come from the non-created (= divine) mind from which the boundaries of things are derived."

The geometrical objects can appear and be real only in concrete things. What this means is explained by Cusanus in the work *De docta ignorantia* [352, Book II, cV, 118]:

> "However, all that really exists, exists in God who himself is the reality of all things. But the reality is the completion and aim of possibility."

7 Mens fecit punctum. [...] [Mens] fecit lineam, considerando longitudinem sine latitudine.
8 Tu nosti [...] quomodo nos exerimus ex vi mentis mathematicales figuras.

How does a human being build in his mind the mathematical objects that are images of objects existing in the divine mind? Cusanus speaks here about the human ability of assimilation and completes what we said above about numbers. The heading of [354, Chapter 7] is: "How does the mind by itself create the forms of things by virtue of assimilation and come in this way in contact with the absolute possibility or materiality".[9] Also here the matter is the abstraction or derivation (cf. [352, Book II, c1, 92 and c4, 114]) and again one notices here certain empiricism.

Important are for us Cusanus' conceptions of the infinite. The problem of the infinite appears by him both in mathematical as well as in theological and philosophical considerations. He claims that the infinite in mathematics is comprehensible by the mind with the help of concepts, however it is inaccessible by sensual perception. Cusanus admits that his aim of considering the mathematical infinite is to approach in this way the "infinity of God". The infinite in mathematics is for him an allegory of God's infinity.

Cusanus begins with the remark that there are no sensually experienced things or processes that could not be increased or prolonged. Hence the infinite cannot be a real object or process. However, in mathematics there are examples that the limit can be achieved conceptually. As such an example provides Cusanus the sequence of regular n-polygons. If n is greater and greater without any bound then the polygons approximate and reach the circle. Among things experienced by senses there is no circle. A circle exists only as a concept in our mind. In [354, c7, 122v] Cusanus writes:

> "It is so when [the mind] conceives the circle as a figure such that lines going from the center of it towards its periphery are of the same length. There can be no material circle of this type outside our mind."[10]

In this way two different figures such as circle and regular n-polygon coincide with each other in the infinite. Cusanus provides many such examples. In all of them one has to do with the same principle, the principle *coincidentia oppositorum* (coincidence of opposites). The completion of a process and consequently for example the limit value of a sequence have the highest form of being and are eternal. Every process trends by itself towards the completion in the infinite.

From the ontological point of view it is important that the infinite cannot get its existence by the existence of its finite parts. The finite cannot secure the existence of the infinite because the infinite will be never achieved in the approximation by finite states. It is just the opposite: the infinite precedesthe finite and is superior with respect

9 Quomodo mens a se exerit rerum formas via assimilacionis et possibilitatem absolutam seu materiam attingit.

10 [...] sicut, dum concipit circulum esse figuram, a cuius centro omnes linee ad circumferenciam ducte sunt equales, quo modo essendi circulus extra mentem in materia esse nequit.

to the latter. So one can see that Cusanus changes here the usual direction of thinking. He states that the finite can be conceived and understood only with the help of the infinite. He writes in [354, cII, 116r]:

> "Hence any finite has its source in the principle of the infinite."[11]

In the work *De docta ignorantia* he applies this principle geometrically and states [352, Book II, cV, 119]: "Now, every finite line has its being from the infinite line, which is all that which the finite line is. Therefore, in the finite line all that which the infinite line – viz., line, triangle, and the others – is that which the finite line is." Hence in the finite one finds the infinite from which it comes.

One sees that the principle of *coincidentia oppositorum* becomes an ontological principle and is used by Cusanus to explain the mathematical knowledge. Let us notice that Cusanus formulated this principle in order to describe the convergence of our knowledge and mathematical thinking to the knowledge of God. And he justifies this procedure writing in *De mathematica perfectione* [353]:

> "My aim is to improve mathematics by concidentia oppositorum."[12]

2.7 Descartes

René Descartes (1596–1650) can be seen as the founder of the modern philosophy. The influence of his thinking can be observed in the whole later development of European philosophy. The scientific contribution of Descartes is immense and it is impossible to describe it adequately. Therefore we will concentrate on his mathematical achievements and his contribution to the methodology of mathematics.

Descartes published only one mathematical work, *La géométrie* – it was one of three appendices to his main work *Discours de la méthode pour bien conduire sa raison et chercher la vérité dans les sciences* (Discourse on the Method of Rightly Conducting One's Reason and of Seeking Truth in the Sciences [95]). However, the meaning of *La géométrie* is enormous – in particular for the methodology of mathematics and other sciences. One finds here the beginnings of the analytic geometry. By consequent application of algebra that was well developed at the beginning of the 17th century to the geometry of the ancient Greeks, Descartes contributed to the unification of both disciplines. Let us note that arithmetic (and algebra) and geometry have been practiced since the time of ancient Greek mathematics as two separate disciplines.

11 Quare omne finitum principiatum ab infinito principio.

12 Intentio est ex oppositorum coincidentia mathematicam venari perfectionem.

Numbers and lengths – represented by line segments – become to be (since Descartes) coordinates in a coordinate system and are in a relation to a given unit line segment. Proportions of magnitudes, the main concept of Greek mathematics, became the relation to a fixed unit line segment. Rational proportions of magnitudes obtained in the course of development of mathematics the status of numbers that is inherited by rational lengths which then adopted again the rational proportions to a unit segment. Lengths that are incommensurable with respect to the unit length became irrational lengths. They moved closer to the status of numbers because they play the same function in the coordinate system as rational lengths. They became – via the calculus of segments augmented by Descartes beside the usual addition also by multiplication – the obvious objects of calculations as numbers were (cf. [318, p. 226]).

A further step made in this context by Descartes forced additionally the proximity of magnitudes, proportions of magnitudes and numbers. Descartes explicitly canceled the restriction accepted by all his predecessors – namely the consequence of the so-called principle of homogeneity. This principle binding relations and proportions with geometrical magnitudes that should always be of the same type had its source in Greek mathematics and in its geometrical algebra. The cancelation of this principle made it possible to consider algebraic equations in which the ideas of a magnitude was irrelevant or even obstructive. It made general considerations of algebraic curves possible and initiated abstraction towards pure mathematics not bounded by geometrical ideas.

The removal of the idea of magnitude from algebra was the beginning of a turn in mathematics that made – after certain time – numbers to become again the foundation of mathematics – as it was the case by the Pythagoreans – in a completely new form, namely as real numbers.

Descartes introduced into mathematics many new symbols – this was the result of a complex development (cf. [219]) – and contributed in this way to the development of symbolic language of mathematics. Some of his symbols are used in mathematics till today (for example = (equality symbol), $\sqrt{b}$ (root), $3a$, ba (for multiplication), a^x (exponentiation)).

Beside the described contribution to mathematics itself that changed it and determined its further development, Descartes contributed also to the methodology of mathematics. One should mention here first of all the general principle saying that the investigation of an object must be preceded by the investigation of the method. A criterion of certainty and safety in science should be clearness and distinctness of its concepts and ideas. Descartes proclaimed a program of a general rational knowledge and of a general rational science that should be developed according to a sample of mathematics (cf. [94]).

Descartes was convinced that only mathematics is able to find proofs and in this way to provide a certain knowledge. The reason for that was according to him the fact that mathematicians consider only quantitative properties. In *Regulae ad directionem ingenii* (Rules for the Direction of the Mind) [96][13] he wrote:

> "[...] strictly speaking the subject of mathematics is what is investigated according to order and measure without considering that this measure is at hand by numbers or figures, stars, voices or any other objects; it should exist an universal science that explains all which can be an object of investigation according to order and measure without knowing about a concrete area to which it must be attributed." ([96, p. 21].)

He proposes to call this discipline "universal mathematics, because it contains all thanks to which other disciplines call themselves mathematical" (loc. cit.).

The reliability of arithmetic and geometry, i.e., of mathematics, is the consequence of the fact that only mathematics

> "deals with such distinct and simple objects that it has nothing to assume which could be insecure by experience. It consists only in pure mental deduction of inferences." ([96, p. 8].)

And sarcastically he remarks that

> "we should not wonder if lots of geniuses prefer to attend other fields or philosophy [in the scholastic manner – authors' note]; that is why everyone presumes to speculate venturously according to vague subjects instead of turning to self-evident matter. For it is easier to follow optional presumptions than coming to truth just in very simple matters." ([96, p. 9].)

From this follows for Descartes the idea of bounding all disciplines to quantitative investigations only and the plan of creating a universal analytic and mathematical discipline (*mathesis universalis*). This idea will appear again for example by Leibniz. Such a discipline should contain all our knowledge about the world.

The conception of *mathesis universalis* was connected by Descartes with the idea that all properties of objects should be deduced from form and motion. The whole nature should be considered exclusively as geometrical and mechanical. This comes again from the conception that the extensive substance has geometrical character.

Mathematics itself – being for Descartes, as we have seen, the paradigm of any discipline – should apply exclusively analytic methods. Descartes permitted here only *intuition* and *deduction*.

---

13 According to the original edition of 1701, published by Artur Buchenau, Leipzig, Verlag der Dürr'schen Buchhandlung 1907.

Under intuition he understood

> "such a simple and clear perception of pure and attentive reason that we never doubt about what we are realizing or what is the same: an inerrant perception of pure reason that arises solely from the light of reason and for being more elementary is more secure than deduction." ([96, p. 12].)

Axioms of mathematics were for him such safe truths that cannot be shook. Under deduction understood Descartes "all that can be deduced with necessity from other things which can be recognized as safe" (loc. cit.).

The analytic method should make it possible to uncover simple components of thoughts. And what was simple was for Descartes exact and clear, hence safe. In *Discours de la méthode* [95, p. 22] he formulated four rules that determine the safety of his scientific work and according to him are generally sufficient:

(i) "Never to accept anything for true which I did not clearly know to be such; that is to say, carefully to avoid precipitancy and prejudice, and to comprise nothing more in my judgement than what was presented to my mind so clearly and distinctly as to exclude all ground of doubt" (intuition).

(ii) "To divide each of the difficulties under examination into as many parts as possible, and as might be necessary for its adequate solution" (analytic method).

(iii) "To conduct my thoughts in such order that, by commencing with objects the simplest and easiest to know, I might ascend by little and little, and, as it were, step by step, to the knowledge of the more complex; assigning in thought a certain order even to those objects which in their own nature do not stand in a relation of antecedence and sequence" (deduction).

(iv) "In every case to make enumerations so complete, and reviews so general, that I might be assured that nothing was omitted."[14]

2.8 Pascal

Blaise Pascal (1623–1662) was highly skilled not only in the field of scientific investigation but also in the literature. He was extraordinarily creative in mathematics and in philosophy. His philosophy was on the one hand tied in with the philosophy of Descartes and on the other it dissociated from it. Pascal in his philosophy starts from a human being in whose life and reason he perceives also irrational elements. He was not a pure rationalist. He interested himself only a bit for ontological problems. He left no closed philosophical system.

According to him the reality disintegrates into two worlds: in the "order of reason" and "the order of heart (*ordre du coeur*)". The reason and the rational method are

14 English translation after p. 22 of http://www.gutenberg.org/files/59/59-h/59-h.htm.

helpless and useless in existential matters. The infinity that surrounds us – as he says – cannot be grasped by reason. Also in the case of ethical or religious questions the reason alone provides no solutions. In those domains clarity and clearness – Descartes strived for – do not offer certainty. Only "heart" can help here. Pascal expresses it by saying that "*Le coeur a ses raisons, que la raison ne connait pas.* (The hearts has its reasons that the mind does not know.)" It should be of course asked what did Pascal understand by the concept of heart. In the philosophical literature one finds various interpretations: from hearts understood as the instrument of getting to know what is supernatural till the heart as a place of intellectual intuition.

In the domain of reason Pascal saw – similarly to Descartes – mathematics and in particular geometry as an ideal of thinking and acting because it is almost the only scientific domain in which there are proofs – it is alone really methodical. On the contrary in all other disciplines there is somewhere disorder. And only geometers are conscious of this (cf. [264]). The new ideal method whose model and prototype is provided by geometry is based on two principles: (1) one does not use any notion whose meaning has not been exactly fixed earlier; (2) all statements should be proved. The usefulness of definitions consists – according to Pascal – in the fact that they make expressions

> "simpler and shorter and contribute to their clarity because they express with the help of a name what otherwise could be expressed only by using many words." ([264].)

Definitions are arbitrary in principle: one can call a given thing as one wants. However, the unique condition should be respected: no two different objects can be called by the same name. Definitions are used only in order to make expressions shorter and not to make clear or to change ideas hidden behind things one speaks about.

Pascal recognized of course that in practice not all concepts can be defined and not all statements can be proved. Hence he accepted some concepts without having defined them as primitive notions that are clear thanks to "natural light", and similarly in the case of general clear fundamental statements that are unproved but provide the fundament of proofs. Among the former are in particular concepts of space, time, motion, number and equality. They can be used without definitions because

> "the nature itself has given us a clearer conception of these things without using words than of those (which) we would get by elaborately explications."

Moreover: the nature has presented all people with coinciding ideas, and similarly in the case of fundamental truths, i.e., axioms. One "perceives" them by heart; all other statements however are deduced from them by proofs, hence within the order of reason. Both axioms and proved statements are certain even when they have different status. Pascal claims that it would be absurd to require from the heart to prove the basic statements, as it would be absurd when the heart required from the reason to "feel" all statements proved by it (cf. *Pensées* [263, p. 479].

Summarizing: On the one hand one has to avoid to use notions when "they are not completely clear and unambiguous" without having defined them beforehand and has to avoid statements without checking and proving them. On the other hand one resigns from definitions of fundamental concepts that are clear by themselves and from proofs of fundamental statements accepted by everybody.

Pascal adds that some of the fundamental statements mentioned by him are evident by themselves and trivial; however, they are not treated as generally applicable because their usage is limited to mathematics only. They are in fact simple but little known because

> "beside geometers who are few – diffused among nations and in uncountably many years – nobody can be found who knows them." ([263].)

However, just those fundamental statements are prototypes and examples of any rational knowledge.

Let us notice at the end that Pascal distinguishes the knowledge about the existence of an object and the knowledge about its nature – as it is the case in the following important example. We recognize the existence and the nature of the finiteness, "because we are finite and extensive as it is" [263, p. 452]; we experience the existence of the infinity since

> "we know that it is false that the sequence of numbers is finite, hence it is true that the infinity exists in numbers." ([263, p. 451].)

However, we do not understand its nature, "because it is extensive as we are but it has no bounds as we have" (loc. cit.).

2.9 Leibniz

Gottfried Wilhelm Leibniz (1646–1716) who worked and researched extremely creatively in many domains and who is treated as the last universal scientist contributed in an essential way also to mathematics and to the philosophy of mathematics. What concerns the philosophy of mathematics one should speak about two of his fundamental ideas.

The first one is the distinction between truth of the reason and truth of the facts. Leibniz distinguished in the domain of all truths, i.e., all true statements, truths of reason on the one hand and truths of facts as well as primitive truths and derived truths on the other hand (cf. [228, Fourth Book, Chapter II]). Primitive truths are truths known by intuition. They do not need any justification because they are clear by themselves and cannot be deduced from anything simpler and more clear. Derived truths are truths which can be reduced to primitive ones – by "the connections of mediate ideas" (loc. cit.); they form the demonstrative knowledge. In fact,

> "often the mind cannot unite, compare, or apply immediately the ideas one to the other, and this compels it to make use of other ideas (one or more) as means to the discovery of the agreement or disagreement it seeks, and this is what we call reasoning." (loc. cit.)

The partition into truths of reason and truths of facts, hence true statements about facts, can be characterized as follows: Truths of reason are necessary, their opposite is impossible because it is contradictory. The truth of them is ensured by logical laws. Leibniz referred here to the principle of identity and to the principle of contradiction. Facts can neither justify nor refute the truths of reason – the reason alone confirms them by itself, quasi *a priori*. Truths of reason are grounded not on facts and they do not concern facts. They concern only the possibility. Hence they are true not only in the actual world but in all "possible worlds" – similarly to the logical laws. They say nothing about any specific type of objects. On the other hand truths of facts are contingent and their opposite is possible. They are based on facts which can either justify or reject them. They are true only in the actual world.

Applying this distinctions and partitions to mathematics Leibniz claimed that not only trivial tautologies but all the axioms, postulates and theorems are truths of reason, hence they are necessary and eternal. They are not related to facts or experience and are true in all "possible worlds".

From Leibniz comes also the idea of treating logic in a mathematical way, as a mathematical discipline, i.e., to develop it by methods characteristic for mathematics. This is connected with his second important contribution to the philosophy of mathematics – the idea of developing a universal logical calculus. One can easily see here the connection of this idea with Descartes's of analytic and mathematical universal science. The source of Leibniz's wide scientific interests was the concern for the strictness and safety of knowledge as well as his clear universal rationalism. However, it is not the rationalism we spoke about at the beginning of this chapter. Leibniz's rationalism is a position that sees the highest, divine reason (*ratio*) acting in the world, in the best of all possible worlds; all things have been harmonically created and ordered according to it. This is illustrated by the following famous dictum:

> "Dum Deus calculat et cogitationem exercet, fit mundus."
> (While God calculates and thinks, arises the world.)

One can see here at the background again the divine numbers that we saw by Nicholas of Cusa.

The universal rationalism of the divine order of the world by Leibniz makes his idea plausible that a general logical language must be possible, a language that would be able to describe what was reasonably ordered. "Being already since his childhood familiar with logic he was fascinated by the idea coming from Ramon Llull (Raimundus Lullus). It is the idea of 'an alphabet of human thoughts' – 'combinations of letters' of it reduce all human notions mechanically to primitive notions, all true statements can be in a mechanical way obtained by them" (quotation after [47, p. 314], cf. [231, Volume 7, p. 185]). This idea appeared finally in the project of a universal and strict symbolic language called by Leibniz *characteristica universalis*. Its universality should be twofold: on the one hand it should express all concepts of sciences and on the other it should serve the communication among members of all nations. The system of signs proposed by Leibniz should fulfill the following conditions:

(1) There should be a one-to-one correspondence between signs of the system (provided they are not signs of empty places for variables) and ideas or concepts (in the broadest sense).
(2) The signs must be chosen in such a way that if an idea (thought) can be decomposed into components then the sign for this idea will have a parallel decomposition.
(3) One must devise a system of rules for operating on the signs such that if an idea M_1 is a logical consequence of an idea M_2 then the 'picture' of M_1 can be interpreted as a consequence of the 'picture' of M_2 (this is a sort of completeness condition).

According to these conditions all simple concepts corresponding to simple properties ought to be expressed by single graphical signs, complex concepts – by combinations of signs. This was based on a fundamental general assumption that the whole possible vocabulary of science can be obtained by combinations of some simple concepts. The method of constructing concepts was called by Leibniz *ars combinatoria*. It was a part of a more general method – a calculus – which should enable people to solve all problems in a universal language. It was called *mathesis universalis, calculus universalis, logica mathematica, logistica*. Leibniz hoped that *characteristica universalis* would, in particular, help to decide any philosophical problem. This is shown in the best way by the following quotation:

> "When this will happen [i.e., when the idea of the universal language will be realized – authors' remark] two philosophers coming into blow would argue like two arithmeticians. For it would suffice for them to take their pens in their hands and to sit down at the abacus, and say to each other (and if they so wish also to a friend called to help): ‚Calculemus!' (Let us calculate)." (The quotation comes from Leibniz's work "De arte characteristica ad perficiendas scientias ratione nitentes" from 1684; cf. [231, Volume VII, p. 125].)

Leibniz assigned an important role to symbolism. He told that "he owned all his discoveries in mathematics exclusively to his perfect way of applying symbols, and the invention of the differential calculus is just an example of it" (cf. [79, pp. 84–85]).

Leibniz did not succeed in realizing his idea of *characteristica universalis*. One of the reasons was that he treated logical forms intensionally rather than extensionally (cf. Section 5.1). This could not be reconciled with the attempt to formalize logic completely and transform it into a universal mathematics of utterly unqualified generality. This would inevitably mean that one must resign from many content domains. Another source of difficulty was his conviction that the combination of symbols must be a necessary result of a detailed analysis of the whole of human knowledge. Hence he could not treat the choice of primitive fundamental notions as a matter of convention. His general metaphysical conceptions induced a tendency to search for absolutely simple and primitive *concepts* (an analogue of monads), the combinations of which lead to the rich variety of notions.

As a partial realization of Leibniz's idea of *characteristica universalis* one can treat mathematical logic developed 200 years later at the turn of the 19th and 20th centuries. It is in fact only a partial realization since it concerns first of all (though not only) the language of mathematics.

2.10 Kant

Immanuel Kant (1724–1804) developed his philosophical systems on the one hand under the influence of the universal rationalism of Leibniz and the empirical philosophy of Hume (1711–1776) and on the other hand in a clear opposition to them.

As indicated above Leibniz distinguished between truths of reason and truths of facts. Axioms and theorems of mathematics are according to him truths of reason, i.e., they are such truths that are necessary, are not based on concrete facts and hold in all "possible worlds". Kant follows this distinction. Propositions and expressions are called judgements by him and he subdivides them into two classes:

- Analytical judgements, i.e., analyzing and deconstructing expressions; they are comparable with Leibniz's truths of reason.
- Synthetical judgements, i.e., expressions connecting concepts; they correspond to Leibniz's truths of facts.

In *Kritik der reinen Vernunft* [199, 201]) Kant characterizes both forms of judgements in the following way [201, B 10]:

> "In all judgments in which the relation of a subject to the predicate is thought (if I consider only affirmative judgments, since the application to negative ones is easy), this relation is possible in two different ways. Either the predicate B belongs to the subject A as something that is (covertly) contained in this concept A; or B lies entirely outside the concept A, though to be sure it stands in connection with it. In the first case I call the judgment analytic, in the second synthetic. Analytic judgments (affirmative ones) are thus those in

> which the connection of the predicate is thought through identity, but those in which this connection is thought without identity are to be called synthetic judgments. One could also call the former judgments of clarification and the latter judgments of amplification, since through the predicate the former do not add anything to the concept of the subject, but only break it up by means of analysis into its component concepts, which were already thought in it (though confusedly); while the latter, on the contrary, add to the concept of the subject a predicate that was not thought in it at all, and could not have been extracted from it through any analysis;"

Kant extended Leibniz's partition further and distinguished two types of synthetical judgements: empirical judgements, i.e., expressions based on experience – *a posteriori*, and non-empirical judgements that are independent of experience and therefore *a priori*. The synthetical *a posteriori* judgements are dependent of experience in the sense that their truth is based on sensual perceptions, for example "this flower is red". To this type belong also general propositions implying single propositions about sensual perception, for example "all ravens are black".

Synthetical judgements *a priori* can be – according to Kant – intuitive and discursive. By Kant, intuitive judgements are connected with the structure of perception, and perceptual judgements and discursive ones with the ordering function of general notions. An example of a discursive synthetical *a priori* proposition is the principle of causality, i.e., the statement saying that every fact has a cause.

Kant claimed – this is characteristic for him and distinguishes him from Leibniz – that all propositions of pure mathematics belong to the intuitive class of synthetical *a priori* statements. In the *Prolegomena zu einer jeden künftigen Metaphysik, die als Wissenschaft wird auftreten können* ([200], Prolegomena to Any Future Metaphysics That Will Be Able to Present Itself as a Science [203, § 6]) he writes about the pure mathematics:

> "Here is a great and established branch of knowledge, encompassing even now a wonderfully large domain and promising an unlimited extension in the future. Yet it carries with it thoroughly apodictical certainty, i.e., absolute necessity, which therefore rests upon no empirical grounds. Consequently, it is a pure product of reason, and moreover is thoroughly synthetical."

And he explains in § 7:

> "All mathematical cognition has this peculiarity: it must first exhibit its concept in a visual form [*Anschauung*] and indeed a priori, therefore in a visual form which is not empirical, but pure. Without this mathematics cannot take a single step; hence its judgments are always visual, viz., "intuitive"; whereas philosophy must be satisfied with discursive judgments from mere concepts,

> and though it may illustrate its doctrines through a visual figure, can never derive them from it."

There is the question: what is pure intuition,

> "*how is it possible to intuit [in a visual form] anything? a priori* [...], how can the *intuition* of the object [its visualization] precede the object itself?" ([203, § 8].)

The answer to those questions is a part of central turn made by Kant's epistemology in the history of philosophy. He sees the condition of the possibility of mathematical cognition in *forms of pure intuition*: space and time. As *forms* of pure intuition space and time are no external realities but *forms of human sensuality* proceeding sensual impressions and ordering them. Space understood in this way forms the foundation of geometry and the pure intuition of time – the foundation of the concept of number and of arithmetic. (In contrast, the aprioristic structure of reason forms the foundation of "apodictical doctrines" of philosophy – it is given by *categories of reason*.) In the pure forms of intuition of space and time mathematics "*constructs*" all its concepts "*in concreto*" and therefore *a priori* [203, § 7]. Mathematical cognition is therefore *a priori* as well as "apodictical and necessary" [203, § 10] – "*my intuition contains nothing but the form of sensuality, antedating in my subjectivity all the actual impressions through which I am affected by objects*" [203, § 9].

Let us quote in this contexts again [203, § 10]:

> "Now, the intuitions which pure mathematics lays at the foundation of all its cognitions and judgments which appear at once apodictic and necessary are Space and Time. For mathematics must first have all its concepts in intuition, and pure mathematics in pure intuition, that is, it must construct them. If it proceeded in any other way, it would be impossible to make any headway, for mathematics proceeds, not analytically by dissection of concepts, but synthetically, and if pure intuition be wanting, there is nothing in which the matter for synthetical judgments a priori can be given. Geometry is based upon the pure intuition of space. Arithmetic accomplishes its concept of number by the successive addition of units in time; and pure mechanics especially cannot attain its concepts of motion without employing the representation of time."

To what extent are mathematical propositions synthetical judgments of amplification and why can they be treated as corresponding to Leibniz's truths of facts? This must be explained.

Synthesis, the power of composing, is "faculty of imagination" that we are seldom conscious of [201, B 103, B 127]. In the developing of our knowledge, in particular of mathematical knowledge, "pure *synthesis*" is proceeded by "manifold of pure intuition"

[201, B 104] that is based on "the synthetical unit a priori". As an example Kant provides calculating and concepts of number under which he understands particular numbers and which belong to the category of quantity in the table of categories [201, B 106]. The synthesis of the manifold by itself is for Kant no cognition. Only the concepts of number which "give *unity* to this pure synthesis" "are the third thing necessary for cognition of an object that comes before us". In today's set theory the concept of a set introduced so aptly by Cantor, represents the pure synthesis of the manifold whereas the cardinal numbers, finite or transfinite, correspond to concepts of number which complete the pure synthesis to cognition.

Exemplary for synthetical judgements in mathematics are elementary arithmetical propositions like $7 + 5 = 12$. For Kant it is an example of amplification judgement in which the concepts "7" and "5" come together and are extended to the concept "12" that is not included in neither of them.

This is knowledge in "concreto" as quoted already above. Synthetical judgements have an *a priori* "objective validity" [201, A XVI]. They are "truths of facts" in quite a new sense. "Facts" are for Kant pure forms of intuition, the manifold in them, the synthesis of this manifold and therefore eventually the knowledge via – in our example – the concepts of number. Kant has put – and this is the epochal achievement – the (earlier extern – empirical or transcendent) foundations of knowledge, in particular of mathematical knowledge into the (transcendental) subject of cognition, into human being.

Let us summarize what is specific here with respect to numbers. According to Kant numbers are *a priori* partial structures of the mind [201, A 142, 143]:

> "The pure schema of magnitude (quantitatis) is number which is a representation that summarizes the successive addition of one (homogeneous) unit to another."

In succession appears the connection of numbers with the pure form of intuition of time. The achievement in the schema "number" is "the unity of the synthesis of the manifold of a homogeneous intuition". Generally, it holds:

Arithmetical propositions are synthetical a priori judgements.

It seems that Kant by grounding mathematics on the pure forms of intuition of space and time had set bounds on mathematics and in this way had not captured the richness of mathematical knowledge of his day. However, it is in our opinion a misunderstanding.

In fact, it is as if the description of space and time as pure forms being conditions of possibility of knowledge contained the claims that space can be only three-dimensional and time one-dimensional and directed. However, Kant never claimed that the structure of time and space would be in this way completely described. Just the opposite, he assumed the activity of the mind. The mind constructs concepts, in particular mathematical ones, in the sense that it creates *a priori* appropriate objects going beyond verbal definitions.

Kant distinguishes clearly between the construction of an object and the postulate of its existence. For example no five-dimensional sphere can be constructed; however, one can postulate its existence. Just this distinction between the assumption of the existence of a mathematical object – for which only the inner consistency is necessary – and its construction – that presupposes a particular structure of intuitive space – is important in order not to misunderstand Kant's philosophy. For example Kant never claimed that it would not be possible to provide a consistent non-Euclidean geometry. The current opinion that for example a non-Euclidean *Geometry of Visibles* described already in 1764 by Thomas Reid (1710–1796) (cf. [299] and [50, 53])[15] or mathematics of the non-Euclidean geometries since Gauß, Bolyai and Lobachevsky rejected Kant's philosophy of mathematics is mistakable if not simply false. Radbruch discusses this problem in detail in [295, pp. 118 ff.].

The conception of the essence of mathematics also enables Kant to explain the relations between pure and applied mathematics. As indicated above theorems of pure mathematics are – according to Kant – synthetical propositions *a priori*. They are objectively valid. On the other hand theorems of applied mathematics are either synthetical propositions *a posteriori* (when they have empirical content) or synthetical propositions *a priori* (when they say something about the pure forms of space and time). The pure mathematics treats time and space independently of the empirical presuppositions, the applied mathematics concerns time and space together with the empirical material in them. But why are theorems of pure mathematics suitable for the description of empirical reality? Kant answers this question in *Prolegomena* [203, Part I, Remark I] in the following way:

> "But if this image, or rather this formal intuition, is the essential property of our sensibility, by means of which alone objects are given to us, and if this sensibility represents not things in themselves, but their appearances: we shall easily comprehend, and at the same time indisputably prove, that all external objects of our world of sense must necessarily coincide in the most rigorous way with the propositions of geometry; because sensibility by means of its form of external intuition, viz., by space, the same with which the geometer is occupied, makes those objects at all possible as mere appearances. [...] But such is the case, for the space of the geometer is exactly the form of sensuous intuition which we find a priori in us, and contains the ground of the possibility of all external appearances (according to their form), and the latter must necessarily and most rigidly agree with the propositions of the geometer,

15 We would like to thank W. Breidert for indicating to us the considerations of Reid.

> which he draws not from any fictitious concept, but from the subjective basis of all external phenomena, which is sensibility itself. [...] It would be quite otherwise if the senses were so constituted as to represent objects as they are in themselves."

In the same remark prior we find:

> "Sensibility, the form of which is the basis of geometry, is that upon which the possibility of external appearance depends. Therefore these appearances can never contain anything but what geometry prescribes to them. It would be quite otherwise if the senses were so constituted as to represent objects as they are in themselves."

The last sentence shows the range of applications that refer solely and explicitly to phenomena of things. This is the next characteristic feature of Kant's "critical" epistemology that concerns conceptions and phenomena and does not penetrate to *things in themselves* (Dinge an sich) and into reality.

At last we come to the problem of the infinity that is essential for mathematics and therefore also for the philosophy of mathematics. Kant distinguishes – as Aristotle did – between potential and actual infinity. However, he does not claim as Aristotle that the actual infinity is not conceivable. According to Kant it is the idea of reason, i.e., the concept that is consistent in itself but that does not concern the empirical experience because its realizations can be neither observed (perceived) nor constructed. One can realize for example the number 7 and perceive this realization by senses, one can even construct the number $10^{10^{10}}$ though one is not able to establish or perceive by senses such a large set of objects. However, the infinite can be certainly neither perceived nor constructed.

2.11 Mill and the empirical conceptions

The empirical philosophy experienced at the beginning of the 19th century especially in England and France a new heyday. The so-called positivism was connected with Locke (1632–1704) and Hume (1711–1776). Positivists saw in "positive facts" the fundament of the philosophy. They saw sources of concepts in constant phenomena and the source of laws in recurring successions. They dismissed as "metaphysical" the question about the essence of concepts and philosophical objects in general as well as the search for the first and real causes. One of the important representatives of this tendency was John Stuart Mill (1806–1873).

First of all Mill developed the methodological version of empiricism. He justified his conception by referring to a psychological logic in a broad sense. A consequence of his conception was the connection of the empiricism with the nominalism – according to it concepts are subjective entities that have no counterpart outside the thinking. Mill looked for the foundation of his epistemology in the psychology, i.e., in searching "positive facts" of the consciousness. Objects were perceptions and combinations of perceptions. The purpose of *logic* was to distinguish constant perceptions from elusive ones and random combinations of perceptions from the constant ones.

To the most important contributions of Mill to logic and the methodology belongs certainly the theory of eliminative induction (i.e., induction that enables to eliminate accidental phenomena) and of inductive knowledge. He treated it as the unique basis of any cognitive knowledge. He systematically described principles of induction. This systems is even nowadays an important instrument of empirical sciences. It describes among others the method of accordance, the method of difference, the method of residues and the method of simultaneous change.

Mill's main work in logic and the methodology is *A System of Logic, Ratiocinative and Inductive, Being a Connected View of the Principles of Evidence, and the Methods of Scientific Investigation* [242] in 1843. This book was thought of as "a handbook of the doctrine that obtains any knowledge from experiment". It should be noted that Mill avoided empirical-inductive extremes and beside induction allowed in his logic and methodology also deduction. Attempts to avoid empirical dogmatism made it possible for him to consolidate different empirical conceptions. This contributed to the convergence of an empirical way of thinking of the 18th and 19th centuries, in particular of the naturalistic positions of the 18th century and the historical ones of the 19th century.

The empirical positions of Mill found their expression also in his philosophy of mathematics. His basis conviction was that the source of mathematics is the empirical reality. Mathematical concepts are obtained by a sort of abstraction from objects of the reality that surrounds us and that are captured by senses: some properties of real objects are omitted in perceptions, others are stressed, generalized and idealized. In *System of Logic* [242, II, V, § 1] he wrote:

> "The points, lines, circles, and squares which any one has in his mind, are (I apprehend) simply copies of the points, lines, circles, and squares which he has known in his experience. Our idea of a point, I apprehend to be simply our idea of the minimum visible, the smallest portion of surface which we can see. A line, as defined by geometers, is wholly inconceivable. We can reason about a line as if it had no breadth; because we have a power, which is the foundation of all the control we can exercise over the operations of our minds; the power, when a perception is present to our senses, or a conception to our intellects, of *attending* to a *part* only of that perception or conception, instead of the whole."

And he adds:

> "But we cannot conceive a line without breadth; we can form no mental picture of such a line: all the lines which we have in our minds are lines possessing breadth. If any one doubts this, we may refer him to his own experience. I must question if any one who fancies that he can conceive what is called a mathematical line, thinks so from the evidence of his consciousness: I suspect it is rather because he supposes that unless such a conception were possible, mathematics could not exist as a science: a supposition which there will be no difficulty in showing to be entirely groundless."

Numbers are for Mill quantities. Here a basis are real sets that are perceived as consisting of units successively put together. Those perceptions are the starting point of the abstraction of numbers. Mill's conception of numbers can be summarized as follows:

> *Numbers have their origin in the reality. Numbers are the result of an abstraction of successively recurring perceptions.*

Arithmetical expressions do not follow from the definitions of numbers but are based on facts that have been observed.

Such considerations yield necessarily the next thesis of Mill claiming that theorems of mathematics are not necessary and safe truths. Their necessity can be rather reduced to the fact that they correctly follow from assumptions from which they are deductively obtained. However, assumptions themselves are far from being necessary and safe. In fact, they are only hypotheses and can be completely arbitrary propositions. Hence the necessity and safety in mathematics is due solely to relations between propositions and not in the very propositions themselves. Theorems of mathematics are necessary and safe only to the extent in which those properties are ascribed to axioms. The latter however can be arbitrary hypotheses. Moreover, in practice they are often simply false because they are only idealizations and generalizations of real relationships. In *System of Logic* [242, II, V, § 1] Mill writes:

> "When, therefore, it is affirmed that the conclusions of geometry are necessary truths, the necessity consists in reality only in this, that they correctly follow from the suppositions from which they are deduced. Those suppositions are so far from being necessary, that they are not even true; they purposely depart, more or less widely, from the truth. The only sense in which necessity can be ascribed to the conclusions of any scientific investigation, is that of legitimately following from some assumption, which, by the conditions of the inquiry, is not to be questioned. In this relation, of course, the derivative truths of every deductive science must stand to the inductions, or assumptions, on which the

science is founded, and which, whether true or untrue, certain or doubtful in themselves, are always supposed certain for the purposes of the particular science. And therefore the conclusions of all deductive sciences were said by the ancients to be necessary propositions."

Recall in this context views of Aristotle who spoke in a similar way.

Elements of positivism can also be found in marxist epistemology. The founder of marxism did not deal systematically with the philosophy of mathematics. However, one finds in their works various ideas that were the base and the starting point of attempts to explain the phenomenon of mathematics inside the dialectical materialism.

Karl Marx (1818–1883) was interested in mathematics and knew for example the differential and integral calculus. *Mathematische Manuskripte* (Mathematical Manuscripts) published after his death contain some remarks on the differential. Friedrich Engels (1820–1895), under the influence of Marx, was also interested in mathematics. He made some remarks on it first of all in his so-called *Anti-Düring* and in *Dialektik der Natur* (Dialectics of Nature). He wrote for example in *Anti-Düring* [104] that the object of pure mathematics are the spatial forms and quantitative relations of the real world.

About the natural numbers Engels spoke in the following way:

"The concepts of number and figure have not been derived from any source other than the world of reality. The ten fingers on which men learnt to count, that is, to perform the first arithmetical operation, are anything but a free creation of the mind." [104, Chapter III, p. 20]

The quote looks almost like polemics towards the "bourgeois" conceptions of Cantor and Dedekind however formulated a bit later – we come to them soon.

In *Dialektik der Natur* [105] Engels claims that the infinity, the concept of the infinite, comes from the nature and only from it, from the reality and cannot be explained as a mathematical abstraction. Other claims made by him, for example concerning negative numbers or complex numbers, indicate that he had no fixed conception of mathematical objects.

Vladimir I. Lenin (1870–1924) spoke out on mathematics only allusively. However, one can find in his works the thesis that mathematics – like other sciences – has a social and class character. He saw the origin of the axioms of logic in the linguistic practice and explained it by the fact that it is permanently applied. According to him the axioms of logic and of mathematics reflect the regularity and repetitions in the nature.

The development of mathematics and in particular the formation of concepts which are more and more abstract and far from applications forced later the dialectical materialists to revise the views of marxist classics. Materialistic views on mathematics are nowadays not uniform. They still agree that mathematics arose from the practical needs of human beings, that its concepts are not inborn but have been developed in the process of exploitation of the nature by human beings. Hence the objects of mathematics are in their essence inferred from the objective reality of the material world. This dependence can have various forms and different character.

Mathematical concepts arise in the process of idealization and abstraction from the material reality that surrounds us. This explains why mathematical statements can be used to describing the world perceived by senses. There are various types of abstraction in mathematics, for example abstraction by generalization, abstraction of the potential realization and the abstraction of the actual infinite. One has to do with the first type when individual objects having a common property are identified. The abstraction of the potential realization consists of dissolving from the real limitations of human being abilities of constructions and overcoming them – this abstraction leads for example to the concept of potential infinity. The third type of abstraction leads – via abstraction from constructibility of elements and the realizability of sets – to the concept of the actual infinite that is fundamental in the more recent mathematics (cf. [308, Chapter 3]). The various methods of idealization and abstraction are interrelated and they supplement each other.

The appropriate method developed and accepted in mathematics is the axiomatic method. It secures not only the clarity but also the possibility of describing unknown – but possible – objects, their properties and possible mutual relations.

However, concepts of mathematics are not free and arbitrary products of the human mind. They are rather taken from the reality. A mathematician discovers relations in the material world and expresses them. Hence objects of mathematics are comparable with objects of natural sciences. There is in principle no separation between *a priori* and *a postertiori* truths. The formal character of mathematics does not suffice to explain the essence of mathematics. Relevant is first of all their content.

The criterion of truth of mathematical knowledge is – according to marxist conception – the social practice in the broadest sense. The essence of mathematics cannot be explained solely by philosophical investigations. Necessary is an additional historical, psychological and sociological reflection on mathematics.

Let us notice at the end that in the constructivism – in particular in the Soviet school of constructivism (see below) – one hoped that mathematics can be integrated into materialism and empiricism.

2.12 Bolzano

Bernard Bolzano (1781–1848) was a theologian, mathematician and philosopher in Prague. His main philosophical work *Wissenschaftslehre* (Theory of Science) [41] was published in 1837. Important for us is the fact that one also finds there some thoughts about logic. They concern the investigation of propositions having special structure. Here the separation of logical and psychological elements was decisive. Bolzano distinguished sharply between the psychological processes and logical contents in judgements – in logic only formal relations should be taken into account. Notice that conceptions of Bolzano were in fact close to ideas of mathematical logic developed at the turn of the 19th and 20th centuries.

From a mathematical point of view Bolzano was scientifically engaged first of all in analysis. He was the author of the first "pure", i.e., not referring to geometrical intuitions, definition of continuity of a function. He introduced the concept of the convergence of a series already before Cauchy. Thirty years before Weierstrass he provided an example of a continuous function that is nowhere differentiable. He formulated also important theorems of analysis. It should be stressed that in all his works in analysis he was an advocate of the so-called "arithmetization" of analysis. He was not satisfied with the geometrical intuitions used in analysis. His aim was to build the theory of real numbers – the fundament of analysis – on the base of the arithmetic of natural numbers. This was realized in a full version later at the end of the 19th century in works by Weierstrass and Dedekind. Bolzano promoted also the development of the tendency in the modern philosophy of mathematics, namely of logicism. Let us also notice that since Bolzano and thanks to him a concept can be introduced into mathematics in a legitimate way when it can be proved that it is "possible" without necessarily proving that it is constructible. Though at Bolzano's time mathematics was treated as a science of magnitudes, he characterized it already as abstract. He wrote that mathematics is "the science investigating general laws that regulates the existence of objects" [42, §13]. It is not the task of mathematics to prove the actual existence of mathematical objects. This is the task of metaphysics.

However, the essential reason why we speak here about Bolzano is different. The reason are his investigations of the infinite that can be found first of all in his in 1851 posthumously published work *Paradoxien des Unendlichen* (Paradoxes of the Infinite) [42].

Bolzano claims there: "Certainly most of the paradoxical statements encountered in the mathematical domain [...] are propositions which either immediately contain the idea of the infinite, or at least in some way or other depend upon that idea for their attempted proof" [42, § 1]. Hence one should investigate this concept.

In *Paradoxien des Unendlichen* Bolzano considers infinite multiplicities – today it would be said "sets" – as well as infinitely small and infinitely large magnitudes. He defines an infinite multiplicity as such

> "that every single finite multitude represents only a part of it" ([42, § 9].)

An example:

> "The set of all numbers manifests itself immediately as an indisputable example of an infinitely great quantity. I say advisedly, as an example of a quantity; and certainly not as an example of an infinitely great number;

> for this infinitely great multitude cannot, as we remarked in the previous paragraph, be given the name of number." ([42, § 16].)

Bolzano speaks here explicitly about the *set* of integers and not about the number; however, he speaks about the *magnitude* of this set and in this way clearly prepares what Cantor will think and do when he will introduce transfinite numbers.

One has to deal with an infinitely large magnitude – says Bolzano – when this magnitude is larger than the sum of any number of magnitudes taken as units. On the other hand a magnitude is infinitely small when every multiple of this magnitude is smaller than the unit.

Bolzano contrasts the mathematical concept of the infinite with the philosophers' concept of the infinity, for example with the concept of infinity by Hegel who – according to Bolzano – claimed that the mathematical infinite is only "the poor infinite" and that the philosophers know another one, "a true, a vastly superior, a qualitative infinity, to be found in God particularly, and speaking generally only in the absolute" [42, § 11]. Bolzano is of the opinion that the problem of Hegel consists here in assuming – as many philosophers and in fact also some mathematicians do – that the mathematical infinity is only potential, i.e., that it is only a variable magnitude that can grow unboundedly. However, the true mathematical infinity is not variable. Hence Bolzano admits here the *actual* infinity. Moreover, he attempts to prove its existence. An example of an actual infinite set should be "the set of propositions and truth in themselves" [42, § 13]. To be able to prove the existence of his actual infinity Bolzano needs an additional assumption. He assumes the existence of God whom he ascribes "the cognitive power that is real omniscience, hence comprises an infinite set of truths, all possible ones" [42, § 11]. The given example of the infinite set is existing, namely in God, i.e., this set has the property of the actual infinity. We see here by Bolzano a specific connection between the mathematical actual infinite and presuppositions of theological type.

Bolzano refers to – as he believes – paradoxical property of infinite multiplicities consisting in the fact that

> "[w]hen two sets are both infinite, they can stand in such a relation to one another that: (1) it is possible to couple each member of the first set with some member of the second in such wise that, on the one hand, no member of either set fails to occur in one of the couples; and on the other hand, not one of them occurs in two or more of the couples; while at the same time, (2) one of the two sets can comprise the other as a mere part of itself [...]." ([42, § 20].)

Recall that about this phenomenon already, for example, Proclus spoke who lived in the fifth century (cf. Section 2.5). He drew the conclusion that potential infinity should be accepted but actual one rejected in order to avoid the paradox. Bolzano did something else. He claims that this paradox is just *the* property which distinguishes finite and infinite sets. Some decades later Dedekind (cf. Section 2.15) will use this property – that from the intuitive point of view seems to be paradoxical – to define in a strict way the concept of the infinite set.

2.13 Gauß

With Carl Friedrich Gauß the history of mathematics reached the threshold to modern mathematics. Particularly Gauß began by his manifold mathematical works to change the mathematical thinking and – what has not been seen at the beginning – to resolve its traditional philosophical ties. This happened in the epoch impressed by Kant's philosophy in which arithmetical and geometrical primitive concepts were included into the epistemology. In this way those primitive concepts have been established ontologically in a new way. This can be seen in the most clear way in *a priori* forms of pure intuition of space that traditionally had Euclidean character. Gauß was the first who considered beside Euclidean geometry also the non-Euclidean one based on geometrical axioms without Euclid's fifth parallel postulate. In this way Gauß released mathematics a bit from ontological bonds and contested the ontological character of Euclidean axioms; recall our reference to the problem of the non-Euclidean geometry in Section 2.10 and notice that neither spatial intuition was meant by Kant mathematically nor non-Euclidean geometry was meant by Gauß epistemologically. However, Gauß sees and criticizes (in [129, Volume II, Remark, p. 170]) the conflict between the pure form of intuition by Kant and "the real meaning" of space "independently of our perception".

In contrast to this, Gauß conception of number seems to be quite traditional – this traditional trend began by Book V of Euclid's *Elements* where the geometrical concept of magnitudes was put before numbers.

According to Gauß the "object of mathematics" are "extensive magnitudes" – to them belong "the space or geometrical magnitudes comprehending lines, surfaces, bodies and angles, the time, the number". He explains that "the proper object of mathematics are relations of magnitudes" [129, Volume X, pp. 57–59]. Numbers are for Gauß abstractions of proportions of magnitudes:

> *"Numbers" indicate "how many times the directly given magnitude should be imagined."*

It is told that towards the end of his life Gauß confirmed those early conceptions (coming from the first years of the 19th century or earlier) (cf. [359, p. 155]). However, Gauß noticed already then that still more and more "priority was given to arithmetical notation in comparison with the geometrical one". He saw the reason of it in our "method of counting (in the decimal system)" that is "infinitely more easy than the systems of ancients" (loc. cit.).

Despite the fact that Gauß deduced his conception of number from the concept of a magnitude, he thought about a purely arithmetical setting of the domain of numbers. On the one hand Gauß contributed by his geometrical interpretation to establish definitely complex numbers as mathematical objects. On the other hand he did not accept the geometrical interpretation as the foundation of complex numbers. In a letter to the mathematician, psychologist and philosopher Moritz Drobisch from August 14, 1834, Gauß wrote:

> "The representation of imaginary magnitudes by relations of points *in plano* does not constitute their essence that should be understood higher and more generally but rather a most pure for us human beings and maybe quite pure example of an application of them. I have many times orally reported on my theory and later recognized that it can be easily grasped and that it contains nothing absurd." ([129, Volume X/1, Briefwechsel [zum Fundamentalsatz der Algebra], p. 106].)

He speaks here about a theory about which also Wolfgang Bolyai asked: "I have waited for a long time for the development of your theory of imaginary [magnitudes] ..." (Bolyai to Gauß on January 18, 1848, cf. [316, p. 129]).[16] Since the construction of domains of numbers begins with natural numbers, one can see here a reference to them as the foundation – detached from the concept of magnitude.[17]

2.14 Cantor

The significance of Georg Cantor's (1845–1918) works for the foundations of mathematics and for the philosophy of mathematics are enormous. We think here first of all about set theory developed by him and introduced into mathematics. This theory turned out to be fundamental for logical and philosophical investigations of mathematics. In Cantor's philosophy one can see all three philosophical basic positions we spoke about at the beginning of this chapter.

According to Cantor the reality of mathematical ideas and concepts can be understood in two contrastive ways: firstly as "intrasubjective" or "immanent realities", secondly as "transsubjective" or "transient realities". He was convinced that mathematical concepts are not only subjective, immanent real elements of the cognition but they possess also transsubjective reality. This implies his thesis that a mathematician

16 In fact, there are authors who ascribe Gauß the representation of complex numbers as pairs of real numbers dating it 1831 and in this way give Gauß the priority before Hamilton (1837) ([212, p. 776], [27, p. 179]). Those authors indicate, without giving the sources, two letters exchanged between Gauß and Bolyai from 1837 – they cannot be found in [316].

17 We would like to thank O. Neumann (Jena) for guiding us through Gauß works.

does not build or invent mathematical objects but discovers them. He often expressed this platonic conviction. One can indicate here the third thesis of his *Habilitationsschrift* or mottos from his fundamental work *Beiträge zur transfiniten Mengenlehre* [65]. The third thesis of Cantor's *Habilitationsschrift* was the following.

> "Numeros integros silimi modo atque corpora celestia totum quoddam legibus et relationibus compositum efficere." (The integers form like celestial bodies a quasi whole determined by laws and relationships.)

The mottos in *Beiträge* are as follows: "Hypotheses non fingo" (I invent no hypotheses) and

> "Neque enim leges intellectui aut rebus damus as arbitrium nostrum, sed tanquam scribae fidels ab ipsius naturae voce latas et prolatas excipimus et describimus." (We give namely the laws of knowledge and of things not by our free judgement but as reliable writers we extract and describe them in the form in which they are born and reported in the language of nature.)

In a letter to Mittag-Leffler (1846–1927) from 1884 he wrote: "What concerns the content of my works I am only a rapporteur and a public officer" (cf. [121, p. 480]). Cantor attributed the real existence to concepts of set theory – not only in the world of ideas but also in the physical world. In particular he was convinced that for example sets of cardinality $\aleph_0$ – like the infinite set $\mathbb{N}$ of the natural numbers – and the continuum $\mathbb{R}$ exist in the real material world.

In the *immanent reality* of mathematical concepts Cantor saw the condition of the possibility of a *pure* or, as he would like to say, of a *free mathematics*. Its freedom consists of "the obligation [...] to check them [i.e., the concepts – authors' remark] also according to transcendental reality." Pure mathematics is in contrast to other disciplines "free of all metaphysical shackles" (cf. [64, pp. 182 f.]).

Cantor introduced the concept of a set not axiomatically but intuitively. Hence his description of the concept of the set was not sufficiently precise and clear. Soon various paradoxes appeared – we describe them below. In Cantor's works one finds two a bit differing circumlocutions of the concept of a set. In the work *Beiträge* already quoted he writes (cf. [66, p. 282]):

> "By an "aggregate" (*Menge*) we are to understand any collection into a whole (*Zusammenfassung zu einem Ganzen*) M of definite and separate objects m of our intuition or our thought (which are called the "elements" of M)."

In *Grundlagen einer allgemeinen Mannigfaltigkeitslehre* (Foundatiomns of a General Theory of Manifolds) in 1883 one finds the following sentence (cf. [66, p. 204]).

> "In general, by a 'manifold' or 'set' I understand every multiplicity [jedes Viele] which can be thought of as one, i.e., every aggregate [Inbegriff] of determinate elements which can be united [verbunden] into a whole by some law."

Cantor introduced the concept of cardinal and ordinal numbers. They were also described only intuitively. This led soon to the discovery of paradoxes in set theory. Two of them, namely the antinomy of the set of all ordinal numbers, called today the Burali-Forti paradox, and the antinomy of the set of all sets have been discovered already by Cantor. He found a solution to those antinomies by distinguishing *classes* and *sets*. A *class* is a multiplicity that cannot be thought as "one", as "a whole" or as "a complete object" and therefore cannot be treated as an element of new multiplicities. Such multiplicities were called by Cantor "absolute infinite" or "inconsistent multiplicities". On the contrary a *set* is a multiplicity that can be thought as "determinate well distinguishable" objects and therefore as being itself an element of classes or sets. Cantor called them "consistent multiplicities" or – as we said – "sets" (cf. letters of Cantor to Dedekind from July 28, 1899, and August 31, 1899, – in [66, pp. 443–448]).

The most important part of Cantor's set theory are his contributions on infinite sets. Cantor distinguished various forms of the infinite. First of all he distinguished – as Aristotle already did – between actual and potential infinity. One has to do with the potential infinity – according to Cantor it was no infinity in the proper sense and therefore he called it sometimes "not proper infinity" – where a undeterminate variable finite "magnitude" appears that either grows above all finite magnitudes or can be smaller than any arbitrary small one. By actual infinity Cantor understood a magnitude that is given as a whole, in all its parts is determinate and is a "constant" and simultaneously exceeds every finite magnitude of the same type (cf. [66, pp. 400–401]). Cantor claimed that potential infinity presupposes the actual one. Notice that Cantor used here the concept of a magnitude what was characteristic for mathematics of the 19th century. This unprecise concept has been later replaced by the concept of a set.

Cantor distinguished also three forms of the actual infinity: (1) the absolute infinity realized solely in God, (2) the infinity appearing in the dependent and created world, (3) the infinity that can be in thought *in abstracto* understood as mathematical magnitude. The absolute infinity is not extendable while the other two types of actual infinity can be extended. In the case of the infinity of type (3) Cantor says about *transfinitum* and opposes it to the absolute infinity (cf. [66, p. 378]). In his works Cantor developed a theory of transfinite ordinal numbers and by using the concept of being equipollent introduced the concept of the cardinality of a set and the hierarchy of transfinite cardinal numbers.

Cantor propagated the existence of actual infinities and at the same time rejected actual infinitely small magnitudes. The later were called by him "paper magnitudes" (Papiergrößen) and he opposed vehemently the introduction of such magnitudes ("the infinitary bacillus of cholera") into mathematics (cf. *Aus dem Briefwechsel Georg Cantors* [66, p. 505]).

Cantor looked for a justification of his theory of infinite sets also outside of mathematics, in particular in philosophy and in theology. He was convinced that set theory is a part of metaphysics. He attempted to prove the existence of the transfinite in mathematics by appealing to the divine absolute. He believed that the real transfinite of his conception is not destructive with respect to the nature of God – just the opposite, it contributes to God's glory. He provided two proofs of the existence of transfinite numbers *in abstracto*. In one of the proofs Cantor concludes the possibility and necessity of creating the infinite on the ground of the concept of God and of the perfection of God. In the second proof he argues as follows: since it is impossible to explain all natural phenomena without assuming the transfinite *in natura naturata*, the transfinite does exist.

Cantor was convinced that set theory is of great importance for metaphysics and for theology. In fact, his works were eagerly studied by philosophers as well as by catholic theologians. Cantor found just by them recognition and faithful readers. On the other hand many mathematicians demonstrated their lack of interest in set theory. They reacted very critically and rejected it.[18] Cantor was in close contact with Jesuit and cardinal Johann B. Franzelin, one of the most important pope theologians of the First Vatican Council, with Jesuits Tilman Pesch and Joseph Hintheim, with Dominican Thomas Esser and with the Franciscan Ignatius Jeiler. Cantor was anxious for their positive opinion about set theory.

Temporarily Cantor was suspected of pantheism that has been officially indexed by Pope Pius IX in 1861. The reason was Cantor's claim that transfinite numbers exist *in concreto*. This led to the suspicion of identifying the infinity *in natura naturata* with God's infinity *in natura naturans*. Cantor refuted this objection by extending the distinction between infinities *in natura naturata* and *in natura naturans* by a further distinction. He distinguished between Infinitum aeternum increatum sive Absolutum, hence the eternal non-created infinite from the Absolute and Infinitum creatum sive Transfinitum, hence the created infinite from the transfinite. This conciliated ecclesiastic theologians and philosophers and Cantor's works received a sort of Catholic *Imprimatur*.

Still let us make some remarks on the concept of natural numbers by Cantor that is not independent of his extension of the concept of number into the transfinite. Cantor regarded the concept of number – both finite and transfinite – as a result of a double abstraction act that he understood as omitting some properties of objects and simultaneous reflection on the similarity of sets – the result was "*Allgemeinbegriff*"

18 One of the opponents of set theory was for example Leopold Kronecker (1823–1891), professor at the University of Berlin, one of the university teachers of Cantor. Among members of the small group of mathematicians who accepted Cantor's set theory was Richard Dedekind. He has early (cf. [85]) recognized the meaning of set theory for the foundations of mathematics.

[65, pp. 281 f.]. He explains:

> "Since every single element m, if one abstracts from its nature, becomes a "unit", the cardinal number $\overline{\overline{M}}$ is a definite set composed of units, and this number has existence in our mind as an intellectual image or projection of the given aggregate M." ([65, p. 282].)

This resembles Euclid's "set composed of units" and holds both for finite numbers as well as for infinite numbers beyond natural numbers. Hence natural numbers are by Cantor finite cardinal numbers. It is clear that the concept of number allowing infinite numbers must exclude links to Kant's forms of intuition of time. Transfinite cardinal number understood by Cantor as actual infinite realities definitely refute the aristotelian dogma concerning the actual infinity. Defending his position – against Aristotle, partially Leibniz, supposedly Gauß, Kronecker and many others – Cantor looked for witnesses in the history of philosophy and found some "points of contact" [64, p. 205] by Plato, Nicholas of Cusa and Bolzano.

Cantor's conception of numbers can be summarized as follows:

> *Numbers are finite cardinal numbers. On the one hand they are given as ideal realities – independently of human thinking, on the other as projections of sets existing in thought and obtained by abstraction.*

Let us close by saying some words about a problem that arose in connection with the hierarchy of transfinite cardinal numbers introduced by Cantor. We mean the so-called Continuum Hypothesis, i.e., the question whether there exists a cardinal number (cardinality) between the cardinality of the set of natural numbers and the cardinality of the set of real numbers. Cantor was not able to solve this problem – and he would not be able to understand the nowadays answer stating the independence of the Continuum Hypothesis. The reason is that this problem did not fit into his empiric-platonic conception. The world of his sets was real and not formal. Cantor doubted occasionally if his set theory has any scientific value. This doubts and the atmosphere of general mathematical antipathy and lack of understanding of set theory led to a nervous breakdown and later psychic illness from which Cantor suffered since the spring of 1884.

2.15 Dedekind

We are writing here about Dedekind even though he spoke out on his philosophical conceptions only in a margin of his mathematical works. His works proved to be extremely important for the foundations of analysis and arithmetic. He was one of founders of modern algebra. He worked in group theory, established ring theory and made researches in the theory of algebraic numbers. At the same time he was – together with Karl Weierstraß and Georg Cantor – one of the representatives of the new trend in mathematics whose aim was to eliminate systematically all ambiguities in basic concepts of mathematics. It was the continuation of efforts of Cauchy, Gauß and Bolzano. Dedekind was a close friend of Cantor and he was one of the first who recognized the importance and the meaning of the set-theoretical works of Cantor. This is documented by their correspondence (cf. *Gesammelte Abhandlungen* [66, pp. 443–451].

From the point of view we are interested in, two works by Dedekind are of special importance: *Stetigkeit und irrationale Zahlen* ([85], 1872) and *Was sind und was sollen die Zahlen?* ([87], 1888). In the first work one finds the development of a theory of irrational numbers on the basis of cuts (today called *Dedekind cuts*) in the domain of rational numbers [85, p. 13]. We will speak about this below. Dedekind's approach recalls definitions of magnitudes of Eudoxus of Cnidus (cf. [109, Book V]). Therefore Dedekind is sometimes called the "new Eudoxus". However, both theories are by no means identical. There are serious differences between them. It has turned out that principles used by Eudoxus and later by Euclid in Book V of his *Elements* in order to develop the theory of incommensurable magnitudes are not sufficient to obtain a complete theory of real numbers – as proportions of magnitudes. Here the principle of continuity introduced by Dedekind is necessary – it was missing in the works of Greek mathematicians. In a letter to Lipschitz from June 10, 1876, Dedekind wrote (cf. [89, Volume 3]):

> "[...] nowhere by *Euclid* or by one of later writers one finds a final formulation of such completion law, nowhere there is the concept of continuity, i.e., the concept of a domain of magnitudes that would be complete in the highest possible degree, whose essential property would be the following: 'if all magnitudes of a domain of magnitudes structured in a hierarchy in a continuous way resolve into two classes such that any magnitude of the first class is smaller than any magnitude of the second one then either in the first class there is the largest magnitude or in the second class – the smallest one'."

And he adds:

> "After all those remarks I still insist in my claim that *Euclid's* principles alone without adding the principle of continuity which is *not* included in them do not suffice as a basis of a complete theory of real numbers as proportions of magnitudes. [...] On the contrary, however, by my theory of irrational numbers there has been created a perfect example of a continuous domain in which every proportion of magnitudes can be characterized by a certain number belonging to it."

The second work by Dedekind that is important here for us, i.e., *Was sind und was sollen die Zahlen?*, contains a set-theoretical foundation of natural numbers, of the complete induction and of the recursion as well as axioms for the arithmetic of natural numbers. Those axioms are known today as Peano axioms. Hence there arises a problem: who was in fact the real author of those axioms, whom – Dedekind or Peano – should be granted the priority? We are of the opinion that it is an ostensible problem. In fact, Dedekind and Peano formulated their axioms in different formal, or better quasi-formal, languages and their aims were quite different. The formulations of Peano were clearly logical and Dedekind was – as stated above – oriented set-theoretically. Hence axioms formulated nowadays in mathematical commonplaceness usually in a set-theoretical way should be called rather Dedekind axioms.

Especially the preface of the work *Was sind und was sollen die Zahlen?* is interesting for us. Here Dedekind summarizes his philosophical conception. It seems that he agrees with Cantor in what concerns the immanent, intrasubjective reality of mathematical concepts in mind or reason, i.e., the rationalistic aspect of his attitude. Dedekind adds a psychological component connected with the conception of a psychological development. Arithmetical truths are – according to Dedekind – "never *given* by inner consciousness" [italics by the authors] but "always *gained* only by a more or less complete repetition of the individual inferences" [87, p. V]. Here Dedekind does not exclude connections with his own mathematical inferences in this famous paper. And he continues: "So from the time of birth, continually and in increasing measure we are led to relate things to things and thus to use that faculty of the mind on which the creation of numbers depends [...]".

The creation of numbers is described by Dedekind in the following way:

> "If in the consideration of a simply infinite system N set in order by a transformation ϕ we entirely neglect the special character of the elements, simply retaining their distinguishability and taking into account only the relations to one another [...] then are these elements called *natural numbers* [...]. With reference to this freeing the elements from every other content (abstraction) we are justified in calling numbers a free creation of the human mind." ([87, p. 73].)

"Simply infinite systems" describe the infinite process of counting in an actually given infinite set. From the structure of counting described by him in three axioms obtains Dedekind the concept of number and sees in the abstraction the core of "creation".

The idea of a "mental creation" is more exact in the work *Stetigkeit und irrationale Zahlen* [85]. Dedekind recognizes gaps in the domain of rational numbers between certain upper classes and lower classes. An example of a "cut" is the upper class of rational numbers whose square is greater than 2 and the associated lower class of rational numbers whose square is smaller than 2. Between them there is a gap because in the domain of rational numbers there is no number $\sqrt{2}$. Those gaps are closed according to him by numbers *created* by us:

> "Whenever, then, we have to do with a cut [...] produced by no rational number, we create a new, an *irrational* number α, which we regard as completely defined by this cut [...]; we shall say that the number α corresponds to this cut, or that it produces this cut." ([85, p. 13].)

Frege and Russell proved later that irrational numbers can be identified simply with cuts (or still simpler with for example their lower classes). In this way psychological creation acts can be mathematically avoided. However, Dedekind defended his psychological conception of "creation" of irrational numbers with great persistence.

In the case of natural numbers there are (mathematically) concrete objects like cuts lacking that could be correlated with natural numbers. The only basis is an abstract structure and the idea of abstract not identifiable positions in this sequence structure. Dedekind says that numbers are obtained by abstraction from counting sequences called by him "simply infinite systems". This is the core of Dedekind's conception of number:

Numbers are abstractions of positions in infinite counting sequences.

It is difficult to understand this kind of abstraction that should create here numbers as "creation". So we try to make "creation" a bit more available.

A concept needs a name, a sign, an image or a word in order to be effective. Dedekind describes the structure of counting. Hence one needs here for example *many* signs that represent not only numbers but also their sequence structure. Counting is then a concept of a sequence *via* a concrete sequence that one must "invent". It is good for a communication to build conventions and to reach an agreement concerning for example the sequence of the usual decimal number symbols. Objects of such sequences are called numbers because they represent the abstract places in abstract structures of counting. One can recognize how directly in such a "structuralistic" concept of number, a number and its symbol belong together.

There are two important and great steps connected with Dedekind's procedure that changed the thinking. The reduction of natural numbers to sets "stops" the process of counting. In fact, set-theoretical concepts are static. However, just in this way the principles of the process – that beforehand were hidden in the intuitive and temporal process – become explicit. The *set-theoretical reconstruction* of the process is needed in order to reproduce the principles of the process of counting in a set-theoretical form.

The second great step is: Dedekind's characterization of natural numbers had already a set-theoretical axiomatic form. It presets the set of natural numbers as an actual infinite set. This forms a firmly given framework. Particular natural numbers are not constructed any more step by step in the process of counting. They do not arise any more, they are *a priori* there. One does only speak in axioms about natural numbers.

Infinite sets are for Dedekind – as they were for Cantor – something obvious and actually given. In *Was sind und was sollen die Zahlen?* one finds the definition of the concept of an infinite set that became classic. This definition is based on a seemingly paradoxical property of infinite sets (cf. Proclus and Bolzano). According to this definition, a set is called infinite if and only if it can be mapped in a one-to-one way onto a proper subset of it. Dedekind formulated this in the following way [87, Item 64]:

> "A system S is said to be *infinite* when it is similar to a proper part of itself (32); in the contrary case S is said to be a *finite* system."

In [87, Section 3, Item 32] Dedekind defined the concept of similarity as follows: "The systems R, S are said to be similar when there exists such a similar transformation ϕ of S that $\phi(S) = R$, and therefore $\overline{\phi}(R) = S$." Today similar transformations are called injective mappings and the symbol $\overline{\phi}$ denotes the inverse mapping of ϕ.

Dedekind not only defined the concept of the infinite set but – similarly to Bolzano – tried to prove the existence of infinite sets. His "proof"[19] is similar to the "proof" of Bolzano (see above). Bolzano's proof uses some theological presuppositions, Dedekind's proof comes from psychologico-philosophical presuppositions. In [87, Item 66] Dedekind formulates the following sentence: "There exist infinite systems". And he justifies this claim in the following way: "My own realm of thoughts, i.e., the totality S of all things, which can be objects of my thought, is infinite." Further he shows that this system fulfills the definition of an infinite set.

The infinite process sawn by Dedekind as realized in the chain of the building of thoughts (the thought of the thought of ...) in his realm of thoughts was later mathematically paraphrased by Zermelo and accepted as a postulate as the Axiom of Infinity in his set theory (cf. Section 4.3.1): There exist infinite sets! What was attempted to justify by Dedekind – from today's point of view "only" philosophically – must be claimed mathematically.

Dedekind attempts to make the foundations of mathematics precise and to arrange them were often criticized. For example David Hilbert criticized Dedekind's idea to found the whole of mathematics on logic – at that time set theory was included into logic. Gottlob Frege and Bertrand Russell did not share Dedekind's conception of the existence of irrational numbers presented above.

19 We put the word "proof" in quotation marks because by today's standards Dedekind's proof cannot be treated as a proof.

2.16 Poincaré

Henri Poincaré (1854-1912) represents in the philosophy of mathematics the rationalistic point of view of apriorism. He was an intuitionist and a constructivist as well as the founder of conventionalism – the latter found its expression in particular in his philosophy of geometry and in his methodology of empirical sciences. There is a connection of Poincaré's views and the philosophy of Kant. His conceptions presented Poincaré first of all in his books, in particular in *La science et l'hypothèse* (1902), *La valeur de la science* (1905), *Science et méthode* (1908) and in the posthumously published *Dernière pensées* (1913).

Poincaré ascribed an important role in mathematical thinking to the creative activity of mind and its ability to construct concepts. This creative ability of the mind is expressed in various forms. One of its manifestations is the intuition. The concept "intuition" appears in Poincaré's works in different meanings. Generally, intuition is for him a native ability of the mind connected with the impulsive activity. It appears both in the unconscious work as well as in the conscious work. In the former it is the activity in the subconsciousness where intuition makes it possible to consider a great number of combinations in a short time. In the latter intuition is able to identify the essential facts among a great number of concrete facts as well as to see the whole beyond many details. Intuition has spontaneous and rational aspects. It provides the feeling of clarity and evidence. It is many-sided and can be independent of sensations. There are various types of intuition: reference to senses or imagination, generalization by induction, the intuition of a pure number etc. In the scope of intuition there is also – according to Poincaré – the possibility (preexisting in mind) of constructing concepts, for example the concept of a group as a pure and not sensual cognitive form. The latter played an important role in Poincaré's philosophy of geometry. We return to this later.

Poincaré connected with intuition the feeling of simplicity, harmony, symmetry and beauty that can be found also in mathematics. He wrote about this for example in *Science et méthode* [280] in the following way:

> "What is it that gives us the feeling of elegance in a solution or a demonstration? It is the harmony of the different parts, their symmetry, and their happy adjustment; it is, in a word, all that introduces order, all that gives them unity, that enables us to obtain a clear comprehension of the whole as well as of the parts. [...] Briefly stated, the sentiment of mathematical elegance is nothing but the satisfaction due to some conformity between the solution we wish to discover and the necessities of our mind, and it is on account of this very conformity that the solution can be an instrument for us. This aesthetic satisfaction is consequently connected with the economy of thought."

Intuition should be supplemented – according to Poincaré – by the discursive knowledge. This means that it must be introduced and closed by conscious activity that should rationally ensure intuitive "epiphanies".

To intuition belongs among others mathematical induction. It is at the base of arithmetic and consequently of the whole mathematics. It enables us to formulate judgements extending our knowledge. The complete induction makes the formulation of general statements possible. It does not result from the experience and it cannot be justified by logic. It is just complete induction (and its applications) that indicates the difference of mathematics and logic. Poincaré opposed the logicism claiming that mathematics can be reduced to logic. He indicated that logicists were not able to dispense with induction – it is in fact a nonlogical principle. From the point of view of cognition logic is – according to Poincaré – empty and tautological and renders only the formulation of analytical judgements possible. Hence mathematics cannot be reduced to logic alone (cf. *Science et méthode* [280, Book II, Chapter III]).

The essence of induction and its applications is summarized by Poincaré in the following way: in a reasoning by induction one formula corresponds to infinitely many conclusions. In the work *La science et l'hypothése* [277, First Part, Chapter I, § VI] Poincaré wrote:

> "This rule, inaccessible to analytical proof and to experiment, is the exact type of the *a priori* synthetic intuition. [...] Why then is this view imposed upon us with such an irresistible weight of evidence? It is because it is only the affirmation of the power of the mind which knows it can conceive of the indefinite repetition of the same act, when the act is once possible. The mind has a direct intuition of this power, and experiment can only be for it an opportunity of using it, and thereby of becoming conscious of it."

And he adds:

> "It cannot escape our notice that here is a striking analogy with the usual processes of induction. But an essential difference exists. Induction applied to the physical sciences is always uncertain, because it is based on the belief in a general order of the universe, an order which is external to us. Mathematical induction, i.e., proof by recurrence, is, on the contrary, necessarily imposed on us, because it is only the affirmation of a property of the mind itself."

As a constructivist Poincaré claimed that mathematical objects are constructed by subject and that there is no domain of mathematical knowledge that would be independent of a knowing subject. Hence it was a position contradicting platonistic realism. One of the consequences of this position was the acceptance only of potential infinity and the rejection of the actual one. In fact, according to Poincaré, mathematical objects are created by a subject and the mind is not able to construct actually infinitely many objects. Hence the actual infinity is mathematically impossible.

Poincaré criticized the so-called non-predicative definitions, i.e., definitions in which an object N is defined by reference to a certain totality E of objects (for example by a quantification over E) to which N belongs. An example of a definition of this type is the following definition of the set $\mathbb{N}$ of natural numbers: $\mathbb{N}$ is the smallest set containing the number 0 and closed under the successor function. One defines here the object $\mathbb{N}$ by refering to the totality of objects to which $\mathbb{N}$ itself belongs.

Poincaré claimed that such non-predicative definitions – being a manifestation of a vicious circle – are the source of paradoxes in mathematics. A standard example of it provided by Poincaré is the following Richard's paradox: let E be the totality of all real numbers given as infinite decimal fractions that can be defined by finitely many words. Hence E is of course countable. Applying now the famous Cantor's diagonal reasoning one can define a real number N not belonging to E. However, the number N is defined by finitely many words. Consequently, N belongs to the set E. A contradiction. Poincaré explains this contradiction by indicating that the definition of the number N as an element of E refers to the set E as a totality to which N belongs as an element. Hence this definition is not predicative.

To avoid paradoxes and contradictions Poincaré suggests to apply the following rules:

> "(1) To consider always only such objects that can be defined by a finite number of words, (2) never to lose sight of the fact that any statement on the infinite must be a "translation", i.e., an abbreviated formulation of statements about the finite, (3) to avoid non-predicative classifications and definitions."
> ([281, Chapter IV, § 7].)

In practice Poincaré not always followed those rules, in particular rule (3). The suggestion to avoid non-predicative definitions has been later adopted by Russell and developed in his theory of types.

As indicated above Poincaré was also the founder of *conventionalism* that formed his methodology of empirical sciences and – what is especially interesting for us – his philosophy of geometry. The conventionalism claims among others that the laws of the natural sciences are not direct and straight descriptions of the reality. They have rather the character of agreements, they are conventions. Hence they can be neither fully justified nor rejected by experience. There are many *pari passu* descriptions of the world. One of them is chosen not because it is true but because it is useful, short, convenient and beautiful.

Poincaré applied this principle to geometry and claimed the following (cf. [277, Second Part, Chapter III, "On the Nature of Axioms"]).

> "*The geometrical axioms are therefore neither synthetic a priori intuitions nor experimental facts.*
> They are conventions. Our choice among all possible conventions is *guided* by experimental facts; but it remains *free*, and is only limited by the necessity of avoiding every contradiction and thus it is that postulates may remain *rigorously* true even when the experimental laws which have determined their adoption are only approximate. In other words, the *axioms of geometry* (I do not speak of those of arithmetic) *are only definitions in disguise.*"

Since the discovery of non-Euclidean geometries in the 19th century there is the following problem: which geometry, the Euclidean one or one of the non-Euclidean geometries is actually right? Poincaré answers this question claiming that it is put in a wrong way. He says: "One geometry cannot be more true than another; it can only be more *convenient*" [277, p. 51]. However, there was the opinion that the Euclidean geometry is and will be most useful. This follows from the fact that it is the simplest geometry and "sufficiently agrees with the properties of natural solids, those bodies which we can compare and measure by means of our senses" [277, p. 52].

Experience plays an important role in the development of geometry. However, this does not mean that geometry is an experimental discipline. In such a case it would be only "a rough and temporary" discipline as Poincaré says.

Poincaré refers to the idea of Felix Klein to characterize geometries as theories of invariants of certain groups of transformations. Note that the very concept of a group – more exactly: the possibility, preexisting in the mind, of the concept of a group as a pure form of cognition – belongs to the domain of intuition. He writes (cf. [277, Second Part, Chapter IV, "Conclusions"]):

> "The object of geometry is the study of a particular 'group'; but the general concept of group preexists in our minds, at least potentially. It is imposed on us not as a form of our sensitiveness, but as a form of our understanding; only, from among all possible groups, we must choose one that will be the standard, so to speak, to which we shall refer natural phenomena. Experiment guides us in this choice, which it does not impose on us. It tells us not what is the truest, but what is the most *convenient* geometry."

2.17 Peirce's pragmatism and the world of symbols

Charles Sanders Peirce (1839–1914) is the actual founder of the philosophical doctrine of pragmatism that was valued in Europe as an American one and not so much respected by European scientists. The central role in Peirce's philosophy was played by the domain of signs lying between thinking and the real world. According to Peirce thinking is a sequence of thoughts performed in signs. Mathematics and logic formed the base of Peirce's philosophy. They are on the top in his hierarchy of scientific disciplines. And vice versa, his philosophy influenced his mathematical works and conceptions.

Peirce has been early and consequently instructed in natural sciences and in mathematics – he was also interested in philosophy. He intensively studied Aristotle and Leibniz and – as he told – at the age of sixteen he knew by heart Kant's *Kritik der reinen Vernunft*. Peirce's theory of three universal categories: the intuitively and plainly given, the simple opposition and the association and universality – arose in a long enduring engagement from these studies.

Peirce was first of all a natural scientist but worked also in other domains: philosophy, mathematics, logic. However, his ingenious contributions were only temporarily or not at all recognized. Many of his scientific works were refused. He made no academic career – this was caused also by social circumstance that nowadays seem to be strange. Peirce himself was, as reported by his few friends, rather a difficult person. He died in 1914 pennilessly and almost isolated. His written *Nachlass* comprises about 100 000 pages – they are till today sighted, selected and edited (cf. [157]).

Peirce followed carefully the development of logic, read works by Boole and De Morgan and contributed to algebra of logic [269]. His ideas have been adopted by Schröder in [320] (cf. [227, 268]). One should mention here first of all his contributions to the logic of relations. He observed the set-theoretical turn of mathematics, followed it positively and considered it as a step from the discrete towards the continuum. In 1900 he wrote two letters to Cantor in which he critically discussed the concepts of an element and of a set. He distanced himself from the conception of the continuum as a set of points and from the identification of the linear continuum with the real numbers.

For Peirce the continuum was a reality and formed the continuous background of thinking, intuition, thought, sign and the temporal-logical connection of concepts. It was unimaginable for him that the continuum disintegrates into points. The continuum was the medium of discrete totalities of any cardinality. Peirce remained in Aristotle's tradition, was tied to the infinitesimal conception of continuum by Leibniz (cf. Section 3.3) and used the concept of a point that can be found today in the nonstandard analysis (cf. Subsection 3.3.3 and Chapter 6). Points were environments of potential points and "without separate individuality" as moments or time points that are always continuously connected with the past by sequences of infinitesimal steps.

Let us summarize some further principles of Peirce's philosophy. Pragmatism is, as indicated by the name, an attitude oriented at acting (in Greek *τὸ πρᾶγμα*). One can briefly say that concepts are determined when their impact on the practice is known. They receive their meaning by the practice, by relations between things. The aim of thinking is to prepare the acting. Propositions refer to intended or real actions and experiments. Propositions and judgements – when cognitively meaningful – are hypotheses. Peirce speaks about "abduction" that starts from single observations and experiments and forms hypotheses. He confronts abduction with deduction and induction that by bare derivations or generalizations provide no real extension of knowledge. Abduction is for Peirce the source of human knowledge given to man from the cradle like instincts to animals.

Propositions are subject to a permanent, even historically, examination and are never certain but only – more or less – accepted. They are hypothetical approximations on the way towards the truth. Hence the question about the *truth* is "sociologized" and put into the infinite process of verification that converges towards the real truth. He told that any "infallibility in scientific context" is for him "irresistibly comic". The degree of truth became "public matter" or radically is measured by its "cash-value" – as is said by James (cf. [257, p. 136]). Such an attitude is relevant also for propositions of mathematics that are obtained by hypotheses, namely deduced from axioms. Their temporary confirmation comes only from applications and actions. Application is the permanent examination of truth.

The old ontological question about concepts is reduced to the practice from which the concepts receive their real meaning. Concepts and their relations are given – and this is new and mathematically important – a second reality in the world of signs. We turn to this below.

Peirce's pragmatism has – and this can now be easily seen – necessarily an evolutionary and empiricist component. Peirce is at the same time a (hypothetical) realist and an idealist. His realistic idealism is manifested in his assumptions concerning the relation between concepts and the reality. The transition from sensations of real things to concepts and judgements is continuous. It assumes an accordance that links concepts as well as things. The connection of real things is like the connection of concepts, namely logical. It takes place at the background of the continuous time in a logical process that accompanies and promotes the hypothetical cognition.

> If there is the reality then it "consists in the following: there is something in the essence of things that corresponds to the reason process, that the world *exists* and **moves** and HAS ITS BEING in a logic of events." ([270, IV, pp. 343 f.], [207, p. 218].)

Everything is embedded in a mental cosmos, in an infinite evolutionary process that is of a mental nature. It rises from a random state and aspires – being directed by repetitions of states and progressive structuralizing to general principles and aesthetics of the world. It seems that Peirce's position lies exactly in the center of the triangle of philosophical basic positions described at the beginning of this chapter.

A central and important feature of Peirce's conception is the linking of cognition with signs and their connections, and consequently the shifting of problems of cognition to the world of signs. Thinking is not possible without signs. Senses provide signs of things. Thoughts are signs – "every thought is a sign", "all thought is in sign". The inner, notional signs are analogies of outer signs. Thoughts are never singular, thoughts precede thoughts and follow them in logico-temporal relations. Signs connect thoughts and the reality. They are the key to connections between concepts and things. In signs the relations of concepts as well as the relations of things became apparent.

One can see that the concept of a sign is by Peirce very wide, even universal. The world of signs connects the world and thinking. For him signs are in the center of an infinite process of cognition. Starting from this fundamental cognitive meaning of signs Peirce considered intensively appearances and functions of signs, distinguished forms of their appearance and described classes of signs. He became the founder of modern semiotics, a scientific discipline with broad applications. Not least it is instructive for manipulating signs and symbols in mathematics and logic, which cannot be imagined without signs, and significant for the conception of mathematics.

Here the base is Peirce's description of the concept of a sign in the triangle sign–object–idea. A sign is a thing whose aim is to communicate the knowledge of another thing represented or connoted by it. This thing is called object of the sign. The idea generated in mind by the sign being the mental sign of the same object is called the "interpretant" of the sign (cf. [213, Volume I, p. 204]). In [270, Volume IV, pp. 20–21] one finds the following abstract "definition":

> "Namely, a sign is something, A, which brings something, B, its interpretant sign determined or created by it, into the same sort of correspondence with something, C, its object, as that in which itself stands to C."

This is the classical description coming from Aristotle. New in it is that in principle it converts into a triangle of signs in which the roles played by sign, object and idea are exchangeable. Important for the world of signs of mathematics and logic are the following distinctions:

Token is reserved for a single sign – as for example the word "sign" just written. *Types* integrate single signs of the same kind, hence all words "sign" in this and all possible texts. A sign, say g, becomes an *icon* when it can potentially appear in a connection with an object. If this object is given, for example a straight line, then it will be an *index*. A sign is a *symbol* if there is a fixed connection between sign and object. For example $\mathbb{N}$ is a symbol. Finally, every sign becomes *rheme*, *dicisign* or *argument* according to the category which the interpretant attributes to the sign's way of denoting its object. The rheme, for example a term, is a sign interpreted to represent its object in respect of quality; the dicisign, for example a proposition, is a sign interpreted to represent its object in respect of fact; and the argument is a sign interpreted to represent its object in respect of habit or law.

A *diagram* is a sign consisting of signs, usually an icon of icons, indices and symbols. They represent relations on the level of signs. Examples are: geometrical figures, formulas, graphs etc. Diagrams form – when treated pragmatically – the world in which mathematicians are acting. They are almost sensual objects of this acting – they denote and mean themselves. Diagrams and their mutual relations are the mathematical reality in which propositions are checked with respect to their validity.

Acting in the mathematical world of signs proceeds according to strict rules – by those rules arise "inevitably" new relations and diagrams. Dörfler sees in this conception a possibility of a "demystification of mathematical practice" that replaces the usual "metaphysical ontology" of mathematical objects by "ontology of diagrams" [98, p. 26].

Diagrams generate propositions that obtain their pragmatical truth by the *inevitability* of acting. Brunner speaks in his work [58, p. 21] about "diagrammatic reality" that spares a mathematician the speculation about "the existence" of mathematical objects "in platonic sense" and of "ideal truths". This pragmatic view seems to describe accurately the activity of a mathematician in his everyday research practice.

2.18 Husserl's phenomenology

Phenomenology, a tendency in the philosophy of the early 20th century established by Edmund Husserl (1859–1938) influenced to a certain degree the modern philosophy of mathematics. Husserl himself came from mathematics. He began his studies of mathematics at the universities of Leipzig and and Berlin under Carl Weierstraß and Leopold Kronecker. In 1881 he moved to Vienna where he studied under Leo Königsberger and in 1883 obtained the doctor's degree on the base of the dissertation *Beiträge zur Variationsrechnung*. He was strongly impressed by lectures of Franz Brentano (1838–1917) on psychology and philosophy which he attended at University of Vienna and decided to dedicate his life to philosophy. In 1886 he went to the University of Halle to obtain his *Habilitation* with Carl Stumpf, a former student of Brentano. The *Habilitationsschrift* was entitled *Über den Begriff der Zahl. Psychologische Analysen*[20] (1887). This work of 64 pages was later expanded into a book (of five times the length) being one of Husserl's major works: *Philosophie der Arithmetik. Psychologische und logische Untersuchungen* ([185], cf. also [186, 188]).

Working as *Privatdozent* at the University of Halle, Husserl came in contact with mathematicians: Georg Cantor, the founder of set theory (which turned to be one of the most important and fundamental theories in mathematics, see Chapter 4) and Hermann Grassmann's son, also Hermann. The former, with whom he had long philosophical conversations when they were teaching together in Halle in the 1890s, told him about Bernard Bolzano. In fact, Husserl was perhaps the first philosopher outside Bohemia to be influenced significantly by Bolzano. Later, as a professor of the University in Göttingen, Husserl had contacts with David Hilbert and, as a professor in Freiburg (Breisgau) where he was appointed in 1916, with Ernst Zermelo.

Cantor influenced in a certain sense the earlier works of Husserl though he is quoted only two times in Husserl's *Habilitationsschrift*. Similarly discussions with Gottlob Frege – the founder of logicism, one of the main trends in the modern philosophy of

20 On the Concept of Number. Psychological Analyses

mathematics (see Section 2.19) – influenced him. Both, Cantor and Frege will appear below when we shall describe Husserl's philosophy of arithmetic. Husserl's philosophy had in fact no visible meaning for the mathematics of his time; however, conversely the mathematics strongly influenced his philosophy.

One of the mathematical motives of his philosophy can be recognized in Weierstraß's program of arithmetization of analysis. Its aim was to found the whole of mathematics on the base of arithmetic and to define all its concepts in terms of arithmetical ones. Quite a lot of mathematicians of the 19th century initialized and supported the arithmetization, among them Augustin-Louis Cauchy, Bernard Bolzano, Richard Dedekind, Georg Cantor and Carl Weierstraß himself. Husserl's aim was to justify by his investigations philosophically and psychologically the Weierstraß's program. In the preface to *Philosophie der Arithmetik* he wrote [185, p. VIII]: Perhaps my efforts should not be wholly worthless, perhaps I have succeeded in preparing the way, at least on some basic points, for the true philosophy of the calculus, that desideratum of centuries".

In order to make our further considerations more clear let us present briefly the main features of Husserl's phenomenology by some keywords. One can see aspects of his phenomenological methods and basic concepts already in his early *Habilitationsschrift*. It, as well as his book *Philosophie der Arithmetik*, was influenced by Brentano and shaped by his descriptive psychology. Later Husserl turned away from this "psychologism" and criticized the psychological point of view in the philosophy of logic and mathematics – for example in the first volume of his *Logische Untersuchungen*.[21] His purely philosophically and *a priori* treated phenomenology that should remove psychology as the foundation has been developed by Husserl for more than forty years. His aim was to establish philosophy as a strict science and to create the universal foundation of all disciplines.

The concept "phenomenology" is at the first sight a bit unclear. It is not about the usual phenomena. They are only the starting point on the way to the "true" phenomena. This way consists in a "reduction" which strips off from the appearance of objects perceived by a subject all that is accidental and not essential or what has been put into them by a philosophical observer. The process of reduction is finished by "*eidos*", one would say in the "essence" or the "idea" of an object. Just here, in Eidos, is located the source of a concept that belongs to the object and precedes the concept.

The word "idea" used today is too abstract to describe the character of *eidos*. Husserl uses the *eidos* similarly to Plato. The Greek word "εἶδος" has a quite different intuitive character – of another, higher intuition."Eidos" *means* intuition. It is – one could say – "deeper", intuitive intuition that belongs by Husserl to *eidos*. From the empirical

21 However, some forms of psychologism which he analyzed there and tried to reject cannot be directly seen in his *Philosophie der Arithmetik*.

observation grows in the reduction the deep intuition of "eidetic phenomenon" that intentionally refers to the reality. Husserl's doctrine is the "eidetic phenomenology".

The first, the fundamental *eidos* is the "*eidos ego*" – as Husserl called the result of the reduction of the empirical I. It forms the basis of knowledge and the "transcendental ego", the faculty of concepts is based on it. It is like a pure consciousness which is necessarily intentional, i.e., consciousness of pure objects. It forms the deepest ground of concepts and of the *a priori* knowledge. The world of real phenomena can be retrieved from it – in its proper "sense of being".

Mathematics was for Husserl a typical example of an eidetic discipline. It studies the fundamental objects, like numbers in the arithmetic, and forms or similar phenomena in the geometry.[22] Husserl claims that one can penetrate in a kind of *Wesensschau* to their essences, their *eidos* – as in the case of physical objects. He makes here no difference. Mathematics as an eidetic discipline studies the abstract objects in which intentionally is more than we can recognize in our normal cognition and to which we will be phenomenologically led back.

Husserl's phenomenology recalls at the first sight Plato's theory of ideas. It seems that it puts the ideas into human beings, in the "*eidos ego*" and the transcendental ego. However, this analogy is inapplicable because it conceals at least one essential aspect: Husserl begins – in a different way than Plato – by the concrete phenomena and tries to get to their essences that resemble rather aristotelian forms in things than the higher ideas. Essential are for him the methods of reduction and intuition, methods which cannot be found in a comparable manner by Plato and Aristotle. However, one can guess that Husserl's phenomenology meets halfway views of mathematicians thinking in a platonic way. By *Wesensschau* and eidetic intuition Husserl addresses also the intuitionistic group of his time, though his concept of number was, as we see now, quite different.

In *Philosophie der Arithmetik* Husserl still referred, as mentioned above, to Brentano's method of descriptive psychology and understood – similarly to Weierstraß and other mathematicians of that time – natural numbers by *empirical* counting, which in the works of Husserl is masked by other principles. Against this Dedekind already in 1888 in his work [87] declared the unclear psychological description of the foundations of numbers to be mathematically insufficient.

In the first part of *Philosophie der Arithmetik* Husserl developed a psychological analysis that started from the *everyday concept* of a number. The analysis begins with the development, application and appearance of numbers and on this base he tries to explain the psychological origin of numbers. He claims that the fundamental concept of a number cannot be defined: "[...] the difficulty lies in the phenomena, in their correct description, analysis and interpretation. It is only with reference to the phenomena that insight into the essence of the number concept is to be won" [185, p. 142]. These

22 Husserl proposed an extension of the geometry in the direction called today topology.

words exhibit Husserl's psychological belief from this period. Here we find already the "reference to the phenomena".

Since our intellect and time are bounded we are able to achieve the comprehension only of a very small part of mathematics. In order to overcome those limits one introduces symbols which accompany and guide our thinking. Almost all we know about arithmetic we know indirectly via the intermediation of symbols. This explains why in the second part of *Philosophie der Arithmetik* Husserl considers extensively the symbolic representations.

Husserl opposed the characterization of natural numbers by axioms as Dedekind and Peano at the same time did. He rejected also the justification of arithmetic on the base of logic (Frege) or set theory (Cantor). In *Philosophie der Arithmetik* he wrote that arithmetic cannot be "a sequence of formal definitions, out of which all the theorems of that science could be deduced purely syllogistically" [185, p. 130]. As soon as one comes to the ultimate, elementary concepts – Husserl speaks here about time, equality, multiplicity, whole, part, number etc. – "all defining has to come to an end" (loc. cit.). Husserl looks for the first origin of fundamental numbers and and attempts to reveal the *phenomena* from which the concept of number arises. The way does not start with abstract numbers. The challenge is to find the sources of the number concept, to comprehend the nature of the abstraction process and to describe the concept formation. According to that one should focus on "our grasp of the concept of number" and not on the number as such.

Husserl understands abstraction in the following way: "to abstain from something or abstract from something means simply: not notice this especially". And he explains: abstraction "does not have the effect that its content and its connections disappear from our consciousness" [185, p. 85]. Here is psychologically indicated what Husserl later included into his method of phenomenological reduction. That are contents that are "not especially noticed" – just they make the "Colligieren" possible, the connecting to a new whole.

This "Colligieren" which leads to "multiplicities" is for Husserl directly connected with the concept of number. This is one of two principles that are fundamental for numbers. The second principle is the underlying everything principle of "something". "The 'something' is no abstract partial content" [185, p. 86] of any "concrete multiplicity", for: "the concept of something is due to the reflection on the psychic act of conception". Again one can suppose that here – in psychological, intentional something – presages the later philosophical *eidos*.

By such copies of something are constituted general multiplicities: "A multiplicity is nothing more than: something and something and something etc.; or any one and any one and any one etc.; or briefly: one and one and one etc." [185, p. 85]. In the word "one" Husserl sees the relation of "partial content" with the whole of the multiplicity that is not expressed in "something".

Multiplicity and *quantity* (*Anzahl*) – and here we are by Husserl's *concept of number* – can be hardly distinguished. "It is *a priori* apparent that they coincide in their essential content" [185, p. 89]. "Quantity" is the "generic term": the concept of quantity distinguishes the "abstract forms of multiplicities", cancels the "vague indefiniteness" of multiplicities and appends to them the "sharply definite how many" (loc. cit.). Multiplicity for Husserl resembles the "something" of number, an indefinable psychological datum [185, p. 130].

The essential element of the abstraction that leads to the just-mentioned concept of quantity is in the concept of "something" [185, p. 129]. It spares – differently than in the case of the set-theoretical concept of a cardinal number – the comparison of "concrete multiplicities" what Husserl explicitly notices. Husserl's concept of quantity comes back to the age-old "definition" of a number by Euclid and corresponds to Cantor's characterization of cardinal numbers stating that it is "a definite aggregate composed of units". We recognize that by Husserl one has to deal here with a process which is like a counting: "One and one and one etc.". He treats numbers as arising and given additively, for example "three" as "one, one and one" [185, p. 87]. The counting process is so explicit and clear in this formulation that it seems that Husserl does not separate quantity and counting. At least in such a way he articulates in his (very sharp and not consistent) critique of Kant's concept of schemata ([185, p. 33], cf. [204, pp. 108 ff.]):

> "Number is the idea of an universal procedure of imagination getting the concept of quantity an image. However, this procedure can only mean counting. But is it not clear that "number" and the idea of "counting" are the same?"

This remark is a bit surprising because just at that time Dedekind (1888) and Peano (1889) provided a clear mathematical separation of number and quantity, the grounding of the concept of number by counting. It seems as if the mathematician Husserl did not want to notice this in his psychological, anti-axiomatic attitude. Let us try to characterize briefly Husserl's concept of number:

> *By Husserl numbers are quantities. Quantities are distinguished multiplicities of abstract units.*

By Husserl just these quantities are primary for the concept of number. On the other hand cardinal numbers as classes of equipollent sets are unfinished and "useless concept formations" [185, p. 129] which state no number but only the equality of number or quantity. This "definition" (Husserl himself puts this in quotation marks) is "considerably appreciable" [185, pp. 130 f.] only for "this wildman" on "that level of mind" for whom the symbolic counting is not available. In our opinion he misunderstands both Cantor (in favor of him) as well as Frege and finally also Dedekind. Note that he criticizes only the decided anti-psychologist Frege.

According to Husserl, a mathematician operates not with abstract numbers but with quantities that are always connected with the idea of special sets via multiplicities.

Mathematics itself is for Husserl a formal ontology. Objects investigated by mathematics are formal categories in various forms – and they are themselves not perceivable. Numbers are an example here. Thanks to the ability of categorial abstraction we can free ourselves from the empirical components of judgements and to concentrate ourselves on the formal categories. In the eidetic intuition and variation we are able to grasp the possibility, impossibility, necessity and contingency of connections between concepts or between formal categories. The categorical abstraction and the eidetic intuition form the base of the mathematical knowledge.

Ideas of Husserl had a not negligible influence on the position of the famous mathematician Hermann Weyl (1885–1955). Weyl's interests in the phenomenology go back to his graduate student days between 1904 and 1908 and his allegiance to it lasted till the early twenties. Some years after the publication of the work *Das Kontinuum* [365] in 1918 Weyl changes his views turning towards the intuitionism of Brouwer and developed his own version of intuitionism. For him the phenomenology and the intuitionism were strongly connected. Later Weyl changed his views again and legitimated Hilbert's program.

The impact of Husserl's ideas on Weyl can be seen in the care with which Weyl treated issues like the relationship between intuition and formalization, the connection between his construction postulates and the idea of a pure syntax of relations, the appeal to the *Wesensschau*, etc. In the preface to the work *Das Kontinuum* Weyl explicitly declares that he agrees with the conceptions that underlie the *Logische Untersuchungen* with respect to the epistemological side of logic. Answering to Husserl's gift of the second edition of the *Logische Untersuchungen* to him and his wife he wrote in a letter to Husserl (cf. [187, p. 290]):

> "You have made me and my wife very happy with the last volume of the *Logical Investigations*; and we thank you with admiration for this present. [...] Despite all the faults you attribute to the *Logical Investigations* from your present standpoint, I find the conclusive results of this work – which has rendered such an enormous service to the spirit of pure objectivity in epistemology – the decisive insights on evidence and truth, and the recognition that 'intuition' [*Anschauung*] extends beyond sensual intuition, established with great clarity and conciseness."

On the other hand Husserl read Weyl's *Das Kontinuum* [365] as well as *Raum, Zeit, Materie* [368] and found them close to his views. He stressed and praised Weyl's attempts to develop a philosophy of mathematics on the base of logico-mathematical intuition. Husserl was pleased to have Weyl – who was a prominent mathematician – on his side. In a private correspondence he wrote to Weyl that his works were being read very carefully in Freiburg and had had an important impact on new phenomenological investigations, in particular those of his assistant lecturer Oskar Becker.

Oskar Becker (1889–1964) studied mathematics at Leipzig and wrote his doctoral dissertation in mathematics under Otto Hölder and Karl Rohn in 1914. He then devoted himself to philosophy and wrote his *Habilitationsschrift* on the phenomenological foundations of geometry and relativity under Husserl's direction in 1923. He admitted that it was Weyl's work that made a phenomenological foundation of geometry possible. Becker became Husserl's assistant in the same year. In 1927 he published his major work *Mathematische Existenz* (cf. [15]). The book was strongly influenced by Heidegger's investigations, in particular by his investigations on the facticity of *Dasein*. This led Becker to pose the problem of mathematical existence within the confines of human existence. He wrote: "The factual life of mankind [...] is the ontical foundation also for the mathematical" (cf. [15, p. 636]). This standpoint in the philosophy of mathematics led Becker to find the origin of mathematical abstractions in *concrete* aspects of human life. In this way he became critical of Husserl's style of phenomenological analysis. This anthropological current played an important role in Becker's analysis of the transfinite. Hence Becker utilized not only Husserl's phenomenology but also Heideggerian hermeneutics, in particular discussing the infinity of arithmetical counting as "being towards death" (*Sein-zum-Tode*).

At the end of his life Becker re-emphasized the distinction between intuition of the formal and Platonic realm as opposed to the concrete existential realm and developed his own approach to the phenomenology called by him "mantic". With this word he referred to the fact that there is a divinatory aspect related to any attempt to understand "*Natur*". In the light of this mathematics appears as a divinatory science which by means of symbols allows us to go beyond what is accessible. Mantic phenomenology will have to replace the older "eidetic" phenomenology.

Becker's works have not had great influence on later debates about the foundations of mathematics, despite many interesting analyses included in them, in particular of the existence of mathematical objects.

Talking about Weyl and Becker one should mention also Felix Kaufmann (1895–1949), an Austrian–American philosopher of law. He studied jurisprudence and philosophy in Vienna and from 1922 till 1938 (when he left for the USA) he was a *Privatdozent* there. He was associated with the Vienna Circle. He wrote on the foundations of mathematics attempting, along with Weyl and Becker, to apply the phenomenology of Husserl to constructive mathematics. His main work in this area is the book *Das Unendliche in der Mathematik und seine Ausschaltung*[23] (1930).

One of the most famous logicians and philosophers of mathematics in whose works one finds Husserl's phenomenological ideas is Kurt Gödel (1906–1978). We write about his philosophy of mathematics in Section 2.23. Here we want to indicate connections between his ideas and Husserl's.

---

23 Infinity in Mathematics and its Suppression

Let us start by noting that Husserl never referred to Gödel. In fact, he was more than 70 years old when Gödel obtained his great results on the incompleteness and consistency, and he died a few years later, in 1938, without seeming to have taken notice of Gödel's work. Also Gödel never referred to Husserl in his published works.[24] However, his *Nachlass* shows that he knew Husserl's work quite well and appreciated it highly.

Gödel started to study Husserl's works in 1959 and became soon absorbed by them finding the author quite congenial. He owned all Husserl's main works. He owned among others the *Logische Untersuchungen* (in the edition from 1968), *Ideen, Cartesianische Meditationen und Pariser Vorträge, Die Krisis der europäischen Wissenschaften und die transzendentale Phänemenologie*. The underlinings and comments (mostly in Gabelsberger shorthand) in the margin indicate that he studied them carefully. Most of his comments are positive and expand upon Husserl's points but sometimes he is critical. One should note that Gödel expressed philosophical views on mathematics similar to those of Husserl long before he started to study them. Views found in Husserl's writings were not radically different from his own. It seems that what impressed him was Husserl's general systematic philosophy which would provide a systematic framework for a number of his own earlier ideas on the foundations of mathematics.

Husserl's post-psychologistic, transcendental view of mathematics is still a live option in the philosophy of mathematics. As Richard Tieszen writes it is "compatible with the post-Fregean, post-Hilbertian and post-Gödelian situation in the foundations of mathematics" [344, p. 335]. The phenomenological approach to the philosophy of mathematics is still being developed by various authors. The starting point for their considerations is, however, not directly Husserl's works but rather Gödel's considerations. Let us mention here, for example, P. Benacerraf, C. Chihara, P. Maddy, M. Steiner, C. Parsons and R. Tieszen. They are not only commenting Gödel's works but are developing their own phenomenological interpretations of mathematics concentrating first of all on the problem of mathematical intuition (cf., for example, [236, 261, 343]).

2.19 Logicism

Logicism is the trend in the philosophy of mathematics claiming that the whole mathematics can be reduced to logic. In other words: mathematics is a part of logic. Logic and its formal character we treat in detail in Chapter 5. The founder of the logicism was *Gottlob Frege* (1848–1925) and its main representative were *Bertrand Russell* (1872–1970) and *Alfred North Whitehead* (1861–1947).

---

24 An exception is his work "The Modern Development of the Foundations of Mathematics in the Light of Philosophy" (cf. [145]) published only after Gödel's death.

Historical sources of the logicism are manifold. One can discover them both in philosophical tradition and in episodes of the history of mathematics. Concerning the axiomatic method, which is a part of this conception, logicists refer especially to Plato, Aristotle and Euclid. They rely also on ideas of J. Locke and G.W. Leibniz. In fact, the logicism can be traced till the philosophical controversies between the universal rationalism (as by Leibniz) and empiricism concerning the essence of mathematical propositions. Logicism refers here to Locke and Leibniz who claimed that mathematical statements are tautological and redundant. It refers also to Leibniz's idea of algorithmization of argumentation both mathematical and generally scientific ones.

The development of logicism as a school of thought in the philosophy of mathematics would be impossible without the development of modern mathematical logic in the second half of the 19th century. This logic was a partial realization of Leibniz's idea of *characteristica universalis* whose aim was the logical analysis of concepts and structures of scientific systems. Mathematical logic was initiated by the project of the mathematization of logic that demanded the development of logical symbolism similar to mathematical symbolism. Simultaneously, the traditional aristotelian logic should be strengthened and extended. Researches of Augustus De Morgan (1806–1871), George Boole (1815–1864), Charles Sanders Peirce (1839–1914) and Ernst Schröder (1841–1902) contributed to the formation of the so-called *algebra of logic*. This term denoted in the second half of the 19th century, and at the beginning of the 20th century the formal logic formed after the pattern of the algebra of numbers. On the other hand logical works of G. Frege and B. Russell should be mentioned – those works represented a non-algebraic approach to logic. Just Frege was the predecessor and one of the founders of modern formal logic. His fundamental logical work *Begriffsschrift, eine der arithmetischen nachgebildete Formelsprache des reinen Denkens* ([122], 1879) prepared the way to the new era in formal logic even though his works were hardly recognized by representatives of the algebraic trend, for example by E. Schröder or John Venn. Frege's *Begriffsschrift* contained the first formal axiomatic systems in the history of logic – the system of propositional calculus with implication and negation as connectives. In such a system every formula arises by a deduction (without gaps) from axioms according to precisely formulated rules. One finds in Frege's work also a formal analysis of propositions with quantifiers as well as appropriate axioms for quantifiers.

All those contributions – both of the algebraical trend and of the non-algebraical one – prepared the formal platform on which the logicism as a philosophical view could be developed.

The development of the logicism was influenced also by the tendency at that time to arithmetize the whole of mathematics, i.e., to reduce it to the arithmetic of natural numbers and to understand it as a uniform whole. This tendency was very strong in the second half of the 19th century. It was closely connected with works of K. Weierstrass and R. Dedekind. The arithmetization of mathematics means first of all the arithmetization of analysis and the latter consists of demonstrating that the theory of real numbers being the base of mathematical analysis can be reduced to the theory of natural numbers. This means that the concept of a real number – and prior the concept of an integer and of a rational number – can be derived from the concept of a natural number (and the elementary concepts of set theory) and further all properties of the real numbers can be deduced from statements of the arithmetic of natural numbers. Dedekind's work *Stetigkeit und irrationale Zahlen* ([85], cf. Section 2.15) quoted above demonstrated how this is possible. One of the ways from the rational numbers to the real numbers has been described in Chapter 1.

In this situation arose the requirement of justifying natural numbers and providing a more elementary base from which the arithmetic of natural numbers could be obtained. A first approach to the solution was provided by Frege in his work *Grundlagen der Arithmetik* [123] in 1884 as well as in two volumes of his *Grundgesetze der Arithmetik* (cf. [124], Volume I in 1893 and Volume II in 1903).

In *Grundlagen der Arithmetik* Frege undertook the attempt of presenting arithmetic as a part of logic: all arithmetical concepts can be explicitly defined by logical concepts and all arithmetical and consequently all mathematical theorems can be deduced from logical laws. This forced at the very beginning a radically new position towards the concept of a number. Frege developed it in a committed dispute with "views of philosophers and mathematicians concerning questions coming here into consideration". His philosophical position can be characterized by:

- Anti-empiricism: theorems of arithmetic are not inductive generalizations.
- Anti-kantianism: theorems of arithmetic are not synthetic *a priori* judgements.
- Anti-formalism: theorems of arithmetic are neither systems of signs nor rules of manipulating signs.

Frege's anti-empiricism contains a definite anti-psychologism. All that is psychic was for Frege necessarily subjective. The main thesis of his studies was "to separate strictly what is psychic and what is logical, the subjective and the objective" [123, p. IX]. Since theorems of arithmetic can be reduced to logical laws, they are – according to Frege – *analytical judgements* and consequently *a priori* per se.

Frege grasped numbers as quantity:

Numbers are quantities.

They were classes of equipollent sets. Sets were understood as extensions of concepts since Frege avoided the concept of a set as a nonlogical one. In a final step concepts as elements of pure thought were objects of logic. This means:

Numbers are elements of logic,

i.e., of pure thought. Concepts existed for Frege – and here he was a Platonist – independently of space, time and human mind. Notice that Frege – in contrast to Dedekind and Peano who proceeded in a structuralistic way – undertook a *mathematical* attempt to settle definitely the *ontological status of numbers*.

Below we shall say more about Frege's definition of natural numbers. Here let us tell only that Frege did not express quite clearly his view on the character of logical laws. He claimed that logical laws are no laws of nature but "laws of laws of nature", they are not laws of thought but "laws of truth".

For a long time works by Frege have remained almost unnoticed or undervalued. One of the reasons was the symbolism introduced by him – it was inventive but very complicated. Behind this symbolism stood his idea of reducing mathematics to logic. Hence it should be completely different from the usual mathematical symbolism – it should help to *see* clearly that mathematics can be really reduced in a complete way to logic.

One of the few scientists who knew Frege's works and was able to value it was the Italian mathematician *Giuseppe Peano* (1858–1932). Peano himself was engaged in developing symbols for mathematics and logic and pleaded for using an axiomatic method in mathematics and its applications. He developed a transparent and useful logical and mathematical symbolism. The symbolism used nowadays comes mostly from him – it was later modified a bit by Russell. Peano demonstrated how mathematical laws can be axiomatically and deductively ordered and systematized by using his symbolism (cf. [247]). In particular he showed how the whole arithmetic of natural numbers can be deduced from five axioms called today Peano axioms (cf. *Arithmetices principia nova methodo exposita*, [266]) (see also the remarks in Section 2.15 devoted to Dedekind).

Bertrand Russell learned about Frege's works from Peano. While studying the first volume of Frege's *Grundgesetze* Russell noticed that the system of logic to which Frege reduced the arithmetic of natural numbers was inconsistent. In fact, one can build in this system the antinomy of the so-called "irreflexive classes" called today Russell's antinomy. Russell discovered this fact in 1901 and communicated it to Frege in the letter of June 16, 1902. He wrote in it:

> "With regard to many particular questions, I find in your work discussions, distinctions, and definitions that one seeks in vain in the works of other logicians. [...] There is just one point where I have encountered a difficulty. You state (p. 17) that a function, too, can act as the indeterminate element. This I formerly believed, but now this view seems doubtful to me because of the following contradiction. Let *w* be the predicate: to be a predicate that cannot be predicted of itself. Can *w* be predicted of itself? From each answer the opposite follows. Therefore we must conclude that *w* is not a predicate. Likewise there is no class (as a totality) of those classes which, each taken as a totality, do not belong to themselves. From this I conclude that under certain circumstance a definable collection does not form a totality." ([159, pp. 124–125]).

This was the first formulation of Russell's antinomy. Frege answered Russell in a letter from June 22, 1902, in which he wrote being worried:

> "Your discovery of the contradiction caused me the greatest surprise and, I would almost say, consternation, since it has shaken the basis on which I intended to build arithmetic. [...] I must reflect further on the matter. It is all the more serious since, with the loss of my Rule V, not only the foundations of my arithmetic, but also the sole possible foundations of arithmetic, seem to vanish. [...] In any case your discovery is very remarkable and will perhaps result in a great advance in logic, unwelcome as it may seem at first glance. [...] The second volume of my *Grundgesetze* is to appear shortly. I shall no doubt have to add an appendix in which your discovery is taken into account. If only I already had the right point of view for that!" ([125, Volume 2, p. 212–215]; English translation [159, p. 127–128]).

Russell's antinomy and attempts of solving it – later in Chapter 4, in particular in Section 4.3.2, we shall return to this – was the starting point of the revision of Frege's project.

Before describing further the history of logicism let us explain Russell's antinomy. It was published in 1903 in Russell's book *Principles of Mathematics*. Today it is formulated in the following way: One can ask whether a given set X is its own element or not: $X \in X$ or $X \notin X$. Let Z be the set of all sets being not their own elements, i.e., $Z = \{X : X \notin X\}$. Now one can ask: is Z its own element or not? If the answer is yes, i.e., $Z \in Z$ then it means that Z has the property of elements of Z. Hence $Z \notin Z$. A contradiction. If the answer is no, i.e., $Z \notin Z$ then Z has the property of elements of Z. This means that $Z \in Z$. Again a contradiction. So both possible answers lead to the contradiction. Such an unsolvable paradox is called antinomy.

In this situation Russell decided to undertook the task of reducing mathematics to logic from the very beginning. He described his philosophical position in the book *The Principles of Mathematics* (1903) in which he also analyzed the antinomy of irreflexive classes. There he presented his views on the philosophy of mathematics that were in fact close to those of Frege. The reduction of mathematics to logic was presented in the monumental work written together with his university professor Alfred North Whitehead and entitled *Principia Mathematica*[25] (Volume I in 1910, Volume II in 1912, Volume III in 1913). There one finds a system of logic fundamentally reconstructed by Russell and called the theory of types. He had this idea of such a system already in 1903 and developed it only in *Principia*.

Russell and Whitehead's *Principia* closed the main period of developing logicism. Later works were in fact completions or improvements of this system and meant the consolidation of the position of logicism.

---

25 One should notice that this title was connected with the title of Isaac Newton's work fundamental for mechanics *Philosophiae naturalis principia mathematica* (1687).

Let us describe in details the conception of Frege and of Russell and Whitehead. Before this will be done one should specify the very doctrine of logicism. According to Russell the main theses of logicism are the following:

- All mathematical concepts, in particular all primitive notions of mathematical theories, can be explicitly defined by purely logical notions.
- All mathematical theorems can be deduced (by logical deduction) from logical axioms and definitions.
- This deduction is based on a common logic for all mathematical theories, i.e., the justification of theorems in all mathematical theories refers to the same basic principles which form one logical foundation for the whole of mathematics. Here is also a thesis included that any argumentation in mathematics should be formalized.

Those theses are not fully clear. In fact, it is not clear what is understood under logic here. The word "logic" can be understood in three ways:

- As a name of a scientific discipline.
- As a name of a formal language in which rules of a logical calculus are formulated.
- As a name of a system of calculus.

It seems that for logicism the third sense is the most appropriate. The theses of logicism given above imply in particular that all mathematical theorems have a uniquely determined content and that all those theorems, similarly to theorems of logic, are analytical judgements – what evidently contradicts Kant's claims.

We wrote above about Weierstrass and Dedekind. The arithmetization of analysis due to them shows that in order to realize the aims of logicism it suffices to develop the arithmetic of natural numbers as a part of logic. This task was undertaken by Frege in *Grundlagen der Arithmetik*.

He used – as indicted above – the concept of the equipollent sets introduced by Cantor. However, to remain strictly in the domain of logic Frege spoke not about sets but about notions and about the equipollency of extensions of notions. One of the consequences was that Frege's formulations were long and complicated but on the other hand this enabled him to use exclusively the language of logic.

Notice that Frege understood all notions in a platonic way, i.e., as real entities. Hence for Frege notions were independent of time, space and human mind. So a mathematician does not construct notions and their mutual relations; on the opposite, he *discovers* them. Frege accepted the actual infinite. All those connected him with Cantor.

The definition of natural numbers proposed by Frege can be described as follows: In order to make it more clear and simpler we will use the set-theoretical language, i.e., we shall use the concept of a set avoided by Frege. One can of course translate what is given below into Frege's terminology and see exactly what complications are induced by this.

(i) The cardinality of a set X is the collection of all sets equipollent with X.
(ii) n is a number if there exists a set X such that n is the cardinality of X.
(iii) 0 is the cardinality of the empty set.
(iv) 1 is the cardinality of the set that consists only of 0.

(v) The number n is the successor of the number m if there is a set X and an element a of X such that n is the cardinality of X and m is the cardinality of the set X without the element a (hence of $X \setminus \{a\}$).
(vi) n is a finite (natural) number if and only if n is the element of all sets Y such that: 0 belongs to Y and if k is an element of Y then also the successor of k belongs to Y.

Having defined the concept of a natural number in such a way one should now introduce arithmetical operations and show that they have the required properties. This task is not difficult.

It should be noticed that Frege's approach was quite different from what was attempted by Dedekind and Peano. It seems that Frege proved the existence of natural numbers and of arithmetical operations in pure thought while Dedekind and Peano only postulated it. In fact, the latter formulated only in axioms the properties of natural numbers and operations on them without showing that such objects actually exist. Frege attempted to secure their existence – as logical and consequently mental objects. His aim was to clarify and explain the nature, existence and place of numbers.

Differences between the approaches of Dedekind and Peano on the one side and Frege on the other are in fact not so deep from the mathematical point of view. Frege's approach comes from axioms, from logical axioms and uses simultaneously a non-axiomatized theory of extensions of notions or sets. His numbers exist on the base of extensions. This means that postulates – described for example by Dedekind and Peano – have been moved to the domain of logic, in particular to the domain of extensions.

There was a serious deficit in Frege's theory as we saw above. It was based on the inconsistent system of logic. In this situation Russell together with Whitehead began to develop a new system of logic – the so-called *ramified theory of types* – in order to establish a consistent base for natural numbers and consequently for the whole mathematics.

The main presumption of this theory is that the totality of all properties that can be considered forms an infinite hierarchy of types: properties of the first type are properties of individuals, properties of the second type are properties of properties of the first type, etc. This hierarchy does not contain properties which could consist of objects of different types, for example of individuals and properties of individuals. Consequently, in this hierarchy there is no uniform equality relation. There is the equality of individuals, the equality of properties of the first type, etc. To avoid the vicious circle of impredicative definitions Russell and Whitehead introduced not only types but also orders of properties (orders depended on the form of a formula describing a considered object or property). By using those means they could eliminate the antinomy of irreflexive classes.

Properties, called by Russell propositional functions, played in the theory of types the same role as concepts and their extensions by Frege. Hence Russell and Whitehead could simply adopt his definition of natural numbers. Some new problems appeared while proving properties of natural numbers. In particular it has turned out that to prove that for every natural number there exists a successor of it one needs an additional assumption being of a nonlogical character, in fact the Axiom of Infinity stating that there exist infinitely many individuals is needed. Consequently, Russell's reduction of arithmetic was not a reduction to logic only, but in fact to a broader base.

Russell proposed a solution to this problem by suggesting to consider in the case of any such theorem not the theorem itself but rather an implication the antecedent of which is just the Axiom of Infinity and the succedent the considered theorem. He used here the Deduction Theorem that states: if a formula φ is the logical consequence of sentences $\psi_1, \psi_2, \dots, \psi_n$ then the implication $\psi_1 \wedge \psi_2 \wedge \dots \wedge \psi_n \rightarrow \varphi$ is the theorem of logic. By this trick one could remain on the ground of pure logic. In a similar way one can treat also other theorems whose proofs require additional assumptions such as the Axiom of Choice. However, there remained the problem that in fact the Axiom of Infinity and the Axiom of Choice are not purely logical assumptions.

The aim of Russell and Whitehead in *Principia Mathematica* was not only the reconstruction of the arithmetic of natural numbers in the framework of the theory of types but the reconstruction of the whole mathematics, in particular of set theory. It is worth noting here that – contrary to Frege – Russell and Whitehead treated concepts not in a platonic way but in a nominalistic way. Hence they did not postulate the independent existence of sets and treated all symbols for sets only as signs denoting nothing. Sets were reduced by them to propositional functions. Russell claimed that from any proposition in which one speaks about sets one can eliminate sets and transform it into a proposition about properties. In this way sets become propositional functions.

On this occasion let us make a remark concerning the problem of geometry. Frege took a radical position there. He claimed that geometry belongs to applied mathematics and consequently ignored it. On the other hand Russell distinguished between pure and applied geometry. The latter was for him an empirical, the former a mathematical discipline. The pure geometry is the theory of abstract spaces that are defined set-theoretically, hence in logic. Axioms of geometry are in such approach elements of definition of an appropriate space.

Logicism was a philosophical project that drew great attention. Since Russell and Whitehead logicism is inseparably connected with *Principia Mathematica* and the theory of types. Russell and Whitehead's work has been developed and modified by many logicians and mathematicians. In particular the very system of the theory of types has been modified. This was done among others by the Polish logician, philosopher, art theorist and painter Leon Chwistek (1884–1944) and by the English mathematician and logician Frank Plumpton Ramsey (1903–1930). They introduced the so-called *simple theory of types*. It was explicitly formulated for the first time by Rudolf Carnap in *Abriss der Logik* (1929). In contrast to Russell and Whitehead's ramified theory of types it was an extensional theory – the former was in fact intensional. This means that in the new theory one took only objects into account propositions referred to, and not the way in which this was done.

In the 1930s the theory of types has been accepted as the best system for the foundations of mathematics; it became a basis which other researches referred to. For example the theory of types was used as basic system in the famous and fundamental work by Kurt Gödel on the incompleteness phenomenon (1931) and in Alfred Tarski's theory of satisfaction and truth (1933). Also for philosophers it was an important point of reference with respect to the image of mathematics. It played this role till the 1950s when its functions were taken up by the axiomatic set theory.

The system of *Principia Mathematica* (and its later simplifications) was the first complete, consistent and natural system of logic. It was in a sense a synthesis of all earlier conceptions in the field of logic and the foundations of mathematics. It indicated the power and meaning of formal methods in logic and mathematics; in particular it showed that the formal principles of logic provide a sufficient tool for deduction of theorems from any given axioms.

Despite of its meaning and its role the theory of types in the formulation of Russell and Whitehead as well as in its later modifications was criticized. Logicism was accused of being metaphysical: in the works of Frege it was platonic and Russell and Whitehead's nominalism was in fact inconsequent – propositional functions they used had again platonic character. It has been said that the theory of types was an *ad hoc* developed theory in which a hierarchical structure of the world reflected in the hierarchy of types is assumed, but this assumption had actually no justification. Also the applications of many postulates which had no purely logical character were criticized. Among them was for example the axiom stating that for any class there exists a class of the same type and "of order 1" having the same elements as the given one. This axiom made the elimination of the so-called impredicative definitions possible, i.e., definitions in which one refers to the object being just defined.

Another criticized feature of the theory of types was the fact that it does not allow "mixed sets", i.e., sets consisting of elements which belong to different types as for example

$$\{\emptyset, \{\emptyset\}\}, \{\emptyset, \{\{\emptyset, \{\emptyset\}\}\}\}, \ldots .$$

However, such sets are needed in the foundations of mathematics. The lack of them makes the investigation of some problems difficult. For example the study of large cardinal numbers was impossible.

A serious weakness of the theory of types is the systematic ambiguity of many notions. It consists of the fact that concepts which are intuitively equal should be introduced separately for every particular type, hence infinitely many times. This occurred in the case of the concept of equality, the concept of an empty set or the concept of a subset.

Beside such objections of a rather technical character also other more general objections of methodic character were formulated. It was claimed that paradoxes discovered in Frege's system and eliminated by Russell and Whitehead were a problem for logic but by no means they were relevant for classical theories of mathematics. In such theories the known paradoxes can be eliminated in a simple and natural way. Hence one can ask a general question whether the reduction of arithmetic to the theory of types is a reduction to a theory which in fact is more problematic than the arithmetic itself?

Logicism based on the theory of types played – despite all faults and imperfections – an important role in the philosophy and the foundations of mathematics. One of its merits is the fact that it indicated the clear way of a systematization of the mathematical knowledge and of specifying the intuitive notion of a mathematical proof. Note that just logicists contributed in an essential way to the development of mathematical logic.

It was shown above that logicism reduced mathematics not to logic but rather to logic *and* set theory. In fact, logicists were forced to accept in their logical systems various axioms of nonlogical character, more exactly: of set-theoretical character. Therefore the doctrine of logicism is formulated today in the form of the thesis that mathematics can be reduced to logic and set theory. However, there exists another version of logicism, a methodological one, which tries to rescue the reduction of mathematics to logic – it is called *if-thenism*. It is based on the finitistic character of the operation of logical consequence and on the Deduction Theorem and claims that theorems proved in mathematical theories should be understood as implications whose antecedents are finite conjunctions of axioms and succedent just the given theorem; such implications are theses of logic. This idea is a principal extension of Russell's attempt to avoid the usage of the set-theoretical Axiom of Infinity (see above).

Note at the end that the idea of Russell and Whitehead influenced the great project of the French mathematicians. In the 1930s a group of mathematicians was formed in France – it used the pseudonym Nicolas Bourbaki. Their aim was the systematization of all mathematical knowledge. Results of their investigations were published under the common title *Éléments de mathématique*. Twenty volumes have been published till today. For Bourbakists the mathematical world consists of structures. The concept of a structure is defined in a set-theoretical way. Three types of mathematical structures are distinguished: algebraical structures, order structures and topological structures. All other mathematical structures can be defined in terms of those three types of structures. However, it is not excluded that in the course of the development of mathematics new types will appear.

2.20 Intuitionism

Intuitionism was developed in the period of 1907–1930 as a counter trend with respect to the logicism. Its founder was the Dutch mathematician *Luitzen Egbertus Jan Brouwer* (1881–1966). Brouwer's ideas have been further developed among others by *Arend Heyting* (1898–1980) and *Anne Sierp Troelstra* (born 1939). One of the essential reasons of the emergence of the intuitionism was the criticism towards the contemporary foundations of mathematics. It concerned two issues that appeared over and over again in the history of mathematics: the concept of the infinite and the relation between the discrete and the continuous, the continuum. Intuitionists criticized the conceptions of Cantor and set theory developed by him, in particular his theory of the infinite and the replacement of the linear continuum by the set of real numbers.

Intuitionists treated as their predecessors those philosophers and mathematicians who claimed that mathematics is a science of real and not only formal contents, who were convinced that mathematical objects are directly given to human and that propositions of mathematics are synthetic *a priori* judgements. They referred willingly to Kant as well as to Paul Natorp (1854–1924), a philosopher from the Marburg School in which Kant's philosophy was studied and his ideas have been developed to the neokantian idealism.

It seems that the ideas expressed later explicitly in works of Brouwer were in a sense "in the air" at the turn of the 19th and 20th century.

For the first time the intuitionistic ideas appeared by the German mathematician Leopold Kronecker (1823–1891) and his students in the 1870s and 1880s. His ideas were in a most clear way expressed in his conceptions concerning the status of mathematical objects. In *Über den Zahlbegriff* (1887) Kronecker formulated a program of "arithmetization" of algebra and analysis, i.e., the program of founding those domains of mathematics on the most fundamental notion of a number.[26] In a number of lectures he developed a unified theory of various types of numbers based on the primitive intuition of a natural number.[27] His scientific credo has been summarized in the best way in his sentence: "*Die ganzen Zahlen hat der lieber Gott gemacht, alles andere ist Menschenwerk*" (The integers were made by God, everything else is the work of man). He claimed:

> "I study mathematics only as an abstraction of the arithmetical reality."

A consequence of this attitude was for example the fact that Kronecker admitted only those definitions of numbers which give a procedure of deciding whether a given number satisfies it or not. He accepted only "pure" existence proofs, i.e., proofs of existential theses giving constructions of the postulated objects.

Kronecker and his students declared war to methods of the theory of real numbers based on Cantor's set theory and new at that time as well as to the theory of functions developed first of all by Weierstraß and his students. They lost this fight – as can be said today. In fact, those theories have been successfully developed – in spite of the discovery of antinomies in set theory. Only Zermelo's proof of the well-ordering theorem (1904) on the base of the Axiom of Choice – we write about it in Chapter 4, in particular in Section 4.4 – caused doubts and opposition, especially in a group of French mathematicians who themselves worked in the theory of functions on the base of Cantor's set theory.

---

26 The restriction of mathematics to algebra and analysis was the consequence of a thesis (Kronecker referred here to C. F. Gauss) that, for example, geometry or mechanics are independent of human mind because they refer to the external reality.

27 After Kronecker's death these lectures were published in 1901 by K. Hensel.

To this group – called today the Paris school of intuitionism or French semi-intuitionists – belonged René Louis Baire (1874–1932), Émil Borel (1871–1956), Henri Léon Lebesgue (1875–1941) and the Russian mathematician Nikolai Nikolaevich Luzin (1883–1950). Their considerations on the foundations of mathematics were mainly connected with the study of the role of the Axiom of Choice – even if they were often concerned with general questions. They did not create a compact philosophy doctrine but formulated several general remarks on the margin of their mathematical investigations in the theory of functions. Their common feature is a certain constructivistic tendency.

Let us mention some of their theses. Lebesgue was convinced that a mathematical object does exist only when it was defined by finitely many words. Borel spoke here about "effective definitions". He claimed that a bare consistency does not suffice that a considered object exists. Real numbers should be given by concrete finite definitions. Hence the set of all such numbers cannot be uncountable. In order to avoid problems connected with the countable continuum Borel suggested to consider the continuum as independently given by the intuition. He called it the geometrical continuum. On the other hand he introduced the concept of the practical continuum consisting of finitely and explicitly defined numbers. His philosophical views can be summarized by the following sentence from his work *Leçons sur la théorie des fonctions* [46, p. 173]: "I do not understand what can mean an abstract possibility of an operation that is not at all possible for a human mind". Lebesgue did not accept the concept of an arbitrary sequence of numbers – he accepted only sequences defined by particular rules or principles. Baire rejected the unbounded operation of forming the power set, for example the set of all subsets of an arbitrary given infinite set.

Many of their views were later adopted by Brouwer and included into his doctrine of intuitionism. It is interesting that – as will be seen later – in many respects the views of the semi-intuitionists were more radical than those of Brouwer. Discussing the problem of forerunners of Brouwer's intuitionism one must still mention two more persons. First of all one should mention Henri Poincaré. We spoke about him already in Section 2.16. His philosophical attitude can be characterized by saying that he was an apriorist, intuitionist and constructivist as well as a founder of the conventionalism. His views strongly influenced Brouwer and his school.

Another person who influenced the doctrine of intuitionism was Gerritt Mannoury (1867–1956), the Dutch mathematician and logician. He was professor of the foundations of mathematics at the University of Amsterdam and the university teacher of Brouwer. Mannoury was connected with the movement of *Signifika*. It was the tendency in the philosophy in which in the framework of strong logical and methodological conceptions also some ethical ideas were formulated. Its starting point was the criticism of the language and their expressions, the denotation in general, the relations between a sign and real facts. Hence it was connected with what is called today semiotics and in particular with pragmatics. Adherents of Signifika were interested not so much in the meaning of expressions but rather in the theory of "psychic connotations forming the basis of applications of human speech". They represented the psychological point of view. Language was treated by them as an instrument with the help of which one tries to influence others.

As an adherent of Signifika Mannoury was convinced that a mathematical formula has a meaning not *per se* but only by and as a result of aims to which it has been formulated and used. He claimed that the selection of axioms of mathematics can be explained only by taking into account psychological and emotional elements lying in mathematics.

In such philosophical-psychological environment Brouwer has developed his views on mathematics. They found their expression for the first time in his doctoral dissertation *Over de Grondlagen der Wiskunde* (On the foundations of mathematics) from 1907. Here he propagated and constituted (and justified) the intuitionistic conception of mathematics. His philosophy of mathematics was a part of his general philosophical views expressed in his book *Leven, Kunst, Mystiek* (Life, Art, Mysticism) from 1905. His dissertation was written in Dutch and therefore its influence was rather small. However, it included already all fundamental theses of intuitionism. Brouwer developed his views later in his inaugural lecture at the University of Amsterdam *Intuitionisme an formalisme* (Intuitionism and formalism) from 1912 (cf. [54]). They can be found also in his paper *Consciousness, Philosophy and Mathematics* [55].

One of the aims of intuitionism was the attempt to avoid the danger of inconsistencies. Brouwer proposed here some means that turned out to be very radical and forced a deep reconstruction of mathematics.

Brouwer rejected platonism that attributed to mathematical objects the existence independent of time, space and knowing subject. Instead the intuitionism proclaims the thesis of conceptualism. According to this mathematics a function is of human intellect and a free activity of mind. Mathematical knowledge is the creation of mind and not a theory, i.e., a system of rules and theorems. Mathematical objects are mental constructions of an (idealized) mathematician. Hence in mathematics there are only objects which have been constructed or are constructible in the human mind. A. Heyting wrote in the work *Die intuitionistische Grundlegung der Mathematik*[28]:

> "that we do not attribute to integers – and in a similar way to other mathematical objects – an existence independent of our thinking. [...] Mathematical objects, even if they are perhaps independent of singular acts of thought, are in their essence determined by human thinking. Their existence is secured as far as they can be determined by thinking; they have only such properties which can be recognized by thinking. However, this possibility of cognition is revealed to us only by cognition itself. The belief in the transcendental

28 Intuitionistic Foundation of Mathematics

existence that is not supported by concepts should be rejected as mathematical proof instrument." ([164, pp. 106 f.].)

As a consequence one should reject the axiomatic-deductive method as a method of developing and founding mathematics. It is not sufficient to postulate only the existence of mathematical objects (as it is done in the axiomatic method) but one must first construct them. The same concerns also properties of mathematical objects. Hence the intuitionism rejected in particular Peano axioms for the arithmetic of natural numbers as well as Zermelo's axioms for set theory. In particular intuitionists attacked the Axiom of Choice. This axiom is – according to them – an example of a pure postulating the existence of a set which can be neither precisely defined nor imagined by our mind.

Another consequence of the conceptualistic view was the rejection of the actual infinite. Human intellect can construct special objects, for example numbers, however it is not able to execute infinitely many constructions. Hence one can understand an infinite set only as a law or a rule of forming more and more of its elements, but they will never exist as forming an actual totality. Such potential infinite sets are countable. Uncountable sets are unimaginable. Hence there are no uncountable sets and no cardinal numbers other than $\aleph_0$. The concept of a set in the works of intuitionists is quite different than the concept used in Cantor's set theory.

The conceptualistic thesis of intuitionism implies also the rejection of any nonconstructive proofs of the existential theses, i.e., of proofs giving no constructions of the postulated objects (cf. [132, p. 4]). Brouwer saw in proofs of this type the source of antinomies, paradoxes and other foundational problems in mathematics. The rejection of nonconstructive proofs necessarily led to the rejection of the classical logic in which every proposition is either true or false. In fact, this logic is the base of such proofs. Consequently, the law of the excluded middle $p \vee \neg p$ (*tertium non datur*) as well as the law of double negation $\neg\neg p \leftrightarrow p$ (*duplex negatio affirmat*) must be rejected. In fact, if they were valid then for any property $\varphi(x)$ we would have

$$\exists x \varphi(x) \vee \neg \exists x \varphi(x).$$

So suppose for example that by assuming $\neg\exists x\varphi(x)$ one gets inconsistency. Then the second half of the above alternative would be valid, i.e., $\exists x\varphi(x)$. Hence one obtains the existence of an object x without having constructed it.

Proofs of this type appear very often in the classical mathematics. Brouwer and the intuitionists did not accept them. For them the existence of an object x with the property φ meant that such an object has been constructed, for there exists only what can be constructed (in thought). The rejection of the proposition $\neg\exists x\varphi(x)$ is in fact no proof of the sentence $\exists x\varphi(x)$.

The fact that the classical logic is not compatible with mathematics in the intuitionistic sense was observed by Brouwer shortly after completing his dissertation. In 1908 he published the first example of an argumentation called today in the literature "weak counterexample". Their aim is to show that some propositions that are accepted from the point of view of the classical logic and mathematics, are in fact not valid when treated intuitionistically. Brouwer presented such examples in order to justify the necessity of the revision of classical theories. In particular they showed that there are definitions which are classically equivalent but from the intuitionistic point of view they do not lead to equivalent concepts.

Brouwer's example was the following: Consider the set

$$X = \{x|x = 1 \vee (x = 2 \wedge \varphi)\},$$

where φ is any undecided mathematical hypothesis, for example the Riemann hypothesis. The set X is a subset of the finite set $\{1, 2\}$. In the classical mathematics X is finite. In the intuitionistic mathematics one can by no means at all speak about it. In fact, in order to be able to claim that X is finite one should first decide what the elements of X are, i.e., decide φ. In this way one obtains a weak counterexample to the thesis that a subset of a finite set is finite. In a similar way one constructs a weak counterexample to the proposition: "For every real number x there holds $x < 0 \vee x = 0 \vee x > 0$."

Let us quote another example. Consider the following two definitions of two natural numbers: (a) Let k be the greatest prime such that $k - 1$ is also a prime or, if there is no such number, put $k = 1$. (b) Let l be the greatest prime such that $l - 2$ is also a prime or, if there is no such number, put $l = 1$. From the classical point of view both definitions have the same structure. In intuitionism this is not the case. In fact, in the intuitionistic mathematics it can be calculated that $k = 3$; however, one cannot say what l is because it is unknown whether there exist infinitely many twin primes. For an intuitionist definition (b) is not a definition at all because to define a number means simply to provide a construction that makes it possible to calculate its value.

The next example will show that in the intuitionistic mathematics the impossibility of the impossibility of a property is not a proof of this property, hence the law of the propositional calculus $\neg\neg p \rightarrow p$ is not valid. So consider the sequence of decimal digits for the number π and write underneath it the decimal fraction $\rho = 0, 3333\ldots$ which will terminate as soon as the sequence $777\ldots7$ of the length k will occur at place k in π. For such k we put

$$\rho = \frac{10^k - 1}{3 \cdot 10^k}.$$

The classical mathematics says that the number ρ is rational. In fact, suppose that ρ is not rational. Then the equality

$$\rho = \frac{10^k - 1}{3 \cdot 10^k}$$

is not true and in the sequence of decimal digits of π there is no place k at which the sequence $777\ldots7$ of length k occurs. But in this situation $\rho = 0, 3333\ldots$ is an infinite sequence and $\rho = \frac{1}{3}$, hence it is rational. However, this contradicts the assumption that ρ is not rational. In this way one proved that the number ρ is rational.

However, no intuitionist would accept such reasoning. One cannot suppose that ρ is rational because this would mean that two natural numbers p and q can be found such that $\rho = \frac{p}{q}$. To this aim however one should either find the place k at which the sequence $777\ldots7$ of length k occurs or prove that there is no such sequence of digits. One can do neither this nor that. Hence it cannot be claimed that the number ρ is rational.

All those examples indicate that classical reasonings partially do not work in the intuitionistic mathematics. Moreover, for intuitionists logic is not the base and starting point of mathematics. Just the opposite. Brouwer claimed that logic is founded on mathematics and not vice versa.

According to the intuitionists mathematics is based on the primitive intuition of *a priori* time. Brouwer adopted Kant's conception of time as pure *a priori* form of intuition (simultaneously he rejected the idea of the *a priori* space). Brouwer says about natural

numbers as "unsubstantial abstractions of perception of time".

The assumption of the *a priori* time leads to the claim that arithmetical judgements are synthetic judgements *a priori.* In *Intuitionisme en formalisme* Brouwer wrote:

> "This neo-intuitionism considers the falling apart of moments of life into qualitatively different parts, to be reunited only while remaining separated by time as the fundamental phenomenon of the human intellect, passing by abstracting from its emotional content into he fundamental phenomenon of mathematical thinking, the intuition of the bare two-oneness. This intuition of two-oneness, the basal intuition of mathematics, creates not only the numbers one and two, but also all finite ordinal numbers, inasmuch as one of the elements of the two-oneness may be thought of as a new two-oneness, which process may be repeated indefinitely [...]. Finally, this basal intuition of mathematics, in which the connected and the separate, the continuous and the discrete are united, gives rise immediately to the intuition of the linear continuum [...]. In this way the apriority of time does not only qualify the properties of arithmetic as synthetic a priori judgments, but it does the same for those of geometry, and not only for elementary two- and three-dimensional geometry, but for non-euclidean and *n*-dimensional geometries as well." ([54, pp. 85–86].)

The claim that mathematics is a free activity of mind and not a system of axioms, rules and laws leads to the conclusion that mathematical constructions are fully independent of any language. Heyting wrote about this in his paper *Die intuitionistische Grundlegung der Mathematik* in the following way:

> "Language, both common language as well as a formal one, is used by him [an intuitionist – authors' remark] solely in communication, i.e., in order to induce somebody or himself to reflect on his mathematical ideas." ([164, p. 106].)

And Brouwer claimed (cf. *Intuitionisme en formalisme* [54]):

> "For this reason the intuitionist can never feel assured of the exactness of a mathematical theory by such guarantees as the proof of its being non-contradictory, the possibility of defining its concepts by a finite number of words, or the practical certainty that it will never lead to a misunderstanding in public human relations."

Hence it is senseless – from the point of view of an intuitionist – to analyze the mathematical language and not the mathematical thinking. This distinguishes the intuitionism not only from the logicism (see above) and formalism (see below) but also from all those philosophers – from Plato via Leibniz till Wilhelm von Humboldt and Ernst Cassirer according to them a language precedes the abstract thinking or the thinking is something linguistic.

The intuitionists claim furthermore that there cannot exist a language for mathematics that would guarantee the real security and exclude the possibility of any contradictions and paradoxes. In this way they turned against formalists who wanted to prove with the help of safe finite methods the consistency of mathematics by reconstructing mathematics as a formal system. Intuitionists claimed even more: to look for the correctness of mathematics not "on the paper" but "in human mind".

Brouwer's theses described above that form the core of his doctrine led him to the requirement of reconstructing the whole of mathematics according to the intuitionistic principles. Sometimes one speaks about "Brouwer's program". In 1912 he began realizing it with full energy. He began by revising the concept of the continuum. Till 1928 he reconstructed a part of topology, of set theory with ur-elements as well as the theory of functions, he developed the theory of countable well-orders and – together with his student B. de Loor – an intuitionistic proof of the fundamental theorem of algebra. Since 1928 Brouwer was in fact rather not very active. The project was continued by his students, first of all by Maurits Joost Belinfante and Arend Heyting, later by Heyting's students. Belinfante worked on an intuitionistic theory of complex numbers, Heyting on an intuitionistic geometry and algebra, and students of Heyting in the intuitionistic topology, measure theory, theory of Hilbert spaces and affine geometry.

An essential and crucial role in the development of the intuitionism played the investigations and works of Heyting. He tried to explain the ideas of Brouwer, to distribute and to popularize them. One can say that without Heyting's endeavour the intuitionism would disappear in the 1930s. The next generation of intuitionists turned prevailingly to metamathematical problems connected with intuitionism. They largely withdrew from the continuation of Brouwer's program Heyting actually strived for.

The main problem that should be solved was the problem of developing the system of logic which would satisfy intuitionistic-philosophical requirements. Andrey N. Kolmogorov (1925) and Valery I. Glivenko (1928) gave a formalization of a part of intuitionistic propositional calculus, and A. Heyting presented in 1930 the first complete system of this calculus. This made it possible, among other things, to compare the classical (aristotelian) logic with the intuitionistic one.

It has turned out that the following formulas are not theses of the intuitionistic system of logic (cf. [132, p. 4]):

$$\neg\neg p \to p,$$

$$\neg(p \wedge q) \to \neg p \vee \neg q,$$

$$(p \to q) \to \neg p \vee q.$$

On the other hand the following formulas can be deduced in the intuitionistic logic:

$$p \to \neg\neg p,$$
$$\neg\neg(\neg\neg p \to p),$$
$$\neg\neg\neg p \leftrightarrow \neg p.$$

The intuitionistic system of propositional calculus is a part of the classical system. Glivenko showed that the formula $\neg\neg\varphi$ can be proved in the intutitionistic system if and only if the formula φ can be deduced in the classical system. Kolmogorov (1925), Kurt Gödel (1933) and Gerhard Gentzen (1933, published only in 1963) proved some theorems on the embedding of classical logic into the intuitionistic logic. The systems LK and LJ developed by Gentzen (1935) turned out to be very important. They made the structural distinction of classical and intuitionistic propositional calculi possible.

Various semantics for the intuitionistic propositional calculus have been developed. In 1932 Gödel proved that this system cannot be adequately characterized by matrices with finitely many values. In 1935 the Polish logician Stanisław Jaśkowski proposed a series of adequate matrices (with infinitely many values) for the intuitionistic logic. In 1938 another Polish logician, Alfred Tarski, proposed a topological interpretation of the intuitionistic propositional calculus.

Interesting is also the proof-theoretic interpretation given at the beginning of the 1930s by Brouwer, Heyting and Kolmogorov. According to it the meaning of an expression φ is given by the proof of φ. Moreover, the concept of a proof of a complex expression is explained by the reduction to proofs of subformulas. In particular:

(i) A proof of the formula $\varphi \wedge \psi$ is a proof of φ and a proof of ψ.
(ii) A proof of the formula $\varphi \vee \psi$ is a proof of φ or a proof of ψ.
(iii) A proof of the formula $\varphi \to \psi$ is a construction that transforms any proof of the formula φ into a proof of the formula ψ.
(iv) The absurdity $\perp$ ("inconsistency") has no proof; a proof of the formula $\neg\psi$ is a construction that transforms any proof of φ into a proof of $\perp$.

Let us mention also the interpretation due to Stephen C. Kleene coming from the 1940s. It indicates the connections between the concept of a recursive function and the intutionistic logic (this interpretation is known under the name of the *realisability interpretation*). There are also Beth models (1956–1959) and Kripke models (1965).

Notice that Heyting's system of the intuitionistic propositional calculus has not been treated by Brouwer as the interpretation of intuitionistic logic that really corresponds to the doctrine of intuitionism. The reason was Brouwer's conviction that it is impossible to grasp in a calculus all potential correct thought processes. Human mathematical activity can never be adequately represented by a system. In fact, the mathematical activity is always dynamic and a system is static. Therefore the intuitionistic mathematics cannot be in a complete and adequate way described by a formal system.

The intuitionistic mathematics is poorer than the classical mathematics. A part of analysis and almost the whole set theory are lost. Additionally it is far more complicated and consequently it is not suitable for applications. Still further criticism is formulated against the intuitionists. It is emphasized that the rejection of the classical mathematics is based on the actually ill-founded variation of the meaning of logical and mathematical concepts. First of all it has been criticized that the fundamental concept used by them, namely the concept of the intuition, is not precise enough, that the derivation of arithmetical laws from the intuition of the two-oneness (see above), the unit and the proportion are in fact pseudo-justifications.

Paul Lorenzen (1915–1994) attempted to refute the latter charge by replacing the psychological manner of describing the series of numbers by *operative construction rules* for counting signs. Christian Thiel (born 1938) describes them in his book *Philosophie und Mathematik* [341, p. 114] in the following way:

$$\mathbb{N}\begin{cases} \Longrightarrow |, \\ n \Longrightarrow n\,|. \end{cases}$$

According to this simple calculus $\mathbb{N}$, the *counting signs* are developed:

$$|, ||, |||, ||||, \ldots,$$

put also by Hilbert (see below) at the beginning of his program. Thiel goes one step further and gets, by the abstraction from counting signs, to "fictitious objects", "numbers" [341, p. 135]. He calls the counting signs n and n' – in equal or in various systems of counting signs – to be "equivalent with respect to counting" (*zähläquivalent*) if it is alone about steps of construction in calculi belonging to n or n', respectively, and their successions in the construction of n or n', respectively, coincide [341, p. 115]. According to this one abstracts from anything that can be in a way specific for systems of signs and makes it possible to speak about the *number* n and to say that n and n' *represent the same number*. This abstraction leads as a "pure logical process" [341, p. 131] from counting signs to numbers and to *arithmetical expressions* that hold in equal measure in all systems of counting signs. Briefly:

> *Numbers are fictitious objects arising by abstraction from counting signs in various systems of counting signs.*

Intuitionism did not find many constant adherents; however, it strongly influenced the foundational investigations. It has turned out that the intuitionistic logic can be a useful instrument in various domains of mathematics, for example in the theory of topoi or in the theoretical computer science.

2.21 Constructivism

Intuitionism is one of the representatives of the extensive constructivistic trend in the philosophy of mathematics. This trend is not homogeneous, there are various constructivistic positions that differ at the meaning of the concept of being "constructive/constructivistic". The constructivistic tendencies appeared in the last quarter of the 19th century as a reaction to the development of highly abstractive mathematical concepts and methods that arose on the base of Cantor's set theory.

Generally speaking, constructivism is a normative position postulating that one should not look for foundations or justifications of the existing mathematics but develop and practice mathematics using particular, more exactly constructive methods and tools. Examples of constructivistic theories are – beside the intuitionism – finitism, ultraintuitionism (called also ultrafinitism or actualism), predicativism as well as the classical and constructive recursive mathematics.

The main principles of the finitism are the following: (1) objects of mathematics are solely concrete (and finitely) given structures – a prototype for this are natural numbers, (2) all operations on such structures should have combinatorial character, hence should be "effective", (3) abstract concepts such as for example the concept of an arbitrary set, operations or constructions are not allowed in the (finite) mathematics. As a founder of the finitism one can see Leopold Kronecker (see above).

The starting point of the ultrafinitism is the observation that already the concept of a natural number is connected with a certain idealization. Usually in mathematics all natural numbers are treated as having equal rights, the same character and being in a way similar. They are objects of the same type – this does not depend on the fact whether one speaks about $1, \ 2, \ 10$ or $10^{10^{10}}$. However, already É. Borel noticed that very large finite objects lead to similar problems as infinite ones and D. van Dantzig asked (in the sense of G. Mannoury) whether "$10^{10^{10}}$ is really a finite number?". In such a way there arose the idea to develop mathematics solely on the base of actual real cognitive abilities of human beings. This explains also the name "actualism" that is used sometimes. By such approach for example the operation of exponentiation is restricted, i,.e., partial. If particular natural numbers are treated as sequences of units then one can doubt whether $10^{10^{10}}$ is really a number. In fact, the number of all atoms in the universe is estimated to be smaller than 10^{80}.

On the ground of such ideas various theoretical projects of reconstructing mathematics are formulated. This is done for example by A.S. Esenin-Volpin, R. J. Parikh, C. Wright, R. O. Rabin and E. Nelson. All such proposals can be called ultraintuitionistic.

Forerunners of the predicativism are H. Poincaré and B. Russell. They were convinced that the source of all problems and troubles of mathematics is the usage of non-predicative definitions (see above). From this comes the idea that in the whole mathematics only predicative definitions and constructions should be admitted. This was performed by Hermann Weyl (1885–1955) in the monograph *Das Kontinuum* [365]. Weyl showed that large domains of analysis can be developed by such restricted methods. This approach has been followed and applied by other authors, for example by P. Lorenzen, M. Kondô, A. Grzegorczyk, G. Kreisel, S. Feferman and K. Schütte.

In the classical and constructive recursive mathematics the concept of a recursive function is used. It was introduced in the 1930s in order to explain the vague concept of effectively calculable functions. In the classical recursive mathematics only the so-called calculable real numbers are considered and studied. They are such real numbers whose decimal expansion can be given by a recursive, effectively calculable sequence. On the base of the set of those real numbers one then develops the analysis. This was done for example by Stefan Banach and Stanisław Mazur in the 1930s.

The second trend, the constructive recursive mathematics is connected with the name of the Soviet mathematician Andrey A. Markov and with the constructivistic school (P. Novikov, D. Botchvar, N. A. Shanin) founded by him. Moreover, this school represented – from the philosophical point of view – the nominalism.

As we saw there is no uniform philosophical base for the constructivistic conceptions. Rather they refer to in fact very different philosophical, in particular ontological, conceptions and assume different presumptions. Taking into account the ontological presumptions one can roughly distinguish the following types of constructivism:

- objectivism claiming that mathematical objects are objective results of construction processes and that they exist independently of the subject who constructed them.
- intentionalism which ascribes an intentional existence to constructed objects, i.e., existence that is typical for cultural objects.
- mentalism which claims that mathematical objects are products of thinking processes and exist only in those acts of thinking.
- nominalism, i.e., a variant of the conception that mathematical objects are nothing more than signs or symbols – they are given as concrete time-spatial objects (for example as particles of chalk on a blackboard).

The constructivistic trends have contributed and still contribute to specify the foundations of various domains of mathematics. Their results are important also for computer science. Hence such trends are important from the point of view of the foundations of mathematical sciences. On the other hand – and this cannot be forgotten – they would lead, if their theses would have been realized, to extensive reductions and restrictions of mathematics.

2.22 Formalism

The founder of the formalism was German mathematician *David Hilbert* (1862–1943). He presented his philosophical views on mathematics in papers published mainly in the period of 1917–1931. Other representatives of this tendency who developed the ideas of Hilbert were *Paul Bernays* (1888–1977), *Wilhelm Ackermann* (1896–1962), *Gerhard Gentzen* (1909–1945), *John von Neumann* (1903–1957) as well as *Haskell B. Curry* (1900–1982) and *Abraham Robinson* (1918–1974).

Hilbert was of the opinion that the attempts to justify and found mathematics undertaken hitherto, especially by the intuitionism, were unsatisfactory because they led to the impoverishment of mathematics and to the rejection of various parts of it, in particular those considering transfinite infinity. He wrote:

> "What Weyl and Brouwer do comes to the same thing as to follow in the footsteps of Kronecker! They seek to save mathematics by throwing overboard all that which is troublesome. [...] They should chop up and mangle the science. If we would follow such a reform as the one they suggest, we would run the risk of losing a great part of our most valuable treasures!" (Quoted after [298, p. 155].)

Hilbert was first of all a mathematician and – in such a way he is characterized by Smoryński – "had little patience with philosophy, his own philosophy of mathematics being perhaps best described as naïve optimism – a faith in the mathematician's ability to solve any problem he might set for himself" (cf. [330]). His aim – presented for the first time in a lecture delivered to the second International Congress of Mathematicians in Paris in 1900 – was to save the integrity of classical mathematics dealing with the actual infinite by showing that it is secure. This should be done by proving that mathematics is certain and reliable. Hilbert was of the opinion that this task surpasses the frameworks of mathematics itself because:

> "the definitive clarification of *the nature of the infinite* has become necessary, not merely for the special interests of the individual sciences but for *the dignity of human intellect itself*" ([173, p. 163], English translation [28, p. 136].)

Hilbert formulated a program – called Hilbert's program – that became the core of the formalism. The purpose of this program was to found and justify the classical mathematics. It is built on the same philosophical traditions as the logicism. Formalists used – this should be noticed – results of logicists, in particular results of Russell and Whitehead. On the other hand Hilbert referred – as Brouwer did – to Kant. The difference is that the intuitionists referred to the Kantian conception of time and space, i.e., to the transcendental aesthetics, whereas Hilbert used Kant's conception of the idea of mind explained by him in the framework of his transcendental dialectics.

In the work *Über das Unendliche* ([173, pp. 170–171], English translation [28, p. 142]) Hilbert wrote:

> "Kant taught – and it is an integral part of his doctrine – that mathematics treats a subject matter which is given independently of logic. Mathematics, therefore, can never be grounded solely on logic. Consequently, Frege's and Dedekind's attempts to so ground it were doomed to failure.
> As a further precondition for using logical deduction and carrying out logical operations, something must be given in conception, viz., certain extralogical concrete objects which are intuited as directly experienced prior to all thinking. For logical deduction to be certain, we must be able to see every aspect of

> these objects, and their properties, differences, sequences, and contiguities must be given, together with the objects themselves, as something which cannot be reduced to something else and which requires no reduction. This is the basic philosophy which I find necessary, not just for mathematics, but for all scientific thinking, understanding, and communicating. The subject matter of mathematics is, in accordance with this theory, the concrete symbols themselves whose structure is immediately clear and recognizable."

Such concrete objects forming the starting point are natural numbers understood by a formalist as number words, digits and number symbols, hence as systems of signs.

In the lecture *Über das Unendliche*, from which both quotations above come, Hilbert said:

> "In number theory we have the numerical symbols
>
> $$1, 11, 111, 1111, \ldots$$
>
> where each numerical symbol is intuitively recognizable by the fact it contains only 1's. These numerical symbols which are themselves our subject matter have no significance in themselves." ([173, p. 89], English translation [28, p. 143].)

> Hence numbers – as symbols – are for Hilbert concretely and immediately given and in fact are nothing more than signs.

Briefly:

> *Numbers are meaningless signs.*

In Hilbert's works one does not find further and deeper explanations of the nature of the natural numbers. At another point [172, p. 18] Hilbert admits:

> "a firm philosophical attitude which is necessary in my opinion for the justification of the pure mathematics as well as for all scientific thinking, understanding and communication is the following: At the beginning [...] is a sign."

If mathematics spoke only about those concrete objects then it would be necessarily a safe and consistent science – in fact apparently given facts cannot be inconsistent. However, mathematics speaks also about the infinite and especially this part of mathematics is indispensable and plays an essential role because

> "the infinite occupies a justified place in our thinking, that it plays the role of an indispensable concept." ([173, p. 165], English translation [28, p. 137].)

However,

> "the infinite is nowhere to be found in reality. It neither exists in nature nor provides a legitimate basis for rational thought – a remarkable harmony between being and thought." ([173, p. 190], English translation [28, p. 151].)

Hence the concept of the infinite is not *a priori* safe. It can lead to paradoxes and antinomies.

In this situation there are two possibilities: either reject the whole classical mathematics in which one speaks about the actual infinite and develop a new one, safe but more poor, – this was proposed by Brouwer and his followers – or try to provide a foundation and justification for the actually existing mathematics. Hilbert chose the second possibility. For him – as for a professional mathematician – it was something unimaginable to resign from a mathematics going beyond the finite. This was stressed by him in a famous sentence: "No one shall drive us out of the paradise which Cantor has created for us." (*Aus dem Paradies, das Cantor uns geschaffen hat, soll uns niemand vertreiben können*) ([173, p. 170], English translation [28, p. 141].)

The clarification and justification of mathematics proposed by Hilbert had an explicit Kantian character. Propositions about the infinite have according to him no real meaning. They are neither true nor false. They cannot be a part of a correct judgement. The actual infinite is an idea of reason in the Kantian sense, i.e., it is a concept that is consistent but is does not correspond to anything in the physical world because it exceeds any experience. On the other hand this concept is indispensable in mathematics (as well as in thinking) because it completes and closes the concrete and the finite.

According to this Hilbert distinguished between the unproblematic, "finitistic" part of mathematics and the problematic "infinitistic" part that needed justification. The finitistic mathematics – to which belongs the domain of symbols and sequences of symbols – is justified *per se* because it deals with concrete clearly and immediately given objects. In contrast to this the infinite mathematics needs a justification. In the finitistic mathematics one deals with so-called real sentences. They are real and meaningful because they refer to concrete objects. The infinitistic mathematics on the other hand deals with so-called ideal sentences that contain references to ideal infinite totalities. Hilbert believed that every true real finitary proposition had a finitary proof. Infinitistic objects and methods played in mathematics only an auxiliary role. They enabled us to give easier, shorter and more elegant proofs, but every such proof could be replaced by a finitary one. He was also convinced that consistency implies existence, hence it is a sufficient condition for the existence of mathematical objects. Moreover, every proof of existence not giving a construction of postulated objects is in fact a presage of such a construction.

Unfortunately the concept "finitistic" and its derivations used by Hilbert are not clear. Hilbert did not give a precise definition or explanation of it. Hence various interpretations are possible. Usually it is assumed that a finitistic reasoning is essentially a primitive recursive reasoning in the sense of Skolem and that it can be formalized in the system of primitive recursive arithmetic (PRA, Skolem's arithmetic). Real sentences are expressions of the form $\forall x \varphi(x)$ where φ contains only atomic formulas, propositional connectives and bounded quantifiers. Such formulas are called Π^0_1-formulas by logicians.

According to Hilbert the infinitistic mathematics can be founded and justified only by finitistic methods because only they can give the necessary security (*Sicherheit*). Hilbert proposed to base mathematics on the finitistic mathematics via proof theory created and developed by him. Its main goal was to show that proofs which use ideal elements in order to prove results in the real part of mathematics always yield correct results.

One can distinguish two problems here: (1) the problem of consistency and (2) the conservation problem. In some of Hilbert's publications both aspects are stressed but usually the one-sided emphasis is put on the consistency problem only (cf. *Grundlagen der Mathematik* [179, Volume I]). The consistency problem consists in showing (by finitistic methods, of course) that the infinitistic mathematics is consistent. The conservation problem consists in showing by finitistic methods that any real sentence which can be proved in the infinitistic part of mathematics can be proved also in the finitistic part, i.e., that the infinitistic mathematics is conservative over finitistic mathematics with respect to real sentences and, even more, that there is a finitistic method of transforming infinitistic proofs of real sentences into finitistic ones. Both these aspects are interconnected. This can be shown when real sentences are identified with the class of $\forall x \varphi(x)$-sentences – G. Kreisel proved that in this case the solution of the problem of consistency provides also a solution of the conservation problem.

Hilbert proposed a program – called today Hilbert's program – to solve those problems. It consisted of two steps. The first step was to formalize mathematics, i.e., to reconstitute infinitistic mathematics as a large, elaborate formal system. It should contain classical logic, infinite set theory, arithmetic of natural numbers and analysis. In order to do this one should fix an artificial symbolic language and rules of building well-formed formulas. Next axioms and inference rules (referring only to the form, to the shape of formulas and not to their sense or meaning) ought to be introduced. In such a way theorems of mathematics become those formulas of the formal language which have a formal proof based on a given set of axioms and given rules of inference. There was one condition assumed for the set of axioms (and for rules of inference as well): they ought to be chosen in such a way that they suffice to solve any problem formulated in the language of the considered theory as a real sentence, i.e., they ought to form – as logicians used to say – a complete set of axioms with respect to real sentences.

The second step of Hilbert's program was to give a proof of the consistency and conservativeness of mathematics. Such a proof should be carried out by finitistic methods. This was possible since the formulas of the system of formalized mathematics are finite strings of symbols and proofs are finite strings of formulas, i.e., finite strings of finite strings of symbols. As such they are concrete, clear and immediately given objects that can be manipulated and investigated with the help of save finitistic methods. To prove the consistency of mathematics it suffices now to show that there are no two sequences of formulas (two formal proofs) such that one of them has as its last element a formula φ and the other $\neg\varphi$ (the negation of the formula φ). To show the conservativeness it should be proved that any (formal) proof of a real sentence can be transformed into a (formal) proof not referring to ideal objects. In order to carry out both those tasks Hilbert developed a theory called "proof theory" or "metamathematics". Its aim is to investigate formalized theories and formal proofs by mathematical methods.

In the metamathematical investigations one abstracts from the contents of all mathematical, hence also real sentences. This is a pure *methodological* procedure of metamathematics. The objection raised by Brouwer that Hilbert treats mathematics solely as a play on symbols and formulas without contents was mistaken. For formalists the procedure of formalization was only a method used in the investigation of mathematical objects. The real difference between Hilbert and Brouwer, and consequently between formalists and intuitionists, consisted in the decision what can provide the foundation of mathematics and which methods of reasoning can be treated as fundamental and legitimate, i.e., the decision which methods can be used: finitistic (as Hilbert proposed) or intuitionistic (Brouwer). The essential difference was and still is that intuitionists strictly reject any non-constructive mathematics.

To be more exact one should add that later various radical versions of the formalism appeared, in particular the so-called strict formalism of H. B. Curry presented in his book *Outlines of a Formalist Philosophy of Mathematics* [80] where mathematics is treated as a science about formalized systems. There mathematics is reduced to the investigation of pure formal theories in which there appear only symbols that constitute a given system. For Hilbert the aim and sense of formalized theories was the defence, clarification and justification of classical, in particular of the infinitistic mathematics whereas for Curry the formalized systems formed a copy of the classical mathematics and should replace it completely. This implies further differences. Proving the inconsistency of a system meant for Hilbert that it is not worth to deal with it. For Curry it was different. He claimed that "a proof of the consistency is neither necessary nor sufficient" (cf. [80, p. 61]) to accept a system and treat it as useful. To justify this claim he provided examples of inconsistent theories that turned out to be useful and could be widely applied. Such theories can be found both in physics as well as in mathematics itself. For example he referred to Leibniz's differential and integral calculus. Leibniz and, first of all, his followers used the basic concept of differential understood as an infinitely small magnitude, i.e., as a positive magnitude which is smaller than any positive real number. For Curry it was an inconsistent concept. Despite of this the differential and integral calculus has been developed and successfully applied. The so-called nonstandard analysis developed in the second half of the 20th century by Schmieden and Laugwitz as well as by Abraham Robinson provided an appropriate theoretical fundament for principles of Leibniz's analysis, in particular for the usage of infinitely small magnitudes. We write about this in Section 3.3 and Chapter 7.

It should be added that Hilbert – differently from intuitionists – saw a language and thinking as a unit. He was of the opinion that thinking in a similar way as speaking and writing proceeds by forming and stringing together sentences.

Hilbert represented a clearly anti-logicistic position. He emphasized that mathematics cannot be deduced from logic alone, that logic does not suffice to justify mathematics (cf. the quotation above). Therefore results of Frege and Russell were for Hilbert of secondary importance.

Hilbert and his students scored some successes in the realization of the program described above. For example W. Ackermann proved in 1924 by finitistic methods the consistency of a fragment of the arithmetic of natural numbers (without induction). However, soon something happened that undermined Hilbert's program (and in a certain sense the whole foundations of mathematics as well). In 1930 the young Austrian mathematician Kurt Gödel (1906–1978) proved that every consistent formalized system containing arithmetic of natural numbers must be incomplete. This means that there are sentences in the language of this system which are neither provable nor refutable. Such sentences are called undecidable. Gödel provided a concrete example of such a sentence. This result of Gödel is called today Gödel's First Incompleteness Theorem. It was published in 1931 in his paper *Über formal unentscheidbare Sätze der 'Principia Mathematica' und verwandter Systeme. I* [136]. At the end of this paper Gödel formulated another theorem[29] called today Gödel's Second Incompleteness Theorem. It says that no consistent formal theory containing arithmetic of natural numbers can prove its own consistency. As a consequence one obtains that the infinite cannot be justified on the base of the finite.

Gödel's incompleteness theorems indicated certain cognitive limitations of the deductive method. They have shown that the whole classical mathematics cannot be included in a consistent formalized system based on the first-order logic (predicate calculus).[30] Even more: one cannot include in such a system all truths about natural numbers. The undecidable sentence constructed by Gödel was an example of a true real sentence speaking only about natural numbers which cannot be proved in the arithmetic of natural numbers. However, it can be proved by infinitistic methods based on set theory and model theory. This indicates also another problem which will be discussed in Section 5.5 – namely the problem of distinguishing between provability and truth.

---

29 Gödel provided neither in this paper nor later in another paper a proof of this theorem. The first proof was given by D. Hilbert and P. Bernays in the second volume of the monograph *Grundlagen der Mathematik* (1939).

30 In the first-order logic quantifiers such as $\forall$ or $\exists$ can bound only variables representing individuals and not properties or sets of individuals.

The undecidable sentence provided by Gödel has in fact no direct mathematical but a metamathematical content. It simply stated: “I (this sentence) am not a theorem”.[31] One hoped that in the domain of strongly mathematical (or better: mathematically interesting) sentences mathematics is complete, i.e., that all such sentences are decidable. However, results of Jeff Paris, Leo Harrington and Laury Kirby showed that that were in fact vain hopes. In 1977 J. Paris and L. Harrington gave an example of a true undecidable sentence of combinatorial contents ([208]) and in 1982 J. Paris and L. Kirby provided an example of a true sentence of number-theoretical contents (Goodstein sentence) that is undecidable in the arithmetic of natural numbers. Hence those sentences are examples of real sentences about natural numbers that are mathematically interesting but have no finitistic purely arithmetical proof. However, they can be proved by infinitistic (using transfinite ordinal numbers) method. – Further comments to those sentences can be found in various works, for example in [250]. D. Hoffmann formulates and proves the sentence of Goodstein in [180, p. 233], presents it in a simple way and explains its arithmetical unprovability.

Gödels results generalized and strengthened by results of Paris, Harrington and Kirby struck Hilbert’s program. They showed that this program cannot be realized in the original form. However, they did not reject the idea of Hilbert’s program. In fact, it was not clear what should be understood as “finitistic” or “real”. Hilbert and Gödel agreed that in the light of the incompleteness theorems the scope of means and methods that can be recognized as finitistic should be extended. On the other hand the fact that Hilbert’s program cannot be realized for a particular formalized system of arithmetic does not mean that it cannot be realized in the case of non-formalized, naïve theory of numbers. It cannot be excluded that the elementary arithmetic can be formalized in a system which can be justified and founded in a finitistic way.

In this situation one tried to continue Hilbert’s program by extending the scope of admissible means and methods. The idea was to allow not only finitistic methods but all “constructive” means – though it was again not quite clear what they are in fact. One of the motivations for such an extension was the proof of the theorem that Peano arithmetic, hence the classical arithmetic of natural numbers can be reduced to the intuitionistic system of arithmetic (called Heyting arithmetic). This was shown independently by Gödel and Gentzen in 1933. Another reason could be Gentzen’s proof of the consistency of Peano arithmetic by using ε_0-induction.[32] This new paradigm forms the so-called *generalized Hilbert’s program*. Many interesting results have been obtain in it. Let us mention here works by Kurt Schütte, Gaisi Takeuti, Solomon Feferman or Georg Kreisel. However, it should be stressed that the extending of admissible methods to all constructive ones (and not only finitistic ones) seriously changed the situation. The finitistic methods and objects had clear physical and concrete basis and they are indispensable for any scientific thinking. The proposed constructive methods cannot be treated

31 Using a coding of arithmetical formulas developed by Gödel and called today the gödelization of metamathematics one can translate this sentence into a sentence about natural numbers (cf. [180, Section 4.2.1]).

32 This is the induction for natural numbers ordered in the type ε_0; the usual induction for natural numbers ordered in the standard way is called ω-induction.

as finitistic. Consequently, the generalized Hilbert's program follows the ideas of the reductionist philosophy of the founder of formalism but results obtained in it do not contribute to the realization of the original Hilbert's program.

Another consequence of Gödel's incompleteness theorems was the idea of reducing Hilbert's program to a "relativized" program. If the entire infinitistic mathematics cannot be reduced to and justified by finitistic mathematics then one can ask for which part of it this is possible. In other words: how much of infinitistic mathematics can be developed within formal systems which are conservative over finitistic mathematics with respect to real sentences? This question constitutes the *relativized* version of *Hilbert's program* (cf. [132, p. 218]).

Recent results of the so-called reverse mathematics developed mainly by H. Friedman and S. G. Simpson contributed very much to this program. Reverse mathematics is a research program initiated by Friedman at the International Congress of Mathematicians in Vancouver in 1974. It can be described as follows: The classical mathematics has not to be formalized in the framework of set theory. Many parts of mathematics, for example geometry, number theory, analysis, differential equations, complex analysis etc. can be reconstructed in a certain weaker system of the second-order arithmetic. Its main axiom is the comprehension axiom postulating the existence of certain objects, more exactly: the existence of sets of natural numbers defined by formulas of given classes (non-predicative definitions are admitted). The main aim of the reverse mathematics is to study the role of the comprehension axiom in ordinary mathematics. In particular one asks the following question: Consider a particular definite mathematical theorem τ. What form of the comprehension axiom is necessary in order to prove τ? To answer this question assume that the theorem τ can be proved in a particular fragment $S(\tau)$ of the second-order arithmetic. Is $S(\tau)$ the weakest fragment with this property? To answer this question positively one shows that that the principal comprehension axiom of $S(\tau)$ is equivalent to τ, the equivalence being provable in some weaker system in which τ itself is not provable.

From the philosophical point of view reverse mathematics is an example of the reductionist program. Many interesting results have been obtained in the framework of it.[33] We can not describe them here in detail because they are usually technically very complicated. It is interesting that they show how large and important parts of the classical mathematics can be finitistically founded and justified. Thus the original Hilbert's program can be partially (though not fully) realized.

2.23 Philosophy of mathematics between 1931 and the end of the 1950s

The main aim of the classical theories in the philosophy of mathematics that arose at the turn from the 19th to the 20th century, i.e., of logicism, intuitionism and formalism, was to eliminate antinomies from the foundations of mathematics and to provide a safe

33 Cf. the book *Subsystems of Second Order Arithmetic* [329] by S. G. Simpson or the papers *Reverse Mathematik und ihre Bedeutung* [248] and *Hilbert's Program: Incompleteness Theorems vs. Partial Realizations* [249] by R. Murawski.

fundament of mathematics. It should be proved that mathematics is consistent and can be safely developed. Gödel's incompleteness theorems from 1931 (we just wrote about them above) were a shock for the foundational studies. After 1931 one can observe a certain stagnation in the philosophy of mathematics which lasted till the end of the fifties. Since the beginning of the sixties a renaissance of interests in it can be observed. We write below about the philosophy of mathematics in the period from 1931 till the end of the fifties and next about its development after 1960. The latter will be preceded by a description of a new philosophical position, namely of the evolutionary standpoint.

Between 1931 and 1960 new conceptions were formulated but they were not so significant as logicism, intuitionism and formalism. One should mention here first of all works by Willard Van Orman Quine, Ludwig Wittgenstein and Kurt Gödel. We wrote already about Haskell B. Curry who worked also in this period – his strong version of the formalism was presented above.

Willard Van Orman Quine (1908–2000) worked mainly in the theory of science. He wrote many works in the domain of semantics and was the author of an original approach to logic and set theory based on the principles of logicism. In the work *New Foundations for Mathematical Logic* [289] in 1937 he proposed a new system of set theory. It is based on two ideas of eliminating antinomies: Zermelo's idea of limitation of size and Russell's idea of the theory of types. In Quine's system called "New Foundation" (NF) one can develop the classical propositional calculus, the theory of relations, arithmetic and some fragments of the classical set theory. In 1940 Quine presented another system called "Mathematical Logic" (ML) which is an extension of NF.

According to Quine the criteria of accepting or rejecting mathematical theories should be similar to criteria used in the case of physical theories. Mathematics should not be treated in a different way than other disciplines only because there are no experiments in it. Mathematics is one of the theoretical disciplines which attempt to explain the reality (cf. the works *Two Dogmas of Empiricism* [292], *On Carnap's Views on Ontology* [291] and *On What There Is* [294]). His holistic view of scientific disciplines led him to the formulation of the so-called *indispensability argument*. Today in ontological discussions in the philosophy of mathematics it is one of the most important arguments for realism. It claims the following: Mathematics should be considered not in separation from other sciences. Since we are realists in the case of physical theories in which mathematics plays a role of an important instrument, one should also be a realist with respect to mathematical objects of mathematical theories. Since mathematics is *indispensable* for example in physical theories, its objects like sets, numbers, functions *do exist* – in a similar way as there are electrons which are indispensable for physical theories. Quine accepted only one kind of existence. Hence one does not find by him a distinction between physical, mathematical, ideal, intentional, conceptual etc. existence but simply existence. He rejects also the possibility of dividing a scientific theory into an "analytical" (hence purely conventional) part and a "synthetic" one (dealing with the reality).

Note that from the indispensability argument there follows the assumption that mathematics is – at least partially – able to describe and to understand the real world especially in such cases where it is an instrument in theories involving the real world.

A similar attitude assumed also Hilary Putnam (cf. his paper *What Is Mathematical Truth?* [287]). Therefore the indispensability argument is called today Quine–Putnam argument.

Quine's arguments against antirealism and antiempiricism prepared the way for a new empiricism in the philosophy of mathematics. An example of such an approach is the conception of Putnam – we write about it below.

The philosophical views of *Kurt Gödel* (1906–1978) were strongly connected with his formal results in mathematical logic and the foundations of mathematics. In fact, the influence was mutual, i.e., his philosophical views were also inspiration for his formal investigations.

The most important works of Gödel which enable us to reconstruct his philosophical views are *Russell's Mathematical Logic* [138] in 1944 and *What Is Cantor's Continuum Problem?* (cf. [139] from 1947, extended and improved version from 1964). Additional information give also three of his works published posthumously: *Some Basic Theorems on the Foundations of Mathematics and Their Implications* [144], *Is Mathematics a Syntax of Language?* [142] and *The Modern Development of Foundations of Mathematics in the Light of Philosophy* [145].

Gödel's philosophical position can be characterized as realism, more exactly: platonic realism. Gödel claimed that mathematical objects exist in the reality independently of time, space and the knowing subject – however he never explained what mathematical objects are and where they exist. Such assumption is indispensable if one wants to develop a conception of the system of mathematics – similarly it is necessary to assume that physical objects do exist if one wants to explain sensations. Gödel stressed the analogy between logic and mathematics on the one hand and natural sciences on the other. He referred here to Russell (cf. [138, p. 137]).

Gödel was convinced that "logic and mathematics [...] are based on axioms with a real content" [138, p. 139] in a way similar to physics which assumes hypotheses about the reality. He rejected the pure linguistic and syntactic interpretations of mathematics. Mathematical propositions are true – according to him – "by concepts that appear in them" even if their meaning is not definable, i.e., if they cannot be reduced to simpler concepts. He objected to the conception that the truth of mathematical statements is based on conventions or on syntactical linguistic rules. Mathematics has real contents, it is a system of mathematical facts (cf. [142, p. 358]).

Mathematical objects are something else than their representations in mathematical theories. They are transcendental with respect to them. This is the consequence of the objective existence of mathematical objects that are similar to "Ding(en) an sich" in Kant's epistemology. Axioms describe the properties of mathematical objects only in an incomplete way. However, contrary to Kant the knowing subject does not impress any structure on them – for example by the organization of his thinking or his perception.

Gödel presented and explained his philosophical ideas mainly in connection with problems of set theory, in particular with the problem of continuum. As a realist he was convinced that the continuum hypothesis has a determined logical value, i.e., it is true or false (though we are not able to decide now which one). Gödel assumed that there exists an absolute universe of sets which we try to describe by axioms of set theory. The fact that we can neither prove nor disprove the continuum hypothesis indicates that the axioms do not describe this universe in a complete way [139]. Hence he saw the necessity of adopting – now and over and over again – new axioms in mathematics, in particular in set theory. They should describe new properties of the universe of sets. The aim should be to be able to decide the continuum hypothesis. Gödel suggested that new infinity axioms postulating the existence of new large cardinals should be considered. He presumed that such axioms can imply not only some corollaries concerning the continuum hypothesis but also some new results in arithmetic of natural numbers. He made this proposal even though no new strong set-theoretical axioms implying a solution of the continuum problem or of new and old problems of number theory have been found (cf. [144, p. 307]). Gödel suggested that looking for the solution of the continuum problem one should look for new axioms based on quite new complex ideas which have not to be immediately plausible.

Gödel claimed that the basic source of mathematical knowledge is intuition. Objects of set theory are in fact far from sensual perception; however, we experience them in a way. He wrote in [139, p. 271 in the version of 1964] that "the axioms force themselves upon us as being true". In [144] Gödel speaks about the mathematical reality which is "perceived [...] by the human mind" and in [142] he writes about concepts and abstract mathematical objects which are recognized by a special type of experience. He does not see any reason to have less confidence in the intuitive mathematical sensation than in a sensual perception (cf. [139, p. 271]).

Mathematical intuition suffices to explain and justify simple basic concepts and axioms. However, one cannot understand it as giving us already a direct, immediate and closed mathematical knowledge. Results of the first intuitive data must be further developed by a deeper investigation which can lead to the adoption of new statements as axioms. Hence mathematical knowledge is not the result of a passive contemplation of data given by intuition, but a result of the activity of the mind which has a dynamic and cumulative character.

Gödel never explained what he meant by intuition. Hence it is difficult to say more about it. In fact, various interpretations are possible.

Our mathematical intuition is being developed by the analysis of concepts and in mathematical practice. Just the analysis of concepts is a basis of our mathematical activity. Especially the analysis of the concept of a set is important here. A deeper understanding of it can be gained phenomenologically. It is worth noting that since

1959 he was interested in Husserl's phenomenology and this probably influenced his theory of mathematical intuition.

As said above intuition provides us with the knowledge of simple concepts and axioms. Theoretical and more complex hypotheses and postulates are justified from outside, for example by interesting consequences, by the fact that they enable us to solve problems which so far were unsolved or by providing the possibility of simplifying proofs. Hence the effect is crucial. Notice that Gödel admits here consequences both in mathematics itself as well as, for example, in physics. This is, according to him, the second (besides intuition) criterion of the truth of mathematical statements.

Another author important for the development of the philosophy of mathematics in the considered period was *Ludwig Wittgenstein* (1889–1951). It should be noticed at the very beginning that it is really difficult to present his views concerning the philosophy of mathematics clearly and definitely. The reason is first of all his manner of writing: ambiguous, imprecise and rather aphoristic. Another reason is the fact that Wittgenstein queried over and over again – especially in his later writings – his own declarations. Therefore his statements can be interpreted in various ways. His philosophy of mathematics can be understood as strong conventionalism as well as a form of behaviorism or even as strict formalism.

Wittgenstein wrote no closed work in the domain of the philosophy of mathematics. His views were expressed in many, often mutually inconsistent remarks made at various periods of his life. One finds them first of all in his posthumous work *Philosophical Investigations* [377] and in the work published on the base of bequeathed manuscripts *Remarks on the Foundations of Mathematics* [378].

Wittgenstein was a firm opponent of logicism, in particular of Russell's attempts to reduce arithmetic and the whole of mathematics to logic. He was convinced that by such reduction the creative character of mathematical proofs and the diversity of technics of proving are not taken into account. A mathematical proof cannot be reduced to axioms and inference rules of a logical calculus, for in the process of proving can arise new concepts and methods (cf. [378, Part III, 41]). Logicism ascribes to logic a fundamental role in mathematics; however, in the mathematical reality it plays only an auxiliary role. Wittgenstein speaks here about a fatal "invasion" of logic into mathematics (cf. [378, Part V, 24]). Using his aphoristic abilities he summarized his criticism against logicism in the following way: Logic is the base of mathematics in the same degree "as the painted rock bears a painted castle".

He underlined the singularity and independence of mathematical knowledge with respect to logic. He claimed that mathematical truths are *a priori*, synthetic and constructive. One can recognize here the clear accordance with Kant's and Brouwer's views. Similarly to Kant, Wittgenstein underlined the necessary character of mathematical knowledge. Mathematicians are not discovering mathematical objects and their properties but creating them. In contrast to Kant, Wittgenstein claims that the consistency is not a priori an indispensable condition of a meaningful construction of concepts.

He queries – similarly to Brouwer – the correctness of applying the principle of the excluded middle in judgements about infinite sets. Wittgenstein emphasized the spontaneous character of creation in mathematics. One can easily see the connections of Wittgenstein's philosophy of mathematics and Kant's ideas as well as the conceptions of intuitionists, in particular of Brouwer.

The classical conceptions in the philosophy of mathematics, that is, logicism, intuitionism and formalism, were not only an inspiration and starting point of new conceptions in the philosophy of mathematics. They caused and influenced also deep and important investigations in logic and the foundations of mathematics. Those investigations were intensified especially after 1931.

One can point here to Alfred Tarski (1901–1983) and set-theoretical semantics developed by him. It gave rise to the very important part of the foundations of mathematics, namely to model theory. Today the latter is a fundamental instrument in studies of elementary and infinitary languages as well as of mathematical theories and structures. Tarski developed it in his investigations concerning the concept of truth. He introduced there the crucial concept of satisfaction which is fundamental for semantics (cf. Section 5.1). His semantics was developed under the influence of logicism and platonism.

Also proof theory, i.e., the finitistic metamathematics introduced by Hilbert, has been developed at that time. After Gödel's incompleteness theorems showing that Hilbert's program cannot be realized in the original setting, the aim of proof theory was to justify the classical mathematics in the spirit of Hilbert's philosophy. To do this an extended and powerful technical apparatus has been developed.

Under the influence of intuitionism various systems of constructive mathematics have been developed – we already wrote about them. The intuitionistic mathematics and logic themselves were also studied.

The philosophy of mathematics influenced the foundations of mathematics and vice versa. Results of logic and of the foundations of mathematics influenced also the formation of philosophical conceptions. One can mention here the development of the theory of recursive functions that influenced the constructive conceptions in the philosophy of mathematics or Gödel's (1938) and Cohen's (1963) theorems on the consistency and independence of the Axiom of Choice and the Continuum Hypothesis. They indicated that the axioms of set theory are not strong enough to decide such questions and that they do not grasp in a sufficient way the nature of sets. One can formulate various quite different set theories and in this way provide diverse foundations of mathematics. We write about this in Chapter 4.

2.24 The evolutionary point of view – a new basic position in philosophy

So far we have presented authors, positions and trends in the philosophy of mathematics, all of which we may call classical. The philosophical views we have seen relate in a different, often complex manner to the classical basic positions, which we briefly characterized at the beginning of this chapter. Sometimes it has been difficult to locate positions and conceptions inside the triangle spanned by the basic positions.

The last-mentioned authors and lines of thought – approximately since Dedekind – belong to a period when a new basic position comes to the fore, getting important in a short period of time. Some new conceptions in the philosophy of mathematics along the lines of the already described philosophical positions can be seen since about the end of the 1950s. They came into being against the backdrop of the new philosophical position generated by the theory of evolution. By this of course the concept of development and evolution gained a certain priority.

We present the evolutionary point of view in order to prepare our discussion of new and related viewpoints, which are associated with the names of various well-known thinkers in the modern philosophy of mathematics. Even if these points of view are new and modify the classical conceptions, they are in our opinion without great importance, rather peripheral to the fundamental questions in the philosophy of mathematics. Since the evolutionary position in philosophy, psychology and other sciences generally is fundamental, in the philosophy of mathematics however it acts rather indirectly as a new basic position. In the philosophy of mathematics it implies a new way of looking at mathematics, namely at its practice. It seems that beyond the concept of number the evolutionary position does not contribute any essential towards relevant fundamental problems in the philosophy of mathematics.

We will first characterize the evolutionary position, and then describe two studies of the concept of number. Through this, in the context of a fundamental mathematical concept, the impact of evolutionary scientific thinking on the philosophy of mathematics will become apparent.

2.24.1 Characterization

First we will modify the triangle of the basic philosophical positions given at the beginning of this chapter; see Fig. 2.1

The theory of evolution, rather the discovery of evolution by Charles Darwin 150 years ago did not only transform biology. It seemed to offer an extreme challenge to philosophy. For a long time it had reached the thinking of humans and changed it essentially within every day life. Considering reality without development is impossible. It is remarkable that the fundamental concept of evolution came into philosophy in its full sense only by biology.

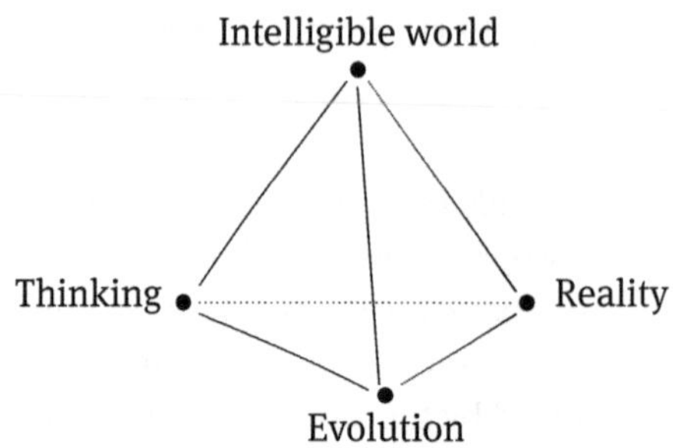

Fig. 2.1. The modified triangle of the basic philosophical positions.

The concept of development was a main factor in the growth of psychology and anthropology separating them from philosophy in the late 19th century. Psychological and anthropological positions are specific conceptions of the world, of reality and of the development of man and thinking as well as of the mechanisms of development. They necessarily start at fundamental philosophical positions questioning and complementing the classical theories of cognition and simultaneously belong to them. Psychology, anthropology and sociology have always reacted on philosophy.

Nearly everyone today, without really being conscious of it, takes over the evolutionary position concerning the world and human beings. Since the evolutionary approach has become more or less obligatory it changed the philosophical environment and influenced the representatives of the philosophy of mathematics in recent decades. We note that mathematics itself has only been marginally affected. Mathematicians see of course the development of mathematical knowledge but normally inside a closed mathematical world – far from any theory of evolution.

Against the backdrop of the theory of evolution in addition to pure conceptual and methodological issues the *conditions* of development emerge – and its actors. The individual mathematician comes to the fore as well as the practice of mathematics, the community of mathematicians, the cultural and scientific environment. Mathematics, which even today one likes to see as absolute, universal, fixed, closed and exclusive, gets a new appearance. Like the other sciences it becomes a historical, cultural and sociological phenomenon.

If we take the evolutionary point of view we get a new perspective on human thinking. From the outside we look at thinking – like at an external object. The traditional inner perspective is replaced. It is no longer only thinking which is obligatory for thinking about thinking, for from the evolutionary point of view we are able to follow the emergence of man in the frame of evolution. Before us we see early products of human intelligence, of art and culture reaching up to nowadays civilization, technology and the sciences.

Epistemologically the concept of evolution means a turning point and challenges the old fundamental positions of idealism and rationalism. For the first time in the history of philosophy it is possible to follow the development of thinking within the biological and psychological development of humans. The way seems to be open to an understanding of the various periods of evolution and the conditions under which the human mind and spirit arise.

We try to characterize the evolutionary conception in brief. The variations of concepts in the periphery of this position and in areas influenced by it are manifold. In the following subsections we describe *tendencies* which, as we see it, predominate.

On basic principles

"Biological" or "evolutionary" epistemology was founded by *Konrad Lorenz* (1903–1989) and is also represented by Rupert Riedl (1925–2005) and Gerhard Vollmer (born 1943). Their fundamental theme is:

- Life is a knowledge-gaining process. It leads continuously from the evolution of molecules up to civilization.

From this results:

- Thinking is like a higher organ which developed in a second stage of evolution. The structures of thinking are copies of patterns in nature.
- Knowledge (in the classical sense) is the actual reflection of reality in structures of thinking.

The old epistemological problem in philosophy changes to a biological, anthropological, historical and finally a psychological and sociological problem.

Life is a process of knowledge. From the evolutionary point of view the old fundamental positions melt together: thinking, mind, spirit, reality are phenomena of one and the same life. The first exercise is the study of evolution and the development of man. This study principally makes it possible to solve the age-old mysteries of philosophy. It levels the artificial abysses of matter and mind, subject and object, thinking and being, idea and reality. When we look at our subject, mathematics, the problem of the application of mathematics seems to vanish (cf. Section 3.4). The distance to other sciences, especially natural sciences, disappears.

What are the consequences – in comparison with the old basic positions – to get the viewpoint of biological evolution?

- Evolutionary epistemology is *empiricist*.
  Biology is a natural science. Thinking in its inner structures reflects the outer structures of the world, which are primary in this process.
- Evolutionary epistemology is *anti-rationalistic*.
  Thinking is the product of nature. Therefore thinking answers to objects and processes in nature. All rationalistic residues vanish, which would isolate thinking and knowledge from objects. The objects of the world are directly given to thinking – by its evolutionary biography. The thing-in-itself (Ding an sich) is metaphysical nonsense.
- Evolutionary epistemology is *anti-idealistic*.

Evolution proceeds by outer processes: chance, mutation, accumulation, selection etc. Higher spiritual and mental principles, which are supposed to guide evolution, are as unnecessary as they are absurd. In fact, for the first time in history it is now possible to look at the development of mind and to say what mind and spirit are. Some cautious attempts can be found in [99]. For example mind is said to be the medium between acquisition and design of the world. Another approach is to connect thinking with the operations and functions of the central nervous system. There is the tendency to look at the mind as a "system" or a "conglomeration" of task-specific cognitive mechanisms or rather modules, that are adaptions to evolutionary problems [333, II 6]. There is a popular comparison not only used by communication media in which the mind is to brain as software in computers is to hardware.

On basic hypotheses

Observing the development of the mind one is compelled to look back far into primeval times. For one knows and will know nothing or little about so early periods, the discovery of evolution becomes necessarily the *theory* of evolution. This theory lives on more or less known basic hypotheses which are partially convincing but often not really noticed as presuppositions. We give some of them in the following short background story and accentuate them by keywords, setting them in *italic* type.

Evolution and development are part of the world. We recognize, observe and experience them every day. Anything beyond however, which one describes or explains, introduces *theory*. A theory of evolution starts postulating the *creation* of life, e.g., primitive *living organisms* arising of dead material. Evolution takes place in mutual *interactions* between organisms and the world. Higher organisms such as plants, animals and humans rise by principles of *chance, mutation, accommodation, selection and cooperation. Humans* are considered to be *higher animals.*

Assuming even primitive organisms one presumes their *inner structures.* Usually an organism is considered as *subject*, which is the active part, the world is assumed as its *object*, that is, as an area of objects and interactions. This is a remarkable *basic hypothesis.* The world and its objects are seen as material, unintelligent, unspiritual and without mind – formed by physical laws and organic mechanisms. At the "end" of all, today, human beings stand opposite the world as the potential owner of mind and spirit. The task is to explain human intelligence and spirituality coming out of an unspiritual and void world. This task is as we think hard to accomplish (cf. [254]).

The process of organic *evolution to higher levels* is found to be an increasingly effective adaptation of organisms, that is, *adaptation* to environments. *Adaptation* is a further basic hypothesis which concerns the biological process as well as the cognitive one. *Adaptation* matches organisms to environment, at first connects organisms to environments. It starts by physical and chemical processes proceeding to organic re-actions and biological patterns of behavior.

Pure matching with environments does not suffice to explain intelligent behavior. The preconditions are *inner structures* which enable organisms to adapt changing environments and to create new behavior. Only here we find the origin of the active part which we already called "interactions". Intelligence probably starts with *inner autonomous regulations of inner structures.* So-called modules of inner structures develop depending on "evolutionary problems" and later on build *conglomerations* or *structures of modules* (cf. [333, II 4]).

Animals and humans differ in complexity of modules and "conglomerations" of modules, in their flexibility, in autonomous regulations of modules and *autonomous regulations of regulations*, and finally in the ability building up gradually meta-modules and meta-regulations. In *regulations of autonomous regulations* one can see aspects of consciousness. Here we may find the beginning of the capacity of planing, of *anticipation* and *acting* in its true sense.

Following the above sequence of hypotheses printed in italic, we are not able to separate clearly human intelligence from animal intelligence. If one trusts in the basic hypotheses humans are considered to be nothing but higher animals. Phrases such as "humans and other animals" nowadays are standard. By this the importance of primatology is explainable, for one assumes that one will find early conditions and elementary characteristics of human behavior and human intelligence. By the theory of evolution, mind and spirit turn out to be intelligence – accompanied by more or less consciousness.

2.24.2 On studies of evolution of number concept

To begin, we will describe two investigations that deal with the development of the concept of number. They come from the psychology and anthropology of culture, the latter in the framework of modern philosophical anthropology (cf. [99]). They rest on epistemological positions which are very close to the evolutionary standpoint and which we may characterize as *genetic.* Their investigational approach shows how the epistemological viewpoint can change into an empirical attitude which substantially needs hypotheses on the early development of individuals. These positions, though generally important, are only marginally acknowledged in the philosophy of mathematics. At the end of this subsection we discuss a project which starts with results of developmental-psychological studies and tries to derive the whole of mathematics.

Where and when in the process of evolution do we find hints about numbers? Again we offer hypotheses (set again in *italic* type). Anthropologically one may suppose signs of numbers in a very early *behavior*, going back to early organisms which search for optimal conditions of surviving and life, e.g., for sufficient foodstuff. Such behavior is something like a *comparison of conditions* starting on a pure biological level. Maybe structures and modules of behavior grow up gradually into autonomously regulating systems of modules enabling oriented actions. To them belong *concrete acts* oriented to *resources* and their quantities. Thus psychologically we have the same *concrete acts*, which are the initial point of development of the concept of number during the early development of children. It is obvious which class of numbers are indicated: numbers abstracted from acts and operations of comparing. The development of numbers which is investigated against an empirical backdrop in such a way will in particular produce cardinal numbers.

Is it adequate to study numbers in such a reduced number environment and to search only for numbers in the development of thinking in abstractions from acts? We will comment this questions in explanatory notes. For doing this in brief we refer to Section 3.1 where we will give an abstract of observations of number concepts in Chapter 2 and where we will offer our own number conception. One can find more details about this conception and its motives in [20, Section 5.1].

A study in cultural anthropology

Not surprisingly, an anthropological study by Peter Damerow [82] begins in a period, when *acts* occur in the evolution of man. It tries to outline the development of the concept of number in evolution.

Well aware that his account is speculative, Damerow entitles his piece carefully "Preconsiderations". Its interests are epistemological. The way to numbers is seen as an interplay between historiogenesis and ontogenesis; it starts with concrete acts and goes through inner operations, representations and abstractions up to numbers as ordered cardinal numbers.

Two "basic hypotheses" determine the study:

> "firstly, logical-mathematical concepts are abstracted invariants and transformations which are realized by acts, [...],
> secondly, such abstractions are handed down by collective external representations [...]." ([81, p. 271].)

To these basic hypotheses a further methodological one is added, namely

> "that Piaget's theoretical reconstruction of the development of number in ontogenesis reproduce this process in a mainly correct way." ([81, p. 255].)

External systems of representatives, e.g., arithmetic symbols or systems of symbols, are local emergences in cultural evolution. On the one hand they are products of psychological *constructions* by single individuals in *social history*. On the other hand they are at every moment in time subjects of *ontogenetic reconstructions* and advancements of individuals. We will speak about Piaget's attempts in the following subsection.

In [81] the development of numbers starts with so-called "arithmetical activities", that are "acts of comparing, correspondence, union and iterating" [81, p. 280]. In a precedent "pre-arithmetical level" such acts are linked to concrete objects in practical situations. Even if they are far from numbers these acts are named "arithmetical", for one psychologically assumes that concrete acts like these will precede in the run-up to cardinal numbers.

In a first "proto-arithmetical level" [81, p. 285] we find "counting objects" that means objective symbols representing single objects like notches or tallies and words or symbolic actions in standard "counting sequences", which by iteration represents quantities [81, p. 286].

At a second level of "symbolic arithmetic" [81, p. 293] representatives of a higher level come to the fore such as abstract but context-sensitive symbols, which beyond pure symbolization are used for calculating by rules. Early arithmetic and technical terms arise from the reference to the handling of symbols and objects. Arithmetical operations and expressions provide a basis of arithmetical activities from where a third and last level in arithmetical evolution is reached: "theoretical arithmetic". This is a domain of expressions of abstract numbers in deductive systems won by mental operations only. If deductions still refer to arithmetical interpretations they are called "linguistic". If they ascend from interpretations and move on to systems of freely interpretable axioms they are called "formal".

The study of Damerow that we have been examining puts

> "the question about the historical or ahistorical nature of logical-mathematical and especially arithmetical thinking" ([81, p. 314], cf. [83, 51])

and assumes in his attempts a basis for an answer. However, he believes that this problem is not theoretically decidable. He speaks about a potential "part of arithmetical thinking" going back to "cultural achievements" and about a potential "extent" of "structures and processes of arithmetical thinking" [...] which are "not variable universals of the nature of *Homo sapiens*". He assumes that external representations are the key to the questions in "historical-culturally comparing" investigations which promise to give further answers.

The concept of number in the study on hand refers to Piaget, whom we speak about below. This concept influences the considerations of Damerow, which are mainly oriented to cardinal numbers. His conception is complex:

> *Numbers are cardinals including ordinal aspects. These are individual and social cognitive constructions and reconstructions which come and came to the fore in the psychological development of individuals as well as in cultural development of societies. External representations are connecting links between individual and social and historical development. The concept of number owns both universal-ahistorical components as well as cultural-historical features.*

We stress that elementary structures of sequence and counting which are fundamentals of the concept of number on the one hand are postulated but on the other they are not objects of the investigation when we are searching for sources and considering early development of numbers. Damerow discusses "iterations" and "repeating acts". They seem to come from action systems belonging to the natural aptitudes of humans which *precede* true arithmetical activities. Does the ability to iterating and repeating and thus of sequences belong to "logical-mathematical universals of thinking", which Damerow considers possible?

Comments

We think that the cardinal imprinting of numbers and the absence of sequence structures in Damerow's study is not surprising, and actually necessary in anthropological studies of this kind. There is dealt with acts without realizing the condition of sequences preceding acts. Basic for acting is the ability to sequence. In counting this ability finally obtains the status of the conceptual. In anthropological studies one usually disregards the correlation between sequence and acting.

According to our outline above we may assume that the development of acts in evolution happens between self-regulation and regulation of regulations. Also the concept of counting points to this early phase in evolution. Our discussion of the concept of number in Section 3.1.2 and the basic hypotheses seen in the theory of evolution and in anthropology suggest assigning the concept of number to ahistorical universals and considering it as independent of cultural influences. Anyway, the principle of counting seems to be fundamental and to cause even anthropologically the concept of number.

Three principal remarks shall close our discussion about this anthropological project.

(a) First an important methodological remark: In fact, it is not correct to ask the question about development of numbers in human evolution as it is usually done. We ask too simple. We have rather to differentiate the question – according to aspects of the number concept. The study above follows the coming out of numbers on the cardinal numeral path, presupposes ordinal features and gains mainly finite cardinal numbers. Other studies could do something like that under the quantity aspect (such studies however we do not know). We ourself stressed the fundamental aspect of counting and of the so-called counting-numbers which seem to point to very early evolution. There are further aspects of the number concept [20, Section 4.2], which are basic and may play a vital role in the development of the concept of number (see (b)). As far as we know these aspects remain uninvestigated.

(b) Second a remark on fundamentals: The phenomenon of time seems to complicate considerations to the process of counting in anthropological studies. There is in a way a kind of self-reference. Evolution and development are not thinkable without time. Reversely thinking and acting happens in time putting points into the flow of time. In the emerging sequences of time points one in fact observes the principle of counting. Such sequences are parts of *reality*. By this, counting and counting-numbers, which represent the concept of sequence, are not only products of the development of mind. Sequences inside the temporal process itself represent the principle of counting and vice versa the principle of counting underlies evolution. By the aspect of counting, numbers and evolution are insolubly interconnected.

Anthropologically however, it is a question of development of *concept*: of becoming aware of time, of setting points in time as basis of acting. It is about the concept of ordering of points in time, of anticipation and recapitulation and in the final stages about the acquirement of counting and counting numbers. By these we finally find the concept of sequence by culturally standardized sequences. – To this concept mathematical intuitionists (Section 2.20) seem to refer when they speak about numbers as “meaningless abstractions of sensations of time”.

Conscious and systematic acting needs operating on sequences and orders in time – in particular the ability to anticipation. Here we realize the condition of order in time and cognitive steps on sequences in time, which point to numbers as ordinal numbers and operators. As structural elements basic for acting they are further fundamental conditions in the evolution of numbers.

(c) Finally, a restrictive remark: After all we are critical of the prospects of anthropological settlings on numbers even though they are of special interest. Interactions are early conditions for the existence of organisms, which somewhen move on to acting. Indeed, at what moment and by which circumstances acting of individuals did start and how did acting run the anthropological way to numbers? Nobody is able to research this. However, the cautious “preconsiderations” of Damerow are instructive and important especially because they ask relevant questions and leave questions explicitly unanswered. These questions point not least to psychology.

Psychological contributions

Fundamental research as we find it in developmental psychology is of epistemological importance. It was Jean Piaget who formed the developmental psychology in its early phases. Piagets experiments, types of problems and evaluations are characteristic elements of his theory. Essential basic concepts and basic statements in developmental psychology come from Piaget and are widely accepted – in spite of some ostensible distance. The basic attitudes today having constructivist features apparently goes back to Piaget. [333, II 4].

Piaget started his academic career as a zoologist. From his biological studies he adopted basic biological notions of the evolution of organisms for his psychological conception of development.

He considered development even of concepts as a constructing process between human individuals and the world. He followed the process up to the beginnings. For Piaget the beginning was the phase where subject and object begin to differ. This meant to investigate development up to the earliest age of infants. Analog investigations are anthropologically impossible. According to Piaget an individual constructs objects and becomes thus an acting subject. At this point cognitive development starts and here Piaget's constructivism established itself. Construction is axiom.

Apart from childish number development this approach is of interest in another way because it tries to understand the construction of objects within cognitive development. Both, subject and object, cause each other and arise in the same process. Piaget questioned the philosophical standpoint which usually starts at the dualism of subject and object. The concept of object is of interest in the philosophy of mathematics, for mathematics starts at special objects, namely at elements of sets being sets. For being objective subjects are eliminated.

Psychologically one looks at the newly born individual, which seems to be "adualistic". It does not differentiate itself from environment and starts to build inner structures and an external world coming out of a state of oneness with environment. It is difficult to have a view of this, for it is unclear where and how the construction begins. Assuming *a priori*-conditions for inner structures seems to be necessary. Additionally these conditions need a far-reaching intentionality. Somewhere Piaget named his theory "dynamic Kantianism".

Hypotheses in developmental psychology

Note that the set of fundamental problems in psychology concerning the early development of human individuals is of quite a different nature than in anthropology. Without a doubt many things, especially in early phylogenesis, must be notional. Piaget characterizes the systematic exposition of fundamentals in his developmental psychology as *Piaget's theory* and *my theory of mental development* [273].

We shall locate again some hypotheses on the psychological way to objects and from objects up to the number concept, which are as basic for psychology as they are for epistemology. Keywords for such hypotheses are printed again in *italics*.

The definitely first and necessarily basic hypothesis, which affiliate to basic anthropological hypotheses, is the assumption of *individuals*. It is presupposed that from the birth on the *intention* to develop inheres in individuals – including modality and orientation to some extent. Cognitive development starts with *elementary reflexes and schemata* and orientates itself to mental abilities. These hypotheses refer to biology and genetics. However, individual development cannot be explained sufficiently by hypotheses of this kind and even no idea of individual development can be seen unless one believes in total genetic and passive destination of humans, including especially the development of the concept of number. According to Piaget individual development however demands individual and active participation from individuals.

By means of *constructing objects* individuals become subjects. Including this process of differentiation Piaget assumed a persistent oneness of subject and the world which is given during the entire development by permanent dual contention. – We think that this idea is fundamental and of great epistemological importance. – Development proceeds by interactions between subject and the world wherein Piaget accentuated actions and constructions by subjects. To this belong *isolating* objects against environment, *differing* features and on the whole increasing *separation* of outer and inner world.

One can speak about "objects" only if objects are independent and persistent. The condition for that is *permanence*, for one is able to identify objects only if they do not vanish or change states and features. It is just this permanence which Piaget investigated. He connected it to inner "transformations" which come from concrete operations. Piaget pointed to a special phase where the separation of objects and subjects is done. This phase is characterized by the formation of a *special inner structure* of inner transformations which must be achieved [273, I 2].

Thus objectification, that means the ability of subjects to have objects, cannot be separated from the ability of acting. The concept of *action*, a central basic concept of developmental psychology, is then widened going back beyond normal conventions and starts in the area of childish sensation and perception, e.g., with motions of eyes, which follow and fixate objects. Recent studies detect clues of activity even by the duration of optical attention. Generally, the concept of action gets vague and very wide. It reaches from technical signals and outputs in cognitive sciences to the intentionally, volitionally, emotionally and freely determined deed in philosophy.

We have already seen the assumptions of interaction between individuals and the world, adaption, action, inner structure, self regulation and regulation as basic hypotheses in anthropology, which however seem to be put partially from psychological notions about ontogenesis into phylogenesis. Anyhow, the study of Damerow cited above relies on the description of ontogenesis by Piaget supposing that it is basically correct. Some of the hypotheses made in anthropology will get more clearness within further basic psychological hypotheses.

As we noticed already we can see a special *intentionality of constructions* as a further hypothesis. This hypothesis cannot be explained by outer adaption or inner self-regulation, accommodation or equilibration. From the beginning intentionality is present in actions, transformations and operations. For instance, having objects or concepts is neither structurally nor naturally given.

The way from outer objects and outer actions on objects to inner *operations* is the permanent outstanding condition within constructing inner structures. The assumption of *internalization* is fundamental for any further detail. But its conditions remain uncleared. Internalization cannot be separated from the concept of development at all and touches even early evolution. Here at the very beginning we see the usually unsaid problem of internalization of outer, dead world into inner structures of living organisms. We stress internalization as an hypothesis because psychologically dressed it comes back within the range of outer actions and inner operations. Internalization is the *main fundamental hypothesis*, which is present everywhere in the background.

There is another psychological hypothesis postulated by Piaget, namely *cognitive structures* themselves in the process of development. Outer actions and inner operations of individuals precede cognitive structures. The emergence of increasingly higher structures is the result and expression of a central principle, the so-called *equilibration* which claims that development tends to inner and outer balance and closeness. "*Reversibility*" of operations in "groupings" is a consequence of an equilibrium in structures and at the same time its condition. – The assumption of balance between inner structures and outer conditions refers back to biological sources. It reminds of processes of emerging from biotopes.

A further substantial assumption is *abstraction* about which very little is empirically known. Anyway, Piaget gave some hints mentioned in the following points.

Piaget's numbers

Higher structures of operations called by Piaget "groupings" are conditions also for the development of the concept of number. Piaget did not include pure counting into his reflections about numbers though counting shows characteristic properties of groupings at special stages. Since pure counting is not connected with outer actions and does not grow out of them it seems to have disappeared from view.

By pure counting, *via* orders of number signs, the formation of sequences becomes a concept. We will speak about this in Section 3.1.2. In Piaget's theory there are only vague descriptions of the building of representations by signs. Representations are seen within perceptions and run through imitations to designations. Here one can clearly see another possible universal restriction of the empirical orientation which is obligatory today. Empirically it is nearly inconceivable to include pure counting into the concept of number because *inner* elements in sequences – expressing the fundamental ability of building successions or orders – are not objects which one can perceive, imitate or indicate like outer objects.

This may explain why in Piaget's investigations counting is not relevant in the development of the number concept. Piaget wrote:

> "To the latter [counting – authors' remark] one does not need to attend for the subject of this book is the investigation of the constitution of number concept." ([272, p. 100].)

Piaget seemed to have overlooked counting as an independent concept building scheme unlike R. Dedekind, who mathematically stated counting as the fundamental structure of numbers and mathematics.

Therefore Piaget oriented himself towards building up sequences coming from actions on outer objects, e.g., comparing sizes. He connected inner operations in this context with classifying sets. The way to "invariant" classes passes through groupings of inner operations which go back to outer actions. These operations are the comparing activities which we already noticed when speaking about the anthropological study above. Finally, such activities lead to stable one-to-one correspondences mathematically interpreted as bijective functions.

Conditions for cardinal numbers are classes. For building up these classes one finds hints to actions and inner operations of comparing and structures of such operations. Beyond that we find out just a little – except usual remarks about emphasizing or neglecting features. The concept of abstraction remains vague. The hypothesis of abstraction by *class building* however was fundamental to the concept of number in Piaget's investigations.

Piaget's concept of number remains somewhat unprecise. He refers to Russell and Whitehead and to cardinal numbers as classes which he found in *Principia Mathematica* [370]. He criticizes the one-sidedness of this conception and claims that there is the direct connection of cardinals to an ordinal inclusion of classes. According to Piaget the concept of number is complete if and only if this connection is given. Piaget summed up his conception of number in the following way.

> Thus finite numbers are inevitably cardinal numbers and ordinal numbers; this follows from the nature of numbers itself which are systems of classes and asymmetric relations melted together to a unique operational oneness. ([272, p. 208].)

From this position, which is mathematically far from being precise, Piaget criticized "the often artificial deductions [...] which fundamentally teared away logistical research from psychological analysis though both are suited for supporting each other like mathematics and experimental physics" [272, p. 11]. We however remember the deep anti-psychological attitude by Frege.

Counting, objects, sets, numbers

Not long after the exclusion of counting in the development of the number concept by Piaget the psychological evaluation changed. Today counting is seen as an important, often as a main condition for the concept of number – in greater harmony with mathematics. However, a really fundamental role – as we see it mathematically in Dedekind or Peano axioms – is as far as we know nowhere dedicated to pure counting. At the most one observes counting only in combination with other number aspects. Counting is not acknowledged as a concept for itself, namely as the concept of a sequence (cf. [20, Subsection 5.1.3]).

Psychologically one continues to depend on cardinal numbers even if meanwhile other aspects of number concept are marginally noted. Counting is primarily seen as an auxiliary factor in determining numbers of elements of sets. One does not see for instance the instrumental meaning of counting and counting numbers. Against that one mathematically knows [20, Subsection 4.2.3] that natural numbers are characterized as universal counting numbers. Usually the pure structure of counting is psychologically characterized as "solid succession" which is in no way adequate. Counting numbers is noticed only by the way and is brought even sketchily into focus.

Without mathematical background descriptions of the development of the full cardinal meaning of numbers by counting are laborious and often unclear. Among other things this problem concerns the idea of measuring sets by comparing them to a section of a sequence of counting numbers. One finds for instance an "accumulator" model (cf. [244, Section 3.2]) which is an only vague figure.

More actual investigations attempt to demonstrate that children have cardinal abilities yet in the first months of life, e.g., the ability of discriminating sets of two or three items. This assumption depends on measuring time intervals of concentration shown by babies when cardinalities of little sets are varied. This is interesting because children as well as adults are able to realize sets of two, three or four elements simultaneously – called "subitizing". It occurs without counting elements or comparing sets element by element.

Here something apparently happens, which takes place between objects on the one hand and sets on the other. The question is: Do infants and little children see small sets simultaneously as objects. Are sets of two, three or four elements realized as integrated objects and their elements are isolated by discrimination afterwards, in the course of further development? An additional investigation attempts to show that this is not the case. It is claimed that it concerns real cardinal registration of sets. The results of investigations in such early phases of development depend only on indirect conclusions and remain strongly hypothetical. However, we may in fact suppose that there exist relations between the perception of number and the sphere of creating and perceiving objects in early childhood.

It is interesting to observe how the area of small sets is won by counting. During the phase of development of counting children count elements even of very small sets. But in doing so these sets are objects for themselves and at the same time sets of elements the number of which children realize simultaneously. Conversely, children build small sets as new objects by counting playfully arbitrary objects. In this way children on the one hand release elements from sets as integral objects, on the other they build sets as new objects from elements by counting. By this, children create counting as a new instrument in the area between sets of elements and sets as integral objects, and experience the cardinal meaning of number words in an early phase of development.

In such sets of few elements, which at an early stage of development are realized as integral objects, and in short processes of counting we may see the source or an aspect of the ability of collecting. This leads to the later ability of building sets as a whole from any distinct objects.

Let us try to look a little deeper into the process of differentiation where finally sets and numbers arise. The gradually increasing differentiation of environment and of features of objects leads to the awareness of details which in the beginning are not independent of these objects. This may happen also in the case of objects which are small sets. Details or elements remain part of the whole – like limbs of a chain. However, mathematical elements of sets are *no* parts. If one tells about "part of an object" it remains intentionally bounded to this object and connected with other parts of the same object. By the mathematical concept of a set mathematics decided upon the total dissociation of parts from sets as objects and upon the complete disconnection of parts among themselves. Parts become abstract elements (abstract in the true sense of the word), the former *connection* of parts becomes an incoherent *summary* of discrete elements (cf. Section 4.2).

In the elementary phase, creating objects, building sets and the development of the number concept seem to be manifoldly interweaved. From this phase of interconnection we see various influences on the later development of arithmetical concepts along individual and complicated inner processes of differentiation and abstraction.

On a project in cognitive science

At the end of this section we make some remarks about a well-known and ambitious project in cognitive science [223]. It starts just in the elementary phase of cognition development considering object building and realizing simultaneously small numbers. The plan is to establish the whole of mathematics and explain it physiologically. The aim is to show that mathematics is "embodied". We are interested in statements about numbers, elementary arithmetic and infinity.

At the beginning we find the psychological interpretations of observations of infants and little children indicated above. One deals with intervals of attention while the number of elements of small sets is changed. The interpretations of such observations are accepted as facts and it is assumed as true that in this way cardinal abilities are proved which are neuronal "embodied". The assumed abilities are taken as innate and figured as elementary numerical abilities belonging to an early number sense. In the following the authors on the one hand postulate so-called "grounding metaphors" being cognitive translations which come from everyday experiences with sets and lead into arithmetic. On the other hand "metaphors" are cognitive instruments extending innate numeric and later on early arithmetic. Additionally they assume that these grounding metaphors are "embodied", i.e., physiologically realized. Combinations of cognitive translations should explain the whole arithmetic which is supposed as starting within the area of material sets. The title of a referring subsection reads as the name of the metaphor: "Arithmetic As Object Collection".

Again we notice what seems to be obligatory in psychology and anthropology: numbers are primarily seen in a cardinal sense. Natural numbers at all are related to sets of objects. We notice the title of another subsection: "Numbers Are Collections of Objects". Counting exclusively means applied counting, i.e., counting objects which in turn starts by counting fingers combined with "subitizing", that is, realizing few elements and fixing small numbers. This accrues by the above-mentioned metaphors to a general instrument of identifying numbers of elements.

We state that in this project

numbers are cardinal numbers arising by metaphors (cognitive translations) from innate and "embodied" numbers which refer to small sets.

It remains unclear where the cognitive translations, the "metaphors", come from. As already said "metaphors" are only postulated and describe nothing but the usual relations connecting everyday situations and arithmetic. There is one exception:

Actual infinity comes from a new cognitive translation. Its name is "Basic Metaphor of Infinity", abbreviated BMI. BMI is assumed to come from the actual end of finite processes which moves on to potentially infinite processes. We ourselves however see the hypothesis of BMI only as an Axiom of Infinity cognitively dressed up. Differently from the authors of the discussed project we think that BMI does not make actual infinity neither mathematically nor in any way cognitively natural or even "embodied" and real. For us the hypothesis of neurological embodying of actual infinity remains strange. Where did BMI originate, when did it appear? Before about 1870 when G. Cantor started to publish his concept of sets BMI did not exist. Do we observe an evolutionary leap into some mathematical brains?

In our opinion the outlined project is not a physical explanation of arithmetic and mathematics as it is engagedly predicted in the introduction of [223]. We think that the project is solely a narrow look onto aspects of mathematics which is motivated and directed by ideas of cognitive science. It seems that the project, which uses the basic concept of metaphors, lives itself on metaphors. Cognitive hypotheses are partially taken for realities. About numbers we read that they are cardinals inborn until 3 or 4 and from this cardinal point "metaphorically" guide the concept of number and arithmetic. Some further aspects of numbers appear in different cognitive metaphors. These aspects however are merely paraphrased in an undifferentiated way as it was in the case of the primary cardinal aspect. Number aspects remain psychologically as well as mathematically vague.

However, we see the concept of "metaphor" as a cognitive transfer to be interesting and important. For instance above we spoke about an early stage of development where we assumed sources of arithmetical fundamentals. We supposed small sets initially being integral objects, which in early phases of childish development gradually differentiate into parts and elements. We considered this as a common source of the concepts of sets and number.

2.24.3 Concluding remarks

We have tried to characterize the evolutionary point of view and to identify its basic hypotheses. Of course we could do this only in a very general way. Today principles which determine the evolutionary position seem to be regarded as obvious and obligatory in many scientific disciplines. By providing examples we presented how these principles have impact on special approaches and investigations, which are of particular mathematical interest, and provoke new hypotheses. The evolutionary viewpoint is empiristic, and scientific disciplines today orientate themselves widely by empiricism.

Outcomes of psychological and anthropological investigations often provide valuable and interesting and sometimes peculiar insights into developments and structures of thinking – even in mathematics. It seems to us that in psychology and anthropology as well as in cognitive sciences there is only little, too little, knowledge of mathematics – too little for the purpose of building hypotheses on mathematical concepts in an appropriate way and of well-founded empiric investigations. Views about numbers are usually narrow and flat. They seem to be stamped *a priori* by the intention to investigate them in an empirical way. Generally, the awareness of the hypothetical character of empirical concepts and so-called results should be present. The study in cultural anthropology discussed above represents this in an exemplary manner, the project in cognitive science does not.

The basic idea of a strict evolutionary standpoint is finally to explain mind, intelligence, spirit, soul, consciousness, behavior etc. by few principles – starting from physics of matter and proceeding to biological and social levels. One believes that only local complexities have to be conquered but that there is in fact no fundamental problem at all. We distrust this common belief – loudly propagated by media – in physical, biological, physiological and electronic reduction, in brief, we distrust the belief in the material reduction of a human being and the world. This idea appears to be somewhat circular, namely to deduce by a theory even the mind which makes the theory. In [254] the project is declared to be "almost certainly false".

Where does the great idea to explain all from one come from? It is the heritage of monotheism and today feeds on physics which is on the way to explain the beginning of cosmos and matter. What follows is biology. "The human will to believe" the american philosopher Thomas Nagel says in the last sentence of his book *Mind and Cosmos* [254] "is inexhaustible".

2.25 Philosophy of mathematics after 1960

In the early sixties a certain renaissance of interests in the philosophy of mathematics can be observed. The old doctrines: logicism, intuitionism and formalism, still prevail but there appear some new conceptions based on other presuppositions and having other aims (as indicated in Section 2.24). The new conceptions are the reaction on certain limitations and the one-sidedness of the classical positions. In fact, logicism, intuitionism and formalism as well as platonism are results of radical reductionist tendencies in the philosophy of mathematics. Hence they have a clear monistic and exclusive character.

Formalism and intuitionism after 1960 did not abandon in principle the ideas of their founders though the development of mathematical logic and the foundations of mathematics influenced them and forced some modifications of the original theses. As indicated above under the influence of Gödel's incompleteness theorems the generalized Hilbert's program as well as the relativized Hilbert's program have been developed (cf. Section 2.22). Recently results of the so-called reverse mathematics developed mainly by H. Friedman and S. G. Simpson contributed very much to this program. In fact, they showed that several interesting and significant parts of classical mathematics are finitely reducible. This means that Hilbert's program can be partially realized. Logicism appears today as so-called pluralistic logicism and is represented by H. Mehlberg and H. Putnam. According to it the main task in mathematics is the construction of proofs in axiomatic systems. By the deduction theorem if a sentence φ is a theorem of an axiomatic theory T then there exist axioms $\varphi_1, \varphi_2, \dots, \varphi_n$ of this theory such that the formula $\varphi_1, \varphi_2, \dots, \varphi_n \to \varphi$ is a theorem of logic. Hence mathematical theories are nothing more than a source of logical laws. It is absolutely inessential what axioms one adopts, there do not exist better or worse theories – important are only the logical interconnections between axioms and theorems. Hence for example there is (from the point of view of the pure mathematics) no difference between Euclidean or non-Euclidean geometry.

The idea of the pluralistic logicism can be in fact found already by Russell in 1900. However, later he changed his mind. One can also see a certain connection of this conception with views of Aristotle. As we indicated above Aristotle claimed that the necessity cannot be found in any single statement about mathematical objects but in hypothetical statements saying that if a certain proposition is true then a certain other proposition is also true. Hence important are not truth, certainty and necessity of single statements but logical connections between propositions.

A typical feature of classical conceptions of the philosophy of mathematics and their versions developed later is the fact that they looked for safe foundations for mathematics as a science. Their aim is to show that mathematics is consistent and consequently its results are irrefutable and safe. Therefore they reduced mathematics practiced by normal mathematicians to various foundations. In reconstructions of mathematics one uses various instruments and methods of mathematical logic and the foundations of mathematics. Therefore those classical conceptions are called *foundational conceptions*.

Those conceptions are reductionist and monistic, i.e., they reduce mathematics with respect to one aspect. Logicism claims that mathematics can be reduced to logic (and set theory). Intuitionism says that mathematics can be based on the intuition of the natural numbers and the latter can be founded on the intuition of the *a priori* time. Formalism sees the rescue in formalized languages in which all mathematical theories should be expressed. As a result there arises an idealized image of mathematics far from the research practice of mathematicians. The foundational conceptions are giving one-dimensional static picture of mathematics as a science and are trying to provide indubitable and infallible foundations for mathematics. They treat mathematics as a science in which one automatically and continuously collects true and proved propositions. They lose sight of the complexity of the phenomenon of mathematics. The historical development of mathematics as a science and the development of mathematical knowledge of a particular mathematician play here no or only subordinated role.

In the sixties there arose conceptions that challenged such views. In English literature they are called *anti-foundational*. They try to reexamine the real mathematics and the actual research practice of mathematicians and not reconstructions of mathematics on the base of various foundations. The starting point of investigations of the anti-foundational trends is mathematics that is really existing and practically done. The aim of those investigations is to describe and to understand this mathematics.

By such investigations one should necessarily take into account the historical aspect of mathematics and its development. Mathematics is no absolute and fixed phenomenon that was and is always the same. It has its own history. It is variable and influenced by the epoch and the culture in the framework of which it is being developed. The research practice of mathematicians is not isolated, just the opposite – it is connected and has many common features with the research practice of other disciplines, in particular of natural sciences. This means that the philosophical questions asked in connection with mathematics are connected with other disciplines in which mathematics plays a role. Additionally it is not appropriate to treat mathematics from the very beginning as a discipline of special character, to separate it from other disciplines and to ascribe its objects a special ontological and epistemological status.

Conceptions taking into account the research practice of mathematicians and trying to grasp the phenomenon of the really practiced mathematics instead of providing "only" reconstructions of it are called quasi-empirical.

2.25.1 Quasi-empirical conceptions

One of the first attempts in this direction was the conception of *Imre Lakatos* (1922–1974). He attempted to apply some of Popper's ideas about the methods of natural science to episodes from the history of mathematics. His results were published in the work *Proofs and Refutations. The Logic of Mathematical Discovery* [220] in 1963.

Lakatos claimed that mathematics is not an indubitable and infallible science – on the contrary, it is fallible as other disciplines are. It is being developed by criticizing and correcting former theories which are never free of vagueness and ambiguity and may include errors and accidents. When a mathematician tries to solve a problem then he formulates a hypothesis and is looking simultaneously for a proof of it and for a counterexample. Lakatos writes:

> "Mathematics does not grow through a monotonous increase of the number of indubitably established theorems, but through the incessant improvement of guesses by speculation and criticism, by the logic of proofs and refutations." (cf. [220].)

New proofs suspend old counterexamples, new counterexamples undermine old proofs. By proofs Lakatos means usual non-formalized proofs of actual mathematics. He does not analyze the idealized formalized mathematics but the non-formalized one practiced by "normal" real mathematicians. The whole text of *Proofs and Refutations* is in fact a critical examination of dogmatic theories in the philosophy of mathematics, in particular of logicism and formalism. The main objection raised by Lakatos is that they are not applicable to actual mathematics. He tries to confront classical conceptions with the Popperian philosophy of mathematics.

Lakatos claims that mathematics is a science in Popper's sense, that it is developed by successive criticism and improvement of theories and by establishing new and rival theories in an evolution-like process. But what are in the case of mathematics the "basic statements" and "potential falsifiers"? These are two key concepts by Popper. One finds no answer in *Proofs and Refutations*. A partial answer can be found in his paper *A Renaissance of the Empiricism in the Recent Philosophy of Mathematics?* (see [221]).

It is said there that informal theories are potential falsifiers for formalized mathematical theories. For example: constructing a system of axioms for set theory one takes into account how and to what extent those axioms reflect and confirm an informal theory used in actual research practice. But what are the objects of informal theories, what do they speak about? What are we talking about when we talk about numbers, triangles or other objects? One finds various answers to this question in the history. Lakatos does not take a definite attitude here. He writes only: "The answer will scarcely be a monolithic one. Careful historico-critical case-studies will probably lead to a sophisticated and composite solution". And just in the history one should look for a proper answer. Lakatos claims that the separation of the history of mathematics from the philosophy of mathematics (what was done by formalism) is one of the greatest sins of formalism. In the introduction to *Proofs and Refutations* he wrote the following (paraphrasing Kant).

> "The history of mathematics, lacking the guidance of philosophy, had become blind, while the philosophy of mathematics, turning its back on the most intriguing phenomena in the history of mathematics, has become empty".

Lakatos indicated that the classical theories in the philosophy of mathematics are not adequate with respect to actual research practice and proposed a new model being closer to that practice. It is a pity that he did not succeed to realize his program of reconstructing the philosophy of mathematics in the framework of his epistemology. One should note that Lakatos' scheme of proofs and refutations does not explain the development of all mathematical theories – for example the origin and the development of the theory of groups can only be explained as a result of certain unification tendencies. Lakatos was conscious of those limitations. Nevertheless, it is doubtlessly his merit that he proposed a picture of mathematics as a science which is alive and which cannot be closed in a framework of a formal system, a picture in which a mathematician developing mathematics is taken into account and in which the actual research practice – and not only idealized reconstructions – is considered.

Another attempt to overcome the limitations of the classical theories of the philosophy of mathematics is the conception of *Raymond L. Wilder*. He proposed to treat mathematics as a cultural system. A source of this conception can probably be seen in his interests in anthropology (his daughter, Beth Dillingham, was an anthropologist and professor of anthropology at the University of Cincinnati).

Wilder presented his ideas in many papers and lectures. A complete version of them can be found in two of his books: *Evolution of Mathematical Concepts. An Elementary Study* [373] and *Mathematics as a Cultural System* [374].

The main thesis of Wilder says that mathematics is a cultural system. Mathematics can be seen as a subculture, mathematical knowledge belongs to the cultural tradition of a society, mathematical research practice has a social character. The word "culture" means here "a collection of elements in a communication network" (cf. [374, p. 8]). Thanks to such an approach the development of mathematics can be better understood and the general laws of changes in a given culture can be applied in historical investigations of mathematics. It enables also to see the reciprocal relations and influences of various elements of the culture and to study their influence on the evolution of mathematics. It makes it also possible to discover the mechanisms of the development and evolution of mathematics. Wilder's conception is therefore sometimes called evolutionary epistemology (compare Section 2.24). One can see here also the influence of Peirce's pragmatism (cf. Section 2.17). Wilder has proposed studying mathematics not only from the point of view of logic but also using methods of anthropology, sociology and history.

In the history of mathematics one can spot various phenomena which develop along general patterns. As an example can serve here the phenomenon of multiples, i.e., cases of multiple independent discoveries or inventions. Probably the best known case is the invention of the differential and integral calculus by Leibniz (1676) and Newton (1671) or the invention of non-Euclidean geometry by Bolyai (1826–1833), Gauss (since about 1829) and Lobatchevsky (1836–1840). Other examples are: the invention of logarithms by Napier and Briggs (1614) and by Bürgi (1620), the discovery of the principle of least squares by Legendre (1806) and Gauss (1809), the geometric law of duality by Plücker, Poncelet and Gergonne (early 19th century). Today such simultaneous and independent discoveries and inventions are very often.

Wilder explains this phenomenon in the following way: mathematicians – who should be treated, of course, as social beings – work on problems which in a given culture are considered as being important, i.e., there are various cultural tendencies which suggest that this or that problem should be solved. If one additionally assumes that the abilities and cultural forces are uniformly distributed then one can expect that solutions will appear simultaneously and independently. It is not an exception but a rule (not only in mathematics but in science in general): "When a cultural system grows to the point where a new concept or method is likely to be invented, then one can predict that not only will it be invented but that more than one of the scientists concerned will independently carry out the invention" ([374, p. 23]).

The conception of Wilder enables also the study of the evolution of mathematics in a given culture or between various cultures as well as to predict in a certain sense the development of mathematics (the latter is possible if one knows the laws and rules of development in a given cultural system). The advantages of Wilder's conceptions can be better seen when compared with other theories.

For example, for E. T. Bell (cf. his book *The Development of Mathematics* [27]) mathematics is a "living stream" with occasional minor tributaries as well as some backwaters. Moslem mathematics was for him a slow spot in the river, a dam useful for conserving the water. It was necessary to conserve the achievements of the Greek mathematics but it brought no new original results.

For O. Spengler (cf. his *Der Untergang des Abendlandes* [331]) mathematics is pluralistic, every culture has its own mathematics. And cultures are organisms like men, they go through definite stages of development. Hence, for example, the Greek mathematics after the decay of the old Greek culture became quite different, quite new mathematics under the Moslems. Greek mathematics had been an abstract and speculative activity of a leisure class, Moslem mathematics was a concrete and practical activity of the descendants of nomads who had been shaped by a somewhat harsher environment. It was simply the mathematics of another culture. Spengler would never condemn the Moslems and their mathematics for not continuing the Greek tradition, as Bell did.

For Wilder a culture, and in particular the mathematical culture, is not an organism which goes through various stages of development from youth to decline and death (as it is the case by Spengler) but a species that increasingly evolves. Hence he does not say that the Greek mathematics died with Moslem mathematics thence born, but rather that mathematics moved from one culture to the other and subjected to different cultural forces. It altered its course of development, but it was the same mathematics.

Wilder formulates certain laws governing the evolution of mathematics. We give here some examples:

(i) New concepts usually evolve in response to hereditary stress (unsolved problems) or to the pressure from the host culture (Wilder calls it environmental stress).

(ii) An acceptance of a concept presented to the mathematical culture is determined by the degree of its fruitfulness. It will not be rejected because of its origin or because it is "unreal" by certain metaphysical criteria (compare, e.g., the introduction to mathematics of negative integers, of complex numbers or of Cantor's theory of infinite sets).

(iii) An important and crucial element in the process of accepting new mathematical concepts is the fame or status of its creator. It is especially important in the case of concepts which break with tradition. Consider for example the non-Euclidean geometries discovered by Lobachevsky and Bolyai. Their epochal discoveries remained almost 30 years unknown and unrecognized. Only when it became known that Gauß had similar ideas they have been accepted. This same applies also to the invention of new terms and symbols (cf. the introduction of π, e and i by Euler).

(iv) The acceptance and importance of a concept or theory depend not only on their fruitfulness but also on the symbolic mode in which they are expressed. As an example can serve here the case of Frege and his sophisticated and complicated symbolism which were not accepted by mathematicians in his day. Therefore his logic initially remained unknown. When his symbolism was replaced later by a better and simpler symbolism due to Peano and Russell Frege's ideas and the epochal importance of his logical work were realized.

(v) At any given time there exists a "cultural intuition" shared by (almost) all members of the mathematical community and affecting their mathematical work.

(vi) Diffusions and connections between cultures or between various fields in a given culture result in the emergence of new concepts and accelerated growth of mathematics. An example can be the Arabic mathematics developed under the influence of Indian and Greek mathematics that influenced the formation and development of western mathematics in the Middle Ages or the emigration of prominent mathematicians from Europe at the time of National Socialism and the connections between the domains of topology, algebra and analysis.

(vii) The discovery of inconsistency in existing theories (or in the existing conceptual structure of a theory) results in the creation of new concepts, theories and methods and accelerated growth of mathematics. As an example one can mention here the crisis of the ancient Greek mathematics that resulted from the discovery of incommensurables and the theory of proportion by Eudoxus of Cnidus as well as antinomies of set theory and the emergence of axiomatic systems of set theory.

(viii) There are no revolutions in the core of mathematics – they may occur in the metaphysics, fundamentals, symbolism or methodology of mathematics.

(ix) Mathematical systems evolve through greater abstraction, generalization and consolidation prompted by hereditary stress.

(x) A mathematician is limited by the actual state of the development of mathematics as well as of its conceptual structure. As an example of this rule let us mention here mathematical logic. It was hardly possible to create it in the earlier stages of the development of mathematics because the new attitude towards algebra and its symbols was required. In fact, one needed the disengagement from intentionality and insight into formality of terms and expressions, e.g., the idea that algebraic symbols do not necessarily represent numbers but that they can represent arbitrary objects of thought satisfying certain operational laws.

(xi) The ultimate foundation of mathematics in any epoch is the cultural intuition of the mathematical community which may be vague and only atmospherical. Without that (hidden arrangement) silent agreement about mathematical subjects and theories mathematical investigations would be chaotic and fruitless.

(xii) As mathematics evolves, hidden assumptions are made explicit and either generally accepted or (partially or fully) rejected. The acceptance usually follows an analysis of the assumption and a justification of it by new methods of proof. As a classical example can serve here the Axiom of Choice used by Peano in 1890 in the proof of the existence of a solution of a system of differential equations. In 1902 Beppo Levi recognized it as an independent proof principle and in 1904 Zermelo used it in his proof of the well-ordering theorem. It was this latter proof that made mathematicians discuss the axiom carefully and study its status.

(xiii) Since mathematics has a cultural basis, there is no such thing as the absolute in it, everything is relative. In particular any mathematical concept must be related to the cultural basis which engendered the mathematical structure to which it belongs. There is also no unique universal pattern of precision and correctness and one cannot say that mathematics in a given period was not precise enough. Exactness is a cultural convention of cultural society at a given time. There are also no absolute objects studied by mathematics. The only existence that the mathematical objects have is the existence of the cultural objects. Mathematical concepts are created by mathematicians, hence of human beings of a given time and culture. Of course it is not being done in an arbitrary way but on the basis of concepts already existing and under the influence of the inner stress (unsolved problems) or of the environmental stress (problems of other sciences or of the praxis). Hence in particular the concept of a number usually seen as universal and absolute is a cultural object and as such it should be investigated.

Briefly: Wilder maintains that mathematical concepts should be located in Popper's "third world" of mental and cultural contents. Mathematics investigates no timeless and spaceless entities. It cannot be understood properly without regarding the culture in the framework of which it is being developed. In this sense mathematics shares many common features with philosophy, ideology, religion and art. A difference between them is that mathematics is a science in which one justifies theorems by providing logical proofs and not on the basis of, say, general acceptance.

An interesting synthesis of the conceptions by Lakatos and Wilder is the paper *Some Proposals for Reviving the Philosophy of Mathematics* [163], by Reuben Hersh. He says first of all that the thesis "that mathematics must be a source of indubitable truth" is simply false. This thesis is not confirmed by the real research practice of mathematicians. In fact, in mathematics we have no absolute certainty – mathematicians make mistakes and later correct them, they are often not sure whether a given proof is correct or not. Mathematicians deal in their research practice with ideas. They are using symbols just to be able to tell about ideas and to communicate their results to others – it is similar to the usage of notes and scores in music. Axioms and definitions are attempts to describe main properties of mathematical ideas. But there are always properties which are not explicitly described in them. Hersh says in [163] that

> "a world of ideas exists, created by human beings, existing in their shared consciousness. These ideas have properties which are objectively theirs, in the same sense that material objects have their own properties. The construction

of proof and counterexample is the method of discovering the properties of these ideas. This is the branch of knowledge which we call mathematics."

The thesis that mathematical knowledge is *a priori*, absolutely certain and indubitable as well as infallible was criticized also in the quasi-empiricism of *Hilary Putnam* (cf. his work *What Is Mathematical Truth?* [287]). He claimed that in mathematics one used always quasi-empirical methods and they are used also today. Those methods are similar to methods of empirical disciplines. The difference is that in mathematics single propositions of a theory are checked by proofs or calculations and not by experiments and observations. However, in mathematics experimental methods play a heuristic role in the context of mathematical discovery. One experiments for example when one formulates a general statement on the base of particular single cases. Experiments can also provide arguments for the choice of axioms or can serve as partial justifications for hypotheses that cannot be proved in a complete way. The criterium of truth in mathematics – similarly to physics – is the success and the applicability of mathematical ideas in the practice in a broad sense, in particular in the research practice of physics or other natural sciences.

The last problem is strongly connected with another problem which interested philosophers of mathematics from the very beginning. It is the problem of applicability of mathematics in describing the real world or the problem of connections between pure and applied mathematics. This problem, discussed by us in details in Section 3.4, is especially interesting nowadays. For example one uses in physics many highly abstract mathematical concepts and theories which have been developed and introduced many years ago without any connection with practice or any practical motivations. In fact, mathematics and mathematization are successfully applied also in disciplines like biology, sociology or psychology and linguistics. It should be admitted that there is no explicit and unique explanation of this puzzling applicability of mathematics. In the philosophy of mathematics various hypotheses are formulated, for example: (1) mathematics is not an *a priori* discipline; there exists a deep ontological relationship between mathematical and physical concepts and the difference between them consists only in the degree of generality. (2) Objects of mathematics are simply mathematical forms of physical phenomena (cf. N. D. Goodman, *Mathematics as a Natural Science* [150]). (3) They are historical-sociological-psychological explanations in which concepts, in particular mathematical concepts are the result of the dissolution from the reality (cf. Lorenz, Damerow, Wilder). (4) The phenomenon of the applicability of mathematics is and will be mysterious and there is no rational explanation of it (cf. E. P. Wigner, *The Unreasonable Effectiveness of Mathematics in the Natural Sciences* [372]). Wigner connects this view with the thesis that mathematics has no real content but is only a "formal game". A mathematician possesses no knowledge, he has only a special ability to operate with formal concepts.

The problem of applicability of quasi-empirical or even empirical methods in mathematics recently got a new dimension. Still more often and in still new domains of mathematics computers are applied. This phenomenon generates new problems and questions in the philosophy of mathematics.

Computers are applied in mathematics in at least six different ways: (1) to do numerical calculations, (2) to solve (usually approximatively) algebraical or differential equations and systems of equations, to calculate integrals, etc. (3) in automated theorem proving, (4) in checking mathematical proofs, (5) in proving theorems (one speaks in such situations about proofs with the help of computers), (6) to do experiments with mathematical objects. Special interests and controversies awake applications of the type (5). One of the main problems considered in this context is the problem of the epistemological status of theorems obtained in this way and consequently – in the case of accepting such theorems as legitimated mathematical theorems – the problem of the character of mathematical knowledge.

This problem is usually discussed in the context of the four-color theorem. This theorem states that four colors suffice to color the regions of the map (on the plane or on the sphere) so that no two adjacent regions have the same color. This theorem was proved in 1976 and is the first mathematical theorem whose proof is essentially based on the application of a computer program and computational power. Moreover, there exists so far no "normal" proof of this theorem, i.e., a proof not using computer. So one should ask whether the mathematical knowledge can be still classified as an *a priori* knowledge and one should choose between the following two options: (1) to extend the scope of admissible proof methods and admit also the usage of computers (hence certain experiments) in proving theorems or (2) to decide that the four-color theorem has not been proved so far and does not belong to the domain of mathematical knowledge.

The first possibility leads to the conclusion that the mathematical knowledge is in general only quasi-empiric and not *a priori*.

Let us notice at the end of these considerations on quasi-empirical trends in the modern philosophy of mathematics that they do not necessarily assume the rejection of classical conceptions in order to take into account the knowing subject and the real research practice of mathematicians. In fact, there are attempts to extend the old theories in such a way that the practice would be regarded. We mean here the so-called "intensional mathematics". It is an attempt to build dualistic foundations of mathematics and to treat it in a way similar to, e.g., the quantum mechanics where the observer must be taken into account. It is proposed to add certain epistemological notions to the classical mathematics.

In particular an idealized mathematician (it should be an idealized picture of a mathematician in a collective sense) and an epistemological operator $\Box\Phi$ (Φ *is knowable*) are introduced. It is assumed that it has the following properties (they are analogous to the properties of the operator of necessity in the modal logic, cf. [149]):

(i) $\Box\Phi \rightarrow \Phi$,
(ii) $\Box\Phi \rightarrow \Box\Box\Phi$,
(iii) $\Box\Phi \;\&\; \Box(\Phi \rightarrow \Psi) \rightarrow \Box\Psi$,
(iv) if $\vdash \Phi$ then $\vdash \Box\Phi$.

Axiom (i) says that everything knowable is true. Axiom (ii) states that anything knowable can be known to be knowable. Axiom (iii) claims that if both a conditional and its antecedent are knowable then the consequent is knowable. Axiom (iv) says that if we actually prove a claim then we know that claim to be true, and so that claim is knowable. One should note that if we take $\Box$ to mean *known* then there arise problems with Axiom (iii). In fact, Axiom (iii) would claim in this case that whenever a mathematician can make an inference, he has already done so. But even an idealized mathematician does not follow every possible chain of inference. Hence we take $\Box$ to mean *knowable*.

The first system constructed according to the described ideas and taking into account the epistemological aspect was the system of arithmetic due to S. Shapiro (cf. [326]). Later other systems were constructed, e.g., a system of set theory, of type theory, etc. (cf. [327]).

Systems of the indicated type make the formal study of constructive aspects of proofs and classical inferences possible. They enable one also to take into account the ways in which mathematical objects are given. From the philosophical point of view one rejects here the platonism.

It is impossible to give here further technical details. Nevertheless, it should be said that systems of intensional mathematics are an interesting attempt to develop classical – as a matter of fact – foundations of mathematics and simultaneously to take into account the real practice of a mathematician. Just this tendency is, as we tried to indicate above – the leading tendency in the modern philosophy of mathematics.

2.25.2 Realism and antirealism

Recently in the philosophy of mathematics some conceptions have been developed whose starting point is the existence problem. It is about the dispute between realism and antirealism or in other words between realism and constructivism. They all refer – in this or other way – to the platonism of Gödel and the realism of Quine and Putnam. Above we wrote about them.

A moderate variant of the modern realism can be found in [375]. T. Wilholt propagates here a partial realism by distinguishing between "realistic" and "formal" mathematics. The "realistic mathematics" goes from the real world via properties in "processes" of concrete "aggregates" and other "physical" objects. Wilholt's thesis, which he tries to justify, is that beside the integers and the rational numbers also the real numbers and consequently also the real analysis belong to the realistic mathematics. In this way the problem of the applicability of such mathematics in "primary" applications is eliminated.

For T. Wilholt magnitudes are real properties of "physical" objects, for example in physics – as quantities are properties of concrete "aggregates" which are exactly so concrete and real as their elements. Magnitudes form the domain of magnitudes whose combinations come from the "physical combinations" of the "supporters" of magnitudes. An appropriate arithmetic of those combinations is introduced and the completeness axiom for this domain is assumed. This axiom claims that it is isomorphic with $\mathbb{R}$ and it seems to allow the identification of the real numbers and "relations of magnitudes" [375, p. 237]. The real numbers inherit from the relations of magnitudes the character of universals. Wilholt recognizes problems that appear in the transition from rational to real numbers. He proclaims the thesis about quasi "real" real numbers treating it as justified. Let us notice that the abstractness of mathematical concepts is especially taken into account in [375], but the construction of the real numbers, the continuum problem, actual infinity and infinitesimals are not considered.

One of the nowadays mostly discussed conceptions is the antirealistic nominalism represented by Hartry Field in his book *Science Without Numbers* [118]. He claims that mathematics is only a convenient and useful fiction. It is a collection of theorems that enable us to formulate and justify statements about the real world which do not allow however a translation into the reality. Hence mathematics is not necessary and nothing more than a useful and beautiful fiction.

Another new trend in the philosophy of mathematics is structuralism – developed mainly in the USA and widely discussed nowadays. We cannot present technical details here. However, we would like to characterize this trend. We refer to conceptions developed by Geoffrey Hellman (in *Mathematics Without Numbers. Towards a Modal-Structured Interpretation* [162]), Michael Resnik (in his book *Mathematics as Science of Patterns* [301]) and Stuart Shapiro (cf. *Philosophy of Mathematics. Structure and Ontology* [328]).

This structuralism has its roots in works and ideas of Dedekind, Russell, Hilbert, Bernays and N. Bourbaki. The latter is in fact a pseudonym of a group of (mainly French) mathematicians who undertook in the 1930s the task of a systematization of the whole of mathematics on the base of fundamental structures (algebraic, order and topological ones). This general idea borrows structuralism and develops it elaborately. The main thesis of structuralism is that the proper objects of mathematics are structures and not single objects. Hence mathematics is the structure of structures. The smallest units are so-called *patterns*. Mathematical objects, in particular numbers, are positions in such patterns. This recalls the description of numbers by Dedekind where numbers were positions in a sequence structure. Mathematical objects have no exterior properties determining them. Their identity is determined solely by their relations with other positions in a structure. New is here also the character of being of mathematical objects. Structuralism ascribes to mathematical objects, in particular to numbers, only a possible, "modal existence". Hellman proposes to treat mathematics – and there at the first place the arithmetic – as a nominalistic theory whose concepts are only names and symbols without any meaning. In the case of numbers this would mean:

> *Numbers are positions in patterns. They have only a "modal" existence and no meaning.*

The "modal existence" of numbers and other mathematical objects are described by second-order logic and modal operators of second-order modal logic. This description uses complex technical details and makes structuralism in a way artificial.

3 On fundamental questions of the philosophy of mathematics

So just that usually produces the enthusiast,
which exclusively was able to form the wise man,
however the advantage of the latter would be less
in becoming former but in not remaining this.[1]

Friedrich Schiller

On the way to the reals described in Chapter 1 there appeared mathematical and philosophical questions we met over and over again in our survey of the philosophy of mathematics given in Chapter 2. They belong to the fundamental questions of the philosophy of mathematics. We now want to consider those questions again and to look for answers to them having in mind what was said about positions, conceptions and tendencies indicated in Chapter 2. Just as different those positions and views were, so too will the attempts of answers differ.

We begin by considering the question of natural numbers that are the beginning of all arithmetic. We got to know various views on them. Further we shall discuss diverse positions concerning the infinite, present historical as well as nowadays conceptions of the continuum in which the question of the infinitely small quantities, the infinitesimal emerges, consider relations between magnitudes and numbers and observe how the intuitive continuum and magnitudes disappear from the pure mathematics. In contrast to the everyday and successful applications of mathematics, the philosophical question why those abstract, formal, transfinite mathematics can be applied to the finite and concrete world is unsettled. In Section 3.4 we shall look for answers to this question from the point of view of various philosophical positions. At the end of the chapter we give some comments – on the base of presented insights – on the particular problems met in Chapter 1.

3.1 On the concept of number

The first fundamental question was and still remains the question on natural numbers. They are at the beginning of the road to real numbers. What are natural numbers, what

1 So bringt gewöhnlich eben das den Schwärmer hervor, was allein imstande war, den Weisen zu bilden, und der Vorzug des letztern möchte wohl weniger darin bestehen, daß er das erste nicht geworden ist, als darin, daß er es nicht geblieben ist.

https://doi.org/10.1515/9783110468335-004

is their nature, how do they exist? We presented various views on natural numbers – as far as they were recognizable – in our survey of conceptions in the philosophy of mathematics given in Chapter 2. We saw a wide panorama of opinions concerning this fundamental mathematical object – ranging from ambiguous mystic to being completely meaningless.

We now look back once more restricting ourselves to some important views, sum up briefly their characterizations and provide a résumé on the base of this. We will not emphasize the rationalistic elements taking into account the structures of thinking because at least traces of them can be found in almost all conceptions.

3.1.1 Survey of some views

For the *Pythagoreans* numbers were elements of a *higher world* that influenced the physical world and that created things according to number, measure and form. In the works of *Plato* they seem to descend a bit and to mediate *between* "the heaven of ideas" and the material reality. The background of both conceptions is a higher, real spiritual world. The concrete meanings of numbers, for example the quantity and magnitude, were secondary. Numbers – in their primary meaning – were independent objects detached from real objects and sets. They formed a closed, higher and "pure" domain in which their properties and mutual relations were determined by themselves. Here is the origin of number theory and of pure mathematics that today still is willingly seen by mathematicians in a platonic way.

Numbers and ideas were higher powers that formed things giving them existence and properties. Indeed, this is an unfamiliar inversion of today's abstraction: forms and properties are not abstracted from objects but granted to them. Numbers are not abstractions of sets, but sets are products of numbers and magnitudes are generated by numbers as measures. Sets are not the base for numbers but numbers form sets.

Aristotle brought numbers to the world and human thinking. Numbers became *forming powers in things* that are recognized by men using a type of abstraction. What is important here is the fact that numbers preserved the old power of platonic ideas creating and shaping the reality. Hence abstraction was – one can see it in this way – in an exchange with creative *interpretation*.

Euclid referred to Aristotle and characterized numbers briefly as *multiplicities* consisting of units. It can be seen that in this "definition" number and multiplicity, we would rather say "set", are not so strictly distinguished as we usually separate them in our refreshed interpretations. The complete differentiation, the separation of numbers and sets seems to be done only by theoretical sets of set theory. The distinction between magnitudes and measures was the end of magnitudes. They have been replaced in the 19th century by their own measures, namely by the real numbers of pure mathematics.

For *Nicholas of Cusa* numbers were *reconstructions* of the divine spiritual numbers obtained by comparison and distinguishing which were put by God into things and according to which God created the world. In this reconstruction – connecting a human being with God – appears in the foreground human thinking that is a precondition of earthly numbers present in the world. One can see here both the platonic and aristotelian conception. The divine numbers remind one of the old ideas forming the reality and the reconstructions in the "comparative distinguishing" – as abstractions by Aristotle.

Nicholas of Cusa attributes the first principle of the emergence of pure numbers explicitly to the divine that creates spiritual numbers in the process of counting. "Only Spirit counts; without the Spirit there is no independent number". Hence the primary aspect of numbers lies in this higher counting process by which the earthly things are determined in their quantity. Since there is nothing in human thinking that has not been before in human senses – as an empiric note of Nicholas of Cusa says – counting understood by human being is connected with things being counted, from which that what is divine is absorbed by a man. Consequently, the earthly "number is a matter of things", hence magnitude and quantity.

For old and new empiricists as *David Hume* and *John Stuart Mill*, world is the unique base on which – via senses – come numbers. "Concepts of a number and of a figure come from no other place than the real world" – as was said by the marxist *Friedrich Engels*. They are outcomes of *impressions* repeating one after another that for Engels were quite close, namely in ten fingers of a hand, on which "human being learned to count". Since those fingers are not "free creations of mind", it is absurd to take numbers as being pure elements of the mind. The idea of abstraction as a background of the process of building concepts was questioned by Hume. One can say that numbers are impressions of nature and nothing more can be stated about connections between concepts and the world. In fact, experience keeps completely silent about those connections, as Hume stated.

Kant transfers – in contrast to this – numbers in a rationalistic way completely into human mind. They are – as schemas – essential elements of its inner structure. Numbers are generated in a process of counting in time.

> Arithmetic by itself forms its number concepts through successive addition of units in time, [...].[2] ([200, I. Teil, § 10], our translation, cf. also [202].)

Kant seems to connect counting – that first is represented as a temporal succession in pure form of the intuition of time – with objectively appearing units. In this way numbers become a *schema of mind* as follows.

---

2 "Arithmetik bringt selbst ihre Zahlbegriffe durch successive Hinzusetzung der Einheiten in der Zeit zu Stande, [...]".

> "The pure schema of magnitude (quantitatis), however, as a concept of the understanding, is number, which is a representation that summarizes the successive addition of one (homogeneous) unit to another."[3] ([199, A, pp. 143 f.], for a translation cf. [201].)

Quantitatively, numbers have by Kant in particular the meaning of quantity and magnitude. Arithmetical sentences about numbers are synthetic judgements *a priori.*

Gauß understood natural numbers still as coming from geometrical magnitudes. They were *multipliers* of them. He noticed the growing dominance of numbers which he attributed to the "method of counting", namely in the decimal system, and saw the necessity of a purely arithmetical construction of the number systems.

In *Georg Cantor's* conception one finds idealistic, empiricist and constructivistic elements: numbers are on the one hand real ideas: "Integers – similarly to celestial bodies – form a whole determined by laws and relations." On the other hand they are *projections* of real sets that are constructed in the human mind by abstraction as cardinal numbers:

> Since every single element m, if we abstract from its nature, becomes a "unit", the cardinal number $\overline{\overline{M}}$ is a definite aggregate composed of units, and this number has existence in our mind as an intellectual image or projection of the given aggregate M.[4] ([65, p. 282], English translation [67, p. 86].)

The background of his threefold conception is set theory, which he has introduced and which for him was philosophically founded and justified. He "discovered" sets and numbers idealistically in a world of ideas, abstracted them empirically from "the language of nature" and formed them in the free human mind.

Richard Dedekind saw numbers exclusively just there, in the human mind – they are free creations of the mind. He thought in a structural way and understood numbers as abstractions of positions in infinite counting sequences that are determined by axioms given by him. Dedekind called counting sequences "simple infinite systems":

> If in the consideration of a simply infinite system N set in order by a transformation ϕ we entirely neglect the special character of the elements; simply retaining their distinguishability and taking into account only the relations to

3 Zahl ist das "reine Schema" des Verstandesbegriffs der Quantität, d.h. die Vorstellung, die "die sukzessive Addition von Einem zu Einem (gleichartigen) zusammenbefasst."

4 "Da aus jedem einzelnen Elemente m, wenn man von seiner Beschaffenheit absieht, eine "Eins" wird, so ist die Kardinalzahl $\overline{\overline{\mathrm{M}}}$ selbst eine bestimmte aus lauter Einsen zusammengesetzte Menge, die als intellektuelles Abbild oder Projektion der gegebenen Menge in unserem Geiste Existenz hat."

one another [...], then are these elements called *natural numbers* [...].[5] ([87, p. 73], English translation [88, p. 68].)

In Section 2.15 devoted to Dedekind we have linked this sort of abstraction with the standard, i.e., decimal sequences of signs representing and realizing the "relations" of elements and the infinite counting structures.

For *Frege*, who was a logicist, numbers were *pure concepts*, hence elements of logic. When his numbers are grasped set-theoretically – what in fact was avoided by Frege himself – then they are finite cardinal numbers possessing for Frege their own existence as ideas by Plato.

The intuitionistic conception of numbers is antiplatonic. Numbers are given as intuitions and "their nature is determined by human thinking" – for *Luitzen E. J. Brouwer* as *void of contents abstractions of time sensations*. They are the "preintuitions" of mathematics belonging, as it was the case for Kant, to the pure intuition of time.

For constructivists this – only psychological – description is undetermined. *Paul Lorenzen* constructs – according to a simple scheme – numbers as counting signs. *Christian Thiel* supplements this conception by treating numbers as *virtual objects* obtained by abstraction from counting signs in various systems of counting signs.

It seems that nowadays the formalistic position dominates. For *David Hilbert* numbers are in fact *meaningless symbols* the meaning of which is determined by their theory, by arithmetic given in an axiomatic way. Numbers do not exist anywhere. Existence is replaced by consistency – however the latter is arithmetically uncertain. Hence numbers are solitary elements of a theoretical arithmetic.

Piaget and *Damerow* treated numbers as *cardinal numbers including ordinal aspects* arising in cognitive constructions that start from acting with concrete objects. Damerow sees in systems of symbols representing numbers that retroact to the development of numbers, elements of social and historical evolution of numbers and of arithmetic.

We would like to try to order the presented conceptions according to their characteristic features and put them together in a scheme. To do this we simplify them a bit and concentrate ourselves on some of their important aspects. We shall distinguish the conceptions by looking at the fact whether they include or exclude historical elements in the concept of number (top line) as well as at aspects that can be related to philosophical basic positions and the status of numbers (first column). One obtains of course no classification. Some conceptions seem to be at seemingly inconsistent places.

---

5 "Wenn man bei der Betrachtung eines einfach unendlichen, durch eine Abbildung φ geordneten Systems N von der besonderen Beschaffenheit der Elemente gänzlich absieht, lediglich ihre Unterscheidbarkeit festhält und nur die Beziehungen auffasst [...], so heißen diese Elemente natürliche Zahlen [...]."

The table below should be read, for example, in the following way: numbers are according to the conception of *Dedekind* (see the body of the table) *structurally* (see the first column) founded, *ahistorical* and *universal* (see the top line).

Tab. 3.1. Characteristics of number concept views.

	Historically qualified	**Ahistorical, universal**
Transcendent	–	Pythagorean, Plato, N. v. Cusa, Aristotle
Transcendental	–	Kant, Brouwer
Abstractions	Mill, Damerow	Euclid, Aristotle, Brouwer, Cantor, Piaget,Frege
Magnitudes, operators	–	Euclid, Gauß
Formal	–	Frege, Hilbert, Peano
Semiotic	Damerow	Hilbert, Weyl, Lorenzen, Thiel
Structural	–	Dedekind, Peano, structuralism
Genetical	Damerow	Dedekind, Piaget, Damerow

3.1.2 Résumé

On this background we now try to formulate a nuanced position concerning numbers that would take into account aspects of the indicated views. We want to draw a *conclusion following from answers* to the question on numbers. Numbers will appear in a certain interdependency between construction, structure and abstraction. We shall use our paper on historical and actual views concerning numbers [24] and sketch our own position.

Let us start by a genetic position and look from this point of view not only at the development of numbers as individual objects but also and first of all at the development of the *structure* formed by numbers. We see – like Dedekind – numbers initially as *counting numbers*. This means:

Numbers are positions in an elementary counting structure,

hence elements of a possibly very elementary, finite counting-like cognitive schema. In the book *Zählen – Grundlage der elementaren Arithmetik* [Counting – Foundations of elementary arithmetic] [23] such structures are called "counting sequences". To numbers understood in such a way belong necessarily internal and external representations, the latter treated by Damerow (see above) as essential elements in the historical development. To be able to understand this approach one must try to explain what *counting* is. From the mathematical point of view this has been done in the book *Zählen* – also in the possibly most elementary finite case.

Already in the case of a simple counting there is something very fundamental that should be taken into account: *the ability of putting into a sequence*, i.e., the ability to form sequences – this is not mentioned in the indicated genetic approach by Piaget and Damerow. Counting – in the most general sense – is

> the concept of succession – via an active conscious succession.

Behind this there is the ability of putting points and intervals into the flow of time in a conscious, anticipating and registering way. It is by the way the necessary condition of any conscious acting. Those points can be connected with acting, perception, points of the space, and eventually signs. Such ability is being developed psychologically and historically. Systems of signs are being handed down and reconstructed. One can say:

> Counting by standardized successions of number symbols is the external collective representation of the concept of succession. ([20, pp. 155 f.].)

The inner construction of the scheme of counting is *in principle* free of abstractions from the reality and any application. In this sense one can speak about a *pure counting*. This construction is closely connected with the inner time – comparable with the form of pure intuition in Kant's sense. Arithmetical contents are *a priori* and analytic because they can be reduced to the simplest principles of pure counting. Numbers are – as for Dedekind and Cantor – "mental creations" or "immanent realities" that arise in the psychological construction of schemas of counting.

However, *in the real psychological* as well as in *prehistorical and historical development* this inner construction of counting and of numbers takes place in principle and from the very beginning in a permanent exchange with the outer reality. It is an instrument of the subject enabling him or her to acquire and to manage the outer world. Dealing with the outer world is for the schemas of counting the source of impulses to be build, differentiated, organized and extended.

The inner construction of numbers meets in this process abstractions that grant to the inner numbers among others the meaning of quantity, ordinal numbers, magnitudes and operators. Thanks to them the inner numbers receive the outer reflection, they become explicit and need external representations by symbols that in an essential way influence the inner construction. Via the internal and external signs the counting can reversely influence itself:

> Numbers count numbers.

This is characteristic for the interaction of counting and numbers as well as a necessary base for the development of arithmetic.

Space conditions are on an equal footing and always connected with time conditions – they influence the forming of the concept of number as well. In the differentiation of the spacial world and the construction of objects begins another way to numbers. Here one finds further elementary origins of numbers. Here begin actions and systems of actions that refer to objects, put relations between objects and become inner operations and structures. They become the basis of *abstractions* and applications that lead to meanings of numbers as quantities, cardinal numbers and magnitudes mentioned above.

Our conception is stamped in an obvious way by the genetic position, and consequently is constructivistic. We do not accept the radical standpoint of the biological epistemology according to which numbers are products of the genetic evolution. The conception just formulated is both rationalistic as far as the construction of inner schemas of counting is at stake and empirical – in stressing the abstraction from outer conditions while forming counting schemas and the meaning of numbers.

We tend to hold numbers as counting numbers to be nonhistorical universals. They have their source in systems of actions lying before "arithmetical activities" as assumed by Damerow. The psychological and prehistorical as well as historical development is – in our opinion – in line with numbers as nonhistorical universals. This is our position. It is supported by the certainty of time that forms a base of any development and the fact that outer actions and inner constructions take place always in sequences of time points. The formation of the concept of number begins with inner building successions and with becoming aware of successions of moments in time.

What about ideas and forms? Does the inner thinking with numbers structure the outer reality, do inner numbers face outer numbers existing independently of thinking, structuring the world and being the condition of the development of counting and of inner numbers? Well, these are *meta*metaphysical questions. We incline towards a conception that integrates aristotelian, platonic and rationalistic elements and that accepts numbers simultaneously as outer forms and inner forms. In the interplay of inner constructions, applications to outer forms and retroactivity on inner constructions we see impulses for the development of numbers and the influence on the interpretation of the outer reality. If one wants to characterize briefly this position then one can say that it is an "objective idealism" in the sense in which for example Friedrich Schiller characterizes it in his aesthetic letters.[6]

Numbers cannot be bare signs or abstractions of them as argued by C. Thiel (see above). The inner construction of counting schemas is dependent of representations via internal and external signs. Additionally abstraction from the outer variety of signs completes the concept of number. We are of the opinion that inner counting schemas and outer symbolic representations are inseparably connected with each other. For the thinking here is an important common point of intersection of inner and outer reality.

Finally, numbers are not bare abstractions (with partially ordinal content) taking place on the background of "arithmetical activities" (cf. Damerow). In fact, the struc-

6 Truth is not something that could merely be received from outside as the reality or sensual being of things; it is rather something that is self-acting produced by the power of thinking and this autonomy, this freedom is just what is missed by a sensual man. ([311, 23. Brief].)

tural aspect expressed by counting and being substantial for the concept of number goes back to those arithmetical activities that always are successions of actions. Successions are basic for ordinal structure of numbers that by abstracting from arithmetical activities become quantities. Therefore the ordinal is not a consequence of abstraction. It precedes abstraction and accompanies it. Succession is a requirement of systems of arithmetical and even "pre-arithmetical" systems of activities (cf. Section 2.24). Its most elementary concept is "counting" – beginning with the most elementary form.

Our considerations suggest that the concept "arithmetical" should be understood in a wider way. In fact, counting and counting numbers are in the best sense arithmetical. From them the whole arithmetic can be in principle derived – without using any other ingredients.

With our formulation of a position that takes into consideration in fact many aspects, necessarily further philosophical questions are connected.

Already in Chapter 2 (Section 2.24) it was mentioned that in the modern philosophy of mathematics little or no attention is paid to the genetic conception of numbers. Contributions of psychology and of the philosophical cultural anthropology were described there. There are various reasons that explain the reservation of the philosophy of mathematics towards those contributions. They were already indicated in Section 2.24.

The object of metamathematics and of the philosophy of mathematics is mathematics itself. Obviously, nowadays mathematics is operating with infinite sets. It is the matter of logical and philosophical investigations and of justification of transfinite concepts and methods of the successful and daily "infinite mathematics". Points of view concerning the origins of the elementary concepts of number are far away from this world of the infinite. The concept of number is a basic one. When in the philosophy of mathematics one speaks about the development then one has in mind the development of higher mathematical concepts, of methods and of mathematics as a discipline.

The reflection on the concept of number begins in the philosophy of mathematics with the concept of a natural number. Natural numbers are described and justified in a transfinite way. Either the base is set theory in which the Axiom of Infinity is assumed: natural numbers are an infinite set. Or Peano arithmetic (PA) is introduced and its models are considered – those models are necessarily transfinite. This is the starting point. It is in fact where the genesis of the elementary concept of number comes to an end.

Additionally there is often a platonic understanding of mathematical concepts. They are nonhistorical universals that determine the aim of every psychological as well as historical development. Only the choice of them, the linguistic shape and their level are influenced by psychological, sociological and historical factors what mathematically is in fact irrelevant. The antipsychologism of Frege is still present.

Is the genesis of the concept of number and of the elementary arithmetic *necessarily* outside the domain of interest of the philosophy of mathematics? We have shown that this is not the case – and we provide now further arguments.

It is not reasonable to connect from the very beginning numbers that themselves are finite with the infinite. The handbook [23] mentioned above is based on a finite foundation (NBG set theory without the Axiom of Infinity, cf. Section 4.3.2). Thanks to that, counting and numbers can be grasped in elementary finite structures [23, p. 9]. The finite approach there is comparable with the transfinite approach by Dedekind [87].

Models of such finite counting structures are usually not isomorphic – contrary to models of natural numbers given by a set-theoretical setting of Peano axioms. The isomorphism of models of those structures is replaced by their mutual embeddability [23, p. 68].

Finite counting structures make the development of a mathematical setting of the elementary arithmetic possible that accompanies the psychological development of the elementary arithmetic. This is the didactic aim of the approach in the handbook [23].

A pure logical axiomatics for finite arithmetical structures – comparable with Peano arithmetic – is by the way not possible as Tarski's theorem shows (cf. Section 5.1). Such approach would be also uninteresting because it would register only arithmetical states and could not describe their development (cf. [23, pp. 234 f.]).

A finite set-theoretical description of finite numbers and the arithmetical genesis of them is interesting for the philosophy of mathematics also for another reason. Finite mathematical standpoints allow in any case to get knowledge about the genesis of the concept of number and to further discuss this concept in a new and advanced way. A well-founded finite foundation provides the possibility to follow the elementary structural development of numbers from the finite into the transfinite. Examples of this can be found in the handbook [23, Chapter 4]. They show how there arises a new position concerning the phenomenon of the infinite – as well as of the finite.

Let us give an example: [23, Theorem 4.5.1] means roughly speaking the following: a counting scheme that enables to count any finite set has the structure of natural numbers. Hence the *universal application* of numbers as tools of counting in order to determine the quantity leads to the infinite.

The approach presented in [23] makes also an elementary definition of the finite by the finite possible. It is quite natural – again roughly speaking: a set is finite if its elements can be *enumerated*.

Let us definitely close our considerations of natural numbers by a prosaic summary that we have in fact already prepared by parenthetical remarks. If one looks back at many possible conceptions of numbers then one comes to the conclusion that there is and will be no final clarification of the nature of numbers. This holds in general for any philosophical, theoretical and empiric approach – in fact insights provided by them are always bounded by fundamental assumptions and hypotheses adopted in them. However, conceptions, explanations and results concerning the status and the development of numbers given by them can be interesting and illuminating.

By no means can and should the nature of numbers be explained by mathematics itself. In fact, it provides only descriptions whose range is in principle obviously bounded by virtue of special axioms. On the other side it should be noticed that mathematics has been developed – unhindered by such deficits – to the nowadays full bloom and that we can understand mathematics of earlier centuries and that our concept of number seems to be at least similar to the earlier ones. Does it mean that the philosophical reflection on the concepts of number and their foundations are irrelevant for mathematics?

It seems to be the case when one thinks about the internal process of the development of mathematics in an isolated way. In no case is the philosophical reflection on the changing foundations and basic concepts irrelevant when one deals with the understanding of mathematics – e.g., in teaching –, with conceptual development and with the *meaning* of numbers. The meaning was in an obvious way fundamentally different in various epochs of mathematics. This was indicated in Chapter 2 and in this section by different characterizations of numbers in various conceptions and tendencies.

We have seen the development:

- from numbers of the Pythagoreans and ideas of Plato where they were the metaphysical powers in the formation of the world,
- from numbers as forms and forming virtue in things by Aristotle,
- via divine numbers by Cusanus and Leibniz,
- via empirical numbers being impressions of the nature,
- via numbers as elements of pure intuition and intuitions of the sense of time,
- to numbers as constructions in their evolutionary and psychological development,
- to numbers on the set-theoretical background,
- till numbers as meaningless symbols.

Conceptions taking into account the development of the concept of number show how the seemingly meaningless symbols can be connected again with a subject and the world.

3.2 Infinities

We are not speaking here about levels of the infinite or about the infinitely small quantities. The latter will be considered in Section 3.3 and Chapter 6, the former in the following chapter. Here we consider the potential and actual infinite and we are interested in a survey of conceptions concerning the infinite proposed by philosophical approaches.

3.2.1 On problems with the infinite

It was said a lot about the infinity in Chapter 2. We met it also in Chapter 1 considering the way from rational numbers to real numbers. The conception of it (beside the conception of the continuum) is the key to the reals.

First of all the infinity is available in the sequence of natural numbers itself. It is also hidden in the concept of a rational number. We will say something about this below in Section 3.5. It appears fundamental, actual and in a differentiated way in extending rational numbers to real numbers as well as in the latter themselves. This was indicated in Chapter 1.

Before turning to the infinities in the extending of rational numbers we want to show the age-old distinction between the potential and actual infinity more vividly than it was done in Chapter 1. This distinction comes from Aristotle.

Counting begins in a finite way. By the use of number words that we *speak out* it has natural bounds. However, from the very beginning – thanks to the principle of the successor – there is the power of extending those bounds and of the cancelation of the limits, and in this way there is the seed of the infinite. Ultimately the symbols of numbers, their system and their projection in the counting process demonstrate the infinity of the sequence of numbers $1, 2, 3, \dots$. The dots "…" illustrate the fact that the counting process will never be closed; it is open in principle. This is the *potential infinity* that we cannot avoid at all. This means that thinking has necessarily to do with the phenomenon of the infinite even when one does not find it in the material nature, in cosmos. It is a phenomenon of thinking. But recall that we have presented philosophical, even empiric conceptions in which infinite sets were real facts. We shall speak about this soon.

It is a risky and big step to think about the open counting process as being closed. Aristotle claimed that such a step is impossible and quasi forbidden. Cantor was the first who made this step explicitly and concretely. One can treat his looking for substantial predecessors as almost failed. For example Plato or Kant allowed the philosophical speculation on the actual infinite. Nicholas of Cusa considered mathematical objects conceptually from the viewpoint of actual infinity. Leibniz praised actual infinite sets to glorify the Creator – simply for reasons of a universal rationalism. In most cases however he condemned them. Only Bolzano can be called the real predecessor of Cantor – he was at least the preparer of the large step done by Cantor. It can be said that he made the first approach. Cantor refers primarily to him. The rebellion against Cantor in mathematics, philosophy and theology was considerable (cf. Section 2.14 about Cantor). Nowadays there is no resistance even at schools. Why? Because the brave step extended the mathematical possibilities infinitely. Hilbert spoke about the "paradise that Cantor created for us".

How should it be done to put bounds to the sequence of numbers $1, 2, 3, \dots$ and to treat it as a closed whole? One begins by collecting the enumerated numbers and it is symbolized by putting the left curly bracket "{". It looks like this: $\{1, 2, 3, \dots$. Eventually, it is acted as if one had them all "in the sack" and one puts the right curly bracket "}" and obtains $\{1, 2, 3, \ldots\}$. The putting of "}" is in practice the implementation of the set-theoretical Axiom of Infinity what nowadays is a daily routine in mathematics. The term "$\{1, 2, 3, \ldots\}$" is in fact paradoxical because the dots "…" say that the process of enumerating is endless; however, } puts just the end to it. The same holds in the case of writing formulas like "$\{2, 4, 6, 8, \ldots\}$" and "$\{1, \frac{1}{2}, \frac{1}{3}, \frac{1}{4}, \frac{1}{5}, \ldots\}$" that are today treated as obvious.

How does this suit with Cantor's own "definition of set":

> "By a set (*Menge*) we are to understand any collection into
> a whole (*Zusammenfassung*) *M* of objects *m* …"

Where does for Cantor the collection of those definite and separate numbers $1, 2, 3, \ldots$ come from? How can he collect them when numbers "run away" by dots "…"?

Though it could seem to be paradoxical, but just "running away", just the dots "…" enable him performing the collection. In fact, the dots indicate the "law" according to which the numbers "can be collected into a whole". They are words used by Cantor in his second definition of the concept of sets quoted by us in Section 2.14. The "law of enumerating" indicated by the dots "…" is just the construction of the successor leading from a number to the next one. This "particular law" is seen by Cantor as realized also in any other set if its elements are properly ordered. This approach is consolidated by Cantor philosophically by attributing to the concept of a set the status of an idea in Plato's sense.

Cantor does *not* see the necessity of the Axiom of Infinity. There is also – besides his platonic attitude – a second reason that Cantor does *not* see the problem of putting } after the dots "…", the problem of infinite sets and the necessity of an *axiom* of infinity, of a requirement. This is connected with another question, with the question on the continuum. Also in this case we share today almost without any exception his attitude.

Let us take an everyday example: Choose two points on a line and denote them as 0 and 1. Consider now the interval $[0, 1]$ of "all" points between 0 and 1. The set of those points is before our eyes. One can *see* this – as is well known – uncountable set! Where is the problem?

It lies just in seeing! We see a continuous interval. Do we see points? – No! We put points and we *think* that they were all put. Also Cantor thought in such a way – and consequently he saw point-sets. Without examination he addresses the problem of the continuum from the concept of point-sets. For him the continuum is – as a perfect point-set – precisely and, as he says, sufficiently characterized. Other approaches are in his opinion a result of an antimathematical "religious dogma" according to which the continuum is a indecomposable "mystery" [66, p. 191]. The deep problem of the continuum, one of the problems mentioned in Chapter 1, will be considered in Section 3.3.

Nowadays problems of infinite sets have almost disappeared from the consciousness. The set-theoretical way of thinking about infinite sets that replace the continuum found recognition. Everywhere sets belong to the self-evident repertoire of mathematicians and teachers of mathematics who often pass them to students and pupils without any reflection.

Poincaré described once this set-theoretical type of thinking as “illness”. For us today it sounds a bit absurd. However, if one looks carefully at problems of the infinite then one can understand better the heated polemics and arguments of the colleagues against Cantor, in particular of his teacher Kronecker. It is not only the matter of mathematics, it was the matter of philosophical, ideological controversies that – as is well known – can sometimes be violent.

One should consider still quite another motive that has prohibited and disturbed through centuries the acceptance of actual infinite sets. Nowadays mathematics is for us stamped mainly by numbers. This is the result of the arithmetization of mathematics that begun in the 19th century and for which set theory formed the base. Earlier, from the moment when the numbers of the Pythagoreans stopped to play the leading role and the magnitudes of Eudoxus of Cnidus and Euclid took it over, mathematics was in principle stamped just by magnitudes and their ideas. This hindered for example the acceptance of negative numbers in mathematics till the 19th century – for there are no negative magnitudes and no negative measures. This is difficult to understand nowadays in time of debts and deficits.

The conceptual separation of sets and magnitudes, i.e., the quantity of its elements, what is clear for us today, has been in fact for a long time not present and clear. Magnitudes were quasi real and first of all *only* intuitive things. Therefore they fell victim in the arithmetization process to “pure” mathematics of numbers that should be free of any “impure” assumptions (cf. Section 3.3.10). Set theory, a domain that until then was unimaginable, is in fact the witness of the final separation of sets and their magnitudes, i.e., their quantities of its elements,.

Sets were closely connected with their magnitudes, their quantity. For magnitudes holds the aged axiom of Euclid: “The whole is greater than the part”. Since this axiom does not hold for the infinite, infinite sets cannot be connected with the idea of a magnitude. Therefore they do not have the level of reality of magnitudes. They are not real and therefore cannot be actual. Such an argumentation can be found already in the works of Proclus. When in teaching one speaks for the first time about the paradoxical property of infinite sets that they are so big as some of their parts then also today one meets resistance by students and pupils because their intuitions are still connected with the classical conception of finite magnitudes.

We would like to approach again problems of the infinite by considering some selected conceptions, in particular fundamental conceptions from the history of the philosophy of mathematics.

3.2.2 Conception of Aristotle

Let us turn first to *Aristotle* who so dogmatically forbid the actual infinity. In Section 2.3 we spoke about this prohibition. We would like to supplement here the remarks on Aristotle's conception by providing three explanations of the infinity given by Aristotle in his *Physics*.

It can be easily understood that in his philosophical approach there should appear problems with the actual infinity. Mathematical objects are connected by Aristotle with abstraction in human thought from real objects; however, they do not exist detached from things. They are forms in those objects that become concepts in thought by abstracting. Since there are no infinite objects in the reality, it becomes clear that the concept of the actual infinite is not possible by Aristotle. He comments this for example in his *Physics* [7, Book 3, Part 6] in the following way:

> "Further, a thing is infinite either by addition or by division. Now, as we have seen, magnitude is not actually infinite."

In [7, Book 8, Part 8] Aristotle says:

> "...and though what is continuous contains an infinite number of halves, they are not actual but potential halves."

The type of existence of the infinite is described by Aristotle in a comparison in the following way: We cannot regard the infinite as an ordinary object

> "such as a man or a horse, but must suppose it to exist in the sense in which we speak of the day or the games as existing things whose being has not come to them like that of a substance, but consists in a process of coming to be or passing away; definite if you like at each stage, yet always different."
> ([7, Book 3, Part 6].)

3.2.3 Idealistic approach

An *idealistic approach* in which ideas possess a real existence has been seen already in the works of Cantor. It is difficult to disclaim that the general concept of sets admitting also infinite sets is an idea. But is it like an idea of Plato, how real is it? Cantor distinguished – as we indicated above – "immanent" and "transient", we would say today "subjective" and "objective" reality. He claimed even that from an immanent reality in principle arises a transient reality and therefore in mathematics there is no obligation to provide evidence for the latter. To justify this conception he used "*unity* of *the universe*" (*Einheit* des *Alls*) (cf. [66, pp. 181 f.]).

All this is very metaphysical. Nowadays such disputes on the levels of reality are rather not understood. We are mostly thinking in an axiomatic way; we set relations between concepts using axioms and look modestly at their consistency as at the sufficient justification of their only theoretical existence. At Cantor's times it was different. By abandoning the obligation to found the realities – what he claimed to be the "free" mathematics – he prepared today's attitude. Bolzano relegated already 30 years before Cantor the confirmation of the existence of mathematical objects from mathematics clearly into philosophy, in fact into theology.

In the case of infinite sets, by Cantor one still has to deal with a very idealistic and even empiric (see below) reality, with "to be or not to be". And this is not only the consequence of the fact that he was engaged in a philosophical dispute with his colleagues. For Cantor the infinite sets and the transfinite cardinal numbers were so real that for example in the case of the Continuum Hypothesis (see Section 4.4) only one alternative was possible: the cardinality of the real numbers is $\aleph_1$ or *not*. He was not able to accept in his platonic attitude the possibility known since 1963, namely that one can take as the cardinality of reals this or another cardinal number. In 1963 Cohen proved that the Continuum Hypothesis is independent of the usual axioms of set theory, and earlier in 1938 Gödel showed that it is relatively consistent with them (cf. [137]).

What are doing adherents of platonism in this case – there are still such philosophers after 1963. For example Gödel was a clear platonic realist and he required – we wrote about this – a new extended or modified set theory that would better describe the world of sets, namely in such a way that the Continuum Hypothesis could be decided in it. This would be also Cantor's attitude. The problem of the ideal reality of transfinite cardinal numbers and their order became the problem of the adequacy, of the "quality" of axioms that should be adequate with respect to particular requirements. In the idealistic conception it is the duty of mathematics to describe axiomatically in an appropriate way the real world of infinite sets that *must* be supposed by a genuine Platonist.

3.2.4 Empiricist point of view

We consider now the *empirical point of view* and summarize how empiricists treat mathematical objects. For them, positivists or materialists, the source of mathematical concepts is the reality. They are extracted from real objects via senses by using a type of abstraction, generalization (induction), idealization or isolation and are in fact simply images of them. From this point of view the infinite exists in mathematics only when it exists in the reality or the infinite can be obtained in mathematical thinking by the mentioned processes.

Does there exist actual infinite in the reality? There are various empiricist attitudes. The usual one, is the standpoint according to which there exists no actual infinite. It was represented for example by Mill. Let us also quote Hilbert who in fact cannot be treated as an empiricist but who described concisely an attitude with which empiricists would agree:

> "[...] the infinite is nowhere to be found in reality. It neither exists in nature nor provides a legitimate basis for rational thought [...]" ([173, p. 190].)

The last sentence is of course understood by a rigid empiricist in a different way than Hilbert did: the actual infinite cannot belong to the range of mathematics because it cannot be found in the reality. It cannot be obtained from finite data either by abstraction or generalization.

What concerns the potential infinite – that for example by Mill can be understood as an abstraction and generalization from enumerating real sets – one can positively answer the question of its place in mathematics. It can be found in determining the cardinality, the number of elements of sets of real objects, hence in applied processes of enumerating. We have noticed above the necessity of potential infinity in thought – and this is accepted by the most empiricists.

However, there are empiricists who find also the actual infinite in mathematics, even in the reality. They belonged and belong, as was briefly described in Section 2.11, to the dialectical materialists. How should it be understood? It seems that there is a sort of a compromise enforced by the development of mathematics. Behind it there is the way of thinking discovered already by Cantor: one considers real continua like space intervals or time intervals and decomposes them into space points or time points. They become real infinite point-sets insofar one can consider points and point-sets as real. From this arise in mathematics via abstraction intervals on a straight line as uncountable, actual infinite sets. On the other hand an appropriate type of abstraction should lead to the actual infinite. It is the abstraction from the constructibility and realizability which can be hardly understood as being empirical.

Also in Cantor's thinking about sets there was an empirical aspect. Besides and behind the ideal sets Cantor indeed saw real sets of concrete objects – he saw them also behind infinite sets. He claimed that the set of atoms in the Universe is countably infinite. Ether in the space treated at that time in physics as the support of the light and considered by Cantor as a set was according to him uncountably infinite.

3.2.5 Infinity by Kant

Now we come to the rationalistic standpoint appearing in differentiated form in the philosophy of mathematics of the modern period. We refer to the classical attitude of Kant.

The basis of mathematics is according to Kant – as we wrote in Section 2.10 – provided by the pure intuition forms of space and time. Using them one orders the "manifold" phenomena, the various things in the world. The mind provides here the mathematical concepts, such as the concept of number, which by "synthesis", i.e., by composition or aggregation of the "manifold" supply synthetic knowledge *a priori*. The synthesis of the various things becomes concrete for example in "manifolds" as Cantor initially called sets. Can those sets be actual infinite?

On the one hand Kant says: No! Because neither in pure intuition form nor in the variety of phenomena there actually exists the infinite. However, in processes, in "regress" (going back) and in "progress" (going ahead) one finds the potential infinite. As an example Kant uses (cf. his *Kritik der reinen Vernunft*, [199, B 552]) a partition of a spatial "whole in the perception that is divisible into infinite". The process of partitioning into the infinite is for him clearly determined. However, one cannot say about the whole that "it consists of infinitely many parts". The parts are namely

> "contained in the given whole as an aggregate, but the whole series of the division is not, since it is infinite successively and never is as a whole, consequently, it cannot constitute an infinite set and no taking together of these same [parts – authors' note] in a whole." [7]

Nevertheless, for Kant there is also the actual infinite on a higher strongly separated level, the level of reason. The reason makes "the concept of the mind [...]" free of "inevitable limitations of possible experience" and looks for it "over the bounds of what is empirical however trying to extend it". Kant speaks in the framework of "cosmological ideas" about the totality that leads to the transcendental ideas [199, B 435 f.]. In the mind and in the perception of phenomena only the potential infinite is possible. The reason however is determined, it "demands" ("fordert", e.g., [199, B 440]) to think the non-whole, e.g., the open series of divisions as a whole, as an "absolute totality". It can be said that by Kant actual infinite sets have an *existence in principle*. A sophisticated presentation of Kant's conceptions of the infinity and its reception is given by C. Kauferstein in [204, Chapter 5].

Thus finite and potentially infinite sets on the one hand and actually infinite sets on the other are by Kant objects of different levels. The former belong to the domain of transcendental schema or are necessarily yielded by them. The actual infinity belongs to the domain of "transcendental ideas" and Kant remarks that "this absolutely complete synthesis is once again only an idea; for with appearances one cannot know [...] whether such a synthesis is even possible" [199, B 444].

7 "...in dem gegebenen Ganzen als *Aggregate* enthalten, aber nicht die ganze *Reihe der Teilung*, welche sukzessivunendlich und niemals ganz ist, folglich keine unendliche Menge, und keine Zusammennehmung derselben in einem Ganzen darstellen kann".

3.2.6 Intuitionistic infinity

In the works of the intuitionists one finds beside other affinities also aspects of Kant's standpoint. Finite sets were of course unproblematic. Infinite sets disintegrate into potentially infinite sets and "infinite non-sets", i.e., such sets that are altogether not conceivable. Examples of the latter are somehow supposed indefinite subsets of the sequence of natural numbers. They are elements of the power set of natural numbers that was intuitionistically not conceivable as well. Such not conceivable sets are for intuitionists not potentially infinite because there is no law according to which they could be constructed.

Intuitionists accept mathematically only such potentially infinite sets that are given as sequences determined by a law. They treated such sets that are given by enumeration as countably infinite. Other sets, in particular uncountable ones were treated as being not conceivable. Potentially infinite sets had for intuitionists no complete objective reality. Therefore they were not mathematical objects that could be elements of sets.

However, potentially infinite sets were not fully unreal. They were – similarly to Kant's ideas of reason – *elements of thought* and consequently somehow given. They have for example the property of being countable. Then they were connected with a law according to which the enumeration could be done. Finally, to certain potentially infinite sets, to sequences, belonged limit values – and this is important with respect to real numbers – that, as the case may be, were irrational or transcendent numbers and consequently were treated as constructible. All this grants to potentially infinite sets certain identity going beyond the pure potentiality.

The thesis that there are no uncountably infinite sets is common for intuitionists and all *constructivists*. In fact, the existence of mathematical objects is connected with their constructibility. To them belong in principle potentially, hence countably infinite sets. All further levels of the infinite are not conceivable. However, the countable infinity is considered by less strong constructivistic approaches in a similarly ambivalent way as by intuitionists what we indicated above. One can say that constructible infinite sets possess a *restricted existence.*

3.2.7 Logicist hypothesis of the infinite

Logicism is a program of reducing the whole mathematics – including real numbers and infinite sets – to logic. The logic came for Frege from the pure thinking. Hence infinite sets should be elements of pure thinking. Frege treated infinite sets in a platonic way. Russell was rather a nominalist and tried to keep them out of logic. Infinite sets were for him pure descriptions that can be replaced by properties. Those properties appearing as logical propositional functions had however again the mental status of ideas.

A major, in fact the most difficult problem for a strong logicism was the infinity itself. It cannot be grasped in a purely logical way but – in order to grasp it – it should be assumed by a type of the Axiom of Infinity. We have just described the conception of such infinite sets or their logical representatives. However, such Axiom of Infinity has no purely logical character – it has set-theoretical character. In Chapter 2 we have described how logicists tried to avoid the Axiom of Infinity in order to save the pure logic. It was eliminated and put as a precedent of an implication where it was necessary. One did not require the existence of infinite sets but it was inserted as a pure hypothesis. Hence the existence character of infinite sets in the logicistic approach is a bit curious: infinite sets have a *hypothetical existence.*

3.2.8 Infinity and the new philosophy of mathematics

Before making some remarks on the impact of formalism on the nowadays attitudes towards the infinite let us say some words about the new philosophy of mathematics. We are faced here with considerations about the development of mathematics. The problem of the infinity was present in the whole history of mathematics and the philosophy of mathematics and is an outstanding example of the significance of this problem. Another perspective of the new philosophy of mathematics is to take into account the research practice of mathematicians. Infinite sets belong today to a natural and self-evident repertoire of mathematical research and work.

Though the new tendencies are against the reductionism of the classical conceptions of platonism, logicism, intuitionism and formalism, they do not contribute something new to the conception of the infinite. We observe there the old conceptions that have appeared in the history of mathematics, in particular in logicism, intuitionism and formalism. One finds there in the foreground the rationalistic tendency taking into account the aspect of the development and of the real mathematical work. Thanks to the everyday application in the mathematical work a *practical existence* is granted to the infinite.

An additional aspect concerning in principle also the concept of the infinite is the cultural dependence of mathematical concepts. Hence to the infinite, in particular to the actual infinite, a type of *cultural existence* should be ascribed. However, one must notice that – similarly to the concept of a natural number – also the concept of the infinite has been in practice mathematically not changed since it has been freed from the mystical by Plato and Aristotle. The infinity seems to be a phenomenon of pure thinking that has – similarly to the concept of a number – at least partially a universal character and that is not submitted to the development. Only the formation of the concept – that thanks to Cantor received in the 19th century a completely new dimension –, the pragmatical attitude of mathematicians towards the phenomenon of the infinite, the treatment of it and the applications of the infinite are of a really historical relevance.

3.2.9 Formalistic approach and nowadays tendencies

Recall the above given quotation of Hilbert, the founder of formalism. For him the actual infinite neither "exists in nature" nor "provides a legitimate basis for rational thought". The stress is put by Hilbert on the *basis*. Hilbert relativises this attitude by the following words:

> "The role that remain for the infinite to play is solely that of an idea – if one means by an idea, in Kant's terminology, a concept of reason which transcends all experience and which completes the concrete as a totality [...]" ([173, p. 190], English translation on p. 151.)

Even if he supplies here the word "idea" by the dilution "solely", the infinite possesses for Hilbert a "well justified place" in thinking and is mathematically "indispensable". The main aim of formalism is to secure the actual infinite and consequently also the infinite mathematics. This should be done by "finitistic" reduction of such mathematics to the "finite" basis of symbols and formulas.

Formalism is strongly connected with the axiomatic method that is implicitly present everywhere in mathematics since Euclid's *Elements* and – in a strict sense (cf. Section 5.2.2) – Hilbert's *Grundlagen der Geometrie*. Nowadays in axiomatized theories the actual infinite has its almost undisputable place. Today mathematics is – as is willingly and a bit sloppy said – set theory. Hence an important background (for many, almost all domains of mathematics) is formed by set theory the axiomatization of which has been accompanied by the development of formalism. Hence the formalistic justification of set theory becomes important – the latter comprises in principle all further domains, in particular arithmetic. Among axioms of set theory there is the Axiom of Infinity in which the existence of actually infinite sets is claimed.

The acceptance of actually infinite sets is associated in the context of this Axiom of Infinity with other axioms of set theory. Due to this actually infinite sets receive not only their authorization but – so to say – their existence. In this way in our days – marked by formalism – in most cases the existence of concepts is considered: consistency of concepts in the framework of a theory based on axioms. This holds also for infinite sets. So one can foremost speak about the *theoretical existence* of infinite sets in the formalistic conception. Whether infinite sets are then additionally understood in a platonic, conceptualistic, nominalistic or similar way is left to the individual mathematician.

Till today no inconsistencies have been discovered in the axiomatic set theory. It is known that the problem of the consistency of set theory is essentially connected with the Axiom of Infinity. Such problematic statements as the Axiom of Choice or the Continuum Hypothesis are independent of other set-theoretical axioms and (relatively) consistent. In the case of set theories without the Axiom of Infinity one can justify their consistency. It will be said more about this in Chapter 4.

Since there are so far no inconsistencies, infinite sets are applied in the formalistic approach without any doubts and one believes in their existence. This is, so to speak, a *pragmatic existence* of the infinite. However, we learned in Chapter 2 that the consistency of set theory cannot be proved. Hence infinite sets possess *no certain existence*.

Finally, infinite sets get their justification by their successes in mathematics and by successful applications of mathematics. Hilbert said that besides the consistency of a theory, "success [...] is the supreme court" for justification of the existence of a concept "to whose decisions everyone submits" (cf. [173, p. 163], English translation on p. 135). Infinite sets are granted – so to say – a *utilitarian existence* that is not completely free of individual and social evaluations.

Nowadays the described situation influences largely – as a background for other possible positions and treatments of the infinite – conceptions of the actual infinite sets. It widely influences also the conception of the real numbers that represent the actual infinite. In the everyday mathematical work the consciousness of this is hardly present and it should not be. However, in the reflection on mathematical research practice and in teaching this fact as well as other facts concerning the infinite should not be ignored from the very beginning.

3.3 The continuum and the infinitely small

The complex of problems concerning the continuum does not consist primarily only of the famous Continuum Hypothesis as it is nowadays often seen (cf. Section 4.4). The latter is rather a result of a preceding and subjacent continuum problem. There is a different and deeper hypothesis, hence an earlier continuum hypothesis. This hypothesis says briefly:

The continuum is a set.

This assumption became soon after Cantor a mathematical daily routine and today it is quasi a mathematical fact. Example: The linear continuum is usually identified with the "number line" that is considered as a set of real numbers. One does not say a word that there is in fact a hypothesis and a problem and that it can be seen in another way. In handbooks this hypothesis is assumed, one usually skates over the problem, does not mention it or dismisses it as metaphysical or even mystic as Cantor already did.

The old, "classical" or "intuitive" continuum should be distinguished from the new continuum of real numbers. The new conception of the continuum as a point-set is beside the actual infinity a fundamental distinguishing feature with respect to the classical mathematics. It is not equally spectacular but in the same way fundamental and – as we will see – causal for the rise of mathematics into the transfinite.

3.3.1 General problem

What is the continuum and what – if not a set – could it be? Really, physically and intuitively seen: The continuum is the space around us, the space with three dimensions in which we are staying and moving, as well as the continuous passage of time. Parts of the space, intervals of time and subspaces like surfaces and lines as well as many other are further representatives of what one can call continuum or continuous.

Considering the hypothesis in the framework of this intuitive conception of the continuum one comes to the reflection: is the space a set of points? Does the space consist of points? Sure, there are points in space if we put them. However, does the collection of points describe the space? By what are points as elements determined and distinguished? By their position? Where? In space? This is the vicious circle.

Maybe still more clear is the hesitation if one considers the continuous passage of time. When we are conscious of time then we experience particular moments or put points in the passage of time. It is done when one plans something or says the word "now". Do we *put* a point of time in the continuous passage of time – or do we *choose* a point of time from the points of time that run in a quasi digitally way before us? Is time a collection of moments, of presents? Is time decomposed into such points of time? How are those points of time distinguished from each other? By time? Augustine of Hippo (354–430) discusses elaborately the problem of now, the present, in his *Confessiones* and says memorably how enigmatic the phenomenon "time" is:

> "What is time? If nobody asks me then I know it; if I want to explain it to somebody who asks me then I do not know." (quoted after [120, p. 251].)

Let us ask another, seemingly naive question. We must ask it. What is in fact a point, a point in space or a point of time? *What are points?* We believe in their existence and we do not know what they are and do not want to know it. For in mathematics one – being trained axiomatically – has refrained from asking this question. One smiles today in mathematics at the old definition of Euclid – in English translation: "A point is that which has no part". However, it throws new light at the question concerning the continuum. What has no part is not divisible. A point in this definition is like an atom.

The intuitive continuum is characterized by the fact that it has parts and those parts are again continua. Points having no parts are representatives of what is discontinuous and what is the opposition of the continuum. They can form only what is discontinuous. Points appear in partitions only as limits of continua and have in this conception no independent existence at all. And it is inconceivable in the case of this conception that just those points whose existence depends on the continua should constitute the latter. The parting (dividing) would become the parted (divided).

Generally, we have an atomistic view when we consider space and time as sets of points, when space or time disintegrate into separate elements. It is transferred from the atomistic view of the matter into continua and therefore seems to us to be familiar. The result is this: The continuum will be *in principle* discontinuous. What we are doing and what the consequences are can be concisely described by a quotation from Kant's *Kritik der reinen Vernunft* [199, B 555]:

> "The infinite division indicates only the appearance as quantum continuum [continuous quantity – authors' note] and is inseparable from the filling of space; for the ground of its infinite divisibility lies precisely in that. But as soon as something is assumed as a quantum discretum [separated magnitude – authors' note] the multiplicity of units in it is determined; hence it is always equal to a number."

Accepting *finitely* many atoms or moments that fulfill the space or time is excluded. This contradicts the unlimited divisibility of the continuous. The infinite necessarily enters here. The "number of units", of atoms, must be infinite. Hence if we want to have the set conception of the continuum, we must – according to Kant – necessarily climb to a fully new infinite number. The "price" that one has to "pay" for the set conception of the continuum is the uncountability of the real numbers – and the *new* continuum hypothesis. The way to the actual infinite begins here and leads to the transfinite numbers. It has been shown to us by Cantor.

The conception of the continua as sets of points enforces the program of set theory (cf. Chapter 4), for the discontinuous given in points must quasi be made good (compensated). Actually infinite sets must be introduced (Axiom of Infinity), procedures from the finite must be transferred into the infinite (Power Set Axiom, Replacement Axiom, Axiom of Choice) and one looks for set-theoretical representatives of properties of the continuous read from the intuition of the continuum (completeness, connectivity). It has been clearly described by Dedekind in his invention of cuts named after him [85].

It is interesting and instructive to study historical conceptions of the continuum. We can do this only sketchy. We will see that earlier conceptions were fundamentally different from the ones today. They enable us to compare conceptions and to confront them with the ones today as well as to scrutinize the latter. Some indications have been made already.

Conceptions of the continuum are not so strongly stamped from the philosophical point of views as it was the case for conceptions concerning the numbers or the infinite (discussed in previous sections). However, the conceptions of the continuum and of the infinitely small quantities are strongly connected with conceptions of the infinity presented before.

3.3.2 On the history of the continuum

The word "continuum" means "the extensive", "the associated", "without gap" or "continuous". The Greek word is "συνεχές" (synechés). Its meaning is also transitive and means "keeping together" or "connecting" and expresses a connection, a relation.

The usage we are following is this: the "continuum" is a superordinate concept of examples given above and of many other examples. Every particular example is "a continuum". However, in case of some particular continua like a body, a surface, a line, an interval one often speaks about "the continuum".

In early times one saw and thought about the continuum first of all in bodies in the space and in the movement in time. Other continua like surfaces, lines, segments were derived from them. Notice that we do today just the opposite: continua of any dimension are built of linear continua.

In the history of mathematics there were very early attempts to explain the meaning of the concept "continuum" and to tell what is characteristic of being continuous in such examples as segments, surfaces, bodies or spaces. They were very different. Conceptions of the continuum that very often were only implicitly developed on the basis of handling geometrical elements have been changed fundamentally in the history of mathematics. We will exemplarily direct our attention over and over again to the linear continuum since it played the essential role in the history of conceptions and will play it in particular at the end.

Pre-Socratic conceptions

A very early author of a conception that one can connect with the continuum is *Anaxagoras* (ca. 500–428 BC). He says:

> For there is no smallest [magnitude] among the small [magnitudes] but always a smaller one. For it is impossible that a being ceases from existence by partitioning into infinite. However, there is always a greater [magnitude] than a given great one. (Quoted after [70, p. 267].)

First of all the middle sentence is important for us. What is the being? Anaxagoras thinks about fully real, physical things in the space.

What is the space? Anaxagoras was a physicist and cosmologist, more precisely meteorologist, dealing with "things in the height" (μετέωρα (metéora)). From that arose his natural philosophy. For Anaxagoras space is never empty. Empty space means for him the same as "not existing". He thought that things in "cosmos" were built of innumerable primary substances that are equally spread in the space, that are moved and ordered according to spiritual principles, that form the physical things and all are in all things. Anaxagoras' philosophy is an early example of a clear dualism of mind (νοῦς) and matter.

The above quotation deals with the unboundedly continued partition which is noted (appointed) there as a characteristic feature of what is spatially physical. A consequence of Anaxagoras' conception of space is that such partitions do not lead into the nothingness. Consequently, there is also no smallest magnitude [70, p. 267]. Since we are interested in mathematics we would like to interpret the continued partition geometrically. So let us think about a "real" segment. The "partition till the infinite" obtained by bisection looks like this:

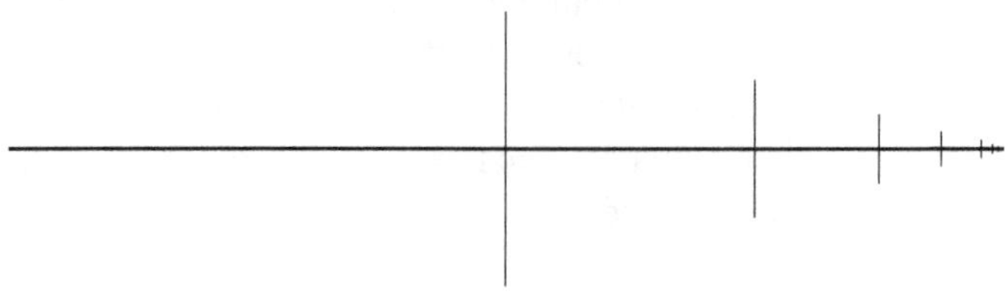

Fig. 3.1. "Partition till the infinite" of a "real" segment (Anaxagoras).

The first and third sentences from the above quotation are illustrated in the following picture – here on an interval on a straight line:

Fig. 3.2. "There is always a smaller [magnitude] than the given small one, there is always a greater [magnitude] than a given great one" (Anaxagoras).

Looking from inside into outside "there is always still a larger [magnitude]", from outside into inside "always still a smaller one". One has here an early formulation referring to the later archimedean property.

These are usual pictures for us. For us the idea of an unbounded partition of segments and of the unbounded extension of straight lines also belong to the idea of a linear continuum. And it belonged already to the geometry of the Pythagoreans. They followed the ideas of Anaxagoras and we pursued them describing in Section 1.2 the discovery of the incommensurability. Also for the Pythagoreans the sequence of the partitions was infinite and did not come to an end.

Democritus (ca. 460–ca. 370 BC) took up a quite different position. We know him as a consequent materialist and as the main representative of old *atomists*. His mathematical works were almost completely lost.

Democritus was a definitive philosophical opponent of Anaxagoras. Atomists had a quite different idea of the continuum than Anaxagoras and the Pythagoreans. Aetius (ca. 50 BC) writes about them in the following way.

> They "claim that division of a material comes to an end by parts having no parts and cannot be continued into the infinite." (Quoted after [70, p. 396].)

So it is again the matter of something material and of partition of it that cannot be unboundedly continued into the infinite but comes to an end by "atoms" having no parts. Again contrary to Anaxagoras who thought in the duality of mind and matter, Democritus puts atoms into the nothingness, into the empty space in which they move and form physical things. His philosophy of nature was of causal nature and monistic with atoms as the last foundation. Also soul and mind, percipience and thinking were explained by him using atoms.

Democritus himself has transferred his atomistic ideas of the material things into geometry:

> "When a cone is cut by planes parallel to the base then how should one think about slice planes arising in this way: as being equal or unequal? If they were unequal then they would make the cone irregular since it would contain many gradual incisions and protrusions; if they were equal then all cuts would be equal and the cone would provide the appearance of a cylinder". (Quoted after [133, p. 232].)

Democritus imagined – the above sentences cannot be understood in another way – a cone as *consisting of layered slice planes*. Those slice planes should have certain heights. Hence they are very thin slices that are not divisible in their height, that are "atomic" and *indivisible*. They are not only flat but have also a certain spatial dimension. Therefore they generate "gradual incisions and protrusions" Democritus speaks about (cf. [238, pp. 19 ff.]).

We would like to transfer Democritus' conception into a *linear* context and to demonstrate it on an example of a segment. We did it also in the case of Anaxagoras's conception and we will do it also for the following conceptions of the continuum. In this way one can confront them and compare them with each other.

Let us translate first Democritus' statement into a flat version. Concepts that have been changed are given in italic:

> When a triangle is cut by *straight lines* parallel to the *base* then how should one think about slice *segments* arising in this way: as being equal or unequal? If they were unequal then they would make the *triangle* irregular since it would contain many gradual incisions and protrusions; if they were equal then all cuts would be equal and the *triangle* would provide the appearance of a *rectangle*.

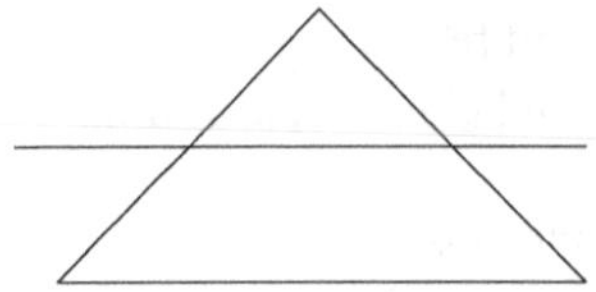

Fig. 3.3. Triangle with lines of intersection.

The cutting would look in the case of inequality “through a magnifying glass” like this:

Fig. 3.4. Part of a side of the triangle and the intersection segment seen “through Democritus’ magnifying glass”.

Segments are here, analogously to slice planes in the case of a cone, very narrow, *indivisible* stripes that are *similar to areas*. Their height is very small, however, of a *fixed positive value*. Otherwise there would be not gradual cuts that we see. There are very many stripes but they can be only *finitely many* because their height is of a fixed positive value. The triangle is composed of those stripe segments.

A further transfer of the formulation into the linear context is only partially possible. “Cut *points*” cannot be parallel. Additionally they cannot be unequal. “Gradual incisions and protrusions” cannot exist. There remains only an analogy whose points have a linear dimension. We are allowed to conclude that for Democritus points are very small, atomic segments. The whole segment is composed of them. Democritus’ statement will become the following (with the complement in brackets):

> When a *segment of points* is cut then what should be the arising intersection *points*? The *segment* would provide the appearance of a segment ⟨composed of intersection points⟩.
>
> It looks – again seen “through the magnifying glass” – like this:

Fig. 3.5. Segment consisting of very small segments (points) by Democritus.

And again segments – being now points – have a *very small length*. Hence there can be only *finitely many* small segments building together the whole segment.

It can be imagined that the conception of the early Pythagoreans – stating that everything, in particular everything being continuous, can be grasped by natural numbers or by proportions of them – has in fact a finite-atomic background. The

Pythagoreans had a procedure – the so-called reciprocal subtraction (cf. Section 1.2) called today the Euclidean algorithm. It enabled one in principle to obtain for any two continuous segments the greatest segment, the greatest common measure with the help of which both segments could be measured and to each of them a natural number could be attached. The proportion of those numbers was the proportion of the segments. Behind the belief in number there was the belief that the procedure of reciprocal subtraction would finish in a finite number of steps – that was probably supported by the belief in smallest atomic segments. Aristotle speculates in an early work [3, pp. 66 f.] about the possibility of such atomic segments and refutes this idea by simple arithmetical and geometrical arguments.

The discovery of the incommensurability by the Pythagoreans excludes the conception of a *finite atomism*. It did not appear again in the ancient Greece as we will soon see when we learn the dominating position of Aristotle. However, it appeared once again in the early Middle Ages. It is said (cf. [117, Volume II, p. 165]) that one treated for example an hour as consisting of 22 560 moments.

We leave now for a short time the chronological progress, skip over Aristotle whose conception will be presented later and come to *Archimedes*. Archimedes (ca. 287–212 BC) connects – as will be seen – the conceptions of Anaxagoras and Democritus, the unbounded divisibility of the continua and of atomism. How does it happen?

Archimedes

Archimedes' conception is split. As a mathematician he stands in the tradition of the Pythagoreans and follows in the *mathematical* conception of the continuum Aristotle and Euclid (see below). On the other hand in his *heuristic* he appears to be an atomist – in a new sense. One can see this in his methodology. From the sketch of his famous quadrature of the parabola – considered here only in a fragment – characteristic features of his mechanical method can be seen.

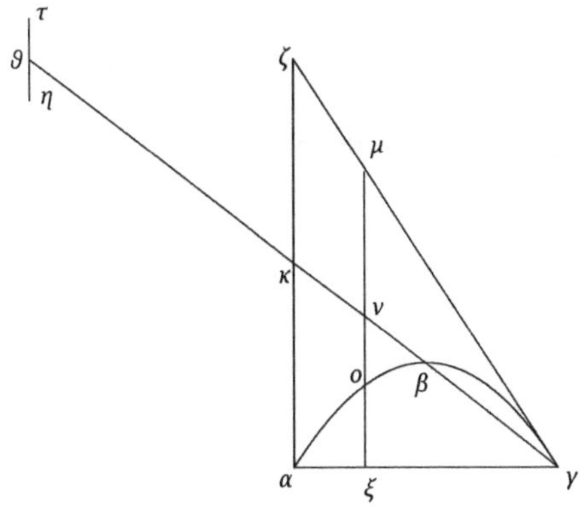

Fig. 3.6. From the quadrature of the parabola by Archimedes.

We do not want to study the geometrical details and to consider his arguments in single statements. However, one recognizes right away: Archimedes *weighs out segments*. We see a steelyard balance with the pivotal point κ. On the left, in ϑ hangs the segment $\tau\eta$ being congruent with respect to ξ. On the right hangs in ν the segment $\mu\xi$.

This indicates the mechanical aspect of Archimedes' heuristic. We get from the following quotation another aspect that is decisive for us:

> "And since the triangle $\gamma\zeta\alpha$ consists of segments in the triangle $\gamma\zeta\alpha$ and the segment $\alpha\beta\gamma$ consists of segments corresponding to ξo hence the triangle [...] is in balance with the parabolic segment, [...]." (Quoted after [317, p. 113].)

What do we read? A triangle is composed in Archimedes' conception of segments – segment by segment. It "consists" of segments:

> A triangle is in Archimedes' conception a composition of segments.

What does it mean when transferred into the linear context?

> A segment is in the Archimedes' conception a composition of points.

It reminds one of Democritus. However, meanwhile a lot of time passed. *Plato* (428–349 BC) wrote 100 years ago clearly and dramatically about the common ignorance of the incommensurability by his contemporaries – as quoted in Section 1.2. Between Plato and Archimedes there are Euclid's *Elements*. The infinite, unbounded partition of a segment was mathematical daily routine. The discovery of the incommensurability had changed mathematics, brought the theory of magnitudes in *Elements* and overcame the mathematical atomism. Hence one should suppose that the conception in Archimedes' heuristic is not the finite atomism of Democritus but an *infinite atomism*:

> The surface of a triangle is in Archimedes' heuristic conception a composition of *infinitely many* segments that are *infinitely thin* and indivisible stripes.

Archimedes himself did not speak out on finite or infinite magnitudes, on small or infinitely small measures, on the status of such atoms, but only operated with them and used their geometrical properties heuristically.

Let us translate this conception concerning a triangle again into the linear context:

> A segment is a composition of *infinitely many* points that are *infinitely small atomic segments*.

A visualization of this situation is not possible. It is mentally only – and has rich consequences.

The transfinite atomism of Archimedes' heuristic is compatible with the conception of Anaxagoras and the practice of the Pythagoreans who came from the unbounded divisibility of continua. If one follows our conclusions concerning Archimedes's conception then it will be noticed that the fact that a segment is composed of infinitely many points presupposes the totality of the latter:

> A segment *consists of* infinitely many points. Hence it is a "whole" of points.

The intuition of a geometrical segment establishes this whole. Hence the *conjecture* is:

> Archimedes considered the *actual infinite*, i.e., the infinite as a given whole.

This has been excluded 100 years ago by Aristotle who excluded the infinitely small quantities and atoms.

Archimedes was again fully aristotelian when he himself held his heuristic proof for not being mathematical. The reason could be the reference to "mechanical" elements in the proof (balance, center of mass) as well as the atomistic conception of a surface. He gave later a strict proof in the style of Euclid. Similar observations as in the case of the quadrature of the parabola can be made also considering his determination of the volume of a ball and of the surface of a sphere (cf. [385, pp. 172–176]).

The atomistic conception of Archimedes was an important element of his heuristic that he probably has taken from the (finite) atomistic idea of the matter in physics spread at that time and that he carried forward infinitesimally. It is also known that Archimedes obtained often ideas for his mathematical theorems from technical ideas, for example from experiments with weighing what is indicated by the word "balance". On Archimedes' heuristic writes B. Zimmermann extensively in [385, pp. 170–221].

The conception of Aristotle

Aristotle has, as we know, refused and blocked the way to the continuum consisting of points. What were his arguments? We know already his arguments against the actual infinite that necessarily apply also to the conception of the continuum as a set. We add to them his arguments coming from his conception of the continuum. They can be found first of all in the sixth and eighth Books of *Physics*. The following quotations taken mainly from [90, 371] come from the sixth book of *Physics*.

Aristotle refers to Anaxagoras and the Pythagoreans' mathematics. To characterize the continuous clearly he opposes it to the discontinuous. We summarize his known conceptions. By Aristotle, time, magnitude, line, spatial bodies and movement fall in the category of continuous. He determines the properties of continuity, literally translated "holding, keeping together" and "being connected" in the following way:

> "It is plain that everything continuous is divisible into divisibles."
> (Physik VI, 1)

This recalls Anaxagoras. Parts that arise in partition are again continua. Important is – again as by Anaxagoras and in the practice of the Pythagoreans:

The potentially unbounded process of partition is not considered to be closed.

In this way the idea of an infinitely small quantity is excluded.

Bounds arising by partitions that bound and connect partial continua determine further the property of being continuous:

"Things being 'continuous' if their extremities are one."

How should one understand this? In spatial and flat examples connectivity is described. In the linear case the "ones" are points arising in partition, bounding segments and connected by segments. Each point, being one, keeps parts together ("συνεχές" (synechés)). If there were two points then the continuum would be split up into two detached parts.

Just points are opposed to continuous segments.

Points are bounds (ἔσχατα, literally "extremities") of linear continua.

Points are plainly not divisible. Aristotle says in *Physics* VI, 1:

"The extremity and the thing of which it is the extremity being distinct."

What is not an extremity is internal of a continuous segment. Points *are* the most external, they have no externals, no bounds. They can be neither bounded nor partitioned. Therefore for Aristotle

points are representatives of what is discontinuous.

From what is non-continuous cannot arise something continuous:

"Nothing that is continuous can be composed 'of indivisibles'."

When points are neighboring then there is "between them a line, hence something of another type". Points cannot touch each other as continua do. Bounds of the latter coincide when they touch each other. Points touching each other are not conceivable. Therefore a continuous setting of points that should touch each other is inconceivable. Moreover, the idea that a line can be potentially infinitely exhausted by points does not hold by Aristotle, for potentiality is understood strictly by him. That means that the process of the partition of a line by points at each point is going on and it is impossible to think it in any way as a whole.

Aristotle excludes definitely that lines and continua can consist of points.

Continua are not sets of points.

Using those arguments Aristotle can easily contradict some of Zeno's paradoxes. For him they are sophisms.[8] Consider for example the arrow paradox. The arrow – according to Zeno – does not move because in every moment it rests. An aspect of this paradox can be simply rebutted if Zeno assumes that temporal continua consists of moments or of smallest time points. This assumption has been discussed and refuted by Aristotle already in the early work [3].

At last it is excluded for Aristotle that a line is composed of "line atoms". Assuming such atoms – done supposedly by Plato – was impossible for Aristotle. They would be on the one hand result of an actual infinite partition process which was excluded. On the other hand a line would be an actual infinite composition of line parts which again was impossible for Aristotle. Finally, the word "line atom" conveys an self-contradictory idea. In fact, "line" is a continuum whose characteristic feature is being divisible and "atom" is something indivisible. Hence ultimately it holds:

Points are not atomic segments,

other than it was the case for Democritus and Archimedes, that is, the atomic or infinitely small is excluded:

There are no line atoms.

Euclid

We make now some remarks concerning conceptions of *Euclid* (ca. 365–ca. 300 BC) that we implicitly and with caution deduce from his formulations and his mathematics.

Conceptions of Aristotle had great influence on the thinking of philosophers in the ancient Greece and among them also on mathematicians, in particular on Euclid (cf. Sections 2.3 and 2.4). This concerns also his conception of the continuum. This is visible in his *Elements* which was a mathematical handbook during the following 2 200 years. For it is difficult, sometimes even impossible, to find consistently adequate translations we have in some cases to take into account the original Greek formulations by Euclid.

The Euclidean algorithm in geometrical form, the procedure of reciprocal subtraction, is the witness of an unbounded divisibility of the linear continuum. It does not stop in looking for a common measure as we can observe in § 2 of Book X of *Elements*. It

8 Compare the paper by H.-G. Bigalke *Über den Unendlichkeitsbegriff* [35, pp. 327 f.].

is a heritage of the Pythagorean mathematics. The unbounded divisibility of continua adopted by Aristotle corresponds to the conception of Anaxagoras.

Notice that from the idea of what is continuous the Greek qualification of the concept of a point arises. What is a point? Points are bounds of a line. This is said in the third definition of Book I of *Elements* by Euclid. The first postulate by Euclid claims the connection of two points by a continuous line, namely by a segment. Both points are connected by this line. So the continuity looked at this way is like a connection, a relation between points. This corresponds also to the meaning of the concept of the continuum both in Greek as well as in Latin and German. The property of indivisibility arising from the opposition to the line becomes by Euclid the first "definition" of a point that "has no part" – as is usually translated.

In Greek there is the word "*σημεῖον*" for "point" that originally meant a "sign". The word "sign" brings to mind points in the works of Aristotle that primary are bounds or dividing points of continua. Points *mark* the bounds of continua or the dividing points that are mental or can be mentally put and given.

How is the relation between points and continua by Euclid? Let us quote some definitions from Book I:

1. A point is that the part of which is nothing. (*σημεῖόν ἐστιν, οὗ μέϱος οὐϑέν*)
2. A line is a breadth*less* length.
5. A surface is that which has length and breadth only.

In Book XI the first definition is:

1. A solid (*στεϱεόν*) is that which has length, breadth, and depth.

In the translation by Heath [110] there is the noun "a solid" for the Greek adjective "*στεϱεόν*" that literally means: "what is rigid", "what is firm, massy". The meaning of "solid" bodies in the spatial geometry is close to concrete bodies, closer than their "bounds", surfaces (XI, Definition 2), and again their bounds, lines (I, Definition 6), and finally their ends, points (I, Definition 3), are with respect to the concrete things.

Definition 1 in Book I of *Elements* says in the translation by Heath: "A point is that which has no part". It is usually quoted just in this formulation. It looks in this translation as if points were something like atoms. Above we have adopted this once. Literally one should translate it just as we did it. This does not suit well to atoms, to what is indivisible by which it is forbidden *per se* to speak about parts. Part comes from partitioning. If one takes the partition of continua continued till points then the Greek formulation of Definition 1 can be understood in the following way: The partition does not lead to continua as it is in the case of continua, but to "nothing".

Hence Definition 1, the beginning of *Elements*, can be understood as a characterization of what is discontinuous – as a consequence and as a contrasting juxtaposition towards the aristotelian characterization of the continuum.

Our interpretation is confirmed and changed a bit when one reads the definitions given above in the reverse order: from bodies to surfaces that have "only length and width" and hence no "depth" through lines "without width" to points that finally have no length, and consequently have participation even in "nothing". "$\mu\acute{\varepsilon}\varrho o\varsigma$" means also "participation".

Points have no size that would make them divisible and not "nothing" as their part. They are neither very small nor infinitely small. In this way in the conception of Euclid, as in the conception of Aristotle, is the atomistic way of thinking excluded. They have no participation in the line, they have no dimension. Points by Archimedes, understood in his methodology as being infinitely small segments that compose segments are not allowed, and Archimedes himself did not treat them as mathematically legitimate.

The conception of a line as an entirety of points cannot be found by Euclid. There are no continua as point-sets. Points, straight lines, surfaces, circles etc. are independent elements in a geometry of constructions. Points are never points "*of* the circle" or "*of* the straight line" but they "lie *on*" them. Between points and straight lines, circles etc. there is uniquely the outer relation of incidence. In connection with for example segments points occur as bounds of them. Entireties of points occur exclusively in a discrete form.

Notice also that contrary to Aristotle, for whom continua are primary phenomena and points as their bounds (ἔσχατα) are secondary, in Euclid's *Elements* the point (σημεῖον) stands at the very beginning. Between points are arising continuous segments, as is said in Postulate 1:

> "To draw a straight line from any point to any point."

Definition 3 of Book I however says:

> "3. The extremities (πέρατα) of a line are points."

It borrows Aristotle's conception of points as extreme bounds. One can see here further evidence of the influence of Aristotle.

Besides the geometrical continua there were also magnitudes found in Books V and X of *Elements* as representatives of what is continuous in the old mathematics. Book V is an amazing witness of the early axiomatic mathematics. The concept of a magnitude is left as undefined.

The concept of relation (ratio) between magnitudes is limited in Definition 4 to magnitudes "of the same kind" which "are capable, when multiplied, of exceeding one another". This formulation is sometimes interpreted as an axiom called later archimedean property neglecting the fact that one has to deal here with a definition and not with a postulate or an axiom. Magnitudes having no ratio are not excluded – they can be "not of the same kind" or multiplied not capable of "exceeding one another". In any case it is remarkable that the archimedean property as a property of the domain of magnitudes has been formulated already here.

The assumption – made implicitly in Book V – of a fourth proportional constructed in Book VI, § 12, for linear magnitudes (segments) demands and describes aspects of the continuity of a continuum.

Incidental remark

Let us stop at this point and state the following: the essential heritage of Aristotle that came to mathematics via Euclid were among others the following principles:

(i) The base of mathematics is build by intuitive geometrical continua and by continuous magnitudes.
(ii) Neither an unbounded partition of continua nor the unbounded supplying of units can be conceived as closed processes: the infinite is potential and not actual.
(iii) Points and sets of points represent what is discontinuous. Straight lines, segments, surfaces etc. are not sets of points.
(iv) There are no indivisible and no infinitely small pieces of lines.

Those principles have been formulated and justified by Aristotle. They determined in principle the Greek mathematics and remained valid through millenia. This concerns the handling of the infinite as well as the conception of the continuum. Since the discovery of the phenomenon of incommensurability the Greek mathematics had to deal with the continuous magnitudes and their proportions because the (natural) numbers and their proportions did not suffice. The way back to numbers would – as we already said and as we will still see – mean another conception of the continuous and another handling of the infinite.

Geometry, the mathematics of magnitudes, the method of exhaustion and the geometrical algebra founded by the Greeks eliminated the problem of the infinite and gave an answer to the question on the continuum. This mathematics presented in main features in Euclid's *Elements*, that till Descartes guided the mathematical thinking, is in fact an admirable scientific phenomenon in the amazing epoch of the history of the human mind.

3.3.3 What is a point?

Considering the problem of the continuum one comes necessarily – as we have seen – to the problem of a point. One must ask the question: "What is a point?"

It is not asked in mathematics. This problem vanished mathematically since there exists the new axiomatics (cf. Section 5.5.2). Nowadays points are *elements* of a set whose relations with each other as well as with other elements of other sets are described by axioms. In this way there arises an implicit definition by the answer to the question "*How* do behave points with respect to straight lines, surfaces etc.?"

The question *what* is a point remains consciously unanswered.[9] The apparent answer "element" does not lead forwards. It simply shifts the answer. In fact, one knows just as little about *what* are elements, again one knows only *how* do the relations of them and other elements look like. Relations between elements are described by axioms of set theory.

Anyway it is interesting how they do this. Objects of set theory are sets, hence elements are sets whose elements are sets whose elements are sets etc. This regress is broken by the Foundation Axiom (cf. Section 4.3). Thus: if points are elements then they are in principle sets. This is strange – and we come to this later. Axiomatically seen it is not interesting – as the title question of this section at all.

When one looks at the issue and is interested in philosophical and historical aspects then the situation is different. Then the question "What is a point?" is relevant. It cannot be separated from the question concerning the continuum. It is not enough to smile for example at the Euclidean "definition" and to put it in quotation marks. The Euclidean conception has been studied above and points were identified for example as opponents of aristotelian continua. Let us briefly summarize what was told so far about points and make some further remarks.

Classical ideas concerning points

In reality points occur in space as for example distinguished positions in spatial objects. Let us imagine for simplicity a concrete cube. A cube has vertices. Vertices will be formed by faces and edges of the cube. They are again real objects. Everybody knows that a concrete vertex is not a point, as a real face of a cube is not a flat surface and an edge is not a straight segment.

In mathematics, in particular in geometry, we are dealing with idealizations and abstractions, with points, segments and straight lines, planes and surfaces, spaces and solid bodies. And the same was done by the ancient Greeks initially in a strong connection with concrete spatial entities and their pictures. It was and is the matter of the characterization of geometrical concepts: What is the space, what are planes and straight lines? Those questions are at least so difficult as the question concerning a point. Let us restrict ourselves to the latter question – for a simple single reason. Nowadays spaces, planes and straight lines are mostly considered as sets of points.

Solid bodies, surfaces and lines etc. were representatives of the geometrical continuum. The aim of early attempts was to understand and to describe them or to operate with them according to their properties. It seems that those concepts have been gradually detached from concrete solid bodies, planes and lines. Points occurred initially only implicitly.

Early atomists – as we quoted – started from the physical bodies and assumed that they are built of smallest, indivisible particles, just of atoms. *Democritus* trans-

9 One can find some progress in answering in [21].

ferred this principle into geometry and treated a cone as built of flat indivisible slices that conceptually seem to be between concrete flat sections and geometrical surfaces. Consequently, lines consist of smallest indivisible segments, of points:

> Points were atomic segments.

Put together they could form segments. All geometrical elements beyond a solid body had themselves – as should be assumed – a bodily dimension.

In *Archimedes' Heuristic* those atoms were infinitely small:

> Points were infinitely small atomic segments.

And segments were compositions of infinitely many of those infinitely small segments.

Since *Plato* and *Aristotle* it has been clear that in geometry one has to deal with idealizations and abstractions. Aristotle contrasted explicitly points with continua. Continua were primary, and consequently

> points were "bounds" of continua, as extremities of segments or as points of partitions,

and as such appeared as dependent.

This was stressed much later by, for example, Kant for whom just the opposite, neither parts and sections of space nor in time can be seen without points as bounds (cf. [199, B 211]). Without the idea of the composition of space by parts there is for him "nothing, even not a point" left over [199, B 467 f.].

Euclid said what points are. His Definition 3 assumes the aristotelian identification of a point as a bound. Postulate 1 of Euclid claims that two points can be connected by a continuous line, namely by a segment. Primary are here the points, and the continuous segment as the connection of points is secondary.

Euclid – and this should be stressed – put the concept of a point at the very beginning of his *Elements*.

> "*σημεῖόν ἐστιν, οὗ μέρος οὐθέν.*" (Semeion estin, hu meros uten.)

This is Definition 1, the first sentence of *Elements* and the first word is "point". Points moved then at the beginning of geometry and of mathematics, the Greek one as well as the following mathematics. Today points became numbers on the real number line and as elements of this fundamental set are present everywhere. We shall describe it below.

Even if the point occurs in *Elements* as primary, the definition of a point lies still in the succession of the characterization of continua by Aristotle. His definition is the consequent continuation of the characterization of continua as unboundedly divisible.

How did we translate Definition 1 above? Quite literally:

> A point is that the part of which is nothing.

Partition of continua leads to continua, a part of a point is nothing. The point is the opposite of the continuum, is a non-continuum.

> The point is the "discontinuum".

> The word "σημεῖον" used for "point" in Definition 1 means "sign".

> Points are signs.

This provides another hint. When a point is a sign then something similar can be assumed for further geometrical elements. Signs stand for what has been thought and denote it. Geometrically seen this points to the level of intuition and pictures which by Plato belong in fact to the domain of material objects but which lead mathematicians to mathematical ideas (cf. [276, p. 60]).

Intuitive geometry is the domain of signs that represent mathematical concepts and "de-note" them. A geometrical picture is a sign of a mathematical context.

> The point is the geometrical basic sign.

In theologically oriented discussions concerning continua in *scholasticism* points became again atoms, in Latin a word for the indivisible. "Nunc" (now), the point in time is the indivisible in time, the moment, the smallest indivisible movement (cf. [49, pp. 23 ff.]).

Here the point appears still in another new way in the foreground before the linear continuum. Albertus Magnus (about 1200–1280) said:

> "Continuum fluit ab indivisibili" (quoted after [49, p. 28])

"The continuum flows from the indivisible." At another place he said:

> The point is the "principium" of a line and "generates it by its flow" ("in processu facit lineam").

The flow, that what is continuous, for example the time, is characterized by the flowing point of time of the present that separates the time into the past and the future and coheres them – similarly to the point of partition in the case of a line that by Aristotle was a "One".

Modern ideas of a point

The indicated ways of thinking of scholasticism certainly influenced the atomism that came back to mathematics thanks to *Johannes Kepler* (1571–1630) and *Bonaventura Cavalieri* (1598–1647) and influenced *Gottfried Wilhelm Leibniz* (1646–1716) through *Blaise Pascal* (1623–1662).

Leibniz himself followed Aristotle in what concerns points and the properties of the continuum. Points were for him no small atomic or indivisible segments that were very small or infinitely small. However, for him there existed infinitely small segments whose notation dx, dy etc. are today still used in analysis – though they had generally no or quite a different meaning.

> Infinitely small segments, the infinitesimals by Leibniz were no points but again continua, hence they were divisible.

For the empiricism the point is a problematic issue. The *idea* of it is for Mill a "minimum visibile" like a line only by means of an areal, width is conceivable. How should one experience points? The geometrical point is – as the geometrical line – "completely inconceivable".

The introduction of coordinates into the space tracing back to Descartes meant that points in space could be presented as triples of numbers. It was based on the old practice of representing numbers as segments or as points on a line, on an axis. However, there were problems with irrational "numbers". In fact, they were no numbers but magnitudes. Nevertheless, they were soon also called numbers and lay as points between the proper numbers on the axis. In this way points on the straight line and "numbers" were moved closer to one another. The one-sided representation of numbers as points on the line seemed to become a reversible relation between both of them. Gradually a quite non-aristotelian conception has been developed: a straight line is a domain of points. This conception of a straight line as a set of points is obvious for Cantor and Dedekind. For Dedekind it were "all points of a straight line" [85, p. 10] that fall into two classes by a cut. The fact that there is *exactly one point* that produces the cut in the set of points, this geometrical "triviality" uncovers for him "the secret of continuity" of straight lines [85, p. 11]. He did not think about infinitesimal segments – and could not think in the framework of the conception of a straight line as a set of points. We shall come to this later.

Since the construction of the real numbers in the second half of the 19th century the following conception has found recognition – it is regarded so obvious that another one seems to be inconceivable:

> Points are real numbers

on the number line or tuples of numbers in the m-dimensional space. "The straight line" by Dedekind became "the number line". If we make the new conception of points plain to us and become conscious of the meaning of the concept of the "number line" then we will give some thought to it. For us geometrical points intuitively are no numbers, the fusion of arithmetic and geometry in the omnipresent "number-line" seems to be disputable. The effect of the fusion is: that what is intuitive is mathematically released from the intuition. And just this was the aim. We come back to this problem.

Also the influence of nonstandard models of real numbers existing for 50 years on the concept of a point is considerable and will engage us. The conception of a point is extended by the idea of an infinitesimal neighbourhood, called monad. This neighbourhood being a continuum accommodates again points that again can be extended by the idea of an infinitesimal neighbourhood.

> To a point belongs an infinitesimal neighbourhood with points. A point is a snap-shot in a process that begins with the usual "real points".

Consequently, this conception stresses again the conception of sets of points and the fusion of arithmetic and geometry. And the concept of a point itself becomes floating: there is no "firm" point anymore, unless we "set" it "firm", interrupt the flow and do not ask anymore. It is similar to the elements of sets that in principle are always sets, sets of sets etc. This was noticed at the very beginning of this section.

One finds this flowing conception of a point already in the pragmatic epistemology of C. S. Peirce (Section 2.17) who referred to the conception of Leibniz. For him the world of mathematical operations was the world of signs and diagrams. Points are geometrical basic signs among signs – they are created by mathematicians in the real continuum like points of time in the time. Points are no isolated objects but like moments infinitesimally surrounded by points and in infinitesimal steps in a continuous way connected with other potential points set by a mathematician. In this conception the continuum can never be grasped by points.

Final remark

In this philosophical conclusion concerning the point we take up the word "σημεῖον" used by Euclid. Points are "signs". Signs lie between the reality and cognition. We set them. They are "ingredients" given by our thinking, additions of our consciousness to the continuum. We express them in connection with the time by the words "now" or "then", in connection with the space by "here" and "there" – and maybe design them as a "point". The point is not a piece and not an element extracted from the continuum. The continuum cannot be – according to this conception – either a composition or a set of points. Nevertheless, the continuum is regarded nowadays as a set of points being numbers or tuples of numbers. We have indicated and we will show that this – also mathematically – is not adequate with respect to the continuum.

What is the thought that denotes a point, what is this add-on, the addition to the continuum? We do not know it. The new axiomatic approach to the concept of a point is – theoretically – inevitable. The question "What is a point" must remain mathematically unanswered. [10]

3.3.4 On the history of the continuum – continuation

Let us close our survey of the historical conceptions of the continuum. We make a jump of about 2 000 years. We make a remark on Leibniz as well as some marginal notes on some of his predecessors and followers.

Leibniz and the infinitesimals

We want to say now some words about a further new conception of continuum that essentially differs from conceptions presented so far. It stamped the analysis of the 18th century and since the 1960s again became important. It was founded and formulated by *Gottfried Wilhelm Leibniz* (1646–1716). It ignores – as Archimedes's heuristics did – in a new way Aristotle's rule forbidding infinitely small quantities.

In the early modern era there arose anew Archimedes's ideas obviously without knowing his methodology. They have been prepared in scholasticism – as indicated above in sections concerning the concept of a point – in a philosophical-theological way. There was then a notable, physically oriented debate concerning the "indivisibles" (cf. [49]). *Johannes Kepler* (1571–1630) and *Bonaventura Cavalieri* (1598–1647) restored the idea of building the continuum by infinitely small quantities, atomic components called, derived from Latin, "indivisibles". It seems that this idea influenced at that time increasingly the mathematical thinking (cf. for example [211, p. 69]) though Cavalieri himself thinking in a traditionally scholastic-aristotelian way recognized problems of the atomistic continuum. He discussed the idea of composing the continuum of the indivisibles and did not exclude it – as indicated by W. Breidert in [49, pp. 68 f.].

The most known occurrence of this conception of the continuum, according to which it is composed of indivisibles, is surely Cavalieri's principle. Geometrically seen indivisibles are here flat slices in solids, linear stripes in surfaces and small line segments in lines. It should be imagined as infinitely thin, narrow and small, hence infinitesimal. Lines we may assume are composed of infinitely many such line segments – points, surfaces of infinitely many lines and solids of infinitely many surfaces. Cavalieri's principle states in the simplest version the following: if all flat slices parallel to a given base of two solids of the same height have the same area then the volumes of both solids are equal.

Leibniz relies in his conception of the continuum both on Kepler and Cavalieri and first of all on *Blaise Pascal* (1623–1662). Pascal got to know Cavalieri's method of indivisibles and applied a linear version of it among others to quadrants. He con-

10 For further thoughts on points see [21]).

nected the idea of infinitely small segments, for example in a circle, belonging to tangents with infinitely many small triangles that showed the slope of the tangent. Steps from indivisibles by Cavalieri to the infinitesimals by Leibniz are described in [60, Section 4.2].

Pascal says about indivisibles the following:

> "An indivisible is one which has no parts. The space is that which has miscellaneous separate parts. Because of these definitions I say that two indivisibles being equal give no space." ([262, R 493–495].)

He deduces again a position that rejects the atomism. Finite magnitudes can be arbitrarily often divisible and cannot be composed of indivisibles. Infinitesimals, infinitely small quantities are a new kind of magnitudes beside the usual finite ones. Leibniz read Pascal's paper from the year 1658 in 1673 during his stay in Paris.

Leibniz adopted Pascal's approach, transferred it into the general case of curves and developed it. Besides infinitely small triangles – with sides dx, dy and ds – he considered together infinitely small stripes under curves and their infinite sum denoted by "$\int$" and out of this developed the differential and integral calculus. In the infinitesimal calculus – as it is nowadays still called – one finds and applies till today his symbols. In his correspondence with contemporary mathematicians and in his works he expresses himself repeatedly in different, sometimes seemingly contradictory way about the essence and role of infinitesimals.

Leibniz used various descriptions for infinitesimals. He speaks about "infinitely small quantities", about "incomparably small", about "indeterminable (inassignabiles) quantities", about "infinitely small parts of lines" etc. At the same time he *distinguishes* points and infinitely small parts of lines.

His conception of the continuum is summarized by him in the following way:

> "One should know however that a line is not composed of points, a surface – of lines, a solid – of surfaces but a line consists of small line pieces (*ex lineolis*), a surface of small surface pieces, a solid of small solid pieces that are unboundedly small (*ex corpusculis indefinite parvis*). This means, as will be shown, that two extensive *magnitudes* can be compared (even when they are incommensurable) if they are decomposed into equal or congruent parts that are arbitrarily small, [...]." (*Mathematische Schriften* [231] and [229, Volume 7, p. 273].)

Notice that Leibniz speaks here about unbounded and indefinite small "line pieces". One should assume them to be infinitely small if they have to give a common measure of incommensurable segments. Incommensurability in the domain of the finite becomes for Leibniz explicitly a transfinite commensurability. Leibniz *distinguishes* points and those infinitely small line pieces, the *infinitesimals*. Contrary to atoms of lines, they are again divisible – by points in those infinitesimal line pieces. Hence Leibniz's conception is in no way an atomistic one.

The fundamental sense of infinitesimals are for Leibniz – as already indicated – the infinitesimal *differentials*. The role of infinitesimals is to describe the relations in moments or points where

> "one notice that dx, dy, dv, dw, dz can be considered to be proportional to instantaneous differences, i.e., increases or decreases of x, y, v, w, z [...]." (Quoted after [16, p. 162].)

As we find relationships between usual magnitudes, also relations in the domain of infinitely small quantities can be considered.

Also *Isaac Newton* (1642–1727) relies – at least in his early works – on the idea of infinitesimals. Instants of "flowing" magnitudes are for him

> "infinitely small increases at which a given magnitude will be increased in an infinitely small interval of time." (Ca. 1670, quoted after [193, p. 97].)

Leibniz characterizes infinitesimals for sceptics in the following way:

> "Infinite and infinitely small lines – even if they are admitted not in a metaphysical precision but as real things – can be after all uncritically needed as ideal concepts with the help of which a calculus will be made shorter as it is in the case of imaginary roots in the usual analysis."

A deviation from this which can suggest the idea of the limit value and the later $\varepsilon - \delta$ approach can be found in the declaration against potential opponents of his infinitesimals quoted above (loc. cit.):

> "[...] so our calculus shows that the error is smaller than any magnitude that can be given because we are able to decrease sufficiently to that end the incomparably small – it can be always assumed to be so small as one wants."

And still more clearly:

> "Instead of the infinite or the infinitely small one assumes so large or so small magnitudes as needed in order that the error is smaller than a given one, [...]." (Quoted after [193, p. 125].)

Hence infinitesimals are for Leibniz at least useful fictions or better idealizations or transfers from finite relationships as can be seen in the last quotation. In any case one does not need "to make mathematical analysis dependent on metaphysical disputes" (loc. cit.).

All in all for Leibniz himself the infinitesimals have the existence yielded from and in his universal rationalism. For him they are "given":

> "There are given also indeterminable magnitudes, namely infinitely small and infinitesimal ones (Dantur et quantitates inassignabiles ...)". ([232, Volume 7, p. 68].)

In a letter to Varignon from February 2, 1702, he writes (quoted after [16, p. 167]):

> "Rules of the finite keep meaning also in the infinite, as if there were atoms – i.e., elements of nature of an assignable firm size – though it is not the case because of the unbounded actual divisibility of the matter, and vice versa, the rules of the infinite are valid also in the case of the finite as if there were the metaphysical infinite, though they are not needed in reality, and the divisibility of the matter never leads to such infinitely small parts. In fact, all is subordinated to the mind and otherwise there would be neither science nor rule and this would oppose the nature of the supreme principle."

By Kant on the contrary one finds a form of transfinite atomism – in the second antinomy [199, B 462 ff.], in purely philosophical context without direct mathematical reference.

For Kant space and time are forms of pure perception that condition phenomena. He distinguishes for example a partition of a matter in space and the partition of a space itself. It is unimaginable to divide a *form* of a perception because "space is not a composition of substances [...]". And

> "if I remove all composition from it, then nothing, not even a point, might be left over; for a point is possible only as the boundary of a space (hence of a composite). Thus space and time do not consist of simple parts." ([199, B 467 f.].)

They are

> "*quanta continua*, because no part of them can be given except as enclosed between boundaries (points or instants), thus only in such a way that this part is again a space or a time." ([199, B 211].)

The situation of the matter *in space* is different. Similarly to the case of infinity there is a conflict between the only potentially infinite partition of a whole and the bare idea of treating this infinite partition as completed "whose absolute totality reason demands" [199, B 440]. In the accomplished partition there is then a first [member] in the series "in regard to the parts of a whole given in its bounds, the simple" [199,

B 446]. It is necessarily indivisible. From the simple "which one could better call the atom" the whole is composed:

> "Moreover I am talking here only about the simple insofar as it is necessarily given in the composite, so that the latter can be resolved into the former as its constituent parts." ([199, B 468].)

3.3.5 Survey of conceptions of the continuum

Our historical considerations show that there were various views concerning the continuum – now we would like to list them. We try to characterize them as briefly and concisely as possible. The list will be closed with today's conception. We shall refer to the linear continuum. To identify roughly the conceptions we name each of them. Its main representative will be indicated in parentheses. Differences between the conceptions as well as the gap with respect to today's conception will be commented at the end.

We distinguish the following views:

(1) *Finite atomism* (Democritus): A line is decomposed into finitely very small line pieces. Those line pieces are not divisible anymore. Points are the atomic line pieces. A segment is a composition of points.

(2a) *Transfinite atomism* (Archimedes, Cavalieri): A line is decomposed into infinitely many infinitely small line pieces. Those small line pieces are not divisible anymore. Points are the atomic line pieces. A line is a composition of infinitely many points.

(2b) *Transfinite atomism – a variant*: A line is decomposed into infinitely many infinitely small line pieces. Those small line pieces are not divisible anymore. Points and the atomic line pieces are objects of fundamentally different type.

(3) *Visualism* (Pythagoreans, Aristotle): Lines are unboundedly divisible. They are neither compositions of infinitely small line pieces nor of points. Parts of a line and points are objects of fundamentally different type. Between them there exists only an exterior relation.

(4) *Infinitesimalism* (Leibniz): A line is composed of infinitely many infinitely small line pieces. Those small line pieces are again divisible. Points and the infinitely small line pieces are objects of different type. Between them there exists only an exterior relation.

(5) *Trans-transfinite atomism* (Cantor, Dedekind): Lines or parts of those lines, respectively, and points are objects of fundamentally different type. Lines and parts of them are (uncountable) infinite sets of points.

Variant (2b) has been introduced because it suggests itself. We have no sure historical witnesses of it. The conception of Pascal could belong here. We are talking about "visualism" in the case of the classical conception (3) because it describes the intuitive continuum of geometry.

Between (4) and (5) there is the border between "old" and "new" continuum. The last conception (5) called the trans-transfinite atomism is just the one that is used today. It is the set-theoretical conception of the continuum dominating today. We call it "trans-transfinite" because it uses the infinity lying beyond the potential or countable infinity and was unthinkable till the 19th century. Conception (5) is a new type of atomism that implies the decomposition into elements and collecting them together and not the composition of parts as it was in the old atomism. Below we shall become acquainted with conception (6) generated by nonstandard models of real numbers – it starts from (3) and continues (4) using set theory as its basis.

As has been noticed above, Aristotle's conception has determined for more than 2 000 years the philosophical conceptions in the background of mathematics. This is true also for his conception of the continuum (3). The atomistic conceptions (1) and (2) seem to belong definitely to the past.

Another line can be drawn between the atomistic conceptions (1) and (2) and the conceptions (3) and (4). The former are in the fundamental opposition with respect to Aristotle's visualism. They assume what Aristotle excluded: the composition of the continuum of atomic components in a similar way as they treat physical bodies. In Democritus's conception it is the explicit philosophical starting point; however, in the transfinite atomism (2) it is only an indirect consequence of an aim-oriented mathematico-heuristical procedure.

Atomistic conceptions are in principle secondary accompanied by a type of set-theoretical conception of continua. However, it is false to think of collecting elements in a set-theoretical sense that would be responsible for the creation of continua as sets of atoms. One should distinguish a *composition* and a *collection*. Continua would be the primary phenomena and their composition of atoms a secondary idea that comes up. In the old atomism the most important thing is – as we see it – first of all the idea of decomposability into atoms and then the composition of atoms. In the case of the transfinite atomism this is clear. How the composition of atomic components goes on is either not considered or – as in Democritus's conception – only required in a philosophical way and vaguely described.

The infinitesimal composition (4) disagrees with the old visualism of Aristotle not in the same way as is done by atomism. It does *not* reduce the continuum to basic components. It *supplements* Aristotle's conception by a new type of magnitudes, namely of infinitesimals. The latter were a powerful invention and an ingenious step never done before. Before Cantor's discovery of the hierarchy of the infinities, the concept of the infinite has been an absolute, at the most theologically or philosophically distinguished one. The indivisibility of infinitely small line pieces was implicitly a consequence of the complete implementation of the infinite partition of a line. Infinitely small lines had to be indivisible. A further partition of indivisibles was unthinkable.

Maybe Pascal but surely Leibniz nevertheless thought about it. One observes a thought to go for the first time *beyond the infinite*. Here is the border between indivisibles and infinitesimals. Leibniz passes in another direction – quasi opposite with respect to Cantor – the limits of the infinite. It appears a bit strange when one considers how Cantor emotionally defamed infinitesimals.

Once again we see just in contrast to other conceptions how difficult it is to conceive the last version (5), the set-theoretical atomism. Old versions are in one or another way connected with the perception. Our today's conception is radically nonintuitive – it seems that we are hardly aware of it. It is incomparable with any former conceptions. Conceptions of sets can be assumed there at best indirectly – as aftereffects.

How can a continuum be considered as a set of elements? The membership relation does not connect elements with each other, just the opposite, elements are in the membership relation distinguished from each other and separated. They disaggregate the continuum. There is no composition. The continuum is decomposed.

In old conceptions the continuum was primary. It was characterized by the infinite divisibility in continua and by its connection or as a composition of atoms or infinitesimal line parts. Points as limits of continua were a secondary phenomenon. Today points as elements are primary. They are the isolated components of sets called continua.

The today self-evident idea to define all fundamental mathematical concepts by using sets and their elements became a program. The mathematical world consists today of sets. Crucial is this: A set – in order to be a continuum – must have properties that are required axiomatically (see Section 1.4). Or (transfinite) constructions are needed in order to make from a set a continuum that again will be a set. In any case a *set* is there that is decomposed into its elements – irrespective whether it is constructed or axiomatically required. In which sense this means an essential progress will be said in the next subsection.

The set-theoretical conception contradicts the classical intuition of the continuum present in the background of mathematics since Aristotle. The new continuum problem, the continuum hypothesis (cf. Section 4.4) and their independence of set theory are the consequence. This indicates the mathematical side of problems with the set-theoretical conception.

How could this set-theoretical conception of continua arise against every intuition? This could happen only by a *historical change in thinking*: the mathematical thinking has turned back from the intuition towards – as was believed – its own elements to which abstract numbers were counted. To build from them the mathematical world, the continuum became inevitably a set.

3.3.6 Notes on the arithmetization of the continuum

How did it come to the change in thinking? This problem was deeply considered and discussed (cf. [193]) and also we have hinted at it in Subsection 3.3.3. We shall try to make this change visible by compacting some events and passing over many details in many deep problems that occurred on the way to today's concept of the continuum. In further subsections one can find supplementary remarks on the fundamental turn of mathematics in the 19th century.

Continuous magnitudes inherited from the old Greeks, infinitesimal quantities of the 18th century and the intuitive geometrical background from the old *Elements* of Euclid caused more and more discomfort for mathematicians in the 19th century. For them they were conceptually unclear and had an element of "impureness" that contradicted growing claims for clarity of thought and for being more scientific. Only numbers seemed to be "pure", i.e., conceptually pure or "logical". This distinction is an old heritage of the Pythagoreans, Plato and Aristotle. By Plato numbers were related to the higher ideas, and magnitudes to the lower material things. Aristotle allocated magnitudes to the empirical world, numbers however to the soul.

Mathematics should be founded on the pure numbers. In particular there was the necessity to clarify the irrational numbers that in fact were called "numbers" but till this moment they were conceived as magnitudes in the old sense. They should really be *numbers*. Like numbers they were long-established in their role of coordinates as points on axes. The border between points, numbers and "numbers" became blurred and the conception of a straight line as a set of points has been prepared. It has been indicated above in the section devoted to the concept of a point that the conception of a straight line as a set of points played an important role in the construction of real numbers.

We make only short remarks on the major problem of the arithmetization behind which tremendous changes are hidden. We indicate the extensive bibliography and refer to further remarks in following subsections. Here our aim is to provide a survey.

Steps and stations in the arithmetization of mathematics were the following:

(a) The representation of numbers and magnitudes as segments (antiquity).
(b) The treatment of irrational numbers like numbers (Stifel (1486–1567)).
(c) The introduction of coordinates into a space since Descartes (1596–1650).
(d) The common interpretation of magnitudes and numbers as points on an axis.
(e) The conception of an axis in a system of coordinates as a set of points (clearly by Bolzano (1781–1848)).
(f) The mathematical legitimization of the actual infinite (Cantor (1845-1918)).
(g) The construction of real numbers (published about 1870).

The consequences were:

(h) The replacement of the linear continuum by the real numbers.
(i) The disappearance of magnitudes from pure mathematics.
(j) The elimination of infinitesimals.

We shall report more extensively about (i) and (j). The desired result was the arithmetized pure mathematics. With respect to the old conception of the continuum particularly striking is:

(k) The geometrical continuum became a set of points.

This consequence of (h), i.e., the *set-theoretical conception of the continuum* is the second fundamental change in the mathematical thinking of the 19th century. It is equally grave as the mathematical legitimation of the actual infinite. Just this set-theoretical conception of the continuum forces the program of set theory, the theory of the infinite.

Practical consequences were:

(l) Geometrical continua can be regarded as copies from $\mathbb{R}^n$.

(m) Points are numbers (or tuples of numbers).

In such a notional world we are living today. The linear continuum is just $\mathbb{R}$, space is an "ether" of tuples of numbers. Permanently present witness and symbol of our conception is the "number line" in which numbers and points are identified. On the base of this mathematical continuum our mathematics is being carried on extremely successfully.

In fact, the progress was tremendous. What so far stands as an impenetrable mist in the background of mathematics – the intuitive geometrical continuum – is now given as $\mathbb{R}$, as an infinite set. Properties of the continuum became explicitly properties of this set following from the construction or from the axioms. The continuum became a well-defined object and a mathematical instrument. Achievements also in applications were exceeding. Hilbert spoke about "a symphony of the infinite". This symphony was and is accompanied by underlying disharmonies, however this does not upset anybody nowadays. Those problems will be discussed in detail in Chapters 4 and 5.

Has *the problem of the continuum* been solved in this way once and forever? We shall see that this is not the case. The crucial invention of infinitesimals by Leibniz and his conception of the continuum are coming back and in a modern way arises a new view of the continuum. However, let us look first at the disappearance of infinitesimals.

3.3.7 The end of infinitesimals and the rediscovery of them

In the dynamic development of mathematics and its applications in physics in the century following Leibniz and Newton – for example by the Bernoulli brothers and by Euler – infinitesimals were broadly used. The idea of them in the background is helpful, they appear to be useful in calculations, they are handled pragmatically, often carelessly, sometimes even in a "breathtaking" way – as Israel Kleiner writes in [211, p. 79] about Euler.

Courant characterized and criticized this epoch of the analysis in the following way:

> "In the mathematical analysis of the 17th century and almost of the whole 18th century the Greek ideal of the clear and precise inference has been disregarded. The noncritical faith in the witchery of the new method dominated then." ([78, p. 303].)

Infinitesimals can be found even in the works of *Augustin Louis Cauchy* (1789–1857) who is rather known as thinking in terms of limits. They were mentioned in the same breath together with the limit values:

> "If the numerical values of one and the same variable arbitrarily decrease in such a way that they become smaller than any given number then one says that this variable is infinitely small or that it is an infinitely small quantity. Such a variable has the limit value 0." ([74, p. 3], quoted according to [235, p. 196].)

It is interesting how on the one hand a variable becomes a firm infinitesimal unequal zero what traditionally is an actual result of an infinite process and on the other hand one associates with it the limit value zero. For Cauchy this situation – that can be often seen by him – seems to be quite natural: the limit process generates infinitesimals, infinitesimals explain the limit value 0. He was accused of this attitude – one spoke about pragmatism or mathematically objectionable liberality.

Elimination of infinitesimals

Already in the 18th century there was a fierce discussion about infinitesimals – however it was led rather not by mathematicians. In it appeared epistemological and theological argumentations pointing out problems of infinitesimals. George Berkeley's *The Analyst* (cf. [52, pp. 99 ff.]) generated in England a wide debate. The basic problem one did not know how to tackle was the fact that differentials or moments ("moments" by Newton) were on the one side treated in calculations as not equal to zero and on the other hand in final results as zero. From the modern point of view this appears to be – as will be seen – a pseudo-problem.

The necessity to provide a foundation of the method of analysis became in the 19th century more and more evident. Cauchy contributed in an essential way to this task. Foundations of analysis have been provided in a such a way that went away from infinitesimals. One looked for finite instruments and they have been found just in the description of processes that before generated also infinitesimals: in processes leading to limit values. Weyl asserted in 1928 that "The limit process won" [367, p. 36] and justified this in the following way:

> "In fact limes is an unavoidable concept whose importance is not touched by acceptance or rejection of infinitely small quantities. If it is grasped one will see that it makes the infinitely small quantities superfluous." (Loc. cit.)

One has taken away the infinitesimals that stood in front of the genuine limit values and covered them. This meant the radical change of conceptions. There appeared new problems and old underlying problems became more clear: what are those infinite limit processes, how should one handle them, what are limit values themselves, what are the connections between processes and limit values? Hence the problem of the infinite appeared vehemently in the foreground and the question of the basis, the continuum, on which the limit processes are running, became louder and louder. The question concerning the status of everywhere present continuous magnitudes turned out to be important. In Chapter 1 we described the way whose beginning in the 19th century were just those questions and that led through various obstacles to the real numbers.

Infinitesimals fell victim. They have been boycotted and finally became forgotten. For example Cantor described them as "Cholera-bacillus" at the organism of mathematics. He polemised with attempts of justifying infinitesimals mathematically. At another place he even proved that if one accepted infinitesimals they would not belong to the domain of the linear magnitudes. At the end it appeared that the archimedean axiom is in fact no axiom but a theorem that follows from "the concept of a linear magnitude with logical necessity" [66, p. 409]. It is clear that his proof could not be correct. In fact, there are domains of magnitudes in which the archimedean axiom is *not* fulfilled.

Hilbert [173, p. 161] was convinced that in the new analysis defects coming from "vague ideas concerning the infinitesimals" were finally put away and in this way "conclusively overcame the difficulties that until then had their roots in the notion of infinitesimal". He speaks about "emancipation from the infinitely small" [173, p. 164].

Courant was convinced that "'differentials' as infinitely small quantities have been definitely discredited and suppressed" [78, p. 331]. Leibniz's infinitesimals led – according to him – only into "mysticism and confusion" [78, p. 330]:

> "Thus the object has been covered up during more than a century by formulations like 'infinitely small quantity', 'differential', 'final ratios' etc. The resistance to dismissing such conceptions was deeply grounded in philosophical attitude of that time as well as generally in the nature of human mind." ([78, p. 329].)

Irrational numbers as proportions of incommensurable magnitudes and infinitesimals had in the mathematics of the 18th century a comparable status. One operated with both of them in a pragmatical manner without knowing in fact with what one has to do. Courant allowed incommensurable magnitudes to be legitimate mathematical individuals (treating them as real measure numbers), however he denied such a legitimacy to infinitesimal differentials and differential quotients. The "natural psychological" inclination to have "a definition of a surface and slope as *Ding an sich* (thing as itself)" should be – according to Courant – given up. The limit process that according to him is remarkably not generating those "Dinge an sich" seemed to Courant to be the unique possible basis (cf. loc. cit.).

Even today one likes to smile at the idea of infinitesimals. In a new handbook of analysis one finds the following remarks emphasized in a box:

> "[...] it should be noticed that the founding fathers [of analysis – authors' remark] in fact imagined in their work such objects as "infinitely small quantities". [...] You are in a good company when you have problems with such an interpretation – it is nowadays hard to believe that infinitely small quantities belonged to instruments of mathematicians till the time of Cauchy and Weierstraß, hence till the middle of the 19th century. You should never (!) use expressions dy and dx as independent magnitudes. (In any case till one comes to the subject of integration [...].)" ([25, p. 237])

The restriction at the end of the quotation is a bit surprising. The symbols dx, dy etc. are used and treated only as *façon de parler*, however one does not give them even an intuitive sense that could be temporarily very useful also in mathematics applying limits. It is astonishing that today everybody believes in the Axiom of Infinity generating infinitely large sets and infinite cardinal numbers but infinitely small quantities and infinitesimal numbers are sometimes indignantly rejected. And this happens even though – as we will see soon – it has been known for a long time that those infinitely small quantities are in fact legitimate mathematical objects.

Critical or even polemical attitudes by Courant or especially at Cantor's time can be understood. In fact, at least Cauchy currently succeeded in making infinite infinitesimals superfluous by using a finite definition of the limit value and Cantor himself introduced on such a basis real numbers via infinite sequences. This was rightly treated as an overcoming of the old big problems generated not only by infinitesimals.

Hence for a considerable time one can speak also in teaching about infinitesimals in a more careful way than in the last quotation. For what has not been done or has not been attempted in the 19th century by looking for foundations of the analysis, namely to justify mathematically the handling of infinitesimals, was successfully done in the 1960s.

Return of infinitesimals

In a paper from 1958 (cf. [315]) *Curt Schmieden* (1905–1992) and *Detlef Laugwitz* (1932–2000) expanded the infinitesimal calculus by just those infinitesimals that have been eliminated. In 1961 *Abraham Robinson* (1918–1974) created by logical means nonstandard models of real numbers [302] and in 1966 published a handbook [303] devoted to nonstandard analysis. In 1978 and 1986 D. Laugwitz wrote the handbooks *Infinitesimalkalkül* [225] and *Zahlen und Kontinuum* [226], respectively. Nowadays there are many, even elementary handbooks of nonstandard analysis. Courses in nonstandard analysis have been prepared and successfully realized.

The nonstandard analysis is based not on limit processes treated by Weyl as inevitable (see above) but on infinitesimals. Instead of considering values of limits one uses the arithmetic of infinitesimals. To this end the real numbers are extended to a non-archimedean field of "hyperreal numbers" including infinitely large and infinitely small numbers. This is done by a sort of an algebraic adjunction of an infinitely large number, set-theoretically by rings of sequences or logically by providing nonstandard models of the reals. Beside infinitely large and infinitely small numbers there are in this field (more exactly: in such fields) next to "standard numbers" also nonstandard numbers at infinitesimal distance.

The conception of an infinitely small quantity is by this arithmetical background not only admissible and useful but mathematically legitimate. Elements of this conception and of the arithmetic of infinitesimals can be without any danger integrated into propaedeutics of the modern analysis. However, one should not be afraid of infinitesimals as it was the case in the handbook quoted above.

In the sequel first remarks concerning the introduction of infinitesimal numbers will be done. More will be said in Chapter 6 where we will calculate with infinitesimals. A propaedeutic introduction – using infinitesimals – to first elements of analysis will be also presented. The latter has some advantages with respect to the usual introduction. We explain also the construction of hyperreal numbers and the extension of real functions as well as discuss problems and backgrounds.

To close this subsection we want to make some general remarks on the recurrence of infinitesimals. In this respect we point in particular to the handbook [226].

The foundations of analysis have been provided for 150 years by limit values. Infinitesimals are present in symbolism but they disappeared from thinking. Since then analysis is practiced and presented everywhere as mathematics of limit values. This mathematics of limits values stood the test of time and this must be taken into account even when one is of the opinion that the infinitesimal mathematics could be the better way to analysis. Books and papers are written in terms of limit-value mathematics and teaching should prepare pupils and students to work with them. Laugwitz asked also critically [226, p. 236]: "Was the resumption of infinitesimal mathematics in the second half of the 20th century justified?"

An optical but important justification can be seen in the infinitesimal notation that is used today in a similar way as it was used in Leibniz's time and that certainly nobody can and will change. Laugwitz notices:

> "The need to justify and to found the infinitesimal notation not only as notation for derivative and integral but also as a meaningful representation of mathematical objects makes concessions to needs of its users." ([226, p. 242].)

Laugwitz discusses strengths and weaknesses of the infinitesimal mathematics. Its advantages that made infinitesimals and "clandestine" calculations with them alive for a long time are obvious. The introduction of infinitesimals and their arithmetic –

which needs only few efforts (cf. loc. cit. and [226], as well as Chapter 1 and Chapter 6 in this book) – can provide insights in teaching and study. Think about the concept of the continuity, about differential quotients and integrals, which by using infinitesimals can be made easier to understand and more available arithmetical objects.

The latter indicates reasons of doing infinitesimal mathematics that are associated with needs of thinking. Infinite processes are difficult to understand and to grasp just because they are infinite. They do not lead to no manageable "Dinge an sich" as Courant said (see above). Let us quote (cf. [367, p. 10]) a remark of Leibniz from a letter and change the words only at one place that we indicate by italics: "However, mind is not happy with limit *process*, it looks for an item, for a thing that would be truthfully identical with it and it imagines it as something outside the subject." Infinitesimals could be such objects.

A new, wider acceptance of infinitesimals would provide a possibility for everybody to make a decision whether one thinks in terms of limit processes and limit values or in terms of infinitesimals or both – according to requirements – as Cauchy did. This seems to lead to some difficulties when seen superficially. We will tell below and in details in Chapter 6 about the options $0.999\ldots = 1$ and $0.999\ldots < 1$ that arise depending on whether infinitesimals are allowed or not.

Further reasons for the enlivenment of the infinitesimal mathematics that can be indicated here rest in itself. It is an important mathematical discipline, it makes the repertoire of mathematical methods, concepts and expressions richer as well as extends and changes the mathematical thinking, it is important with respect to applications and foundations of analysis and finally contributes to the understanding of the historical mathematics.

Laugwitz shows how the ways of thinking of Leibniz and his students as well as mathematicians of the following centuries can be legitimated and how controversial methods used by them could be justified. An example: Laugwitz shows [226, pp. 77 ff.] that the addition theorem of Cauchy on continuous functions can be seen as correct in the infinitesimal mathematics (though from the point of view of limit-value mathematics it is often qualified as false). One accused Cauchy of not knowing the uniform continuity – though he argued in his proof in an infinitesimal way. However, for theoretical needs the necessity of the definition of the uniform continuity comes no longer into consideration on the basis of an infinitesimal definition of continuity.

The infinitesimal, "qualitative" definition of continuity is extremely more intuitive than the usual "quantitative" ε-δ-definition – provided one is ready to imagine infinitesimals: f is continuous if an infinitesimal change of x produce at most an infinitesimal change of $f(x)$. This definition of "Cauchy continuity" comes from Cauchy.[11]

---

11 Laugwitz describes the ε-δ-definition of continuity as "quantitative" because it is based on a quantitative estimation by real "measure numbers" and the infinitesimal definition as "qualitative" because it refers to other, qualitative aspects of the continuum and it is sufficient for "pure" mathematics

The seemingly paradoxical phenomenon of infinitesimals consisting in treating them once as not equal to zero and at another time as equal to zero can be easily solved. Consider for example the function $f(x) = x^2$ and calculate the differential quotient

$$\frac{dy}{dx} = \frac{(x + dx)^2 - x^2}{dx} = 2 \cdot x + dx.$$

How can it be said that the *derivative* at x is $2 \cdot x$? From the point of view of measuring dx plays in fact no role since it is infinitesimal, and $2 \cdot x + dx$ is infinitely close to the real number $2 \cdot x$. Distinguish the hyperreal differential quotient and the real derivative. This is the simple solution to the polemic philosophical-theological debate of the 18th century we wrote about.

It should be noted that the actual foundation of the infinitesimal mathematics is relatively complex and assumes elements of logical as well as strong set-theoretical and algebraic instruments. This shows that mathematical foundations of the infinitesimal mathematics were in fact "improbable" at Cauchy's time, i.e., even more far away than the concept of limit and the construction of real numbers. There were no set-theoretical and algebraical methods that only then were forced by limit-value mathematics and developed in the following decades.

Does the infinitesimal mathematics belong to teaching subjects? In our opinion the answer is positive in the case of propaedeutic and by introducing elements of analysis. Providing foundations is – as we said – difficult. Laugwitz explained it shortly as:

> "Not everything that one can teach should be taught." ([226, p. 242].)

The propaedeutic maxim is already applied when one does not teach the construction of real numbers. The actual foundations need methods indicated above and they belong to further courses in nonstandard mathematics. However, this mathematics should not be left unsaid or even defamed. An occasion to speak about it is provided when dx, dy etc. occur that otherwise remain unnecessarily mysterious.

Deeper reasons to teach and learn elements of the foundations of the infinitesimal mathematics are fundamental and of the philosophical type – they are connected with the concept of the continuum and of a point as well as with the conception of a magnitude. We speak about this in the following two subsections below.

[226, pp. 29 ff.]. There is a problem of ε-δ-continuity of a given Cauchy-continuous function f. It consists of the extending of real functions to hyperreal numbers – to solve it one introduces so-called 'inner" sets of hyperreal numbers and "inner" functions (cf. Chapter 6) and the consideration of extensions is limited to them.

3.3.8 Nonstandard numbers and the continuum

The nonstandard analysis is based on nonstandard models or set-theoretical extensions of real numbers. Fields arising here are called *hyperreal numbers*. The logical approach to nonstandard models was used by A. Robinson 1961 in the work *Non-standard Analysis* mentioned above and in the following handbook *Non-standard Analysis* from 1966. Schmieden and Laugwitz used another approach. In the paper *Eine Erweiterung der Infinitesimalrechnung* (1958) they brought back again the "infinitesimal calculus" (as it was and is still called), the lost infinitesimals. Their method was *set-theoretical*. Laugwitz proposed in [226] beside the set-theoretical also a logico-algebraic approach and compared both of them.

The logical approach proposed by Robinson is formal. It begins with nonstandard models of real numbers. Such models – there are many of them – do exist as a consequence of the completeness theorem ([135], cf. Section 5.4) and model-theoretic theorems following from it. Their existence is only "formally" justified on the base of the first-order logic – in a similar way as nonstandard models of natural numbers. (They were studied already by Skolem in the 1930s – cf. Section 5.4). From the mathematical point of view the application of those models is of interest to complete and to extend the classical analysis as well as the possibility of an alternative analysis in general. The reference of nonstandard models to the arithmetical continuum of real numbers from which they come is primarily arithmetical. Logical nonstandard models seem to be far from the intuitive geometrical continuum.

Also the logico-arithmetical approach developed by Laugwitz in [226] does not come organically from the intuitive continuum. It is the matter of a sort of adjunction of a transfinite element that seems to blow up arbitrarily the intuition of the continuum. We make below a remark concerning this.

It is different in the case of the set-theoretical construction described in [225, 226] and since then used in many handbooks of nonstandard analysis (see, among others, [10, Chapter 7], [101, pp. 211–233] and [224, 350]). One begins here with $\mathbb{R}$ and considers arbitrary sequences of real numbers. There arise infinitesimals and infinite, infinitely large, numbers in a similar way as real numbers do from sequences of rational numbers.

To present it in a vivid way let us choose a popular example. The following question is asked over and over again:

Is (a) $0.999\ldots = 1$ or (b) $0.999\ldots < 1$?

Though (a) is permanently asserted and naturally proved, possibility (b) does not disappear. Why? It is connected with the problem of the actual infinite.

It is generally difficult for a non-mathematician or a mathematical beginner to connect actually the limit value 1 that in fact belongs to $0.999\ldots$ with $0.999\ldots$ because it is assumed here that the series, i.e., the infinite sequence of partial sums $(0.9; 0.99; 0.999; \ldots)$ should be thought to be closed and given as a whole. A mathematician has simply learned or adopted this. A mathematical beginner cannot imagine the finished and complete series, the actual infinite.

A beginner is doing something naive – however not false. He compares every *single* member of the sequence with 1. This means that he compares the – potentially or "restricted actually" (cf. 3.2.6) infinite – sequences $(0.9, 0.99, 0.999, \ldots)$ and $(1, 1, 1, \ldots)$. And in fact the first sequence is in a sense smaller than the second one. Their difference $(0.1, 0.01, 0.001, \ldots)$ can be specified and it is a candidate for something infinitely small.

Hence a mathematical beginner speculating about $0.999\ldots$ operates with infinite sequences. He compares – seemingly in a naive way – those sequences instead of thinking about limit values. He goes beyond real numbers remaining intuitively in the linear continuum in which real numbers are represented by points. However, in the continuum there is apparently – beside real numbers – also place for new numbers, for example numbers between $0.999\ldots$ and 1. If we consider both numbers as sequences then $(0.95, 0.995, 0.9995, \ldots)$ lies between $(0.9, 0.99, 0.999, \ldots)$ and $(1, 1, 1, \ldots)$ – it is the arithmetic mean of them. The difference of $0.999\ldots = (0.9, 0.99, 0.999, \ldots)$ and $(1, 1, 1, \ldots)$ is the sequence $(0.1, 0.01, 0.001, \ldots) > 0$. It is infinitely small, namely smaller than $\frac{1}{n} = (\frac{1}{n}, \frac{1}{n}, \frac{1}{n}, \ldots)$ for any $n \in \mathbb{N}$. Those relations between such sequences remain valid when one moves to hyperreal numbers. In Chapter 6 we consider such questions in detail.

Hence alternative (b) of the above question is not false. It is nonstandard – and different than taught today.

Sequences of rational or real numbers – now of course grasped as *actually* infinite – form a ring. This ring, more exactly a factor ring of it, is the starting point of the set-theoretical construction of a non-archimedean field with infinitely large and small elements described by Laugwitz in [225] and [226, pp. 91 ff.]. This construction will be presented in Chapter 6 entitled "Thinking and calculating infinitesimally". There arises a field of "hyperreal numbers" – with infinitely large and small numbers, among them ω and Ω given by the sequences $(\frac{1}{1}, \frac{1}{2}, \frac{1}{3}, \ldots)$ and $(1, 2, 3, \ldots)$, respectively.

Notice already here that the construction requires the Axiom of Choice – hence the evaluation of nonstandard numbers seems to depend on it. Later we say more about it.

The set-theoretical construction has a very concrete and intuitive beginning being immediately close to the continuum. Laugwitz says the following:

> "For me, and this should be repeated, the continuum is not identical with the set $\mathbb{R}$, [...]". ([226, p. 223].)

And:

> "We have seen that the intuitive linear continuum leaves place for the Omega numbers and does not need to be regarded as being exhausted by real numbers." (Loc. cit.)

The set-theoretical construction described in Chapter 6 begins with sequences of *real* numbers and leads to hyperreal numbers. An analogous construction that uses only sequences of *natural* numbers leads to the "hypernatural", nonstandard natural, numbers $^*\mathbb{N}$. Beside standard natural numbers represented here by constant sequences $(n, n, n, \ldots)$ one has here infinitely large "hypernatural" numbers. The "simplest" such number is the number Ω mentioned above represented by the natural sequence $(1, 2, 3, \ldots)$.

3.3.9 Consequences for the conception of the continuum

Let us notice at the beginning that the term "nonstandard" used by Robinson in [302] is a bit unfortunate. It could suggest that one has to do here with something that is not standard, hence extraordinary, not natural, perhaps mathematically absurd or even "not normal". But this is not the case. The new nonstandard conception of the continuum that arises by nonstandard models of real numbers is as natural as the conception of sets being standard today. No! It is "more natural", namely it is more intuitive than the set conception of the continuum. It is tied with the old aristotelian, classically geometric conception of the continuum and extends it by infinitesimal quantities as Leibniz did. Has it been only the caprice of the history that nonstandard is not standard? Above we already speculated about this and will come back to it later.

New view of the continuum

In fact, for us standard mathematicians it is a certain effort to draw consequences for the new conception of the continuum from the indicated constructions or logical models. Our thinking habits should be overcome. The latter are determined by $\mathbb{R}$. Today $\mathbb{R}$ *is* the continuum from which all other continua are derived and which makes a line to become the number line.

What makes thinking in terms of nonstandard objects so difficult? Certainly not the real numbers themselves, it is rather the impact of them that made a line to be the number line and consequently a set of points, an *a priori* given finished set of points. This hypothesis of the set of points became such a mathematical fact that one finds it difficult to question it. If one does not experimentally drops this primary *continuum hypothesis*, as we described it at the very beginning of this section, it will be impossible to accept and understand the nonstandard approach. We do not mean here the mathematical constructions described and indicated above but the geometrical intuitive starting point. Consider the following words standing at the very beginning of the work by Schmieden and Laugwitz [315]:

> "The concept of Cantor-Dedekind real number and consequently the whole today's analysis are based on the assumption that the same limit value should

> be ascribed to two sequences of rational numbers whose members finally differ at less than ε, where ε is arbitrary rational number. Dedekind's approach is equivalent: any "cut" in the domain of rational numbers is generated by unique real number.
> One can admit that there is an element of arbitrariness in this. Consider a cut, i.e., a partition of rational numbers into two classes, [...], then – as is well known – Dedekind justifies the postulate of the existence of a "real" number separating both classes by the 'continuity' of the number line. Another unjustified assumption made by Dedekind is the assumption that for every cut there is unique such separating element." ([315, p. 1].)

Consider the last sentence. If one assumes the linear continuum as the set of points then it is false. In fact, in the problem of the uniqueness of a separating number identified with a point the issue is not the unjustified assumption but a necessary consequence. How can one think about two points that divide *one* and the same cut of the line into equal parts?

Hence one should accept again, at least temporarily, the old classical conception that the linear continuum is *no* set of points.

The new situation generated by nonstandard numbers looks like this:

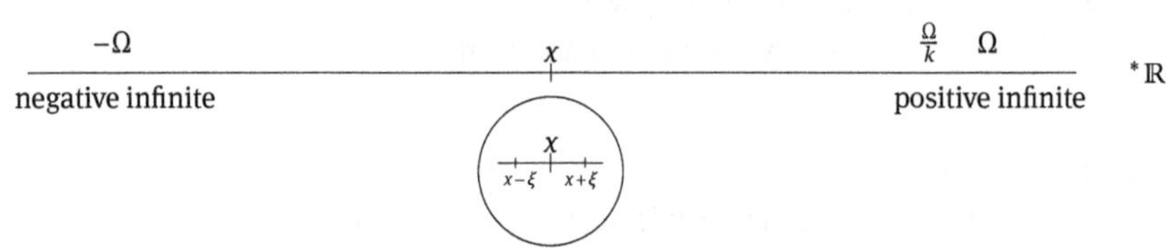

Fig. 3.7. Infinite numbers and infinitely close numbers on the „number line".

One obtains a longer, a "long" number line with infinitely large positive and negative numbers and a "more dense" number line on which every real number has an infinitesimal neighborhood of nonstandard numbers. We shall soon see that one must resign from a usual conception of a "number line", i.e., of a line captured by numbers. In the picture x is a real number. In an infinitesimal neighborhood, in the so-called monad of x one finds nonstandard numbers $x + \xi$ infinitely close to x. The intersection of two monads of two different real numbers is necessarily empty.

What does it mean for example for the old conception that a point that we put and that represents the number x does divide the line? – Every infinitesimally neighboring nonstandard number $x + \xi$ divides the line "in almost the same manner"! This means that one cannot distinguish both numbers quantitatively, hence by *measuring* – and similarly this holds for the points representing them. Consequently, – as we experimentally illustrated – it is not a point but rather an infinitesimal continuum, in which again there are points, a neighborhood that divides the line. Or – formulated in a slightly paradoxical way – seen in a nonstandard way a point is a symbol for the conception of an infinitesimal neighborhood of points on the line.

However, there arises another difficulty: even if we mathematically smile at Euclid's definition stating that a point is that which has no part, it is in fact somehow present. The new infinitesimal conceptions come into conflict with our Euclidean approach. One should think now in the following way: a point is a sign for a point together with a monad. Such a point has possibly "parts" – for example infinitesimal intervals – and the latter are not nothing.

The conception of a point became fluent: a point is an infinitesimal neighborhood of points that again are neighborhoods of points that again etc. This reproduces the potentially infinite extensions of real fields of nonstandard numbers. Every such field of numbers is – like $\mathbb{R}$ – a *model* of the continuum and none of them *is* the continuum. Real points are followed by hyperreal, the latter by hyperhyperreal etc. Every of those points, also the "first", the real point, is a snap-shot.

The indicated conceptions supplement our quantified way of thinking that is stamped by the idea of measuring and is documented by $\mathbb{R}$. The new infinitesimal conceptions are extensions of the old quantitative ones. However, they are not easy for our intuition stamped by magnitudes.

It is obvious: Cantor and Dedekind started from the intuition that a line is a set of points. The aim to fill the line by numbers became a precondition that to every point there correspond a number. Infinitesimal quantities were excluded from the very beginning. This explains also the highly emotional polemics of Cantor against infinitesimals.

What are in fact the essential elements of the nonstandard conception of the continuum?

One should imagine infinitely small and infinitely large numbers and nonstandard numbers that are infinitely close to standard numbers. One must – and this is especially mathematically difficult – take leave of identifying the linear continuum and $\mathbb{R}$. One must – and this is perhaps even more difficult – return to the old intuitive geometrical continuum. In the geometrical continuum, in the continuum of lines, one represents new numbers again as points – as the "old" real numbers. One should accept that there is in the continuum beside and after the real numbers still "place" for new numbers, a place that has been earlier rejected by identifying the continuum with $\mathbb{R}$.

One should take leave of the way of thinking and speaking about the "number line" in which points and real numbers are identified and consequently other numbers are eliminated. A "longer" and "more full" number line must be accepted. *Even more*: according to our experience one should be and one will be careful in identifying again the new numbers with *the* points of the geometrical continuum to get a new "number line". We shall speak about points *in* the continuum and should take leave of the conception that the continuum is a set of points and the line consists of numbers. One cannot speak any more about "all points of the line".

The new situation has many advantages. In fact, the real numbers did not disappear. All we know about real numbers can be entirely transferred to nonstandard numbers: all theorems about real numbers hold in the domain of nonstandard numbers that is a model of real numbers. Mathematics becomes richer by this – *external* – distinction between standard and nonstandard numbers – it is even more comfortable because one has now infinitesimal and infinite numbers. This makes many things simpler and available that otherwise could be described only in the language using limit values.

If we had learned early to think in terms of nonstandard models they now would not be alien to us. The nonstandard models are called "nonstandard" only because we are usually thinking in another way. It was the destiny, possibly inevitable destiny that we have learned just in such a way. The constructions indicate that the set-theoretical, algebraic and logical assumptions of them were not given or maybe more away than the concept of a limit value and the real numbers when solid foundations of analysis were provided in the 19th century. It was just the concept of a limit value that stood at the very beginning of the development that led to the real numbers and to the theory of the infinite, to set theory. The logical assumptions were formulated in the first half of the 20th century when a new axiomatic approach had been developed.

3.3.10 Medium of free evolution

A first look into the history of mathematics showed us the intuitive geometrical continuum replaced in the 19th century by the real numbers. Being richer at some insights we now again come back to the "old" continuum. We have indicated how in the history in many different ways one tried to characterize the continuum or even to fix it. We came to the conclusion that it cannot be arithmetised or grasped as a set. It is and remains a medium for points that can be put into the continuum and a medium for presenting numbers – standard or nonstandard.

Laugwitz distinguishes qualitative and quantitative aspects of the continuum. Real numbers represent its quantitative aspect. They arise in the continuum by the transfinite extending of concrete quantifying measuring. This is instantiated in the constructions of real numbers as classes of nested intervals or as fundamental sequences. Real numbers are the *universal measure numbers.*

However, the continuum has also qualitative properties that go beyond the purely quantitative ones in the most genuine sense and that are demonstrated by infinitely large and infinitely small quantities. They are implemented by the set-theoretical constructions of the infinitesimal and infinite numbers. The continuity of the linear continuum is described in a new way by what is infinitesimal. This makes it also possible to grasp in a qualitative way for example the continuity of functions f. The continuum cannot be grasped by pure measuring.

The continuum is – as nonstandard models tell us – in the truest sense of the word measureless.

The continuum can also not be grasped by the set-theoretical or quasi-algebraical constructions of hyperreal numbers. This is shown for example by the construction using the logico-algebraical adjunction that can be repeated. In fact, the obtained nonstandard field has all properties of the initial field whose model it is. Standard and nonstandard numbers can be distinguished only "externally". The continuum *cannot be exhausted* by infinitesimal and infinite numbers.

This is clearly demonstrated by the nonstandard conception of the continuum based on the logical foundations – it shows even more. Logically seen it is the matter of nonstandard models $^*\mathbb{R}$ of real numbers. One nonstandard model is as good to the same extent as any other.

– There are nonstandard models of *arbitrary large uncountable cardinality,*

bigger than the cardinality of the real numbers [237, p. 239, Example 7.3.11.3]. The continuum is – from this point of view – not only measureless, rather

– the continuum is *inexhaustible.*

Atomism that began as finite, became transfinite and nowadays determines the mathematical everyday life set-theoretically as trans-transfinite seems to be not adequate with respect to the phenomenon of the continuum. It can be said:

– Any type of atomism fails to grasp the continuum.

Summarizing considering the situation on a straight line, we obtain:

– $\mathbb{R}$ is *not* the straight line.
– Not all relations in the continuum can be approached by measuring – even not by transfinite extensions of them.
– The linear continuum is *measureless.*
– There are nonstandard models of $\mathbb{R}$ of *arbitrary large uncountable cardinality* – beyond the cardinality of the real numbers.
– The linear continuum is *inexhaustible.*

Aristotle was right:

– The linear continuum is *not* a set of points.

The mathematical consequence of the existence of nonstandard models is:

– The number line is a theoretical entity.
– $\mathbb{R}$ is a *model* of the geometrical straight line – among others.

There is however no doubt that this model is extremely effective.

There remains the old intuitive and geometrical continuum. We will not get rid of it. It is the whole time present as an incomprehensible "space sauce" at the background – as said by Brouwer (quoted after [16, p. 346]). Weyl expressed it in a more philosophical way: the continuum is a "medium of free evolution" (*Medium freien Werdens*) [366, p. 49]. In fact, the continuum is not graspable as additionally indicated by problems connected with the set-theoretical Continuum Hypothesis (cf. Section 4.4). It remains – also mathematically – transcendent.

3.3.11 The disappearance of magnitudes

The matter of this subsection is the disappearance of magnitudes in pure mathematics. In applied mathematics as well as in mathematical physics they are still present as they used to be. However, the concept of a magnitude is reduced to the real numbers (cf. [152]) that have adopted their role. Magnitudes are not fundamental notions anymore. Now we shall tell how this happened. We will also speak about circumstances that were directly connected with the problem of the continuum.

From the moment when Eudoxus and Euclid replaced the Pythagoreans' numbers by magnitudes till the middle of the 19th century mathematics was the science of magnitudes. At the background of the continuum the basis was formed by continuous magnitudes. However, there still was the old desire for numbers. And it were and remained just the natural numbers that had the reputation and brightness of purity and abstraction. Only they seemed to be appropriate for the genuinely pure mathematics. Since the Greek mathematics connected the concept of a number with the concept of the natural number, for a long time it was unimaginable to extend them to rational and then to real numbers – as genuine numbers. A Greek substitute for the rational numbers were proportions of natural numbers, and for the real numbers proportions of geometrical magnitudes. Mathematical knowledge was expressed in the form of equations of proportions. Leibniz criticizes in a letter to Clark the approach of Euclid

> "who – being not able to define properly the concept of a geometrical proportion in the absolute sense – determined what should be understood under equal proportions." (Quoted after [367, p. 10].)

The mathematical pressure grew as there appeared problems connected for example with the necessity of constructibility of proportions in the old algebra. The quotation says that this algebra was able to recognize proportions of two magnitudes – being commensurable or not – as equal. However, there was no possibility to determine the quality in the case of incommensurable magnitudes "in an absolute sense", i.e., through an independent mathematical object. As far as one speaks about proportions of magnitudes one thinks in fact about the principle of measuring magnitudes. When they were incommensurable there was no common measure and no measuring procedure that could attach a measure number to the magnitudes. As a consequence many problems had to remain unanswered – for example the Delian problem of doubling the cube because one was not able to give the length of the segment $\sqrt[3]{2}$.

Leibniz describes the problem of the method of proportions and the growing requirement of mathematical individuals, of new numbers, in the following way:

> "However, the mind is not satisfied by this accordance, it looks for identity, for an object that would be truthfully the same and it imagines it as being outside of the subject." (Loc. cit.)

This problem existing for thousands of years and becoming more and more urgent with the development of mathematics required new numbers whose admission was in fact controversial. For a long time there has been a type of a transitional stage, a correspondence between numbers and magnitudes. Once they were rather numbers – when representable as proportions of numbers, hence rational – at another time rather magnitudes – when proportions were irrational. Here is for example the statement of Stifel (1486–1567) symptomatic who helplessly oscillated between the admission and rejection of irrational numbers:

> “It is rightly disputed whether irrational numbers are genuine numbers or fictitious ones (*ficti*). Since irrational numbers appear to be useful in proofs concerning geometrical figures and rational numbers fail [...], we are induced and forced to admit that they do really exist, namely by effects they bring and which we sense as real, safe and established.
> However, other reasons induce us to the contrary claim, [...]. [...] An entity that is not exact enough and that has no relation to genuine numbers cannot be called a number. As an infinite number is in fact no number, so an irrational number is no genuine number because it is hidden by the fog of the infinity; the relation of an irrational number to a rational one is as indefinite as the relation of an infinite and a finite one.” (Quoted after [134, p. 245].)

If it concerns proportions of magnitudes and of measuring then rational numbers are used in practice. The latter got in the history their status of numbers as proportions of natural numbers and that they suffice for applications. However, geometrical phenomena and mathematical problems required real numbers – with emphasis put on “numbers”. Where should they come from? The basis of natural numbers was not enough. Sequences of rational numbers, i.e., of pairs of natural numbers, that go to irrational magnitudes were not comprehensible. They disappeared in the infinite. They could be understood only as open processes and not as proper mathematical individuals.

The requirement of real numbers was in fact the requirement of a new foundation that will be able to grasp the infinite. The potential infinite could not remain potential. The conception of the infinite as a whole, as a proper mathematical object was requested but not at all suspected at the beginning of the 19th century. It was Georg Cantor who risked the step towards the actual infinite – we wrote extensively about him and his great work. It was not a step in an evolutionary development of mathematics, it was a revolutionary step.

Together with infinite sets and transfinite cardinals, sets in general and the concept of them entered clearly into mathematics. It became a mathematical task to justify them. The first steps in what today is called set theory was made by Dedekind in his famous work *Was sind und was sollen die Zahlen?* [87] in 1888. Natural numbers – so far being an uncontroversial base of *pure* mathematics – have been given a base by sets or "systems" as Dedekind used to call them. Frege provided a logical base for them in [123] where he described sets by using concepts and their extensions. Just the parting from the base of natural numbers caused a heavy resistance that via Kronecker led to intuitionism.

The excitement caused by Cantor's step is clear and understandable here. It meant – and this was more or less clear for any mathematician at that time – in effect the cancelation of the foundation of the whole mathematics accepted so far: the rejection of fundamental magnitudes. The ground on which one stood was in danger – not so much methodically but rather "ideologically". Conceptions and patterns of thinking began to totter.

The young Hermann Hankel, who did also deep research on the historical development of mathematics, published in 1867 a book about new higher number systems (complex numbers and Hamilton's numbers) that theoretically blew up the classical concept of magnitude. However, at that time, when set-theoretical constructions of real numbers and the formulations of the completeness had not been published yet, he did not think about a resignation from continuous magnitudes.

His attitude is probably typical for the majority of mathematicians of those years. Infinite sequences used to define irrational numbers were for him

> "in their completion incomprehensible" and led "at any time to a contradiction". ([154, p. 59], quoted after [117, Volume II, p. 163].)

"Such an instrument" to grasp irrational numbers – and Hankel was convinced about this –

> "can be provided only by geometry through its operations on magnitudes independent of any numbers but only if the concept of continuity – in which is hidden the contradiction – is seen as given. The pure thinking, free of any intuition is not able to capture the infinite and the formal number theory – the irrational." (Loc. cit.)

How strongly the conceptions of the infinite and of infinite sequences have been changing is indicated by Hankel's encyclopedic contribution [155] from 1871.

Frege refused later firmly to accept for example Dedekind's construction of real numbers. He held on magnitudes and tried to introduce real numbers on the logical base as their proportions. Griesel writes about this in [152]. The attempt remained unfinished.

How difficult it was to resign from magnitudes is also documented by correspondence between Dedekind and Lipschitz concerning Dedekind cuts (cf. [340, pp. 86 f.]). For R. Lipschitz (1832–1903) the real numbers seemed to be completely unimaginable without geometrical ideas and he treated the rejection of the Euclidean position concerning magnitudes done by Dedekind as superfluous, incomprehensible and not as going beyond the Euclidean theory of magnitudes. However, Dedekind was conscious that he had done a historically important step and used all his powers and all mathematical and historical arguments to convert Lipschitz.

Dedekind presented in proper words the latent idea "away from magnitudes and go towards real numbers" in the preface to his work *Stetigkeit und irrationale Zahlen* [85] in 1872 and he recalled there a lecture on differential calculus in 1858 after which he already drafted Dedekind cuts.

> "For myself this feeling of dissatisfaction was so overpowering that I made the fixed resolve to keep meditating on the question till I should find a purely arithmetic and perfectly rigorous foundation for the principles of infinitesimal analysis." ([85, p. 4], English translation on p. 2).

It seems that he knew how important the mathematical step done by him was. In fact, he gave the exact date of his discovery – November 24, 1858, hence fourteen years before the publication of [85].

The fundamental aim of Dedekind was to introduce irrational numbers by which incommensurable magnitudes can be measured and which would be genuine *numbers* and whose concept "is not based – as used to be before – on magnitudes" [85, p. 9]. This old correspondence between numbers and magnitudes – being only geometrical – "could not claim to be scientific" for him (loc. cit.).

His justification then could not be only arithmetical. Natural numbers and rational numbers did not suffice. The arithmetical justification demanded a new thinking and new concepts, new foundations. His construction was – similarly to those of Cantor and Weierstrass – set-theoretical and transfinite. The matter were actual infinite sets. He considered – already in 1858, long before Cantor's defense of the actual infinity – the domain of rational numbers as an actually given set. Using it he constructed via cuts – which again consisted of actual infinite sets of rational numbers – the domain of real numbers – again as an actual infinite set whose elements correspond to points of the straight line. Speaking at the beginning about continua we noticed how big the step from the idea of a continuous straight line to the straight line as a set of points was.

Cantor, for whom anything else than the conception of a straight line as a set of points was mystical, obviously assumed exactly this in his construction of real numbers. He never doubted that by real numbers all points of a straight line are taken. The same holds for Dedekind – however, he clearly and constantly *distinguished* points on a straight line and numbers. Real numbers were for him mental creations generated by cuts of rational numbers. They *correspond* to points on the straight line.

Dedekind attempted to make the transition from the intuitive straight line to the number line plausible. It is interesting how this was done. The effect of his approach like that of Cantor's should be the elimination not only of the finite magnitudes but also – from the very beginning – the infinitesimals.

Dedekind anticipates – as Cantor did in his conception of a line – the concept of a linear continuum as a point-set. He speaks about “all point of a straight line” [85, p. 10]. For him this conception is a part of “the triviality of the secret of continuity” [85, p. 11] of a straight line as he imagines it.

It is a hypothesis that a straight line is a point-set. This has been noticed at the very beginning of this section. It became mathematical reality when the real numbers have been constructed and it was *postulated* that real numbers capture all points of a line. For how should the set of points of a straight line be formed other than by coordinates? The required result $\mathbb{R}$ that fills the line influenced beforehand Dedekind's geometrical conception. A line filled by numbers – represented by points – was presupposed for the purpose of his idea of cuts. This conception of a straight line as a point-set led for Dedekind inevitably to the statement that there is a unique point that produces a cut of the line: “there exists one and only one point which produces this division of all points into two classes” (cf. [85, p. 10], English translation on p. 11). This uniqueness makes infinitesimal quantities impossible from the very beginning because points being infinitesimally close are unimaginable. This was explained above.

We repeat: the conception using sets contradicts the intuitive conception of the continuum. It considers the continuum as consisting of isolated elements. This is just – as noticed above – the crucial and revolutionary step. Properties of the continuum being beforehand hidden in magnitudes become – *just thanks to this* – explicit. In the set-theoretical *reconstruction*, in this attempt to connect again what was decomposed, as points connect parts of a decomposed straight line, important properties become first apparent. How was this said by Courant? It was, it is a necessary mathematical destiny to make this vast step towards a set-theoretical conception if one wants to understand properties of the continuum. However, he indicates also some gaps that remain even after this giant step towards a set-theoretical approach.

This is the same destiny that a bit later (1888) caught natural numbers. It was again Dedekind – we wrote about this (cf. Section 2.15) – who gave them a certain set-theoretical frame. Through this, principles of the series of natural numbers hidden previously in intuitive temporary counting processes became explicit.

Moreover, $\mathbb{R}$ became the universal domain of measure numbers for magnitudes. Its properties, in particular the archimedean property, are generated – in processes of extending – by natural numbers via rational numbers, which primarily was governed by needs of measuring. Laugwitz [226, p. 18] stresses and emphasizes in italics:

> “*The archimedean property that excludes both infinitely small and infinitely large numbers is a consequence of properties of a process of measuring and therefore one of properties of the real numbers of the calculus inherited from measure numbers.*”

The archimedean property of the reals has an obvious but in no way necessary or even logical retroaction on our concept of magnitudes which we usually consider measured and in the same way ordered. But if we take impartially the position of continuous magnitudes the infinite small and infinite large is *not* excluded *per se*. If one allows the non-archimedean conception different magnitudes may get the same real measure if the difference is infinitesimal. Infinite small means *unmeasurable* small.

However, the retroaction was completely different and stronger. When after all set-theoretically founded real numbers were well-established in mathematics the retroaction entered definitely. They eliminated the intuitive classical continuum from mathematics and took its place. The new set-theoretical conception of the continuum seemed to be the liberation of analysis from the geometry that so far could not be separated from the intuitive continuum. The arithmetization of mathematics had begun by Descartes by algebraic "numbers" as coordinates took its course. Geometry now moved to number spaces generalized to spaces of arbitrary and infinite dimensions on the base of set-theoretical formation of concepts and on the new algebra.

Also magnitudes became necessarily the victims of the elimination of the intuitive continuum and the arithmetization of mathematics. The way – that long ago in the Greek mathematics led from numbers to magnitudes – led now back to numbers. The concept of a number became definitely separated from magnitudes. The latter were on the fringes in applied mathematics. It is not plausible that they will sometimes turn back into pure mathematics as fundamental entities.

However, the philosophical questions left when excluding the magnitudes and the intuitive continuum from the foundations of mathematics were and are unanswered. As at the turn of the 19th and 20th centuries there were often fierce disputes, especially challenged by Intuitionism, for a long time as well as nowadays there prevails silence. M. Epple comments in [107, p. 410] the disappearance of magnitudes from analysis as follows:

> "The elimination from mathematical research and teaching of any philosophical discussion that broke out at the end of the theory of magnitudes meant and means till today the elimination of a fundamental and multipolar dissent from the modern mathematical practice."

Looking back it seems to be appropriate that a new discussion should begin. It receives new impulses by the return of the infinitesimal magnitudes and the cancelation of the identification of the continuum with real numbers in nonstandard models of reals.

3.3.12 Final remarks

We started by our fully intuitive conception of the continuum and have observed that there were various attempts to understand the continuum. We saw that it was at the beginning coherent, unboundedly divisible and not consisting of simple parts. Later it was decomposed into finite or transfinite atoms and thought of as being formed by them. Points were atoms. However, the history has been dominated since Aristotle and Euclid first of all by the continuous magnitudes and the continuous unboundedly divisible continua. Only in the Middle Ages the atoms as indivisibles came back and they were in the 17th century again replaced by divisible infinitesimals that Leibniz added to the aristotelian continuum. In the 19th century the end of the aristotelian continuum and of the magnitudes came. The continuum became a set of points, magnitudes became real numbers and the straight line became the number line $\mathbb{R}$. For 50 years now, in the background the old classical continuum is returning – via nonstandard models of real numbers in a deep shadow of the standard continuum. The status quo however is: the mathematical continuum *is* $\mathbb{R}$.

But $\mathbb{R}$ is not the intuitive classical continuum. The latter has not mathematically but intuitively, physically and philosophically survived. The difference was always clear for leading mathematicians – both at the hour of birth of $\mathbb{R}$ as well as later. Richard Dedekind, who found, as he claimed, the "trivial" characterization of the continuity of lines, was thoughtful and cautious in the evaluation of his discovery:

> "[...] for I am utterly unable to adduce any proof of its correctness, nor has any one the power. The assumption of this property of the line is nothing else than an axiom by which we attribute to the line its continuity. If space has at all a real existence [...]" ([85, p. 11], in the English translation on p. 12.)

Hermann Hankel excluded still in 1867 (cf. [154]) that the concept of a real number can be grasped by infinite sequences. For him the geometrical continuum was the unique possible foundation of the continuity – he claimed this while Dedekind, Cantor and Weierstraß already began to construct the reals.

Let us recall the motto at the beginning of Chapter 1 and carry on the quotation from the handbook [77]. Courant describes there the relation between the intuitive and the mathematical continuum in the following way:

> "The idea of a continuum and of the continuous flow is entirely natural. However, one cannot refer to it when one wants to explain the mathematical situation; there will always be a gap between the intuitive idea and the mathematical formulation that should describe the scientifically important elements of our intuition by precise expressions. Zeno's paradoxes indicate this gap." ([77, p. 46] in 1927.)

The matter is the conception of the continuum as set. As long as the continuum consists of isolated elements, Zeno's paradox of the impossibility of motion will be unsolved. Many well-meant popular persuading attempts should not mislead here. Something different are the counter-arguments coming from nonstandard mathematics.

It were just Zeno's paradoxes that challenged Aristotle and led him to the classical characterization of the continuum. He succeeded in solving some of Zeno's paradoxes because he recognized their source just in the point conception and rejected it. The old Greek mathematics based just on continuous magnitudes was so far not affected by this type of paradoxes. However, the old mathematics was finally not sufficient any more.

Courant's remark indicates concisely the mathematical dilemma between intuition and mathematical precision. The problem of the continuum is closely connected with the problem of the infinite. The conflict between intuitionism and logicism or formalism can be seen in the case of the latter as well as in the case of the continuum. The intuitive conception of Brouwer was confronted with the conception using sets. In Chapter 2 we wrote about "Brouwer's program" whose aim was to reconstruct in particular analysis by intuitionistic methods. Weyl ([367, p. 44] in 1928) complains about this dilemma in the following way:

> "Mathematics gained by Brouwer the highest possible clarity. He was able to develop the beginnings of analysis in a natural way and to keep contact with the intuition much closer than it was the case before. However, it cannot be denied that in further development of it" the intuitionistic mathematics with its restrictions "ultimately leads to hardly tolerable heaviness. A mathematician sees with pain how the greater part of his tower built, as he thought, of solid building blocks dissolves in fog."

The gap between the mathematical description of the continuity and the intuition of the continuum indicates that only some aspects of the continuum can be mathematically described. Successes of applications show after all that important or *the* important aspects are captured by $\mathbb{R}$. However, the merit of the infinitesimal mathematics is that it has been proved that also *further*, namely *qualitative* aspects can be represented in a mathematical way. Laugwitz says:

> "Careful distinction between numbers for measuring ($\mathbb{R}$) and numbers for calculus ${}^{\Omega}\mathbb{R}$) that takes into account mathematical consequences is for me an essential contribution to the clarification of mathematical aspects of the continuum. The latter are now better seen than it was by the coarse identification with $\mathbb{R}$." ([226, p. 237].)

Here ${}^{\Omega}\mathbb{R}$ are hyperreal numbers or Omega-numbers as Laugwitz calls them. His summary of the continuum problem was that with the introduction of $\mathbb{R}$, numbers for measuring, the intuitive continuum has by no means been eliminated. The continuum is – and it was said above – immeasurable.

Let us give an example. A qualitative aspect not caught by real measure numbers will be visible when one considers Zeno's paradoxes of motion. An arrow does not move because at any instant of time it is in the state of rest. However, – thinking infinitesimally – any given real point has an infinitesimal continuum as its neighborhood in which an infinitesimal though not measurable motion can be assumed. It proceeds by infinitesimal intervals. This relativizes Zeno's statement that the arrow is in the state of rest, however it does not reject it. The infinitesimal motion transfers only the problem. In fact, the state of rest returns when one asks about the motion from a point to a point in infinitesimal intervals.

Another interesting and advanced nonstandard solution of Zeno's paradox of motion via – internally – *finite* number of *infinitesimal* steps collecting all real numbers of an interval is given in [239]. The author McLaughlin argues on the basis of axiomatic internal set theory (IST in [255]) which we did not present here. This argumentation can be difficult for someone that used to think in a standard way and can seem to be a bit exotic. However, it indicates in detail new aspects of the continuum that are not seen in the standard approach.

Is the intuitive classical continuum mathematically grasped by mathematics of infinitesimals? No. Infinitesimal intervals by Leibniz are continua and hence divisible. Higher differentials d^2x are infinitesimal with respect to the infinitesimal dx. Today it is obviously shown by nonstandard models of real numbers that themselves can be basis for further extensions. They all can be *models* of the linear continuum, however not *the* continuum itself. The intuitive geometrical continuum is – as said above – *inexhaustible*.

The method of infinitesimals, the basic contribution of the infinitesimal mathematics indicates that the intuitive continuum escapes from being grasped just at the moment when one attempts to grasp it. Hyperreal numbers as well as other nonstandard models are sets. The classical continuum cannot be understand and grasped by mathematical, at least not by set-theoretical means. The meaning of this insight for the conception of the continuum is rarely or not at all noticed.

Finally, Observations made by us require some meta-remarks: about mathematical approaches to the continuum and a possible mathematical distance to nonstandard conceptions.

Till the late 19th century there were only rudiments of approaches grasping the continuum in a mathematical way. There were magnitudes that resisted the explicit mathematical approach. The continuum remained intuitive in the background of mathematics – both for geometry and for magnitudes, finite or infinitesimal. Since that time the mathematical approaches varies widely.

The first approach undertaken in the 19th century was the (first naive) set-theoretical construction. It emerged from the measuring and constructed the real numbers as unique outcomes of infinite approximating sequences in the infinite continuation of measuring processes. One believed and believes to grasp the continuum in this way.

In the set-theoretical construction of nonstandard numbers due to Schmieden and Laugwitz the measuring sequences are distinguished more exactly, and infinite small and large differences are indicated. The Axiom of Choice turns out to be crucial here – it is needed to obtain a field and a linear order of the new numbers that enable to transfer the infinite and infinitesimal relations into the linear continuum. Those new numbers cannot be obtained – even by methods of the transfinite set theory – in an equally "constructive" way as the real numbers as classes. This new domain of numbers seems perhaps to be less "concrete" just because of the usage of the Axiom of Choice. Is it difficult to recognize it as an alternative to $\mathbb{R}$? Let us mention our remarks on the Axiom of Choice made in Sections 4.4 and 4.5. It is said there that the Axiom of Choice (AC) can be seen as a plausible, almost necessary consequence of the Axiom of Infinity. The acceptance of the Axiom of Choice is growing despite some paradoxical consequences that arise in a combination of it with the Axiom of Infinity.

Let us notice that the assumption of nonstandard models of real numbers does not lead to contradictions even when the Axiom of Choice is eliminated (cf. [350, p. 2]). This is clear. In fact, the consistency of nonstandard models with ZFC implies the consistency with ZF – without the Axiom of Choice. There is a regular analogy to the situation of the Axiom of Choice. The assumption of nonstandard models is *independent* of ZF and *consistent* with ZF – similarly to AC (cf. [12, p. 851]). Hence the decision to work with nonstandard models is reliable. Nonstandard mathematics cannot be dismissed, as nobody dismisses mathematical domains in which AC is applied.

In the logico-algebraic adjunction proposed by Laugwitz in [226] one postulates simply the infinitely large and small, as is done in the internal set theory (see above). This approach seems to be appropriate with respect to mathematical practice – except for the logical component, i.e., the transition of arithmetical expressions to the new number domain that can be seen as only a formal postulate.

Nonstandard models are in the purely logical approach of Robinson simply formal logical phenomena. Both approaches, logico-algebraic adjunction and the logical determination of nonstandard models are based on the first-order predicate calculus. In the purely logical approach just the expressive weakness is used – namely the fact that the class of archimedean fields cannot be axiomatized (cf. [102, pp. 119 ff.]). The existence of nonstandard models becomes in a certain sense relative, and the nonstandard models themselves additionally get a formal character. There can arise an impression of a logical game of levels of language decreasing the acceptance of nonstandard models and nonstandard methods in mathematical everyday practice. Similar reservations can be formulated also with respect to other approaches not discussed here.

In Chapter 6 we try to weaken such and similar doubts and show how plausible and elementary the new infinitesimal calculus is. It is the matter of extending $\mathbb{R}$ to ${}^*\mathbb{R}$, hence of the extending of the basis and the work in analysis that could not be ignored. Their advantages should be checked. There can only be advantages – in fact there are no losses since ${}^*\mathbb{R}$ is a model of the same arithmetic as $\mathbb{R}$.

There remains an important question: to what extent do the nonstandard methods really extend the standard analysis. In [226, pp. 65 f. and 161 f.] δ-functions are indicated that are equal to 0 in $\mathbb{R}$ and are arithmetically defined infinitely close to 0, hence are not standard (cf. [350, Section 10.2]). M. Väth in his book (loc. cit.) describes how further implicit elements become explicit and concrete in the nonstandard approach.

Since the early 1960s nonstandard analysis has been the mathematical reality. At the beginning the reactions were controversial and cautious – however they became soon open-minded and positive. Nowadays there are very good handbooks, old and new (see, e.g., [226, 303, 350], cf. the Bibliography for more examples). Today nonstandard analysis means a break for fundamental mathematical concepts. The infinitely small – the conception of which was present through the history of mathematics – was rehabilitated in it. It changes decisively the conception of the continuum and the point.

It is in a sense astonishing that the nonstandard analysis is still a special domain and on the fringes. Despite all its contributions it is in fact not present in the mathematical everyday practice, in mathematical conceptions and in teaching. This is indicted by a small letter exchange in *Mitteilungen der DMV* – we report in Chapter 6 about this. Also a look into handbooks of analysis shows this (cf. for example [25, p. 237]). Even all possible formal reservations which may exist and which we spoke about do not justify the fact that the nonstandard mathematics and its implications are rather not really present and accepted. This concerns also the elementary teaching.

However, what remains – and this is perhaps an "ideological" obstacle – is the principal influence of nonstandard models on the *conception of the continuum*: the clear relativization, even the abolition of the identification of real numbers and the continuum. Briefly: the set-theoretical atomism is rejected.

But the powerful *model* of the real numbers seems to be unrivalled. The status quo of the conception of the continuum is today determined by real numbers. They *are* – in the standard mathematics – the continuum. Nonstandard concepts are still what their name says – they are not standard.

3.4 On the problem of applications of mathematics

The title question of this section appeared – implicitly or explicitly – already in Chapter 2 when various philosophical positions have been presented: how is it possible that abstract and theoretical mathematics is practically applicable in the concrete world? In the case of some positions we indicated an answer or an attempt of an answer. In those various attempts one sees clearly the differences of the philosophical approaches described at the beginning of Chapter 2: does mathematics come from the world by experience, does mathematics build its own higher world or is it a rational construction of human beings? According to the given positions there arise divergent attempts and problems connected with the question concerning the applicability of mathematics.

We would like to consider in this section the problem of the applicability of mathematics systematically: at the beginning we describe some aspects of the problem and then we discuss some important positions and conceptions.

3.4.1 Aspects of the problem

At first the problem of the applicability of mathematics in the concrete world seems to be irrelevant. In fact, mathematics is developed in the world. From the very beginning it is obtained by interacting with the reality. The world of mathematics is in a sense a world over the world to which it is related in a natural and reversible way. Those relations begin by the geometrical intuition and by simple numbers. Relations of magnitudes and simple numbers to real, "physical" things are considered in detail in [375] in order to get a "realistic mathematics". Already in such elementary relations deep epistemological problems are hidden as described by C. Thiel [341, Chapter 2, pp. 43 ff.]. Below we indicate such problems when we report on historical conceptions concerning the applicability of mathematics.

Let us consider the elementary numbers: they are – as observed in Chapter 2 and discussed in Section 3.1 – among others abstractions with manifold meanings according to the domain from which they come. Important meanings of numbers are quantity and magnitude.

Already here, at the very beginning, the simplest mathematics, mathematics of "natural" numbers, commences its own life that is detached from the world: in counting. The pure counting by numbers is a proper world being developed without reference to the real world. In the counting process towards the infinity it easily exceeds any reality. From the counting mathematics of *pure* numbers is developed – without any further reference, fully independently –, the complete elementary arithmetic. However, there is never a problem to refer this mathematics via meanings of numbers back to the reality. Quantities or magnitudes are for example *applied* numbers, and the putting together of sets or magnitudes is *applied* addition. Concrete magnitudes are measured when numbers are related to their representatives by special procedures.

A particular side of numbers is number theory being first of all a pure, "beautiful" domain, a "playground" of pure thinking having various connections with other disciplines. Even though one does not expect any application, one finds some of them (e.g., cryptology). There are also domains of mathematics in which one does not ask about applications since they are intentionally far from any reality and any idea of being applied (e.g., large cardinal numbers).

However, even in the case of such "pure" domains applications can not be excluded in principle. A famous example is the non-Euclidean geometry that as epistemologically unthinkable, as courageous rejection of the Euclidean intuition took leave of the "reality" and that some time later found oneself in the new reality, the "reality" of relativistic physics of the 20th century. It should be noticed that the concept of the reality was changed by the modern physics. However, we shall naively speak over and over again about various realities - physical, concrete, material etc.

Briefly: we are going to discuss today's everyday mathematics that we learn and teach and whose applicability is – despite permanent applications – a fundamental problem.

Degree of abstraction

We consider the "normal mathematics" beyond the elementary arithmetic which is not deliberately out of touch with the real world. Almost any such mathematics begins with real numbers or assumes them. On the way from natural numbers to real numbers – indicated in Chapter 1 and mentioned over and over again in this chapter – essential things happen. Mathematics exceeds a boundary there. It is not only the matter of the abstraction from the concrete. Real numbers arise in the world of abstract pure numbers according to new principles among which an essential role is played by the actual infinity (see above). Real numbers are set-theoretical *abstractions of the abstract*.

This abstraction is connected in an abstract way with a very concrete application of numbers: with measuring of magnitudes. Real numbers recall this measuring. They are – i.e., only "few" of them – infinite sequences of rational measure numbers. Real numbers come from measuring processes extended into the transfinite. In this way they become universal measuring numbers being "only" theoretical, namely *set-theoretical* numbers. They are abstractions of abstract sets of abstract numbers. Those highly abstract real numbers replace finally the continuum and concrete, intuitive, geometrical magnitudes (cf. Section 3.3.7) which they are measuring and make also the latter to be *abstract elements* of pure mathematics. Consequently, the problem of application with respect to abstract magnitudes in an abstract world disappears of course. Reality and application itself become abstract. However, there is a problem of overcoming the distance between the abstract-abstract world and concrete applications in very concrete world – it is not a mathematical but a philosophical problem. It indicates an aspect of the problem of applicability that is characteristic for today's mathematics.

Apriority

Are mathematical concepts *a priori*? This is a widespread conception lying in the still strong tradition of the Kantian philosophy. It is assumed that mathematical concepts are not empirical, that they in fact precede all experience and stamp the experience. If this is the case then there is a distance between objects of experience and mathematical concepts. Mathematical concepts are according to this conception additions to real things, i.e., better to say to their appearances. In fact, things can approach us only as phenomena because in our conceptually dependent experiences they are not as they really are. Mathematics provides in this conception only models of reality that never reach the reality. There arises a problem of reference to the reality, of the applicability.

Set-theoretizing

We indicated several times – and we will come again to this in Chapter 4 – that mathematics is nowadays in principle set theory. It is more serious than one wants it to be. Our mathematical thinking has been deeply influenced by set theory in the last hundred years. A standard example is provided by real numbers – which themselves are sets – as well as by what they induce.

Without actually thinking of sets real numbers are elementary elements of our mathematical thinking. However, they have a strong set-theoretical impact on another elementary object of our mathematical thinking: they have in the 19th century almost "occupied" the linear continuum, then replaced it and thus in many domains they became the *fundamental set*. The paradoxical concept of the *number line* determines or influences, mostly implicitly and unconsciously, our mathematical consciousness. The continuum that before – also mathematically – was seen quite differently became a set.

The *continuum* of real numbers is, because itself not abstracted, not really abstract. It is "only" theoretical and demands to be the intuitive, classical, physical continuum. Does it work? The countless successful applications indicate that it works quite well. However, it is often mathematically overlooked that already here the problem of the applicability of mathematics appears. It has been forgotten that there is still another continuum than $\mathbb{R}$, namely the intuitive, real, physical continuum. In Section 3.3 we have seen that $\mathbb{R}$ is "only" a model and – briefly – not the reality. We know that $\mathbb{R}$, a set, is a very good model of the geometrical continuum that is not a set. However, one should ask why?

Similarly to the continuum also other elementary mathematical objects became sets on which one operates without especially paying attention to this. We spoke already about numbers. It is in particular remarkable that one sees geometrical objects, motions, functions and relations quite differently than they are grasped really, intuitively, practically. We have learned and internalized to think them as sets. This changes the thinking and removes it set-theoretically from the concrete world. Why does mathematical thinking still capture it apparently?

Infinity

Actual infinity – being earlier unimaginable – became since Cantor a self-evident element of mathematics. The influence of the infinite on mathematical thinking is universal. Hilbert spoke about analysis as of a "symphony of the infinite". Set theory is the theory of the infinite and mathematics is set theory – as is willingly said.

In set theory a new world has been created that nowadays – instead of the real, intuitive world – forms the background of mathematics or, as just said, is this mathematics. New laws determining this new world had to be found. This was a task that did not appear ever before. Earlier the physical reality and the intuition coming from the philosophy formed the background of mathematics (usually it was not mathematically realized, even less considered). Laws of intuition as far as they were geometrical have been thought of as being recorded in *Elements* of Euclid. The fundamental laws of the physical world were and are puzzles of natural sciences and of philosophy – and they will remain incomprehensible even if there are projects that impose the claim to explain the world.

Before the turn of the 19th and 20th centuries there was a situation that was similar to that before the appearance of set theory – it was in the infinitesimal mathematics of the 17th and 18th centuries. There were the infinitely small quantities, the infinitesimals that formed the new world in the background of analysis. Historically one did not manage to understand and to describe this world in a mathematically appropriate way – in contrast to later set theory. The foundations of analysis in the 19th century went via limit values into the world of infinite sets and into the universal conception of sets. Recall that the conception of the continuum in the historical development of the infinitesimal mathematics was contrary to the set-theoretical conception of today. Only in the 20th century one has succeeded to found and to justify the nonstandard analysis, the infinitesimal calculus. Its continuum looks – as we noticed – necessarily quite differently than the standard continuum $\mathbb{R}$.

Mathematics with its new world of infinite sets has been *separated* from the old finite world and stands today vis-à-vis it. Here necessarily arises – with respect to another aspect – the question concerning its applicability. How can a mathematics of the infinite be applied to – by all appearances – finite world? How can it be that the infinite fits the finite?

The infinite exceeds the finite that should be grasped by it. The natural first exceeding of the finite is the potential infinite that was mathematically important but not manageable. Before Bolzano and Cantor this transcendence was still expressed in this purely potential form. The actual infinite can be seen as a kind of idealization of the potential infinite that accomplishes the imperfect infinite. It is radical from the philosophical point of view and revolutionary in the history of mathematics. It appears in the form of the Axiom of Infinity of set theory and stamps decisively the world of sets (cf. Chapter 4). The Axiom of Infinity grants to the formerly transcendental mathematically not manageable infinite ostensibly the manageability of the finite. The *actual infinity* is a special mystery and seems to be a puzzle also in the framework of the problem of the applicability of mathematics.

Formality

The essence of the new axiomatic approach in comparison with the old Euclidean one is that it dispenses entirely from the reality. We shall report in Section 5.2 on the history of the axiomatic approach till formal axioms. The *formality* being nowadays the fundamental feature of mathematics is a further aspect that provokes the question of the applicability.

Concepts are axiomatically reduced to primitive concepts which are represented as symbols having in principle no meaning. Axioms are not – as has been earlier assumed – evident truths but – again in principle – formal expressions consisting of symbols. This grants concepts and expressions the possibility that they can be freely interpreted – just here the new power of theories is located. A given theory can be applied in various *mathematical* domains. The most known example is of course the group theory. *Why* formal theories despite their formality can also be practically applied is another aspect of the problem of applicability of mathematics. How is it that those formal mathematics removed from the world can be applied just to this abandoned world?

Remark

Mathematics is a particular way of thinking. The question concerning the applicability of mathematics is therefore a sub-question of the question of the applicability of thinking in general. How is it that human thinking suits the reality? For this general question, answered in various ways by various philosophical positions, the sub-question concerning the applicability of mathematics – which is so heavily exposed in the scientific thinking – is of a special meaning. Concepts, statements and methods of mathematics are very special instruments and products of thinking which has been developed in the history and especially in the last 150 years in an extraordinary way, and they call in question the relation between thinking and reality in a new and provocative way. The question concerning the applicability of mathematics can be seen as the test case for the general question on the relation between thinking and reality. Answers – if there are any – have a great meaning for the general question on the applicability of thinking.

Questions about applicability and answers to them vary – both historically as well as currently – according to various philosophical trends at which we shall look now. In our report we take into account principal statements of given positions and generally disregard sometimes numerous different specifications of them.

3.4.2 The problem of applicability in the historical setting

The indicated aspects show that the problem of applicability is first of all a problem of modern mathematics and does not concern to the same extent the historical mathematics till the 19th century. In the old mathematics the same conditions were applied as those described by us at the very beginning of the case of the elementary arithmetic. In the consciousness of mathematicians it referred always – with the exception of number theory – to the real world and to the intuition. It proved itself in practice. A free theoretical development beyond the real world as it happens today did not virtually occur. The question concerning the applicability of mathematics is historically in principle connected with the problem of relations between ideal or abstract concepts and real or intuitive objects. We indicate again the work [341] in connection with epistemological problems that are here. Problems of formality and of the infinity are historically rather rare margin aspects of the problem of applicability, and the aspect of set-theoretizing does not occur at all.

Platonism and Aristotelianism

For the old Platonism there was in fact no problem of applicability. Mathematics and its applications are parts of the platonistic philosophy.

By Plato, ideas, and to them belong – at least in a higher intermediate world – mathematical concepts, are forming a higher world. The physical world is only a projection, a "shadow" of it. The world of ideas is the extended heritage of the Pythagoreans' higher world of numbers from which the material world was formed by numbers. Physical things of the material world are in the Platonism imperfect and unstable copies of real ideas and mathematical concepts. Things result as shadows of ideas and they approximate the laws of the world of ideas and in particular the laws of mathematics.

Possible deviations of things from laws are a characteristic feature of the uncertain material world and not a phenomenon putting the question of the applicability of the ideal mathematics. The ideal impact of higher mathematical concepts on earthly things is like a higher, ideal application of mathematics. A mathematician investigates the real world of ideal mathematical concepts and laws, follows them into the lower material world and captures by them the physical things.

Aspects of the applicability question arising from the formality of mathematics or from the infinity in mathematics do not play any role in the historical platonism. Infinity was only a philosophical concept, an idea. Laws, axioms were ideal truths determining the real world.

In Aristotelianism – as indicated in Section 2.2 – the relations were in a certain sense reverse. There is also no problem of applicability. Mathematical concepts are – as in Platonism – realities. They are abstractions, mental images of forms that are in the real things. Mathematics is the science of properties and of relations between those concepts that are applicable to the empirical world through forms of things. Aristotle excludes the actual infinite both philosophically as well as mathematically. His conceptions of the continuum prohibited the set-theoretical conception of the geometrical objects till the 19th century. Thus aspects of the infinity and of the set-theoretical view are – similarly to the problem of formality – irrelevant with respect to the problem of applicability of mathematics in the case of aristotelian philosophy.

Nominalism

Nominalism is radically opposite to Platonism with respect to the view of concepts. It refutes the platonistic realism of concepts and treats mathematical concepts – similarly to universals in the philosophy – as pure names, i.e., as descriptions, as bare linguistic signs. Concepts become fictions without any real meaning and have in a radical fictionalism no factual reference. Strong nominalism comfortably refuses any further meta-mathematical reflection on the status of mathematics and its concepts. Among fictions there is in nominalism also the infinity as long as it is mathematically given.

In nominalism Mathematics builds at most a fictitious domain of names – there is no way from their emblematic isolation to the reality. Mathematics is only a formal science, a fictitious game of signs whose relation to the reality is denied. Consequently, the applicability of mathematics is from the point of view of nominalism quasi a *postulated puzzle*.

Nicholas of Cusa and Leibniz

For *Nicholas of Cusa* – similar to Aristotle – mathematical elements were in things. This is justified in a theological manner: God created the world according to mathematical laws and has put mathematical ideas such as numbers or forms in things. Relations and elements of divine creation are registered by human beings and they are simulated and reconstructed in mathematics. This mathematics belong to physical things. Hence there is no problem of applicability by Nicholas.

In the conception of Nicholas of Cusa mathematical infinity is a fundamental concept. It is an image of the divine infinity that can be seen in processes as potential and in its completion as actual. Infinity is the core of the principle of *coincidentia oppositorum* in which the opposites are dissolved – as between polygons and a circle treated as an infinite polygon. The finite gravitates in a process towards completion in the infinite. The finite can be understood and grasped only with the help of the divine and mathematical infinite. By Cusanus the infinity was a philosophical-theological fact whose impact on the world is included. Hence the problem of applicability was not relevant.

"God is a mathematician. The world is an applied mathematics." In this way can Leibniz's following sentence be translated very freely (for mathematical usage only): "Dum Deus calculat et cogitationem exercet, fit mundus". This is the core of a universal rationalism by Leibniz – it plays a crucial role in mathematics. Mathematics is a divine source of divine creation. The applicability of mathematics is not a problem – in fact it is a requirement – in the cosmological dimension.

This explains and towards this are addressed Leibniz's great mathematical projects – *mathesis universalis* and *characteristica universalis* – referring to the philosophy, language and logic. We wrote about them in Section 2.9. They indicate the earthly task of a human being – namely to penetrate by thinking the divine order being in fact mathematical. As in the reasoning of Nicholas of Cusa, mathematics means to "reflect" on and to reconstruct the "mathematical" creation of God, the divine ratio in the creation. The world would not be the best of worlds, a perfect creation of the perfect creator if the mathematical reconstruction were not possible and the mathematics of a mathematician were not applicable. The mediation between the divine and human, spiritual and physical justifies Leibniz in monadology. The applicability of mathematics is also here on earth a presupposition.

For Leibniz there was no problem of applicability of a *formal* mathematics having distance to the reality. He did not see the necessity of a formality of *characteristica universalis* – it was one of the reasons why it could not been realized. What concerns the *infinity*, Leibniz's views were not uniform. He was swinging between the recognition and rejection of the actual infinite. However, his conception of infinitely small was initially clear. Infinitesimals belonged to those mathematical elements whose applicability was for him, as he himself proved, unquestionable. *Indirectly* he operated thus with the actual infinite claiming that a line consists of infinitely small segments – there must be infinitely many such segments and they are actually given as a collection by the unity of a line. The usage of those infinite collections was – indirectly and noteless – as unquestionable as the usage of the infinitesimals.

Pascal and Kant

For Pascal mathematics is a part of "the order of reason" (*ordre de la raison*). A sort of intuition coming from heart (*ordre du coeur*) is metaphysically fundamental for it. On the one hand it justifies convictions in "the order of heart" and on the other hand it provides mathematics with fundamental concepts and axioms in a "natural light". Mathematics is not formal. Nature itself has given mathematics the clear view of elementary concepts. Since mathematics is in this way intuitively connected with nature, its applicability in the nature is explainable. Pascal does not pick applicability out as a central theme.

Infinity is real for Pascal – mathematically in the concept of a number. It is boundless, hence potential. Actual infinity presupposing the bounds cannot be conceived. It is not a concept of reason that would be mathematically accessible. Hence there exists no problem of the applicability of the actual infinity.

For *Kant* mathematics is pure intuition. It is the creation of the imagination based on a pure, *a priori* fundament of spatial and temporal sensuality of the human being. Mathematics inherits the purity of intuition and became a "formal" character far from the empirical world. We quoted Kant (*Prolegomena*) in Section 2.10 in order to explain how for him the applicability of mathematics was directly given. Pure intuition composes on the base of sensual perceptions of things phenomena that are shaped spatially or arithmetically *a priori* by pure forms of intuition and therefore "must be in accordance with" mathematical laws coming from pure forms.

However, "It would be completely different if the senses had to represent objects as they are in themselves" [202, First Part, Note I]. This indicates the natural limits of Kant's epistemological concepts – as well as the problem of the applicability of mathematics. Mathematics does not reach the reality, "things in themselves" (*Dinge an sich*). It is directed to subjective phenomena and constructs – as any cognition by Kant – *models* of the objective world whose accordance with the world we are not able to understand. Those conviction of limits of knowledge at general has since Kant been spread over all domains of science.

Infinity arises – as other mathematical concepts – from pure intuition. It cannot be met anywhere in the reality. However, it is a potential element of thinking. Actual infinity is only a principle claiming that the potential, open infinity can be ideally completed. For Kant actual infinity is a philosophical idea of pure reason and not an element of mathematics. Hence Kant's critique does not give any information on the fundamental phenomenon of the actual infinity in modern mathematics and on its mathematical application.

Empiricist conceptions

Empiricism appears in various forms that in different ways react to the question concerning the applicability of mathematics. A common feature is the conception – indicated in Chapter 2 – that the knowledge, hence also mathematical knowledge, must come alone from experience. The world appears to us through senses that provide us with concepts. In a positive case – as for example by *John Locke* (1632–1704) – they are directly connected with real things. This first step from things to concepts is in fact a positive starting position when the problem of applicability of mathematics is considered.

However, already in the second step there is a serious conflict between mathematics and empiricism. In fact, the latter shifts even the relations between concepts established by thinking into experience. They come from experience and become statements and knowledge by iteration in experience. This empirically necessary conception contrasts with the systematic-logical methods of mathematics that obtain the relations and its statements in its own way independent of any experience.

The particular way of getting knowledge, the precise and safe classification in mathematics is empirically recognized, however its value is sometimes more, sometimes less denied – or even entirely negated. The truth value of mathematical statements lies for empiricists only in the correctness of deduction of consequences from axioms. Their real value being dependent of assumed axioms is hypothetic. By Mill – about whom we wrote in Section 2.11 – geometrical axioms are in no case necessary assumptions and "not even true". In this conception mathematical statements are products of internal mathematical procedures that are irrelevant to reality or even contradict the reality.

From the empirical point of view there are serious problems already in the first step leading from experience to concepts. It is difficult to disassociate concepts from things. *David Hume* (1711–1776) denies the possibility of abstraction [184, Section 12]. If they must be connected with things then for mathematical concepts remains only the possibility to be bare names whose connections to things are challenged.

This nominalistic tendency in the case of general concepts and complex ideas can be asserted already by the early empiricist Locke. Hume discusses the problem of connections between things and concepts and states that we cannot know anything about it.

> "Are the perceptions of the senses produced by external objects that resemble them? This is a question of fact. Where shall we look for an answer to it? To experience, surely, as we do with all other questions of that kind. But here experience is and must be entirely silent." ([184, Section 12, Part 1].)

A consequence for him is a sharp scepticism not only towards metaphysics alone but also towards mathematics. However, he appreciates the latter as an activity of the reason and puts it as a second pillar beside chains of thought "about facts and being". However, theorems of mathematics are and remain "abstract" [184, Section 12, Part 3].

Mill views for example mathematical concepts such as "point" and "straight line" as formed in a special, eliminative way – no real things correspond to them. They are "entirely unimaginable". Implicit definitions of concepts via axioms, being for him neither necessary nor evident, have any value only inside mathematics. It can be said that in the works of Mill one finds a formal picture of mathematics having a life of its own, released from the real world. The applicability of this mathematics, never denied, remains however a puzzle in this empiricism.

Infinity as a concept justified by experience is rather impossible. In fact, there is nothing in experience that could lead via senses to the actual infinity or to infinitely small quantities. Hume declaims intensely against both infinities. Even the doctrine of the infinite divisibility of extension is absurd for him [184, Section 12, Part 2]. In contrast to this Engels claims that even the actual infinite is derived from the nature and in the dialectical materialism an abstraction and idealization are introduced that should lead to the concept of the actual infinity. It looks a bit compulsively: the aim is to earn all concepts from the reality to be able to remain materialistic even when it has to do with mathematics. However, the actual infinite cannot be seen as "image, photography, copy of real quantitative relations and spatial forms" as it should be from the materialistic point of view [308, p. 193].

There is a danger for the consequent empiricism to degenerate to a kind of dogmatism. However, empiricism tends very often – as indicated – to nominalism or it represents a deep scepticism towards cognitive abilities of human beings. Hume is an early, Mill a late skeptic – also in what concerns the applicability of mathematics.

Problems that empiricism has with mathematics can be avoided only by compromises – in fact met by various empirical positions. Also the concept of the applicability is modified in this context. The applicability of mathematics is treated not epistemologically but "socialistically". Mathematics is characterized as an indirect "production procedure" having economical roots and effects [304, pp. 69 f.] and evaluated by its social gains. The latter, and consequently the applicability of mathematics, are indubitable.

Pragmatical positions

The basic pragmatical concept is "action". An essential mathematical action is application. Hence applicability is from the point of view of pragmatism a requirement for mathematics that cannot be doubted. What pragmatism in fact does not explain is the question how the applicability of mathematics is possible. Indeed it is epistemologically picked out as a central theme but answers are presuppositions. A continuous transition from things to concepts, a transition between signs of sensual impressions and signs of concepts is *postulated*. In principle an equivalence of logical connections between conceptual and sensual signs that lies at the base of an open process of cognition is assumed.

The relation between mathematics and the reality in Peirce's pragmatism is noteworthy. Mathematics generates concepts, develops methods and makes statements obtained on the base of primitive notions and hypotheses, i.e., axioms, that consequently are themselves approximations or hypotheses. They are proved to be of value in applications or are falsified. If they are proved to be of value they are in no case correct or certain. They remain hypothetical. They will be in a further process falsified or they remain in an open process of probation and approximation to the reality.

It is explicit for example in the case of real numbers which for Peirce are not the linear continuum, but a model, a "diagram" on the way of approximating the real continuum. Points are signs in the real continuum, hence individuals of another, alien kind – it is excluded for them to form the continuum. Infinities of point-sets of arbitrary cardinality are signs of further steps towards the continuum – the latter itself is never a point-set but it provides a continuous background for real as well as for conceptual signs. Hence one finds actual infinity – in an epistemological sense. Infinity never possesses the reality of things or of ideas. It possesses the reality of signs whose connections with the reality is set *pragmatically* in applications and remains – just in the sense of pragmatism – *hypothetical.*

3.4.3 Classical positions

Three positions called by us classical, i.e., logicism, formalism and intuitionism are trends in the philosophy of mathematics developed in the 19th and 20th centuries when mathematics was detached from philosophy and the task of providing foundations of it became a mathematical task. Assistance from the philosophy itself to discuss the philosophical problems related with the foundations of mathematics has been rare. The classical positions were stamped by influential mathematicians.

Logicism follows so to say an elitist program that concerns only mathematics and nothing else. It is the matter of reducing mathematics to pure thinking for the purposes of the real purity of mathematics. Pure thinking should be logic – and there arose the task of developing the latter in a *mathematical* manner. One of the consequences was the distance towards intuition and reality as well as the demarcation of mathematics of all other “impure” sciences among which there are – according to some representatives of logicism – also applied mathematics and geometry. Hence the problem of the applicability of mathematics plays no role in the program of logicism. However, the idea of this program and the conception of mathematics as a part of logic that became more and more formal do provoke *a fortiori* such questions – in a purely philosophical way of course.

Since the problem of the applicability of mathematics was not formulated in logicism, one cannot expect an answer from logicism itself – one can do this in another way, namely by analyzing philosophical views of representatives of logicism and those views varied from platonism (Frege) till nominalism (Russell). They could not be a-priori-synthetic and empirical because mathematics should be analytically reduced to logic and must be free of impure empiricism. Logicism is not in a broader sense a philosophical position from which we can expect information on the problem of applicability of mathematics. This pertains equally and yet more to the aspect of infinity within this problem too. Actual infinity was a problem for logicistic foundations of mathematics that could not be managed in a pure logical manner. There were attempts to exclude it by admitting the actual infinite only in a hypothetical form (cf. Section 2.19 and Section 3.2.7).

Similarly to logicism one can speak about the approach of *formalism*. Its aim is to justify the consistency and certainty of mathematics – to accomplish this one uses *formalization*. The matter is to reduce mathematics to a safe finite mathematics of symbols. Again in the focus is only mathematics for which formalism developed a special conception. It seems to dominate today largely the picture of mathematics. The formal conception of mathematics provokes again the question about its applicability and adds to it the problem of the gap between formality and the reality.

One should notice that the original aim of formalization was not to develop a view of mathematics but to be an instrument that should help to establish mathematically the certainty of mathematics. The proper problem that made for formalists the formalization necessary was the *infinity*. The matter was to secure the new infinite mathematics with its big richness and to defend it against foundational doubts and attacks coming in particular from the constructivists. The formalists saw that the infinite cannot be encountered in the reality and that mathematics with the actual infinity removes itself from the reality also when applicability is concerned. The purpose of formalization, i.e., of the reduction of mathematics to the finite mathematics of symbols, was to show that the actual infinite maintained to be indispensable for mathematical research is in fact quasi eliminable. It should – and one wanted to prove this – play only the role of an instrument making relations and proofs mathematically more transparent and not changing the connections between mathematics and the reality. Against this the idea became prevalent that mathematics is – in principle (!) – formal.

Also formalism is in fact no particular philosophical position that would give us information on the applicability of mathematics. Generally, it presupposes the applicability, namely for the finitistic part and its concepts and “real sentences” that refer to objects understood as concrete. Formalism attempts to reduce by formalization "infinitistic" mathematics to the concrete applicability of these "real sentences". In Section 2.22 we reported in detail on those not quite clear connections between finitistic and infinitistic mathematics.

It seems likely that the radical representatives of formalism tend to the philosophical position of nominalism. Hilbert, the father of formalism, was a rationalist, i.e., explicitly Kantian seeing the infinite concepts of mathematics as being ideas of pure reason that complete and perfect the finite methods. Intuitionism and constructivism were the vehement opponents of formalism.

Logicism and formalism contribute to the problem of applicability only indirectly. When they eliminate from mathematics all uncertain intuition and reality and in this way save its certainty, they ensure the certainty of mathematics also in applications – if the puzzle of applicability is solved.

Intuitionism – and this is typical for this approach – has a conservative side. It attempts or better attempted to connect itself to traditional foundations. Its aim was to preserve the connections with intuition and to tie in with abstract elements associated with the reality such as natural numbers. Its vicinity with clarity and concrete reality and in this way with philosophy was thus evident. Its critique concerned just the removal of mathematics from all the philosophy, intuition and reality.

Intuitionism had quite a different view of mathematics than formalism and logicism – in it the problem of applicability of mathematics actually does not emerge. For formalism and logicism mathematics was a closed domain of knowledge whose truth should be justified and protected. There was necessarily a distance to the intuitive and real world. On the other hand for intuitionism mathematics was a vivid, permanently changing, open process whose sources were the intuition of time and space as well as the thinking of mathematicians. The development of mathematics from the intuition of numbers secured – even when it was not explicitly asked – the return path of mathematics to the intuition and applications. The development and practice of mathematics – as the intuitionists wanted to arrange it – reduced the degree of abstraction of mathematical concepts. The problem of the actual infinity excluded by intuitionism and the formality negated by it were not present in the only hypothetical question of applicability.

The apriorism of the intuition of time and space and of concepts in intuitionism alone give occasion to ask the question of the applicability of mathematics. We refer here to what was said about Kant's views. By him the applicability of mathematics ranged – in a stringent way – till phenomena, however not to things. Intuitionists did not perceive the gap between their mathematical concepts and the real things and therefore did not answer the question on the applicability.

Some occasion to ask the question concerning the applicability is finally given by ontologies of mathematical concepts in constructivistic tendencies that appeared in various forms. Those are questions we already met in nominalism or considering platonic conceptions. The latter will be considered in the retrospect to this section.

3.4.4 New conceptions

Also in the new conceptions following the classical ones and discussed in Section 2.25 the question concerning the applicability of mathematics is not in the foreground. The new positions are either following the classical ones and as there the question on the applicability plays no role in them , or they are under the influence of the evolution theory, cultural studies, psychology or sociology. For those historically or sociologically oriented positions the development of mathematics, the "evolution" of mathematics is on the one hand basic and on the other hand the individual mathematician and its research is put in the foreground. From that follows a quite different picture of mathematics ranging from quasi-empirical sciences in the vicinity of physics till the proximity of religion and art. Mathematics is freed from its elitist being and brought back to the community of "ordinary" disciplines.

The fact that the question on the applicability is for historical-psychologically oriented positions not so urgent is connected with the evolutionary point of view in their background. This point of view being nowadays very common among mathematicians was described in Section 2.24. Briefly: Thinking – so is claimed there – and in particular mathematical thinking has been developed from the very beginning in the interchange with the reality. In this interchange thinking forms concepts and structures that suit the reality in the natural way. Applicability is not a question but a *consequence*. The application of mathematics is an essential element of its development. Thus it is far away from an absolute, pure or formal science. At any moment it is a special phenomenon of the contemporary culture. In its applications and in its use its era-specific "truth" is demonstrated.

In our opinion such characterization concerns today's mathematical thinking only seemingly (cf. Section 2.24.2). There are at least two puzzles concerning mathematical thinking: sets and their actual infinity. Neither the set-theoretical way of thinking in terms of mathematical concepts – so it seems to us – can be in fact understood as historically explicable detachment from the nature nor is the actual infinity explainable in an evolutionary way. The latter for example is at the most a detachment from the thinking itself, an abstract meta-detachment so to say, that makes mathematics an abstract theory and can be hardly evolutionary, sociologically or psychologically interpreted. Modern mathematics is theoretical and determined by transfinite concepts. Its way via the applications to the reality is evolutionary dark.

3.4.5 Retrospect

Why is the question of the applicability of mathematics so clearly asked today?

It seems to be discovered *anew*. In fact, this question has not been asked so clearly and so urgently in the history of mathematics and of philosophy though it was always present. The applicability of mathematics was in the historical positions often an unexpressed obviousness, a presupposition or even a part of the conception. In the classical positions the problem was not of high priority. The matter was first of all the foundations of the new, set-theoretical and highly abstract mathematics – the intention was to set them, to criticize them or to defend them. The question on the applicability was not asked or at best it was a further question in the sequel of the problem of foundations.

Besides the classical positions and new conceptions developed at the background of the cognitive and cultural evolution of human beings and putting an individual mathematician and his mathematical research practice at the center, there are also positions in the philosophy of mathematics connected with historical philosophy. They are today still strong.

Important is the position of today's *mathematical platonism* derived from the classical platonism – we discovered it already in Cantor's and Gödel's conceptions. Even today it can be noticed in the mindset of many mathematicians in the (seldom) case of making philosophical remarks. They establish ties with the real existence of ideas in the historical platonism transferred to modern mathematics and its concepts (cf. [48]). Mathematical concepts are reality and form a particular independent mathematical world investigated by mathematicians. Connections with the real world play no or only a subordinate role here – today it has nothing to do with connections of ideas by Plato or forms by Aristotle within physical things.

The question of the applicability is seldom asked in mathematical platonism. A mathematical Platonist lives in most cases in a pure mathematics of pure mathematical ideas. His main aim is to describe adequately this real world. Among realities of the modern platonism belong infinite sets. However, infinitely small quantities or infinitesimal numbers seem to find only little generosity by mathematical platonism. The distance of mathematical platonism from the reality provokes again – just in the case of a modest, concrete mathematician – the question of the applicability of his mathematics.

This provocation is answered by *mathematical realism* represented by W. V. O. Quine and H. Putnam (cf. Section 2.23). It claims that mathematical concepts in applications are as real as they are in mathematical platonism, however not in a pure world of ideas but as in the concrete physical world. This follows from the *indispensability principle* in realism (cf. Section 2.23) that seems to solve at one stroke the problem of applicability.

How problematic the reality of concepts is can be seen in a modified form, in partial realism presented by T. Wilholt in [375] where he explicates it on the example of real numbers as a part of a "realistic mathematics" in such a way that difficulties become visible in detail. This position was briefly described at the end of Section 2.25.

Magnitudes and proportions of them are treated in [375] as real properties connected with real facts. Wilholt constructs an arithmetic of proportions of magnitudes and shows – under the old Euclidean assumption of the existence of the fourth proportional element and the postulate of completeness – that it is isomorphic to $\mathbb{R}$. This isomorphism grants to the real numbers the character of universals and the reality of proportions of magnitudes with which they are *identified*.

Identification *by* isomorphism is in mathematics a commonly used method. However, here the matter is not an inner mathematical relationship but explicitly the overcoming of distance between mathematics and the reality. Such an identification is philosophically dubious unless the "realistic mathematics" is not assumed as a part of the reality – but this is a vicious circle.

Wilholt recognizes the difficulties connected with the transition from rational to real numbers exactly at the place which is for us the crucial step towards highly abstract, set-theoretically stamped and transfinite mathematics. However, he sticks at his thesis concerning "real" real numbers. Neither the actual infinity and the construction of real numbers nor infinitesimal relationships and the property of being non-archimedean are picked out in [375] as a special theme.

This gives the impression that artificial theoretical products such as real numbers are confounded with the (supposed) reality of proportions of magnitudes. This confusion points at a partial lack of historical account that neglects the 19th century revolutionary step from magnitudes to real numbers. Just this step led out of the intuition, of the domain of universals and the old "realistic" mathematics into the new higher mathematics.

Another historical position, the *apriorism* is still alive. It is connected with intuitionism and compatible with constructivism. According to it mathematics is a particular domain of *a priori* concepts and statements, more exactly a domain constructed from *a priori* elements. Also in this conception there is a gap between the real world and the world of mathematics. The *a priori* mathematics provides models of which one does not know why they work universally in the reality.

Some marks were also left by logicism and formalism. We have adopted the purity and systematics of mathematics from the logicism and the formal character from the formalism that influence today our view at the perception of mathematics. However, all those features are far away from the "impure", unordered and concrete reality. Why does a pure, formal and systematic mathematics capture the unordered, chaotic reality?

There is at least one more reason to ask anew and attentively the question of the applicability: there appeared and still appear more and more situations that mathematical theories dissociated from any practice and developed without taking into account any practical applications, but they can be and are suddenly applied just in practice. How can it happen that a pure, theoretical mathematics developed from pure, inner-mathematical reasons according to theoretical, inner-mathematical principles does describe the reality?

One can see such applications in modern physics. Usually physics is an orderer of applied mathematics. Today it seems to be sometimes just the opposite – mathematics gives to physics an occasion to describe the realities and maybe even to see them for the first time. Mathematics quasi becomes an eye by which a physicist perceives. Werner

Heisenberg reports that the following statement of Einstein has opened his eyes on the way to discovering the uncertainty principle:

> "The theory decides what can be observed."

Heisenberg quotes this sentence three times [161, pp. 92, 111 and 112]. The theory meant here was the quantum mechanics. Through the theory the reality shines.

It should be noticed that the concept of the reality has been drastically changed by modern physics. The physical macro-cosmos as well as micro-cosmos seem to be almost "dissolved" in mathematics. Both cosmology as well as quantum world provide examples. Mathematics and reality seem to blend.

We have seen that regardless of the chosen philosophical position one cannot obtain a convincing answer to the question of the applicability. We are far away from answers proposed by historical platonism or aristotelianism or by scholastic theology. Do we come – via the mathematical physics – closer to the universal rationalism as it was the case for Leibniz? The clear and substantial question of applicability has persisted since there is the infinite in mathematics and space and time became sets.

The applicability of mathematics is and remains unexplained. At the same time it is the daily, hourly constant fact. It cannot be more "factual". From this perspective the problem of applicability is not relevant. An active mathematician does not ask the question of applicability. The applicability of mathematics is no problem in practice. If there were problems then mathematics would be false. It is only a theoretical, philosophical problem. The fact that it is formulated is typical for the modern mathematics. This problem does not hold true for former mathematics in the same way. The relevance of the question of applicability is a consequence of the fact that today's mathematics has left any real, intuitive and philosophical connections to such extent as no other science has ever done. It seems that theoretical physics is in a similar way taking leave of intuition, the reality and the philosophy. It confides more than ever in mathematics.

At the end of this subsection let us formulate our own conception of applicability. At the beginning of this chapter when the concept of natural numbers has been considered we spoke about our "objective idealism" and quoted Schiller. What can our aristotelian-platonic-rational conception, as we sometimes called it, contribute to the problem of applicability?

The objective idealism – so can be briefly stated – quasi lost "the object" in modern mathematics. The degree of abstraction became so high that the objective connection with the concrete reality has been broken. We stressed it many times. What was still present by the former building of natural numbers – the impact on the interpretation of the objective world and its development by the exchange with it – has been lost. Both interpretation and impact have been principally relocated into a distinct, purely mathematical world. The reference to what is real and objective became indirect,

indirect-indirect etc. An objective idealism cannot rebuild the bridge. However, it can observe.

We spoke above about an *idealization* that turns the potential infinite into an actual one. This revolutionary mathematical step induces certain consequences. Let us quote Schiller again [310, p. 210]:

> "By idea the ennoblement moves always to the infinite because the reason does not link its challenges to the necessary bounds of sensory world and does not stand still unless before the absolute perfection. Nothing about which greater is thinkable does suffice; no need of finite nature excuses before its rigid court: it accepts no limits other than those of thought about which we know that it flies over all limits borders of time and space."

Some lines below in the same paragraph Schiller reminds:

> "That which he [the idealist] allows himself beyond those lines is eccentricity and just to this he is far too easy misled by a misconceived concept of ennoblement."

The last sentence of the paragraph from which we have taken these quotation is just the motto of this chapter.

Above we have relativized the "ideal" of the idealization to the actual infinite. In fact, we have mentioned – and this will be considered in detail in Chapters 4 and 5 – that this type of idealization leads to inner difficulties in mathematics. The terrain is shoreless and not safe. There appear problems of truth, incompleteness and consistency as well as the problem of the Axiom of Choice and the Continuum Hypothesis having directly its source in the abstract real numbers. Behind all of this there is the problem of the actual infinity.

We have learned – as we believe – that we have to deal with a mathematics that celebrates great successes in applications even if it provides only models. The latter are and remain – so teaches the objective idealism – nothing but models. They are not the reality and they do not achieve it. The real numbers are not the linear continuum, they are a model of it. The objective idealism does not know why the mathematical models are applicable. It indicates only that the modern set-theoretic and infinite mathematics is hypothetical and together with Friedrich Schiller warns about eccentricity and enthusiasm.

The old intuitiveness of concepts and the historical evidence of axioms that came from the reality disappeared. They have been replaced by generality, purity, formality of mathematics and – beside all the beauty – the danger of "emptiness" [310, p. 215]. It is just the application from which mathematics retrieves the whole wealth of reality. The fact of its applicability is its permanent verification. Here mathematics experiences and receives its everyday justification – however always, so to say, “on probation”, for its applicability remains a mystery.

3.5 Conclusion

Let us now look back at the way to real numbers described in Chapter 1 and consider – taking into account insights we got by now – problems identified in Chapter 1. At the beginning we make some fundamental remarks.

Problems connected with the concept of real numbers are the source of many questions considered nowadays in the philosophy of mathematics. These are questions that do not directly affect the everyday mathematical practice and research. The matter is the background, the development, the awareness of mathematics, of its methods and concepts, hence things concerning first of all the reflection, the understanding, the teaching and learning of mathematics.

The basic challenges with respect to real numbers can be seen in the development that began in the 19th century and primary led to the fact that real numbers have been established as a purely arithmetical number domain and have replaced the geometrical continuum. This establishing of reals was the precondition of the arithmetization of mathematics that dominates its picture today. Old fundamental questions and new problems connected with the arithmetization did not disappear. However, from the fact that real numbers are put as the foundation follows a definite decision with respect to many foundational problems that could have been decided in another way. The real numbers *are* – as we want to clarify – *the* answer. Other answers, for example that coming from the nonstandard mathematics, are hardly taken into account.

In Chapter 2 we have followed and seen in the previous sections how then a change of a thousand year old mathematics of magnitudes into number mathematics took place. This happened not in a step by step process but it was partially a dramatic, in principle revolutionary event. So far in magnitudes was hidden the continuity that in the course of justifying the analysis required an explicit formulation. The precondition of this was – to say it in a pointed manner – the “demolition” of the continuum and the “abolition” of the continuous that caused the conception of the continuum as a set and the introduction of actual infinite sets. In the case of such transitions it is often spoken – here a bit harmlessly – about “the change of the paradigm”. The transition from mathematics of magnitudes to number and set-theoretic mathematics was a *revolution* comparable (and taking place at the same time) with the change from the biology of creation to the evolutionary biology. It was not only a revolution with respect to the method but a revolution in *thinking*.

This thinking is a challenge both for learners learning mathematics (we took their point of view in Chapter 1) and for tutors who should recognize it and should take it into account in teaching. Its essence is the *leaving of intuition*. It was necessary to be able to make explicit properties of the continuum – in almost the same manner as leaving temporal counting processes was necessary in the case of natural numbers to be able to recognize their principles. A tutor knows that he must "pick up" a learner from the intuition as is often said today. Dedekind maintained that "even now such resort to geometric intuition in a first presentation of the differential calculus [is] exceedingly useful from the didactic standpoint" ([85, p. 4], in the English translation on p. 1). However, in teaching there is often no such bare resort but an irreducible mixture of abstract number domains and the intuitive geometrical continuum that generates by Dedekind "a feeling of dissatisfaction" (loc. cit.). Today such a commixture is often claimed to be indispensable and didactically necessary and the reference to "basic ideas" of the intuitive continuum treated as "phenomenological" or even "genetical" (cf. [22]). We have questioned all this providing clear arguments. H. Boehme describes in [38] briefly and clearly how axiomatic, constructive and geometrical elements cooperate. However, the background not considered there is the conception of the geometrical continuum as a set of points and its completeness.

Beside the arithmetization of the continuum there is also another basic turn in thinking: the axiomatic method. At first it sounds a bit surprising because in principle since Plato, Aristotle and Euclid mathematics is determined by an axiomatic method. However, in association with sets the axiomatic method got a new stimulus. This was indicated in Section 2.15 in connection with Dedekind's essay *Was sind und was sollen die Zahlen?* [87]. A new step into this new quality was the axiom system for real numbers given by Hilbert [166]. In Section 5.2.2 and in Section 5.3 we shall say more about this new degree of abstraction of modern axiom systems.

New in the new (set-theoretical) axiom systems is their beginning. It is usually as follows: Let a set be given. This set is readily denoted – in the case of real numbers – as a rule by "$\mathbb{R}$". In it there are elements called "real numbers", which axioms address. What is new here in comparison with the old axiomatic approach is the fact that from the very beginning by specifying a set one puts a sort of borders. One is not dealing here – as it was in the old approach – with constructions and processes in an open domain from which objects, for example magnitudes, are *arising* and a domain of objects is being *developed*. An open domain of construction becomes a closed "domain of verbal discourse".

Here, too, is a challenge for a learner here. It has not been constructed so far by him, he knows nothing about its existence but, however, it is already spoken about real numbers, even about *all* of them. The challenge is here the abstraction from the construction and consequently from active experience and from intuition.

3.5.1 From natural to rational numbers

We look back at a topic a bit before Chapter 1 began. It has been often indicated that the way to the real numbers does not begin by rational numbers but by natural numbers. At least one marginal note should be made in connection with the extension to rational numbers. A short historical note concerning this will be made at the beginning of the next subsection.

Let us consider the rational number $\frac{2}{3}$. What it is? Instead of the fraction $\frac{2}{3}$ one can put the fractions $\frac{4}{6}$, $\frac{6}{9}$, $\frac{8}{12}$ as well. What does a today's mathematician immediately do who has at his disposal the actual infinite sets? He writes:

$$\frac{2}{3} := \left\{\frac{2}{3}, \frac{4}{6}, \frac{6}{9}, \frac{8}{12}, \ldots\right\}.$$

It does not look well because the fraction and the "fractional number" are not distinguished. A fraction is a pair of natural numbers. Therefore it is written more exactly:

$$\frac{2}{3} := \{(2,3), (4,6), (6,9), (9,12), \ldots\}.$$

On the left-hand side there is the fractional number, on the right-hand side there are fractions forming an infinite sequence thought of as being given as a ready actual infinite set. Is this in principle not the same as defining $\sqrt{2}$ as a sequence (a_n)? Not at all.

First of all actual infinite sets are not needed for the concept of rational numbers. An extensional step is not needed. Intentionality and consequently the potential infinity suffice. Members of the potential infinite sequence $\frac{2}{3}$, $\frac{4}{6}$, $\frac{6}{9}$, $\frac{8}{12}$ are identified by their meaning. Every element of this sequence represents $\frac{2}{3}$ and operations on rational numbers are operations on such representatives. Representatives are pairs of natural numbers. This is the reason why one often does not distinguish at all between the reduction of real numbers to rational and the reduction to natural numbers.

In the case of $\sqrt{2}$ and real numbers the situation is of course completely different. Here $\sqrt{2}$ is not represented by members a_n of the sequence but by *the sequence at a whole*, hence by the sequence as an actual infinite set. If this step towards the actual infinite is made then it is coherent to define $\sqrt{2}$ as actual infinite set of all sequences equivalent to (a_n).

3.5.2 Incommensurability and irrationality

After natural numbers and proportions of them historically there appeared magnitudes that did not have any relations or proportions given by natural numbers and thus *no proportions* at all. In fact, an attempt to determine their proportion vanished in the infinity of an geometric Euclidean algorithm. Segments incommensurable with a given, verbally "determined" segment [109, Book X, Definition 3] are called by the Pythagoreans and later also by Euclid "ἄλογοι" (alogoi), which means "segments *without any relation*".

The big mathematical step made by the Greeks was to accept proportions of such magnitudes and to move from a mathematics of numbers to a mathematics of magnitudes. Subordinate or better aside from the mathematics of magnitudes remains the mathematics of *natural numbers and their proportions.* Calculational and algebraic manipulation of proportions of natural numbers and proportions of magnitudes being in a relation of natural numbers granted step by step a number character to proportions. They became *rational numbers*. Proportions of magnitudes incommensurable with respect to a *given unit magnitude* were considered as *irrational*. Calculational and algebraic manipulation of such irrational proportions granted also them more and more the character of numbers – however, the concept of them remained obscure. The mathematical concept of irrational numbers given in the 19th century is the result of the development of 2 300 years.

The real numbers ℝ are the answer to the problem of the incommensurability. In fact, it disappeared when the real numbers have been constructed and a *second big mathematical step* back to the domain of numbers has been made. "Incommensurability" became "irrationality" and can be deleted from the mathematical vocabulary. And it has been deleted. It is spoken today for example about irrational proportions even when one speaks about magnitudes by Euclid. This is legitimate but fatal for the understanding of Greek mathematics, of the development of concepts and for teaching. By one hit one puts the long and great development do death and omits a didactic necessity.

It is important to note here that in mostly used translations of Euclid's *Elements* (e.g., [109, Book X, Definition 3]) one speaks about "irrational" segments or areas instead of translating ἄλογος (alogos) faithfully as *being without proportion*. The concept "rational segment" can be found in translations to describe segments that are commensurable with a "given segment". Here "ῥηταί" (retai) is used which does not mean "rational" but "definite" segment. The translator possibly impressed by the advancement of real numbers chose here the unfortunate actualizations that have naturally influenced the reader of Euclid as well as the teacher. In this way there can appear the idea that the old Greeks were quite close to introducing the concept of the irrationality or even that they already had it. It has been indicated several times how far it was from the truth, from the real development.

In Chapter 1 we claimed that in teaching and tuition one should speak about irrational numbers only when before it was spoken about incommensurability. Steps of the historical development should be closely connected with steps of the systematic introduction of number domains. Similarly to the history of mathematics also in teaching it can and should be indicated that at the beginning of the way from rational numbers to real numbers there was the incommensurability. It would help to see what has happened by the introduction of ℝ and how specific it was. For a learner a bridge of incommensurability is needed on the way to real numbers, he should be *astonished* by recognizing that *not all is a number*, he needs the challenging question: How does it go further, what can be done mathematically, what has been done? Only in such a way the new, big mathematical step, the experience of a cognitive adventure and theoretical achievement standing behind the construction of real numbers, in axioms for them and in the algebraic adjunction will be recognized.

Since the number line is indispensable in teaching, the incommensurability should be picked out as a central theme. However, there prevails – both at school and university – a convenient, regrettable and permanent mix-up of numbers and points or magnitudes, of number domains and geometrical straight lines. In teaching we are often not further than at the time of Stifel whom we quoted above. We stay with an intuitive, unconscious correspondence of numbers and magnitudes. Stifel clearly distinguished both of them and he struggled against mixing magnitudes and numbers. Today nobody struggles anymore. Such problems are not noticed at all or are treated as solved because real numbers are there and there is no incommensurability. Moreover: magnitudes did disappear. So what should be incommensurability? Real numbers are *the answer to the problem of the incommensurability* – by dismissing it.

Instead of considering the conceptual development one has decided to choose a quicker and more effective way and presents irrational numbers to be on hand. However, it is abstract, technical and without any fundament to assume real numbers as given. In addition there is the identification of real numbers and points of a straight line and vice versa. Both have sad consequences. It would be invisible and hidden that real numbers are in fact a *work of art* – in the genuine sense of the word, that by introducing $\mathbb{R}$ a giant step has been made by mathematics – and the humankind. Even taking into account all successes of standard mathematics based on real numbers one should not forget that only some aspects of the continuum are met by $\mathbb{R}$. It has been said already above: $\mathbb{R}$ is *not* the continuum (see below). $\mathbb{R}$ is a *model.*

3.5.3 Adjunction

If there are only rational numbers, if *all arithmetic is a rational number,* then $\sqrt{2}$ has no sense. It can be only said: there is no number whose square would be 2, there is no measure number being the measure of the diagonal of a unit square. The term "$\sqrt{2}$" remains a *formal term without any meaning*. It is important to state this. Only when this is done, the important question can be asked: What can be done?

To be able to calculate one adjuncts $\sqrt{2}$ to $\mathbb{Q}$. What this means and how then the calculating proceeds cannot be understood unless one does not *excogitate* it. The altogether *formal construction* must be visible. First of all it should not be forgotten that $\sqrt{2}\cdot\sqrt{2} = 2$ is set. Then the calculations with formal terms like $b\cdot\sqrt{2}$ and $a+b\cdot\sqrt{2}$ should be set. (If $\sqrt{2}$ is a formal term then for example the symbol of multiplication put in front of it has no sense.) This is accompanied by intuitions that are geometrical *interpretations* and *not* definitions. It can be recognized that there arises a new calculating domain, a field including $\mathbb{Q}$.

Such a consciously formal construction is possible in teaching mathematics – it is also important. Only in such a way one is able to comprehend that there happens something purely theoretical, that an active mathematician in a learner is addressed. In academic studies the formal construction without the term $\sqrt{2}$ should be made.

The reality in academic studies and in teaching is usually different. Formal adjunctions appear of course in teaching algebra, however they are seldom in the course of introducing real numbers. It is justified by axiom systems describing arithmetically real numbers – and providing them in a complete way. Calculating with real numbers is assumed without any comments and the adjunction in $\mathbb{R}$ disappears. We indicated above what this in fact means with respect to the consciousness of what has been mathematically done.

The presence of real numbers usually put right away is a radical *answer to the problem of adjunction*: there is no such problem.

3.5.4 On the linear continuum

The biggest "trick" behind the artwork of real numbers is the following: $\mathbb{R}$ is as a set, the "opposite" of what the continuum intuitively is, for $\mathbb{R}$ decomposes into elements but the continuum is homogeneous. However, $\mathbb{R}$ *is* today the continuum.

More exactly: points can be put into continua, however they cannot be distinguished by themselves. It is possible only when numbers are put on points or coordinates are introduced into the space. However, this is *nota bene* a *projection*. Coordinates do not form the continuum. The continuum is a medium being given for such projections. Even when $\mathbb{R}$ as a set has originally nothing continuous in it, its elements should behave as *the* points of the linear continuum. This is required by axioms presuming that *the* straight line consists of points. Or this is provided by the artificial constructions of the set $\mathbb{R}$.

In spite of what has been said $\mathbb{R}$ is today not only the *answer to the question of the continuum*. Again: $\mathbb{R}$ *is* the continuum.

The medium has been replaced by that for what it was a medium.

This has far reaching consequences. Let us mention for example the problem of the continuum hypothesis presented in Section 4.4.

Nonstandard models – called a bit unfortunately "nonstandard" – refute the equalization of the linear continuum and $\mathbb{R}$. This as well as the meaning of the intuition are usually omitted. The remembrance of the mathematical development in which the intuitive continuum stood in the foreground till the 19th century seems to be buried alive. This is a large obstacle for teaching. The consciousness of the mathematical fundament of real numbers, of what it is and what it provides can arise only if the continuum and $\mathbb{R}$ are distinguished (cf. [18]). This is done in teaching if ever unwillingly. The number line being mathematically fixed seems to be a didactically proven and in any case convenient instrument. It was indicated in the Introduction how problematic this orientation of teaching is. It is explicated in [19, 22].

When $\mathbb{R}$ is the continuum then this means its identification with the geometrical straight line. In this way numbers and points easily become the same in teaching. Straight lines become geometrically sets of points. In this context problems indicated in Chapter 1 often are invisible. They are buried. For a long time now we are going round in circles: the set-theoretical conception of the continuum, the postulate of the actual infinite, the indistinction of numbers and points, the conception of the straight line as number line goes unnoticed from lecturers to students and future teachers of mathematics, from them to learners some of whom will become lecturers etc. Already for Hilbert, the lecturer, "the points of the interval 0 to 1" were "the totality of real numbers between 0 and 1" (cf. [173, p. 167], in the English translation on p. 139).

3.5.5 Infinitely small quantities

The idea of the continuum was from the very beginning accompanied by small atoms, then soon by infinitely small quantities. Both were forbidden by Aristotle what however did not prevent Archimedes from thinking the latter. Only in the 19th century after the full bloom of infinitesimals in the analysis the infinitely small has been regularly "relegated" from mathematics. A solution – the limit value – has been found that made infinitely small quantities superfluous. The fact that since about 1960 the infinitely small quantities have been mathematically rehabilitated (by Schmieden, Laugwitz and Robinson) is today commonly not noticed or treated as not serious. This is not present in the world of ideas being stamped by the real numbers. However, this world of ideas meets in the case of learners conceptions and ways of thinking for which the infinitely small quantities are not necessarily foreign but possibly related to (cf. Chapter 6). What for centuries could be and was mathematically thought cannot suddenly disappear from the thinking of human beings. When one is looking at the sequence of great mathematicians who worked with infinitesimal elements then it seems to be inappropriate to treat infinitesimals from the perspective of today's limit mathematics as not important or even as senseless.

Mathematically in the Completeness Axiom for real numbers, seemingly based on the situation on a geometrical straight line, exactly one number is put – for example inside a nesting of intervals. Other conceptions are excluded and pretendedly proved wrong – by arguments from the limit-value mathematics. It is not taken into account that other representations of numbers and other conceptions of ordering connected with the infinitely small quantities are possible. It is not known or not believed that for example $0.999\ldots$ can be smaller than 1. It will be seen in Chapter 6 how such ideas are denied possibly because one does not know or ignores the infinitesimal mathematics.

The handling of infinitesimals has been for a long time comparable with the handling of irrational numbers. One did not know what they both in fact are. Nevertheless, one calculated with them. Practically today the situation with irrational numbers is similar to earlier. In instruction one does not know what they are. In the everyday mathematical work one does not want to know it. Dedekind said: "... because the issue is only little fruitful" [87, p. 4]. This could have been a justification of the fact that he published his Dedekind cuts not in 1858 but only in 1872.

When this is the situation then why is it today (still) strongly forbidden and quasi improper to think in terms of the infinitely small? After some propaedeutic preparations one can calculate with infinitesimal numbers – in a non-archimedean way – with the same warranty as with unexplained irrational numbers. It is presented in Chapter 6. However, it is usually attempted to expel the infinitesimals like Beelzebub.

It can be easily understood. In fact, the answer to the *question on infinitely small quantities* provided by the everywhere present $\mathbb{R}$ is: There are no infinitely small quantities. They cannot exist because there is no place left for them beside $\mathbb{R}$ on the straight line, for $\mathbb{R}$ *is* the number line.

3.5.6 Construction, infinity, infinite non-periodic decimal fractions

Just the constructions of real numbers apply the actual infinite in a visible and substantial way as an instrument. There would be no $\mathbb{R}$ if there was not the actual infinite. Hence the indisputable fact $\mathbb{R}$ is the definite *answer to the question on the actual infinite.*

The constructions are mathematically hardly noticed because at best one registers them once and then goes to the everyday calculations that are axiomatically justified. Also the axiom systems refer to the actual infinite – practically however in a hidden way as framework of the formalism of limit values. The question what real numbers are is irrelevant.

The construction of real numbers appears in teaching at most marginally, at schools, possibly for good reasons, by no means at all. It remains in the background. Still just here – by introducing the quite new world – are the requirements of the actual infinite towards the thinking very clear. They are connected with the concept of the limit value. It is not the matter of the logically a bit complicated and sophisticated concept of the limit. The matter is the requirement to think about infinite sequences as being completed, as actual given wholes. Even the geometrically intuitive specification of the limit of a sequence, for example $\sqrt{2}$, does not help. This will be clear in unavoidable discussions whether $0.999\ldots$ is equal to 1. It is dictated but not convincing. It is learned but not understood. We report on this in Chapter 6. Another difficulty is the latent treatment of the problem of the limit value of a sequence as a solution of it (cf. Section 1.5). What is the limit value of a sequence, what is $\sqrt{2}$? It is the infinite sequence itself, the nested intervals, more exactly: it is represented by this sequence that therefore *must* be necessarily treated as an infinite whole. It is mathematically with a certain elegant style, psychologically however difficult. One hundred years ago it really had to rack mathematicians' brains over this problem and generated moral conflicts.

Today in teaching – maybe without noting it and despite intuitive support – using the concept of real numbers, we *implicitly* claim from learners two things: to think about the actual infinite and to accept sequences as their own limit values. We omit the fact that thinking in the domain of the infinite can be quite different than usually done.

Even more difficult and more puzzling are the so-called "non-periodical infinite decimal fractions" with their inestimable immense sequences of digits. Words are said, decimal fractions are categorized, all real numbers are here. A learner is blindsided. There is no possibility for him to notice how big the requirement in those few words is. "Non-periodic" is a negation and offers nothing concrete. Processes following a pattern as it is the case by identifiable sequences and their limit values are just negated here. There hardly arises a concept. Even a vague idea of such "non-periodic infinite decimal fractions" cannot be generated. How can "..." be written and "and so on" said when "non-periodic" in fact says that it does "not go so on". How can a sequence of digits be given as completed and determined when it is by definition indeterminable (cf. [22, Section 4.1])?

What does in fact arise is a truthfully irrational fog of "non-periodic infinite decimal fractions" in which the irrational numbers are hidden. We have seen how mathematical minds are divided at this place. Sequences without a background of constructions are not accepted by intuitionists and constructivists. It has been noticed in Chapter 1 which decision problems arise by assuming that infinite sequences of digits of such "non-periodic infinite decimal fractions" are given as a whole. The digits and sequences of numbers lying behind them are neither connected by a law nor determined as it was expected by Cantor in his explanations of the concept of a set. The requirement to consider to such an extent the indefinitely infinite as a given whole is a further challenge lying in the unbounded usage of the actual infinite. In the case of π or $\sqrt{2}$ one can refer to sequences and series or the constructive processes of approximation. However, there are only few – in fact only countably many – sequences of this type in comparison with completely arbitrary sequences of digits in "non-periodic infinite decimal fractions" that are not connected with constructive processes.

Also here the answer is given by $\mathbb{R}$: $\mathbb{R}$ would not be uncountable if not all possible infinite non-periodic sequences of digits – even sequences without a construction in the background – were allowed. The decision depends in principle on the Power Set Axiom of set theory that allows even for infinite sets to form and to put together all possible subsets (cf. Chapter 4).

3.5.7 Concluding remarks

The "most philosophical" question has not been asked yet in this retrospective, namely the question of the essence and existence of the real numbers. The essence of real numbers is hidden in their construction that reduces the question on their essence and existence first to the question on the natural numbers. In Chapter 2 and in the summary at the beginning of this chapter we reported about many conceptions of natural numbers. The methods of construction suggest further questions.

Actually it seems a bit whimsical to ask the question of the existence of real numbers, for they are used everywhere and there is no better reason for their existence than this fact. Nevertheless, just here in connection with what seems to be everyday and self-evident, such a philosophical question should be asked. Since the concept of real numbers is grounded essentially on the actual infinite the question of the existence of real numbers is connected with the question of the existence of this actual infinite. In Section 3.2 devoted to infinities various conceptions have been indicated and we noticed that today – stamped by formalism – the existence problem is in principle understood as the question of the consistency of set theory. One thinks about a kind of *theoretical existence* of infinite sets and therewith of the real numbers. The consistency problem however is and remains in principle unanswered when Gödel's second incompleteness theorem is taken into account (cf. Chapter 5). Consequently, also the problem of the existence is open – for real numbers too. Since so far no inconsistencies have been discovered in axiom systems the work with infinite sets is carried forward without any doubts. The existence of real numbers receives in this way a sort of in principle *pragmatic* feature. However, there is also a strong *utilitarian* argument that confirms this type of existence: the success in applications. But we want to stress: in the case of a strict, the most exact science of all disciplines this is rather a strange situation when its own foundations are not secured. Hermann Weyl wrote in 1921:

> "In fact: Every serious and honest reflection must lead to the conclusion that any drawback in frontier domains of mathematics [i.e., set theory and logic – authors' remark] should be valued as alarming symptoms; they indicate clearly what is hidden by externally brilliant and smooth business in center: the inner groundlessness of foundations on which rests the structure of the empire." ([366, p. 1].)

Hilbert – believing in his program (cf. Section 2.22) – saw this in another way and spoke in the paper on the infinite [173, p. 166] about analysis as of a "symphony of the infinite". This symphony begins with real numbers. The "paradise" of the infinite from which "no one shall drive us out" was created by Cantor. Only in it the real numbers could become pure numbers. However, – as we have seen – there are problems connected with the infinite in them. The real numbers have quasi been "bought" together with problems of the actual infinite. Almost all questions asked in Chapter 1 in connection with the development of real numbers and considered in this chapter have directly or indirectly to do with such problems.

We have seen that behind today's $\mathbb{R}$ – by nobody seriously questioned – there are distinct decisions concerning many philosophical foundational problems. Therefore the willingness in mathematics and in teaching to ask questions is generally rather small. Would it not mean that mathematics – such as it is – would be questioned?

No, this is not the case. We have shown that questions asked in the philosophy of mathematics are for many reasons important for the teaching of mathematics, for the reflection and the consciousness of what is mathematically thought and done.

4 Sets and set theories

I imagine a set as an abyss.[1]

Georg Cantor

Why do we occupy ourselves with set theory in an introduction to the philosophy of mathematics, more exactly: with different set theories? We have seen in Chapters 2 and 3 how set theory did arise as a distinct mathematical discipline and how it developed as a framework for almost all mathematical branches. It was just the step towards the actual infinity and a boundless extension of the possibility of building sets that required a hundred years ago a mathematical theory of sets. Nowadays set theory together with logic constitutes the foundations of mathematics – as a distinct mathematical discipline and as the foundation of practically all mathematical branches. In this chapter we want to reflect on set-theoretical foundations of mathematical thinking and talking and to indicate the set-theoretical framework within which mathematics is nowadays developed. We point to limitations and problems as well as ask questions of foundational type.

An important object of set theories are infinite sets. Set theory is – so to speak – a mathematical theory of the infinite. The concept of the infinite was from the very beginning an important notion of mathematics and caused big difficulties. All the time one tried to study it in detail and to explain it. This has been done – as we have seen – till the 19th century – within philosophy and theology. Then the infinite became – through the concept of a set – an object of mathematics. The concept of a set replaced the unclear concept of a magnitude that so far was the basic one. Since the concept of a set is nowadays fundamental for the mathematical concepts of natural and real numbers, it is in fact the basic notion of mathematics.

Set theory provides a precise language in which (almost) all mathematical notions can be formulated. This has not only methodological but also ontological consequences. Since all mathematical concepts can be defined by set-theoretical notions, the question about the existence and the nature of mathematical objects can be reduced to the question concerning the existence and the nature of objects of set theory, hence of sets. This shows how important set theory is with respect to the philosophy of mathematics. It is in particular the matter of the reduction of the ontological questions.

We will consider here neither the history of the infinite in mathematics nor the development of the concept of a set and of set theory itself. We limit ourselves to general remarks. We did it already in Chapter 2 while writing about philosophical views on

1 Eine Menge stelle ich mir vor wie einen Abgrund.

https://doi.org/10.1515/9783110468335-005

mathematics and its objects as well as in Chapter 3 when we wrote systematically about opinions concerning the infinite.

Recall that from the antiquity till the end of the 19th century a great majority of mathematicians rejected the actual infinite or at least were in opposition to it. It seemed that the potential infinity would suffice the aims of mathematics. The infinity of processes, sets or magnitudes consisted only in the possibility that they can be prolonged or enlarged.

One of the consequences was for example the fact that the mathematics of the ancient Greece was far from constructing the concept of an irrational number. The theory of proportions and the method of exhaustion of Eudoxus of Cnidus (ca. 408–355 BC) are characteristic for a method avoiding the concept of a value of the limit and of the actual infinite. Note that the method of exhaustion did not in fact mean that infinite sequences of polygons can be grasped as a whole and that a curvilinearly bounded figure can be finally exhausted by them. To show for example that the area of a circle is proportional to the square of the radius one used such sequences to give then an indirect proof and to reject in finitely many steps the supposition that they are not proportional (cf. [385, pp. 120–169]).

4.1 Paradoxes of the infinite

One of the reasons for rejecting or objecting the actual infinite in the past were paradoxes connected with it. Some of them are attributed to Zeno of Elea (ca. 490–ca. 430 BC). We know them from *Physics* by Aristotle. Among them are: the dichotomy (continued bisection), the arrow (impossibility of motion), Achilles and the tortoise (Achilles cannot overtake the tortoise) and the moving rows or stadium (relativity of motion). They concern the difficulties with the infinite in case of a partition into infinitely many parts and infinite summation. They also indicate problems with the idea of a combination of infinitely *many* infinitely *small* objects.

Another paradoxical property has been indicated by Proclus Lycaeus (Proclos Diadochos) in his *Commentary on the First Book of Euclid's Elements* (cf. Section 2.5). He has noticed that an infinite set can have exactly as many elements as a proper part of it:

> "A diameter divides a circle into two equal parts. However, if *one* diameter generates *two* semicircles and if infinitely many such diameters will be drawn then there will be twice as many semicircles as infinitely many diameters." ([285, p. 158].)

For him this was the reason to reject the existence of actual infinite magnitudes and to accept only the potential infinity. He wrote the following.

> "the magnitudes can be partitioned into infinite but not into infinitely many parts (*ad infinitum, sed non in infinita*). The latter would mean that there were actually infinitely many parts, the former – that only potentially; the latter admits the infinite a substantial existence, the former only the rising."
> (Loc. cit.)

The paradox indicated by Proclus was probably known already to Plutarch (ca. 46–120 BC). Later it appeared over and over again. It occurred in the works of scholastics of the 14th century, for example Thomas Bradwardinus (ca. 1290–1349). In the 13th century it was used as the base of argumentations that should show the impossibility of the existence of an eternal world. In the year 1638 Galileo Galilei (1564–1642) formulated a paradox named today after him – it was based on the same pattern. He noticed that on the one hand the square numbers $1, 4, 9, 16, \dots$ form a part of the natural numbers but on the other there are exactly as many square numbers as natural numbers. He evaluated this in the following way:

> "This is one of the difficulties that arise when we try to grasp the infinity by our finite minds and to attribute to it the same properties as those ascribed to the finite and the bounded; by me this is not correct – one cannot speak that one infinite quantity is greater, smaller or equal to the other." (*Discorsi* [128, p. 33].)

In a similar way also Isaac Newton (1642–1727) argued in a letter written in 1692:

> "The infinities, when considered without any restrictions or limitations are neither equal nor unequal nor are they in any relation to each other."

Gottfried Wilhelm Leibniz (1646–1716) who knew the paradox of Galileo Galilei wrote:

> "There is nothing more absurd than the idea of an actually infinite number."

At another place however he affirmed:

> "I accept the actual infinite – instead of assuming that the nature detests it as is sometimes stupidly claimed – in the sense that I am of the opinion that the actual infinity is available everywhere in the nature to stress the perfection of the creator."

Bernard Bolzano added in his *Paradoxien des Unendlichen* (1851) to paradoxes of the Proclus–Galilei-type a variant saying that a one-to-one correspondence between the reals of the open interval $(0, 5)$ and the reals of a greater interval can be established (cf. Section 2.12).

All those paradoxes contradict the principle found by Euclid and saying that the whole is greater than its part [109, Book I, Axiom 9]. Richard Dedekind and Georg Cantor turned this seemingly negative property into a positive one. They treated it as the *property* of infinite sets distinguishing them from finite sets and raised it to be the core of the definition of infinite sets (cf. Sections 2.14 and 2.15). This opened the way to today's set theory – now we will write about it.

4.2 On the concept of a set

We make now some remarks on the word "set" and the status of the concept of a set.

4.2.1 Collecting together versus putting together

The word "set" can have two meanings: the *collective* one and the *distributive* one. In the collective meaning a set is – as a collection of objects – a certain given *whole* consisting of those objects. Its elements are parts of it. According to this a library can be seen as a set of books/texts and a chain is a set of links. In this meaning a set of concrete sensually perceptible objects is itself a sensually perceptible object quasi going ahead its elements. The expression "x is an element of the set A" means here "x is a part of A". A theory of sets in the collective sense has been developed by the Polish logician Stanisław Leśniewski (1886–1939) – it is called "mereology".

Mathematics decided for the concept of a set on the total separation of parts from the whole and on the complete separation of parts among themselves, hence the second meaning of the word "set". Parts become "abstract" elements. Set theory understands sets not as putting together but as collecting together. This leads to extensionality. Relationships between elements should be reconstructed – extensionally – as sets, i.e., as set-theoretical relations. The connection to the whole is an abstract membership relation.

The sentence "Venus is an element of the set of all planets of the solar system" means – in the distributive meaning of "set" – simply: "Venus is a planet", and the sentence "3 is an element of the set of natural numbers" means "3 is a natural number". Planets and numbers are primary here and they build the set of planets and of natural numbers, resp. elements are ahead of sets. Hence a set in the distributive sense is not a sensually perceptible object even when its elements are concrete objects. A set is here an abstraction.

Both described meanings are principally different. This can be clearly seen by noticing that there are sentences which are true by one of the meanings and false by the other. Since the word "set" is nowadays used almost exclusively in the distributive sense, the term "totality" or "system" is sometimes applied to see this better. Consider for example the proposition: "A tenth part of Venus is an element of the solar system".

This can be seen as true in the collective meaning but is false in the distributive meaning. The membership in the collective meaning is transitive: If x is an element of a totality y that is an element of a totality z then x is an element of the totality z". By the distributive meaning it is generally not transitive.

Set theory deals only with sets in the distributive sense – Cantor characterized them in his famous "definition of sets" (cf. Section 2.14) as collections. This concept of a set enabled the step towards the actual infinity and is fundamental for today's mathematics. The idea of putting together which historically – starting from the primary continuum – stood in the foreground was one of the obstacles on the way to the infinite (cf. Section 3.3.5).

The philosophical question that can be asked now is: how and in what sense are there those sets – meant in the distributive sense. This is connected with the old problem of universals.

4.2.2 Sets and the problem of universals

One meets the problem of universals – that in the Middle Ages generated the quarrel of universals – for the first time in the works of Plato. It concerns the question: what does in fact correspond to general notions like "line", "number" or "man", "beauty", "goodness" etc. Answers given in the history of philosophy can divided into four groups. The following characterizations describe simultaneously the four main positions in the quarrel of universals.

Plato is the representative and the founder of the *radical realism*. It is assumed here that universals are self-dependent realities and they exist independently of objects and subjects. In Plato's ontology universals are ideas. They form an own world over the material-physical one. Consequently, for example, beside a single man there exists a man *as such*, i.e., a man in general, by Plato the idea of a man.

A bit less extreme is the *moderate realism* dating back to Aristotle. Also here universals have their own existence. However, they are not self-dependent and independent of the material world. They exist objectively only as properties of concrete single objects. For example a man as such, a man in general has no independent existence – he is a summary of essential properties specific for concrete men. "Man" is the whole of the characteristic properties of men.

One should stress that both those positions towards universals award the universals an external reality independent of the knowing subject.

In the Middle Ages within the Christian philosophy emerged a new position – the *conceptualistic* one. Its founder was Jean Roscelin (ca. 1050–ca. 1120), a monk from Compiégne (better known by his Latinized name Roscellinus Compendiensis or Rucelinus). According to this position universals exist only in the human mind, they are only concepts. Therefore the name of the position – it comes from the Latin word *conceptus*, concept. Hence there are no real objects like *a man as such* or *a line as such*. Concepts "man" and "line" are constructs of the mind and do exist only in it.

Still further goes William of Ockham (also Occam, from Latin: Gulielmus Occamus; before 1300–1349/1350). His position is called *nominalism*. According to him universals exist neither as such nor as general concepts in mind. Only single concrete objects do exist really. The source of the assumption that there is something outside of objects is the misinterpretation of the language. There is neither a man as such nor the concept of a man but only a general name (Latin: *nomen*) "man" in the language – it is a linguistical abbreviation for the description of single men. Universals, general concepts are fictions. They have no existence.

The quarrel of universals in the Middle Ages – especially together with other problems and questions of a theological nature – were sometimes intense. It is actual also today in a moderate form and is still not settled. Mathematics provides an example: what does it study? Does it study a real world – independent of things that are sensually perceptible, consisting of ideal objects like numbers, functions etc, is it a rational world of concepts or even a fictitious language game?

Since nowadays practically all mathematical concepts can be defined in set theory, they are reducible to the concept of a set. Consequently, the problem of the ontological status of mathematical objects has been reduced to the problem of sets, and the mathematical quarrel of universals to the universal "set".

Three of four positions in the quarrel of universals described above have their counterparts in the problem of the way of existence of sets. They are: platonism, (neo)conceptualism and (neo)nominalism.

The main thesis of *mathematical* platonism is: everything that is not contradictory does exist. In the case of sets this means: for any well-formulated and consistent property there exists a set of all objects fulfilling this property. The set of those objects possesses an independent existence just as well as its elements. Its existence is independent of the existence of its elements. To avoid contradictions one introduces some restrictions concerning properties when necessary. An example for this is the Comprehension Axiom in the Zermelo–Fraenkel system ZF of set theory – we write about this later – or in the theory of types (cf. Section 2.19). So set theory, and consequently also the whole mathematics, becomes a science about such legitimate and for itself existing objects, and its aim is to describe the world of those objects like the aim of zoology is to describe the world of animals.

This platonism in mathematics does not mean – and this should be stressed – that sets are treated as ideas like by Plato. The way of existing of sets does not correspond there to the way of existing of their elements. Their existence comes up to the existing elements. Sets possess another independent existence – just as *sets* of elements. Example: The set of natural numbers exists apart from its elements, i.e., natural numbers. Sets are not forms of some real concrete objects as it was the case for ideas by Plato.

Cantor – as we saw in Chapters 2 and 3 – was a representative of this platonic position. His firm conviction of the independent existence of the infinite contributed in an essential way to risk the step towards the actual infinity and to enforce it. In today's philosophy of mathematics the platonic position is represented for example by Kurt Gödel who in *Russell's Mathematical Logic* wrote:

> Classes and concepts may, however, also be conceived as real objects [...] existing independently of our definitions and constructions. It seems to me that the assumption of such objects is quite as legitimate as the assumption of physical bodies and there is quite as much reason to believe in their existence. ([138, p. 137].)

Today's mathematical *conceptualists* ascribe the existence only to such sets – and generally such mathematical objects – that are constructible, more exactly: such objects that can be constructed on the base of recognizably existing objects. Axioms postulating the existence of non-constructible sets are rejected. The attitude is strongly connected with constructivism. This position is represented for example by intuitionism (cf. Section 2.20). Moreover, in the ontology of conceptualism four versions of constructivism can be distinguished: (1) the objective version in which the existence of constructions is independent of the constructing subject, (2) the intensional version in which the constructed objects have an intensional existence, (3) the mentalistic version in which constructions exist as mental products in the mind of a constructing mathematician, and (4) the finitistic version in which the existence of constructions is given only in the real, material signs, i.e., in the inscriptions on a table or in printed characters (cf. Sections 2.21).

Nominalism, the third new position, assumes the existence only of special individual single objects. Statements about other objects, for example about sets, are statements about those individuals. Hence the question about those individuals and their existence arises. In answers to this question formal and factual nominalism can be distinguished. The former is represented for example by N. Goodman (cf. *A World of Individuals* [148, p. 17]). The formal nominalism rejects complex sets that cannot be directly comprehended. It approves individuals of any type. It is only required that everything which is accepted as existing should be treated as an individual object. The factual nominalism distinguishes certain objects and treats only them as existing. Other objects – eventually also individual objects – are excluded. As an example of such a type of nominalism can serve reism of the Polish philosopher Tadeusz Kotarbiński (1886–1981) in which only physical objects (bodies) possess the existence.

Notice that in the modern philosophy of mathematics there is no counterpart of the moderate realism. According to it a set could be intensionally treated as a property, as a property common to all its elements. Consequently, sets would exist as properties of their elements. Hence their existence would be derived from the existence of their elements and therefore not independent. However, there is a difficulty connected with so-called coextensive properties. They are properties that are intensionally different but extensionally determine the same sets of objects. For example the property Φ_1 of natural numbers to be divisible by 5 and the property Φ_2 of ending in 0 or 5 in the decimal expansion. If the position of the moderate realism were accepted, all coextensive properties should be found and treated as being equal.

Let us mention that there is still at least one more, comfortable position towards the existence of sets and mathematical objects. One can treat the problem of existence simply as a philosophical pseudoproblem having in fact no influence on teaching and research in mathematics. Such a position was represented for example by Rudolf Carnap (1891–1970) and it is probably popular among mathematicians. Recall – in contrast to this – Georg Cantor whose platonism and the firm conviction of the real existence of infinite sets was not without influence on the origin of set theory, on the development of mathematics and on the transformation of mathematical thinking. Notice also that it was just this transformation that ultimately made that modern and comfortable – in connection with problems of existence – position possible.

4.3 Two set theories

The founder of set theory as a mathematical discipline was Georg Cantor (cf. Chapter 2). In his works published between 1874 and 1897 he formulated the base of set theory. He described fundamental properties of sets and proved main theorems.

The beginnings of set theory, i.e., some of its aspects can be found earlier. We find an early text on relations and relata in the works of Joachim Jungius (1587–1657) in his *Logica Hamburgensis* [195] in 1638. Elements of the algebra of sets can be found already by Gottfried Wilhelm Leibniz. His ideas have been later carried on by Johann Heinrich Lambert (1728–1777) and Leonhard Euler (1707–1783). In particular Euler has given the geometrical interpretation of connections between sets (Euler's diagrams). William Hamilton (1788–1856) and Augustus De Morgan (1806–1871) attempted to study relations in a formal way, and George Boole (1815–1864) was the founder of the purely formal algebra of sets called today Boolean algebra. This theory allows various interpretations not only as an algebra of sets. Also Charles Sanders Peirce (1839–1914) and Ernst Schröder (1841–1902) developed a theory of relations and the algebra of sets. Until the first half of the 19th century one does not find explicitly an independent concept of a set being not only a formal, linguistic instrument. One finds it by Bernard Bolzano in the first volume of his *Wissenschsaftslehre* (1837) and later in his *Paradoxien des Unendlichen* (1851) (cf. Section 2.12). Some considerations concerning infinite sets one finds in the works of Richard Dedekind (cf. Section 2.15) and Paul Du Bois-Reymond (1831–1889) devoted to analysis. In [87] Dedekind developed a theory of "chains". All those investigations were in fact – from the point of view of set theory – fragmentary. Only Cantor developed set theory – in the real sense.

Cantor's set theory however was founded on an intuitive unclear concept of a set. Various characterizations of it (cf. Section 2.14) show this in detail. Aside from the philosophical way of speaking which uses other not defined concepts to explain the very concept of a set, there appear also concepts that should be made precise: what do "determine", "well-differentiated", "law" or "collecting together" mean?

Associations and intuitions connected with the concept of a set were different for different authors. O. Becker wrote in *Grundlagen der Mathematik in geschichtlicher Entwicklung* [16, p. 316] that Emmy Noether reports on an anecdote attested by F. Bernstein:

> *Dedekind* spoke out with respect to the concept of a set: he imagines a set as a closed sack containing quite specific things which one does not see and of which one knows nothing except that they are given and determined. Some time later gave *Cantor* his imagination of a set: He raised his immense figure, described a circle with his arm and said looking far away: 'I do imagine a set as an abyss'. [89, Volume III, p. 449]

The usage of intuitive and imprecise concepts of a set that thanks to the step into the infinite enabled an unregulated and unlimited building of sets led to antinomies in Cantor's development of set theory. The most important antinomies are: the antinomy of the greatest ordinal coming from Cantor and C. Burali-Forti, the antinomy of the set of all sets of Cantor and the antinomy of irreflexive classes of Russell. We wrote about them in Chapter 2.

The discovery of those antinomies at the turn of the 19th and 20th centuries shook Cantor's set theory. It became clear that set theory cannot be based on intuition alone. The concept of a set must be made mathematically precise. This could be done only by developing an axiomatic set theory or by founding it on logic. One looked for systems that would eliminate the known antinomies. The results can be divided into two groups: axiomatic theories and theories in the framework of the theory of logical types. We present two set theories that are very different from the point of view of the intellectual approach. We wrote in Section 2.19 about the logical theory of types developed by Russell and Whitehead – now we make only remarks about it here and there and indicate its connections to other approaches.

4.3.1 Set theory according to Zermelo and Fraenkel

Cantor's explanations of the concept of a set link the concepts "set", "collecting together" and "element". They indicate the challenge of an axiomatization: to describe how sets are built of elements and in this to implicitly explain the very concept of a set. In his early explanation of the concept of a set from the year 1883 Cantor spoke about "totality (*Inbegriff*) of certain elements that can be collected by a law as a whole". This type of collecting together (Latin: *comprehensio*) by a "law" is the general *principle of abstraction* according to which Cantor and others built sets without clear rules, bounds – and without any doubts.

Principle of abstraction or comprehension. Let Φ be a property (determining a law of collecting together). Then there exists a set x of elements y such that $\Phi(y)$.

In a formal way:

$$\exists x \forall y [y \in x \leftrightarrow \Phi(y)].$$

By such building of sets they look like this:

$$x = \{y \mid \Phi(y)\}.$$

Ernst Zermelo (1871–1953) saw in such a generous unbounded way of collecting, the reason of antinomies. In such a way huge collections could arise that lead to antinomies. He avoided antinomies by restricting the formation of sets and limiting the size of them. Restrictions of such a type are characteristic for the approach to the first axiomatization of set theory given by Zermelo [383] in 1908. Zermelo replaced the principle of an unbounded abstraction by the following axiom.[2]

Axiom (of Comprehension).

$$\forall z \exists x \forall y [y \in x \leftrightarrow y \in z \wedge \Phi(y)].$$

This means:

Let Φ denote a property. *If a set z is given* then there exists a set x *of elements y from z* such that $\Phi(y)$.

The restriction of the building of sets is immediately recognizable. Only in a given set z one can build sets by singling out elements y. It should be clear what one speaks about. Sets are a priori subsets. Set terms now look like this:

$$x = \{y \in z \mid \Phi(y)\}.$$

By such a selection from given sets neither Russell's antinomy of a set of sets being not their own elements nor consequently the set of all sets can arise. This can be easily seen: Consider the property "$y \notin y$". When one wants to build a set r by the property "$y \notin y$" according to the Comprehension Axiom then there is a set z given from which one selects elements y, i.e. $r = \{y \in z \mid y \notin y\}$. We ask – as in Chapter 2: Has r the property $r \in r$ or $r \notin r$? One gets the following contradictions: (a) $r \in r \to r \notin r$. In fact, r as an element of itself has the property $y \notin y$ of its elements. (b) $r \notin r \to r \in r$. In fact, because $r \notin r$ hence r has the property $y \notin y$ of elements y of r, it belongs to its elements. However, now the situation is not antinomic – we conclude that $r \notin z$. Indeed assume that there is a set m of all sets. We build the subset $r = \{y \in m \mid y \notin y\}$.

2 The Comprehension Axiom is frequently called *Separation Axiom* distinguishing it explicitly from the unrestricted comprehension principle above.

For this r one gets the above contradictions (a) and (b) and one concludes $r \notin m$. But this contradicts our assumption that m is the set of all sets.

The system of Zermelo formulated in 1908 has later been extended by Abraham H. Fraenkel (1891–1965) and Thoralf A. Skolem (1887–1963) and it is called today Zermelo–Fraenkel set theory (ZF). If the Axiom of Choice is added then the system is denoted as ZFC. There are eight axioms of ZF.

We shall describe now the axioms of ZF and comment on them. They set in a pragmatic way the requirements for manipulating sets familiar to the concept of a finite set and commonly used by handling such sets. They should hold also for infinite sets from the moment when the Axiom of Infinity will be formulated.

Note that the system ZF is formulated in the language of the first-order predicate calculus (cf. Section 5.1.1). For example instead of the usual symbol $\Longleftrightarrow$ common for the mathematical practice we use the symbol of logical equivalence $\leftrightarrow$. Small Latin letters denote set variables. The only nonlogical predicate is the membership relation "$\in$". We try to avoid fully formal formulations but we give axioms in formal setting that we shall often prepare and simplify and always explain.

The first axioms says that sets are determined by their elements and only by them. This means that sets are equal if and only if they have the same elements – so to say: if they "range" over the same elements, i.e., when they have the same "extension".

Axiom (of Extensionality).

$$\forall z \forall y [\forall z (z \in x \leftrightarrow z \in y) \rightarrow x = y] .$$

This axiom determines the fundamental principle of set-theoretical thinking that is not intensional but purely extensional, quantitative. If an intensional component were added then sets would be determined not only by their elements but also by their meaning. For example The sets $\{a\}$ and $\{m\}$ where a is the evening star and m is the morning star would be different though $a = m$. Or: The set of divisors of 27 and the powers of 3 smaller than 81 can be distinguished by their meaning though they have the same elements. However, such differentiations are impractical and inexpedient. One should distinguish for example various empty sets.

The extensional element in the concept of a set grants the set-theoretical and consequently also the mathematical thinking in principle a static and discontinuous character. The latter has been discussed in detail in Chapter 3 in connection with the concept of the continuum. The static character is expressed when for example a process is mathematically presented. Think for example about the process of counting. To grasp it one starts from a set whose elements are provided in a static way (usually this set is called $\mathbb{N}$). The process is simulated set-theoretically when a construction of a successor of a given set is added. This construction of a successor being a correspondence, a function, is again a set, a static "register" of pairs. In this set of pairs the movement, the activity originally included in the concept of correspondence and function is canceled. Or think about geometrical mappings, e.g., about rotations in the space defined as functions from $\mathbb{R}^3$ into itself, i.e., as (infinite) tables of values. It is far from the intuitive idea of a rotation.

Notice that we speak in the sequel *exclusively* about sets – this is proper for *set theory*. Sets are *determined* by their "definite and well-differentiated" elements. This is expressed by the Axiom of Extensionality. And just sets themselves suit best to be again elements. Set theory avoids the problem of unclear "objects of our intuition or our thinking" by allowing as elements of sets only sets. So called "urelements" that can be introduced to set theory make the language more complicated and – from a mathematical point of view – are an unnecessary ballast. In nonmathematical applications however the acceptance of them can be useful.

The following axiom states something simple: it claims that any two objects can be collected to a set, i.e., to a new, abstract object, a pair.

Axiom (of Pairing).

$$\forall x \forall y \exists z \forall u (u \in z \leftrightarrow u = x \vee u = y).$$

Such sets look as set terms as follows:

$$z = \{a, b\}.$$

If $a = b$ then one gets the singleton $\{a\}$ that should be distinguished from a. This distinction can be in a nice way illustrated by a hat that once is outer and later inside a box { }. (Try to put the latter on your head.)

The Union Axiom states that if there are two or more sets given then a set y, the union of them, can be built – it collects elements of all given sets.

Axiom (of Union).

$$\forall x \exists y \forall z [z \in y \leftrightarrow \exists u (z \in u \wedge u \in x)].$$

If $x = \{a, b\}$ then one writes $y = a \cup b$. If x consists of many sets then the term looks like

$$y = \bigcup x = \{z \mid \exists u (z \in u \wedge u \in x)\}.$$

The following Power Set Axiom allows to collect all subsets of a given set x to a new set $\mathcal{P}(x)$. It is a powerful axiom. If the number of elements of x is n then $\mathcal{P}(x)$ has 2^n elements. Imagine this in the case when x is infinite. Here occur the problems of the actual infinity that we see again and again. In Chapter 2 some remarks have been made, in particular in connection with the continuum hypothesis.

Axiom (of Power Set).

$$\forall x \exists y \forall z [z \in y \leftrightarrow \forall u (u \in z \rightarrow u \in x)].$$

The variable y denotes here the power set

$$\mathcal{P}(x) := \{z \mid z \subseteq x\}.$$

The Comprehension Axiom (called sometimes: Separation Axiom) mentioned already above, which plays an important role in the system ZF, is in fact a *schema* of axioms: for every property – given by an expression φ – there is an appropriate axiom.

Axiom (Schema of Comprehension).

$$\forall y \exists z \forall u (u \in z \leftrightarrow u \in y \wedge \varphi(u)).$$

Properties are connected with relations that can have many variables and are usually expressed by complicated expressions. We give this schema again in full generality for $(n+1)$-ary expressions φ.[3]

Axiom (Schema of Comprehension).

$$\forall x_1 \forall x_2 \ldots \forall x_n \forall y \exists z \forall u (u \in z \leftrightarrow u \in y \wedge \varphi(u, x_1, x_2, \ldots, x_n)).$$

Here φ is a formula of the language of set theory in which the variable z is not free (cf. Section 5.1) and such that the variables $x_1, x_2, \ldots, x_n$ of φ are different from u. Note that this axiom does not exclude the impredicative definitions of sets. They are definitions that refer to totalities to which the set being defined belongs itself. We wrote about such definitions of sets and problems connected with them in Chapter 2, e.g., in Section 2.16 devoted to Poincaré.

The following *Axiom of Infinity* elevates the common description of handling finite sets and the forming of new sets to a higher level, into the realm of the infinite.

The Axiom of Infinity is an existential axiom. The existence of a set x with the following two properties is claimed:

(1) The "first" and the simplest element of x should be the empty set y. It is determined by the property ψ of having no element:

$$\psi(y) := \neg \exists z (z \in y).$$

One defines

$$\emptyset = y :\leftrightarrow \psi(y),$$

and $\emptyset$ is uniquely determined, hence it is *the* empty set. $\emptyset$ is a subset of every set.

(2) In the set x should a type of a counting process be possible: If $u \in x$ then also the "successor" $u \cup \{u\}$ should be in x.

Axiom (of Infinity).

$$\exists x \big[\emptyset \in x \wedge \forall u (u \in x \rightarrow u \cup \{u\} \in x)\big].$$

Infinite sets having this property are called *inductive*. Together with x there exists the empty set $\emptyset$, further the element $\emptyset \cup \{\emptyset\} = \{\emptyset\}$, then $\{\emptyset\} \cup \{\{\emptyset\}\} = \{\emptyset, \{\emptyset\}\}$ and $\{\emptyset, \{\emptyset\}, \{\emptyset, \{\emptyset\}\}\}$ etc. The "smallest" inductive set is denoted by ω. Intuitively it consists of the just given sequence of sets:

$$\omega = \big\{\emptyset, \{\emptyset\}, \{\emptyset, \{\emptyset\}\}, \ldots\big\}.$$

3 The Comprehension Axiom Scheme is frequently called *Separation Axiom Scheme*.

If one puts

$$0 = \emptyset, \quad 1 = \{\emptyset\} = \{0\}, \quad 2 = \{0, 1\}, \quad 3 = \{0, 1, 2\}, \quad \dots, \quad m + 1 = m \cup \{m\}, \quad \dots$$

then it is clear that ω is a sequence of sets that proceeds like the natural numbers.

ω is *the* set of natural numbers in the set theory.

In the following schema the so-called Replacement Axiom will be formulated. Its necessity has been recognized by Skolem, D. Mirimanoff and Fraenkel only in 1922 or shortly before. The Replacement Axiom describes the transition from given sets u to a new set w that arises when functions, more exactly *set-theoretical operations*, F are applied to elements x of u and the images $F(x)$ are collected together. Operations are unique assignments on the universe of all sets (that itself is no set) given by expressions $\varphi(x, y)$ that assign in a unique way sets y to sets x. An example is the construction $\mathcal{P}$ of the power set that assigns to a set x its power set $\mathcal{P}(x)$.

The uniqueness of such operations is in a formal way expressed in the first part of the following formula.

Axiom (Schema of Replacement).

$$\forall u[\forall x \forall y \forall z(x \in u \wedge \varphi(x, y) \wedge \varphi(x, z) \rightarrow y = z) \rightarrow \exists w \forall v(v \in w \leftrightarrow \exists x(x \in u \wedge \varphi(x, v)))].$$

It looks a bit confusing. So we denote by F the operation given here by φ and use set terms. Then one gets the following formulation.

Axiom (Schema of Replacement).

$$\forall u \exists w(w = \{F(x) \mid x \in u\}).$$

The meaning of the word "replacement" can be easily seen here: If F is given then one can move from the given set u to a set w of images of elements of u and quasi "replace" in transition u by w. For example $\mathcal{P}$: The Replacement Axiom gives for the set ω the set w whose elements are power sets $\mathcal{P}(n)$ of elements $n \in \omega$. Or: one can define recursively a function f on ω by putting $f(0) = \omega$ and $f(n+1) = \mathcal{P}(f(n))$. This gives the set $\{\omega, \mathcal{P}(\omega), \mathcal{P}(\mathcal{P}(\omega)), \dots\}$. Just by the example of this set Fraenkel and Skolem have recognized the necessity of the Replacement Axiom.

We come now to the last axiom of ZF set theory, to the *Foundation Axiom* formulated by Zermelo in 1930 and connected with some ideas of Mirimanoff and von Neumann. It expresses a further natural expectation of sets that can be built by using the hitherto introduced axioms. We expect that the construction of sets begins by "simple" objects (that are already sets) and rises from sets to sets, then from sets of sets to sets etc. A set can then contain objects, sets, sets of sets and arbitrary "higher" sets as elements. If a set is given then one can trace back the complexity of its elements and in a finite number of steps find elements that are of a relatively "simplest", "lowest" type. This means that there is *no* not interrupting chain of membership relations of the type

$$\cdots \in x_n \in x_{n-1} \in x_{n-2} \in \cdots \in x_1 .$$

Since objects, elements, are always sets, one can say this in such a way: in every nonempty set x there are elements y such that no element of y is an element of x. This is expressed by the Foundation Axiom.

Axiom (of Foundation).

$$\forall x[x \neq \emptyset \to \exists y(y \in x \wedge \forall z(z \in y \to \neg z \in x))]$$

or

$$\forall x[x \neq \emptyset \to \exists y(y \in x \wedge y \cap x = \emptyset)].$$

Since $\in$ can be treated as a relation restricted to x, this means: Every nonempty set contains a $\in$-*minimal* element.

What has been positively said in connection with this axiom, can be also restrictively formulated in such a way: sets are sets of elements, and in a real set theory those elements are again sets. They again are sets of elements being sets of sets of etc. Hence in a concept of sets a non-interrupting, infinite regress of membership is included. The Foundation Axiom says "finished", the regress is interrupted – "ultimately". The foundation axiom – though it actually looks "harmless" and rather technical – is in fact theoretically strong.

This will become "palpable" in the following equivalent formulation of the axiom. One simply requires that the formation of sets begins by the simplest set $\emptyset$ and the world of sets is constructed according to the formulated principles of building sets. There arises a hierarchy of sets – the so-called *von Neumann hierarchy* in which a not interrupted regress of membership is prevented *per se*.

The von Neumann hierarchy plays an important role in set theory. It makes the above-indicated composition of sets explicit from stage to stage. The hierarchy begins with $0 = \emptyset$ in the first level V_0. Then there is the set whose element is just the empty set: $\{\emptyset\}$. This is the power set of the empty set, hence $\mathcal{P}(\emptyset)$. It forms the next level V_1. By applying the power set construction one moves to the level $V_2 = \mathcal{P}(\mathcal{P}(\mathcal{P}(\emptyset)))$ and so on from level to level. If V_n is reached then $V_{n+1} = \mathcal{P}(V_n)$. The ascent in the hierarchy V_n for finite n is exemplified by Fig. 4.1. Recall the set-theoretical natural numbers $0 = \emptyset$, $1 = \{0\}$, $2 = \{0, 1\}$ etc.

The first transfinite levels after all finite n are

$$V_\omega = \bigcup\{V_n \mid n \in \omega\}, \quad V_{\omega+1} = \mathcal{P}(V_\omega), \quad \ldots, \quad V_{\omega+(m+1)} = \mathcal{P}(V_{\omega+m})$$

etc., till the transfinite level

$$V_{2\omega} = \bigcup\{V_{\omega+n} \mid n \in \omega\}, \quad V_{2\omega+1} = \mathcal{P}(V_{2\omega}), \quad \ldots, \quad V_{2\omega+(m+1)} = \mathcal{P}(V_{2\omega+m})$$

etc., unlimited to high and always higher transfinite levels.

$\vdots$	$\cdots\cdots\cdots\cdots\cdots\cdots$
V_4	$\{0, 1, 2, 3, \{1\}, \{2\}, \{\{1\}\}, \{0, \{1\}\}, \{1, \{1\}\}, \{\{1\}, 2\}, \{0, 1, \{1\}\}, \cdots\}$
V_3	$\{0, 1, 2, \{1\}\}$
V_2	$\{0, 1\}$
V_1	$\{0\}$
V_0	0

Fig. 4.1. Start of the hierarchy of sets.

It is claimed that all sets can be grasped in this way, i.e., that every set appears on one of the levels and that there are no "exotic" sets not originating from $\emptyset$ and built by those systematic construction of sets. This is expressed in an equivalent way by the foundation axiom referring to this hierarchy.

Axiom (of Foundation). Every set x belongs to a level V_α for some ordinal α.

Already the first steps of the hierarchy V_n for $n \in \omega$ are a strong demand for our imagination. Further steps to V_ω and even further require the actual infinite and go beyond the usual imagination. The universe V of sets arising by applying axioms of building new sets is enormously large. The unlimited process beyond the finite into the transfinite indicates radically the explosive power of the Axiom of Infinity together with the Power Set Axiom.

Axioms listed so far form the system of ZF set theory. If the Axiom of Choice (AC) – usually treated with a certain reservation – is added then the resulting theory is denoted as ZFC. We shall write in a separate section about some problems connected with this axiom that are reasons of this reservation. Here let us say only few words.

The Axiom of Choice describes something usual: If several sets are given then it is unproblematic to choose one element from each of those sets. The Axiom of Choice allows also that infinitely many sets are given. How should one choose an element from each of infinitely many sets? It can be laborious even if the number of sets is finite but great. In the case of infinitely many sets it is *practically* impossible. It will again become visible how big the notional step up to the Axiom of Infinity is.

Axiom (of Choice).

$$\forall x\big[\forall y \forall z(y \in x \land z \in x \to y \neq \emptyset \land z \neq \emptyset \land (y = z \lor y \cap z = \emptyset))$$
$$\to \exists w \forall v(v \in x \to \exists u(w \cap v = \{u\}))\big].$$

This means:

> For any system x of nonempty pairwise disjoint sets v there is a set w having exactly one element from every set $v \in x$.

4.3.2 Von Neumann, Bernays and Gödel set theory

Also in the case of BG or NBG set theory formulated by von Neumann[4] [355, 356] in 1925 and 1928, Bernays [30, 31] in 1937 and Gödel [137] in 1940, the point is to avoid the known antinomies. According to von Neumann the reason of them is not the existence of "very large" sets but the fact that "large sets" can be treated as usual sets. The point is not to bound from the very beginning the process of building new sets and the size of set-theoretical objects as it is done in ZF set theory by the Comprehension Axiom but to grade the arising totalities.

Consider for example the domain V of all sets – and assume that it is an object, i.e., a set as any other. Then obviously $V \in V$. This is strange. In fact, the formation of V assumes that V is already given. Further $\{V\}$ and $\{V, \{V\}\}$ are also elements of V and should be built before V would be "ready". This does not suit our conception of forming sets. It is difficult to accept V as "definite" and as a "whole" as Cantor required with respect to objects of set theory. One should be – at least initially – careful and treat V not as a usual set and handle it as an object similar to others. What is the consequence of this? One should not allow V without further ado as an object in the process of building sets, as a potential element of a set.

What is the source of Russell's antinomy? The point is the totality $R = \{y \mid y \notin y\}$. The contradictions (a) $R \in R \Rightarrow R \notin R$ and (b) $R \notin R \Rightarrow R \in R$ arise only when the question $R \in R$ is asked. If it is not provided then there is no antinomy. This means that one is *requested* to exclude R as an object from the process of building sets.

Objects of NBG set theory are called *classes*. Among them are *sets*. *Sets* are those classes that can be members of other classes. *Proper classes* are those classes that cannot be members of other classes – as for example R.

Definition. A class x is called a set if there is a class K such that $x \in K$. We write "$\mathrm{mg}(x)$". A class is called *proper* if K is not a set.

We use small Latin letters as set variables, and Latin capitals as class variables. This has been used already in the definition and when we spoke about V and R. For R it is clear by our above considerations: if R were a set then by (a) and (b) one would get a contradiction. In the case of V this follows from the comprehension axiom of NBG set theory formulated below.

First however we secure the building of classes. Since classes are collections of sets only, we restrict properties used in forming classes to properties of sets. If in a property a quantifier occurs then it should refer to sets only (set theory in which there is no such restriction is KM set theory – named after Kelley and Morse). To each such

4 It is interesting to note that von Neumann chose as primitive notions not the concept of a set and the membership relation but the concept of a function. The formulation of axioms in [355, 356] were difficult to comprehend and did not correspond to the intuitive understanding of sets. Bernays and Gödel formulated then the axioms in a simpler way by using again the concept of the membership.

property φ corresponds an axiom. Hence one has again a schema of axioms describing the collecting, the "comprehension" of objects into classes.

Axiom (Schema of Comprehension).

$$\exists Z \forall u(u \in Z \leftrightarrow \mathrm{mg}(u) \wedge \varphi(u)).$$

To indicate more exactly set and class variables in properties and expressions we write this axiom schema explicitly again for $(n+1)$-ary properties φ. Note that the unbounded variables in φ can be class variables denoted here – according to our convention – by capital letters X_i.

Axiom (Schema of Comprehension).

$$\forall x_1 \ldots \forall x_n[\mathrm{mg}(x_1) \wedge \ldots \wedge \mathrm{mg}(x_n) \rightarrow \exists Z \forall u(u \in Z \leftrightarrow \mathrm{mg}(u) \wedge \varphi(u, x_1, \ldots, x_n))].$$

For the class Z we used the "class-term" as it was done above for sets:

$$Z = \{u \mid \varphi(u)\}.$$

The comprehension schema ensures, e.g., the empty class. It suffices to consider an unsatisfiable property – for example $x \neq x$ – and one gets

$$\emptyset = \{x \mid \mathrm{mg}(x) \wedge x \neq x\}.$$

The class of all sets is

$$V = \{x \mid \mathrm{mg}(x) \wedge x = x\},$$

and Russell's class is

$$R = \{x \mid \mathrm{mg}(x) \wedge x \notin x\}.$$

Notice that the originally set curly brackets "{" and "}" in the class term become "class curly brackets".

The distinction between sets and classes is the *fundamental idea* of NBG set theory. The fact that to each property corresponds a class makes the notation easier. Instead of speaking about properties φ and expressions $\varphi(u, x_1, \ldots, x_n)$ that belong to the domain of logic one can use classes K that are *objects of the NBG set theory*. We have just seen that for example the class of all sets V and Russell's class R are objects of this set theory being proper classes. In ZF set theory one can speak about them only indirectly as about domains outside set theory. Notice that NBG set theory is an extension of ZF set theory – in fact the domain of sets is extended by classes.

The following axioms of NBG set theory require that particular ways of building classes should lead to sets. This is the reason why we speak here not about NBG class theory but about set theory. With regard to contents the axioms differ not so much from the axioms of ZF. However, in thinking, speaking and writing there are – as a consequence of the fundamental idea – differences. We give here the axioms respecting the difference “classes – sets” without detailed comments and explanations (they are similar to those made in the case of ZF). We finish with some remarks on the connection between NBG and ZF.

The Axiom of Extensionality is similar to the case of ZF. The only difference is that here one speaks about the equality of classes whose elements are sets. The different variables make it clear.

Axiom (of Extensionality).

$$\forall X \forall Y [\forall z (z \in X \leftrightarrow z \in Y) \rightarrow X = Y].$$

One can collect only sets to pairs. This should be explicitly formulated. If x and y are sets then one can without further ado build the class $\{x, y\}$. It is required that $\{x, y\}$ is a *set*.

Axiom (of Pairing).

$$\forall x \forall y [\mathrm{mg}(x) \wedge \mathrm{mg}(y) \rightarrow \exists z (\mathrm{mg}(z) \wedge \forall u (u \in z \leftrightarrow u = x \vee u = y))].$$

No axiom is needed to build a union of two classes or a union of classes.

Definition. Let A and B be classes. Then the class

$$A \cup B := \{u \mid u \in A \vee u \in B\}$$

is called a union of A and B.

Definition. If $\mathfrak{A}$ is a class of sets then the class

$$\bigcup \mathfrak{A} := \{u \mid \exists x (x \in \mathfrak{A} \wedge u \in x)\}$$

is called union of $\mathfrak{A}$.

It should be ensured that if a and b are sets then also $a \cup b$ is a set and that if $\mathfrak{A}$ is a set then also $\bigcup \mathfrak{A}$ is a set. We do this generally for union of sets. Since it concerns only sets the appropriate axiom is the same as it was in ZF set theory.

Axiom (of Union).

$$\forall x [\mathrm{mg}(x) \rightarrow \exists y (\mathrm{mg}(y) \wedge \forall z (z \in y \leftrightarrow \exists u (u \in x \wedge z \in u)))].$$

Briefly:

$$\forall x \Big(\mathrm{mg}(x) \rightarrow \mathrm{mg}\Big(\bigcup x\Big)\Big).$$

The Comprehension Axiom Schema ensures that for any property φ restricted to sets there exists a class Z of elements fulfilling this property. If the elements u come from a given *set* v from which one separates u then this class should be a set. This is required – as in the case of ZF set theory – by the Comprehension Axiom (Separation Axiom).

Since elements $u \in v$ having the property φ form in any case a subclass Z of v one can formulate the Comprehension Axiom simply as follows.[5]

Axiom (of Comprehension).

$$\forall z \forall v(\mathrm{mg}(v) \land z \subseteq v \rightarrow \mathrm{mg}(z)).$$

It should be added that a subclass of a class A is a class such that all its elements are also elements of A.

Subclasses of classes X can be proper classes. So "power classes" of classes – analogue to power sets – cannot be considered. However, if x is a set then – according to the above-formulated Comprehension Axiom – subclasses y of x are sets. So one can make the following definition.

Definition. Let x be a set. Then $\mathcal{P}(x) := \{y \mid y \subseteq x\}$ is said to be the *power class* of x.

Now the Power Set Axiom states that power classes of sets are sets.

Axiom (of Power Set).

$$\forall x[\mathrm{mg}(x) \rightarrow \exists y(\mathrm{mg}(y) \land \forall z(z \in y \leftrightarrow \forall u(u \in z \rightarrow u \in x)))].$$

Briefly:

$$\forall x(\mathrm{mg}(x) \rightarrow \mathrm{mg}(\mathcal{P}(x)).$$

The Axiom of Infinity of NBG set theory corresponds literally to this axiom of ZF. However, the property $\mathrm{mg}(x)$ must be inserted into the requirement of an inductive class. About inductive classes we can speak exactly as it was done above with respect to inductive sets.

Axiom (of Infinity).

$$\exists x[\mathrm{mg}(x) \land \emptyset \in x \land \forall u(u \in x \rightarrow u \cup \{u\} \in x)].$$

Briefly: *There exists an inductive set.*

Another formulation will be given in order to use the expression possibilities of NBG set theory. One can speak about the *class* ω of (set-theoretical) natural numbers – it can be described like the set ω of ZF indicated above. It is not necessary to require first by an Axiom of Infinity the existence of an inductive set.

Definition. $\omega := \{x \mid \forall X(\text{is inductive class} \rightarrow x \in X)]$.

This is a typical case of an impredicative definition (cf. Section 2.16 about Poincaré). One can easily define ω predicatively as a class by properties of elements, namely as a class of special "finite" sets (cf. [23, p. 75]). The Axiom of Infinity is now short and concrete: the (set-theoretical) natural numbers form a set.

5 The Comprehension Axiom is frequently called *Separation Axiom*.

Axiom (of Infinity).

$$\mathrm{mg}(\omega).$$

Turn now to the variant of the Replacement Axiom Schema for NBG set theory. It is about operations F and images under those operations. In the framework of classes this can be easily formulated. Operations are definite objects of NBG, namely simply functions on the universal class that do not have to be described by properties φ as it was in the case of ZF. The following formulation demands generally for a function F that images of elements of sets x are again sets: If x is a set then $\{F(y) \mid y \in x\}$ is a set. In symbols:

Axiom (of Replacement).

$$\forall F[F \text{ is a function} \to \forall x(\mathrm{mg}(x) \to \exists v(v = \{F(y) \mid y \in x\} \wedge \mathrm{mg}(v)))].$$

One can see that the expression possibilities of NBG allow to formulate a single Replacement Axiom instead of the Replacement Axiom Schema as it was in the case of ZF.

The Foundation Axiom of NBG is a literal translation of the Foundation Axiom of ZF set theory.

Axiom (of Foundation).

$$\forall X[X \neq \emptyset \to \exists y(y \in X \wedge y \cap X = \emptyset)].$$

Above we spoke about the meaning and plausibility of this axiom.

Also the Axiom of Choice in the formulation for ZF set theory can be translated into the case of classes: For every class X of nonempty disjoint sets y there exists a class W containing exactly one element from every set $v \in X$ and nothing more.

Axiom (of Choice (AC)).

$$\begin{aligned}\forall X[\forall y \forall z(y \in X \wedge z \in X \to y \neq \emptyset \wedge z \neq \emptyset \wedge (y = z \vee y \cap z = \emptyset))\\ \to \exists W \forall v(v \in X \to \exists u(W \cap v = \{u\}))].\end{aligned}$$

To exploit the expression possibilities of NBG set theory a stronger formulation is often used which secures the choice of elements from all nonempty sets. This is done by a function F on the universal class V assigning to every nonempty set x a (not exactly determined) element. This axiom is denoted by GC (Global Choice).

Axiom (of Choice (GC)).

$$\exists F[F \text{ is a function} \wedge \forall x(x \neq \emptyset \to F(x) \in x)].$$

The extension of NBG obtained by augmenting one of the Axioms of Choice is denoted by NBG + AC or NBG + GC, respectively.

4.3.3 Remarks

Though the fundamental ideas of ZF and NBG set theories are different, they are neverteheless in a certain sense equivalent:

- sentences about sets provable in ZF are also provable in NBG and vice versa (cf. [102, pp. 214 ff.]).
- ZF and NBG are equi-consistent, i.e., ZF is consistent if and only if NBG is consistent.

However, generally the expressing possibilities of NBG are more convenient for the mathematical everyday practice than those of ZF.

In NBG one can speak about classes that are not objects of ZF set theory. So one can speak for example about the class G of all groups, which is a proper class – as can easily be shown – hence no object of ZF. To say that g is a group one can write $g \in G$ in NBG, and if ZF forms the base of considerations then this must be paraphrased. NBG provides for example a (partial) basis for the theory of categories and functors – not provided by ZF set theory.

If one forgoes the Axiom of Infinity then in the resulting NBG set theory the infinite is not excluded. It turns out that in such, say, NBG_e set theory (in which one can resign further axioms except the Foundation Axiom) sets are just finite classes. The domain of the infinite is the domain of proper classes [23, p. 88]. In ZF set theory, without the Axiom of Infinity the infinite is eliminated. There are no infinite objects there.

It has often been stressed that we do not know whether both set theories that form today the foundation of almost all domains of mathematics are consistent. Moreover, some concrete essential problems as for example the continuum hypothesis (see below) cannot be decided in those set theories. They are in a certain sense too weak to describe all ideas concerning sets and on the other hand too strong to secure their consistency. Nevertheless, they seem to correspond to the intuitive concept of a set though they come from purely pragmatic considerations and are based on probably random axioms describing the formation of sets. Evidences of the acceptance of those systems of axioms provides another system, namely the system of Scott (1967) that first looks about for conceptual necessities and the hierarchical structure of sets. For example W. Felscher wrote about it in [117, Volume III, pp. 83 ff.]. Scott's system is comparable with both axiomatic systems presented above. Ebbinghaus develops in his handbook [102] explicitly the set-theoretic axioms as open and understands the formulating of ZF set theory as a step-by-step approximation of the intuitive concept of a set that in principle cannot be axiomatically grasped.

Besides the set theories ZF and NBG described above there exists a third important, logically formulated approach to a theory of sets trying to avoid antinomies in another way. B. Russell outlined it already in *The Principles of Mathematics* in 1903 and developed it together with A. N. Whitehead in the famous *Principia Mathematica* in 1910–1913. It is based on the hierarchy of logical types. We wrote about this conception in Chapter 2 in Section 2.19 devoted to logicism. Recall here only that the theory of types consists of distinguishing properties according to "types" forming levels of an infinite hierarchy. The key to avoiding antinomies is the fact that properties belong to exactly one level of the hierarchy and that there are no "mixed properties", e.g., properties being simultaneously properties of individuals of the lowest level and of the level of properties of such properties. The substantial expression problems of this approach for mathematical practice have been indicated.

The axiomatic set theories ZF and NBG coming from Zermelo and von Neumann as well as the type-theoretic approach of Russell play in the discipline "theory of sets" an important role. They are, say, the standard in set-theoretical investigations. We refer here to Chapter 2, in particular to the sections devoted to logicism, formalism and intuitionism as well as the following sections of this chapter. Till the 1950s the version of set theory in the framework of the theory of types in Russell's setting was popular. Today it seems that the Zermelo–Fraenkel set theory with the Axiom of Choice (ZFC) – in spite of its smaller expressing power – enjoys much more popularity as a background of the mathematical practice and teaching than NBG set theory. This is certainly a consequence of the fact that it exists since 1908 while the NBG set theory, more exactly a variant of it, became known outside the very "theory of sets" only in the 50s and 60s thanks to the popular handbook of topology by J. L. Kelley [206].

4.3.4 On modifications

Set theory denoted by QM developed by Quine and Morse is based on the same axioms as NBG. However, in the Comprehension Axiom are now properties allowed that are *not limited to sets* in which also classes as variables may occur and stand in the scopes of quantifiers. It is just the set theory that has been used in the handbook of topology [206] mentioned above. The QM set theory is stronger than the presented ones, i.e., there are propositions about sets that are provable in QM but neither in ZF nor in NBG. Also the problem of the consistency of QM is different. It cannot be obtained from the proof of consistency of ZF even if this would be given.

In 1956 Wilhelm Ackermann (1896–1962) proposed another approach to avoid antinomies. His set theory follows slightly modified principles. A weaker version of the limitation of size is taken into account in its axioms. Ackermann's set theory, denoted by A, speaks – similarly to NBG set theory – about classes. It is assumed like in the Comprehension Axiom of NBG that every subclass of a *set* is a set. But contrary to NBG not every element of a *class* is immediately a set – as it was *per definitionem* in the case of NBG. The primitive notions of Ackermann's theory A are the predicate "set" and the membership relation $\in$. The aim of Ackermann was to describe the step by step composition of the world of sets by a theory of sets in which it is not assumed from the very beginning that all sets are given. It was shown that exactly the same propositions about sets are provable in Ackermann's set theory A as in the Zermelo–Fraenkel theory ZF. Hence the theories A, ZF and NBG are equivalent with respect to the domain of sets.

In the 60s a group of Czech mathematicians around Petr Vopěnka proposed a theory in which so-called *semisets* occur (cf. *The Theory of Semisets* [358]). In this theory the universe of objects is larger than in ZF, NBG or QM. Sets, classes and just semisets are considered there where a semiset is a subclass of a set being not a set. This is an essential difference with respect to NBG in which the Comprehension Axioms claims that a subclass of a set is a set.

In the 70s Vopěnka and his collaborators proposed still another set theory called *alternative set theory* (cf. [357]). Its aim was to provide a framework the starting point of which are concrete phenomena of the reality and which would be a basis for the reconstruction of mathematics. It is a non-formalized theory though some important parts of it are axiomatizable. It is similar to the theory of semisets. In some aspects it is connected with Robinson's and Laugwitz's approach to nonstandard analysis (cf. Sections 3.3.7 till 3.3.9 and Chapter 6) and with the ideas of ultraintuitionism (cf. Section 2.20). There are two sorts of objects in this theory: sets and classes. Sets are clearly defined, well-differentiated, unchangeable *finite* objects. Classes are in contrast considered as idealizations of properties that therefore can be infinite. Large parts of the mathematics can be reconstructed in the alternative set theory. At the same time it allows to formulate conceptions that cannot be adequately or only with difficulties formulated in other set theories. As examples can serve the differential and integral calculus in Leibniz's formulation with the infinitely small or the paradox of a bold man[6], the old paradox of moving[7] and generally domains between being continuous and being discrete.

Also Russell's proposal to avoid paradoxes has been developed and modified. Let us mention systems of Quine known as New Foundations (NF) in 1937 and Mathematical Logic (ML) in 1940. The former was an attempt to combine Zermelo's idea of limitation of size and Russell's idea of types. In the latter von Neumann's idea of distinguishing between classes and sets (as in NBG set theory) is applied.

4.4 The Axiom of Choice and the Continuum Hypothesis

The Axiom of Choice is perhaps the most considered and discussed axiom in set theory and in mathematics. The interest in it is similar to the interest in the parallel postulate of Euclid that accompanied mathematics till the 19th century. The axiom AC has been formulated above. It says:

> For every family of nonempty disjoint sets there exists a "selection set" containing exactly one element of each of those sets.

This statement seems to be highly plausible and it is often unconsciously applied. It is for example usual to assume that equivalence classes have representatives and that they form a set. The choice of such representatives from the classes of a partition provides an equivalent formulation of the axiom. There are many such equivalent formulations. Some of them will be given below.

The plausibility of the Axiom of Choice has its source in the finite that is characteristic for our thinking. If the Axiom of Infinity is assumed then it is consequent, even necessary to transfer the self-evident choice of elements from sets to infinite collec-

6 When does a man loosing his hair become bold?

7 Arrow paradox – the moving arrow rests at any moment and hence it does not move.

tions. However, if a family of sets from which elements should be chosen is given and infinite then there is a problem. This indicates again the challenge hidden in the Axiom of Infinity. The problem of the Axiom of Choice – and simultaneously its strength – consists of its nonconstructivity. It is not said – what in the finite case is no problem – how the choice should be done. If one recalls Cantor's explanations of the concept of a set (cf. Section 2.14) then it is clear that the formation of the choice set is out of the ordinary. There is neither a "law" through which the elements "can be united to a whole" nor are the elements that should be collected "definite".

The necessity of an axiom of choice has been noticed for the first time apparently by Giuseppe Peano in a work from 1890 in which the problem of the existence of solutions of special systems of ordinary differential equations was discussed. He came in a proof to the point where an element from every set of a family of sets $A_1, A_2, A_3, \ldots$ of reals should be chosen. He wrote there:

> "However, since one cannot apply infinitely many times an arbitrary law by which one assigns (*on fail correspondre*) to a class an individual of that class, we have formulated here a definite law by which, under suitable assumptions, one assigns to every class of a certain system an individual of that class." [267, p. 210]

In 1902 Beppo Levi applied explicitly the Axiom of Choice in a work devoted to the cardinality of the set of all closed sets of reals. Prior it was applied by various mathematicians, among them by Cantor, but without being conscious of doing something special. So far it was not treated in the classical mathematics and in logic as a special principle of forming sets.

In 1904 Zermelo gave the first explicit formulation of the Axiom of Choice and used it in his proof of the *well-ordering theorem* (see [381]). In 1906 Russell gave the formulation of the axiom used today. He called it the "multiplication axiom".

The Axiom of Choice plays an important role in mathematics. This role is greater than thought by mathematicians not working in the foundations of mathematics. Many mathematical statements use it or some principles equivalent to it. Among those equivalent principles are among others: (1) Zermelo's Well-Ordering Theorem: Every set can be well ordered. (2) Tukey's Lemma: For any property Φ of subsets of a given set A being of finite character[8] the following holds: Every subset of A having the property Φ is contained in a maximal set with property Φ. (3) Kuratowski–Zorn Lemma: A partially ordered set such that every chain (linearly ordered subset) of it has an upper bound containing a maximal element.

8 The property Φ is of finite character if and only if $\Phi(\emptyset)$ and for arbitrary sets $B \subseteq A$ one has $\Phi(B) \leftrightarrow \forall C(C \subseteq B \wedge C \text{ finite} \rightarrow \Phi(C))$. This means: the property Φ is of finite character if it transfers from sets B to its finite subsets and vice versa.

Now we give examples of some mathematical theorems in proofs of which the Axiom of Choice plays an essential non-eliminable role. In set theory itself the Axiom of Choice is needed for the following statements: (a) Every infinite set contains a countable subset. (b) The union of countably many countable sets is countable. (c) The set of reals is not a union of countably many countable sets. (d) The Cartesian product of nonempty sets is nonempty. (e) Every two sets are comparable with respect to their cardinality. In topology the Axiom if Choice is needed in particular in the proof of Tichonov's theorem stating that the product of compact topological spaces is compact. It was shown that Urysohn's Lemma[9] is equivalent to the Axiom of Choice. In measure theory AC is used in the proof that there exist Lebesgue nonmeasurable sets of reals. In analysis the Axiom of Choice is needed in order to prove that the definition of continuity in the sense of Heine[10] is equivalent to the ε-δ-definition of continuity of Cauchy. Moreover, the implication that if a function is continuous in the point x_0 in the sense of Heine then it is also continuous in the sense of Cauchy, is equivalent to the "countable Axiom of Choice", i.e., to AC for countable families of sets. In functional analysis one uses the Axiom of Choice (in the form of Tukey's Lemma) in the proof of the existence of a basis for an arbitrary vector space and in the proof of the Hahn–Banach Theorem. In algebra AC is applied in the proof of the theorem stating that every field F has exactly one (up to isomorphism) algebraically closed extension. Also this theorem is equivalent to AC for countable families. The Axiom of Choice is also needed to show that a subgroup of a free group is free or that every group has a maximal Abelian subgroup.

This is only a small selection of theorems in the proofs of which the Axiom of Choice is needed. There exists an extensive book [182] where hundreds of such examples are given and classified.

The Axiom of Choice seems to be indispensable for the mathematical practice. On the other hand it leads to some paradoxical consequences. Its best known paradoxical consequence is certainly the Banach–Tarski theorem on the paradoxical decomposition of a sphere from 1924. They have shown – using some ideas of Felix Hausdorff – that a sphere can be decomposed into a finite number of parts which then can be put back together to yield two identical copies of the original sphere.

Let S be a 3-dimensional sphere of diameter 1 and let $X, Y \subseteq K$. The sets X and Y are said to be congruent (briefly: $X \equiv Y$) if and only if there exists a rotation δ of K such that $\delta(X) = Y$. They are said to be congruent by decomposition (briefly: $X \approx Y$) if and only if there exist disjoint sets $X_1, X_2, \ldots, X_n$ and $Y_1, Y_2, \ldots, Y_n$ such that $X = X_1 \cup X_2 \cup \cdots \cup X_n$, $Y = Y_1 \cup Y_2 \cup \cdots \cup Y_n$ and $X_i \equiv Y_i$ for every $i = 1, 2, \ldots, n$. The Banach–Tarski theorem states now that there exist disjoint subsets X and Y of K

9 Let X be a topological space in which closed disjoint sets can be separated by open sets. If $A, B \subseteq X$ are disjoint and closed then there exists a mapping $f : X \to [0, 1]$ with the value 1 on S and the value 0 on B.

10 The definition of Heine says: a function f is continuous at x_0 if and only if for any sequence x_n such that $\lim x_n = x_0$ it holds $\lim f(x_n) = f(x_0)$.

such that $X \approx K$ and $Y \approx K$. It can be also shown that the same holds for a closed or open ball without the middle point. R. M. Robinson proved that 5 is the minimal number of parts into which a sphere can be decomposed in such a way that they form (in the sense of the relation $\approx$) two spheres. Banach showed that there exists no paradoxical decomposition of a figure on the plane.

The decompositions from the Banach–Tarski theorem seem to contradict our intuition originating from our experience with measuring. However, this experience is based on the contact with and perception of "normal" sets, first of all polygons and polyhedrons. Sets from the theorem are mostly pathological, in particular Lebesgue nonmeasurable – their existence is proved by using the Axiom of Choice. How the proof looks like and how one ball can be transformed into two balls is sketched for example in [217] by K. Kuhlemann. He tells there a short story of "Downfall of Mathemagica" (*Der Untergang von Mathemagika*) with many mathematical and philosophical allusions.

The reason for our doubts towards the Axiom of Choice are not only its paradoxical consequences. In fact, it has a completely different character than other axioms of ZF or NBG: it is not constructive. It postulates the existence of a set without giving any information about it. Moreover, the set postulated by the axiom is in no way uniquely determined – in contrast to sets required by other axioms of ZF or NBG. In the case of equivalence classes for example there are generally various sets of representatives. This is the reason why the Axiom of Choice is strictly rejected by constructivists and intuitionists.

Since the Axiom of Choice generates some problems (primarily of the philosophical nature) and, on the other hand, it is indispensable in many branches of mathematics, some mathematicians treat theorems proved with the help of it in a special way and explicitly mark this fact.

The outstanding position of the Axiom of Choice in mathematics is the reason of great interest paid to it in the foundations of mathematics. The situation concerning the Continuum Hypothesis is similar. Recall what it says. By a theorem of Cantor the cardinality of a set A is smaller that the cardinality of its power set $\mathcal{P}(A)$, briefly: $\overline{\overline{A}} < \overline{\overline{\mathcal{P}(A)}}$. The operation of a power set generates an infinite sequence of cardinals, the hierarchy of beths ($\beth$). There is also another hierarchy of cardinals, the hierarchy of alephs ($\aleph$). It is the hierarchy of cardinals of well-ordered sets. By the Axiom of Choice, more exactly by the Well-Ordering Theorem of Zermelo equivalent to it, it follows that this hierarchy contains all cardinals. The first elements of those hierarchies are equal by definition, i.e., $\aleph_0 = \beth_0$. The Continuum Hypothesis, briefly CH, states – what can be expected – that also the second elements are equal, i.e., $\aleph_1 = \beth_1$, hence $\aleph_1 = 2^{\aleph_0}$. The Generalized Continuum Hypothesis, briefly GCH, is the statement that the hierarchy of alephs and the hierarchy of beths are equal. This means that for all ordinals α it holds $\aleph_\alpha = \beth_\alpha$, i.e., $\forall\alpha(\aleph_{\alpha+1} = 2^{\aleph_\alpha})$. Sometimes the Continuum Hypothesis is given in another form. It can be shown that the set of all subsets of the set $\mathbb{N}$ of natural numbers, hence the set $\mathcal{P}(\mathbb{N})$ and the set of reals $\mathbb{R}$ are equipollent. So one can formulate the Continuum Hypothesis in the following way: every infinite set of real numbers is equipollent either to the set of natural numbers $\mathbb{N}$ or to the set of all real numbers $\mathbb{R}$.

The Generalized Continuum Hypothesis then looks as follows: every family of subsets of an infinite set A is equipollent either to a subset of A or to the whole set $\mathcal{P}(A)$. Note that the latter formulations are equivalent to the former if the Axiom of Choice is assumed.

The last remark indicates that there are interconnections between The Axiom of Choice and the Continuum Hypothesis. In 1926 Alfred Tarski and Adolf Lindenbaum conjectured and in 1947 Wacław Sierpiński proved that the Generalized Continuum Hypothesis GCH (in the second formulation) implies the Axiom of Choice. It is worth noting that this proof was purely combinatorial and did not use any transfinite principles.

A particular problem connected with the Axiom of Choice and with the Continuum Hypothesis is their status in the set theory ZF. What is the connection of those principles and the axioms of set theory? If it were shown that they are consequences of other axioms – that are not so problematic and not connected with any controversies – then their philosophical and methodological position would be strengthened and they would be seen less critically. However, it has turned out that the situation is different: in 1938 Kurt Gödel proved that GCH (thus also CH) and AC are relatively consistent with other axioms of ZF. This means that if ZF is consistent then the theories ZF + GCH and ZF + AC are also consistent. Or: if GCH or AC added to other axioms of ZF led to inconsistency then the inconsistency would be provable already in ZF alone. Hence neither the negation of GCH nor the negation of AC can be proved in ZF. Gödel proved this by constructing a model of ZF in which AC and GCH are satisfied. This is a model of the so-called "constructible sets" – they are generated from the empty set by given set-theoretical operations iterated infinitely many times. In Section 4.4.1 we give the universe of constructible sets in another but equivalent formulation.

Gödel's result does not solve the problem completely. Only in 1963 Paul J. Cohen completed the answer to the question by a new method called the method of forcing. He proved the following:

(i) It is not true that (CH $\rightarrow$ GCH).
(ii) It is not true that (AC $\rightarrow$ GCH).
(iii) AC and CH (and *a fortiori* GCH) do not follow from other axioms of ZF.
(iv) Neither AC implies CH nor CH implies AC in ZF.

Hence the consequence of Gödel's and Cohen's results is: AC and GCH are relatively consistent with and independent of Zermelo–Fraenkel set theory. Consequently, a set theory ZF with the Axiom of Choice and the Generalized Continuum Hypothesis is as possible as a set theory without them or even with the negation of one or both of them. If one takes into account that set theory is the base and foundation of the whole mathematics and that many theorems of analysis and algebra are dependent on the Axiom of Choice then one should realize that various different mathematics, in particular various different forms of analysis are possible. Which of them would be or is correct? How should this situation be handled? This can be compared with the situation in geometry after the discovery of non-Euclidean geometries. The point is here whether the parallel postulate or its negation should be assumed. Today both

Euclidean or non-Euclidean geometries are developed and applied in mathematics. The case of set theory is more serious – here the point is the foundation of the whole mathematics.

If one treats Cantor as the author of the Continuum Hypothesis and as the first who applied the Axiom of Choice then one speak about Cantorean and *non-Cantorean set theory* as well as about Cantorean and *non-Cantorean mathematics*. However, the analogy with geometry does not help any more. One cannot simply choose this or that mathematics according to these or those conditions or requirements. One expects – independently of our philosophical views – a clear situation. One can be calmed down by the hope – indicated above as well as in Section 2.23 about Gödel – that we are still on the way to a set theory that would adequately describe our intuitions of a set and solve problems connected with the Axiom of Choice and the Continuum Hypothesis in this or another way. After all there were and are researches in the foundations of mathematics.

4.4.1 Search for new axioms

Since the position of the Axiom of Choice in set theory is not quite clear and its consequences are ambivalent – beside "positive" there are also, as we have seen, unwanted consequences of AC – and its role in mathematics is not beyond dispute one looks for other alternative principles. One of them is the Axiom of Determination (AD) introduced by Jan Mycielski and Hugo Steinhaus in the paper *A Mathematical Axiom Contradicting the Axiom of Choice* (1962).

To formulate this axiom consider a play between two persons. Let a set M, an ordinal α and a set $A \subseteq M^{2\alpha}$ be given. Hence elements of A are sequences of elements of M of the length 2α. We define a play $G^{\alpha}_{M}(A)$ in the following way: Players I and II choose elements from the set M and player I begins. He chooses an element $v_0 \in M$, player II chooses an element $w_0 \in M$. Then player I chooses an element $v_1 \in M$, player II chooses an element $w_1 \in M$ etc. After α steps they will build a sequence $\langle v_0, w_0, v_1, w_1, v_2, w_2, \ldots \rangle$ of length 2α, hence an element of $M^{2\alpha}$. The first player wins if this sequence belongs to the set A, otherwise wins player II.

A *strategy* is a mapping τ assigning to every sequence $\overline{u} = \langle u_0, u_1, \ldots \rangle$ with $u_i \in M$ of length $< 2\alpha$ an element $\tau(\overline{u}) \in M$. A strategy τ is said to be a *winning strategy* for player I in the play $G^{\alpha}_{M}(A)$ if and only if for any sequence $\overline{w} = \langle w_0, w_1, \ldots \rangle$ of length α chosen by player II the sequence $\overline{v} * \overline{w} = \langle V_0, w_0, v_1, w_1, \ldots, \rangle$ of length 2α belongs to the set A. The sequence $\overline{v} = \langle v_0, v_1, \ldots \rangle$ of elements chosen by player I is determined in the following way $v_0 = \tau(\emptyset)$, $v_1 = \tau(\langle v_0, w_0 \rangle)$, $v_2 = \tau(\langle v_0, w_0, v_1, w_1 \rangle)$, etc. In a similar way a winning strategy for player II is defined. The play $G^{\alpha}_{M}(A)$ is said to be *determined* if and only if one of the players has a winning strategy.

The *Axiom of Determination* (AD) now says the following.

Axiom. For any set $A \subseteq \omega^{\omega}$ the play $G^{\omega}_{\omega}(A)$ is determined.

Recall that ω is here the sequence of natural numbers and ω^{ω} is the set of all infinite sequences of natural numbers. Note that by theorems of the arithmetic of ordinals it holds: $2\omega = \omega$. So one can say briefly:

(AD) For any set A of infinite sequences of natural numbers there is a winning strategy for the play $G^{\omega}_{\omega}(A)$.

What are the consequences of the Axiom of Determination? In the system ZF + AD the following statements can be proved:

(i) Every set of reals is Lebesgue measurable.
(ii) The Axiom of Choice for countable families of real numbers.
(iii) There exists a pre-well-ordering (i.e., a transitive and founded relation) on the set $\mathbb{R}$.
(iv) Every filter on the set $\mathbb{N}$ of natural numbers is principal.
(v) $\aleph_1$ is a measurable cardinal.
(vi) $\aleph_3$ is a singular cardinal.
(vii) For any set X of real numbers either $\bar{\bar{X}} \leq \aleph_0$ or $\bar{\bar{X}} = 2^{\aleph_0}$. This means that the Continuum Hypothesis holds.

Again the question about the status of the Axiom of Determination in the system ZF can be asked. Note that the main problem of the consistency of AD with respect to the axioms of ZF is so far unsolved. R. Solovay proved that if the theory ZF + AD is consistent then the theory ZF + AC + "there exist uncountable measurable cardinals" is also consistent. This result indicates how difficult the problem of relative consistency of AD is.

It has been noticed above that the Axiom of Choice leads to the construction of not Lebesgue measurable sets. This indicates a conflict with the first of the consequences of AD listed above. So AD and AC are contradictory. Which of them should be chosen, what should be used as the foundation of mathematics – ZF with AC or ZF with AD? In looking for the – philosophical – arguments for a decision one should take into account that both, i.e., the Axiom of Choice and the Axiom of Determination hold in the case of the finite. They are formulated in the language of set theory by formulas of a similar form (structure), i.e., as formulas of the form $\forall[\forall\exists \rightarrow \exists\forall]$. Add that the Axiom of Choice AC can be formulated as a special form of the Axiom of Determination. In fact, Mycielski proved that in ZF the following sentences are equivalent:

(a) The Axiom of Choice.
(b) $\forall M \forall A \subseteq M^2$ [the play $G^1_M(A)$ is determined].

Note that the theory ZF + "$\forall M \forall \alpha \forall A \subseteq M^{2\alpha}$ [the play $G^{\alpha}_M(A)$ is determined]", hence the determinancy of described plays in all generality, is inconsistent! Even for any single $\alpha \geq \omega$ the theory ZF + "$\forall M \forall A \subseteq M^{2\alpha}$ [the play $G^{\alpha}_M(A)$ is determined]" is inconsistent. It is worth noting that the Axiom of Determination speaks only about the play $G^{\alpha}_M(A)$ for $M = \alpha = \omega$ and that in this case $2\omega = \omega$.

Hence AD is not a general set-theoretical principle. It does not concern arbitrary sets but only ω and ω^{ω} and subsets $A \subseteq \omega^{\omega}$. So it is an axiom that concerns only the Baire space ω^{ω}, and it cannot be generalized without inconsistency. This distinguishes it from the Axiom of Choice that concerns generally the concept of a set. "Visually" this induces acceptance of AC and the rejection of AD. However, AD has some "beautiful" consequences – for example: all sets of real numbers are Lebesgue measurable. So we see that today there are no convincing philosophical arguments towards the acceptance or rejection of one of those axioms.

The results of Gödel and Cohen on the consistency and the independence of the system ZF of the Axiom of Choice and of the Continuum Hypothesis described above still show something else. They show that the characterization of sets given by axioms of ZF is too weak to decide, for example some properties of sets expressed by AC or GCH. Hence the question about a stronger systems of axioms arises. Some new strong axioms of infinity have been proposed – they postulate the existence of large cardinals. The simplest one claims the existence of inaccessible cardinals, i.e., cardinals closed with respect to the operations of the power set and the union of sets.

More exactly: A cardinal $\mathfrak{m}$ is said to be *inaccessible* if and only if (a) $\aleph_0 < \mathfrak{m}$, (b) if $\mathfrak{n} < \mathfrak{m}$ then $2^{\mathfrak{n}} < \mathfrak{m}$, (c) if $\overline{\overline{A}} < \mathfrak{m}$ and F is a mapping of A whose values are cardinals smaller than $\mathfrak{m}$, then $\Sigma_{x \in A} F(x) < \mathfrak{m}$. One can iterate the inaccessible cardinals into the transfinite and obtain in this way the sequence of Mahlo cardinals. It was described for the first time by Friedrich Paul Mahlo in 1911.

The study of large cardinals has been intensified since 1960. Various types of such cardinals have been introduced, for example measurable, compact, supercompact etc. cardinals. All this leads to the following two questions:
(1) How can those axioms postulating the existence of large cardinals be justified?
(2) What are their consequences with respect to the Continuum Hypothesis?
In the literature there are various answers to question (1). Gödel suggested to rely on the intuition. He wrote in the paper *What is Cantor's Continuum Problem?* [139, p. 265 in the version from 1964]:

> "also there may exist, besides the usual axioms, the axioms of infinity [which assert the existence of large cardinals – authors' remark], [...] other (hitherto unknown) axioms of set theory which a more profound understanding of the concepts underlying logic and mathematics would enable us to recognize as implied by these concepts."

However, he said neither *how* such a intuitive "profound understanding" can be achieved nor whether this or that definite axiom should be accepted or rejected.

Gödels set-theoretical visions have been of a large extent and at the same time optimistic. We read (loc cit):

> "There might exist axioms so abundant in their verifiable consequences, shedding so much light upon a whole field, and yielding such powerful methods for solving problems (and even solving them constructively, as far as that is possible) that, no matter whether or not they are intrinsically necessary, they would have to be accepted at least in the same sense as any well-established physical theory."

A. Kanamori and M. Magidor suggest in the paper *The Evolution of Large Cardinal Axioms in Set Theory* [198] introducing new axioms of infinity according to two principles: either in a "theological" way, i.e., metaphysical-philosophically motivated and justified or in a purely formal way taking into account only their "aesthetic value" seen in "beautiful" consequences and interesting interrelations. A clear position towards this proposal has been formulated by P. J. Cohen. He rejected any platonic realism and recommended pure formalistic arguments in set theory (cf. *Comments on the Foundations of Set Theory* [75]).

There are attempts to justify the existence of large cardinals by general principles, in particular by the reflection principle. The latter has been formulated by Azriel Levy for properties of the first-order logic and later generalized by Paul Bernays for properties of the second order. It states that any property that holds for the universe V of all sets holds also for a certain level V_α of the cumulative hierarchy of sets (see above). Recall that the universe V is a union of the levels V_α over all ordinals α. From the reflection principle for the second-order logic follows for example the existence of Mahlo cardinals.

And now about question (2): Do new axioms of infinity together with axioms of ZF enable to solve the Continuum Hypothesis? It seems that the Continuum Hypotheses CH and GCH are on the one hand consistent with and on the other independent of *every* Axiom of Infinity proposed so far, which means: if K is a new Axiom of Infinity such that ZF + K is consistent then the theories ZF + K + GCH as well as ZF + K + ¬GCH are also consistent. Hence the axioms proposed so far do not give anything new with respect to the Continuum Hypothesis. The expectations and hopes of Gödel have not been realized so far.

In this situation one can ask which levels of the infinity are really needed for the "finite mathematics", i.e., for the mathematics about finite objects as natural numbers or finite sets. Gödel claimed in his famous paper *Über formal unentscheidbare Sätze der 'Principia Mathematica' und verwandter Systeme. I* (cf. [136, footnote 48a, p. 191], English translation in [140]) the following: "For it can be shown that the undecidable propositions constructed here become decidable whenever appropriate higher types are added (for example, the type ω to the system P)[11]. An analogous situation prevails for the axiom system of set theory." Since Gödel's undecidable sentences are of finite character, i.e., they speak about finite objects, in particular about natural numbers, one can state that unbounded transfinite iterations of the power set operation are necessary for the completeness and justification of finite mathematics. New results of J. Paris, L. Harrington and L. Kirby on new "independent", i.e., undecidable sentences of a mathematical, more exactly combinatorial or number-theoretical and not – as it was the case in the proof of the First Incompleteness Theorem of Gödel – of a metamathematical content, show that Gödel was right. The new independent sentences can be proved with the help of transfinite methods that exceed the arithmetic. They

11 *P* is the system from the *Principia Mathematica* [370] together with constants for natural numbers – authors' remark.

show also that at least the first level of the transfinite in Cantor's set theory should be necessarily assumed and accepted for the mathematics of finite objects.

Let us still mention a theorem due to Kreisel. It states that a sentence φ of the language of Peano arithmetic, hence a sentence about natural numbers provable in the theory ZF + AC + GCH is also provable in the set theory ZF itself. Hence the set theory ZF + AC + GCH is a *conservative* extension of ZF with respect to natural numbers. This means that the extension of ZF by the Axiom of Choice and by the Generalized Continuum Hypothesis does not prove any new arithmetical theorems. Briefly: AC and GCH have no meaning for our knowledge about natural numbers.

Another approach to extend the axioms of ZF goes back to Gödel. Above we already hinted at it briefly. The point is to prove the relative consistency of the Axiom of Choice and of the Continuum Hypothesis. Gödel constructed to this end a model by *limiting* the formation of sets. The limitation is done from outside. Consider how it works in the case of building a new hierarchy by a limited power set operation. By building new subsets, only so-called "constructible" or "definable" sets are allowed. Let x be a set. Then a subset y of it is said to be *definable* if and only if there is an expression φ and elements $z_1, \ldots, z_n$ such that $y = \{z \in x \mid \varphi(z_1, \ldots, z_n)\}$. Instead of the power set operation $\mathcal{P}$ a new operation $\mathcal{D}$ is introduced now – it collects the definable subsets of a given x:

$$\mathcal{D}(x) = \{y \mid y \subseteq x \wedge y \text{ definable}\}.$$

The relation $\mathcal{D}(x) \subset \mathcal{P}(x)$ holds. With the help of $\mathcal{D}$ and $\bigcup$ the *constructible* hierarchy L arises:

$$\begin{aligned} L_0 &= \emptyset, \\ L_{\alpha+1} &= \mathcal{D}(L_\alpha), \\ L_\gamma &= \bigcup\{L_\beta \mid \beta < \gamma\}, \end{aligned}$$

where α is an arbitrary ordinal and γ is a limit ordinal, i.e., an ordinal without a predecessor. The hierarchy L is the "universe" of all definable sets as V is the universe of all sets. The *Axiom of Constructibility* states now that both these universes coincide, i.e., that every set is definable.

Axiom (of Constructibility). $V = L$.

If ZF is consistent then ZF + $V = L$ is also consistent and the Axiom of Choice, even the Axiom of Global Choice (GC) as well as the Generalized Continuum Hypothesis are provable in ZF + $V = L$.

The Axiom of Choice holds because all sets are now "constructible" and in this way the "non-constructive part" of this axiom is reduced. This was the aim of Gödel's construction. One of the consequences of the limitation of the universe expressed by $V = L$ is the fact that now the Continuum Hypothesis also holds.

Note that $V = L$ is *independent* of the axioms of ZF together with the Axiom of Choice since ZFC + $\neg V = L$ is consistent if ZF is consistent. And again the following question can be asked: which set theory ZF + $V = L$ or ZF + $\neg V = L$ should be chosen and accepted? And again the answer is difficult. In fact, on the one hand only the limitation of building sets to definable sets seems reasonable. It is in a sense an extension of the fact that in the finite every set is in principle built by a certain property. On the other hand the limitation expressed by the Axiom of Constructibility seems to be a bit artificial. The axiom $V = L$ and the Power Set Axiom do not fit together. The requirement that all subsets of, say, ω or even of $\mathbb{R}$ are definable does not satisfy our intuitive expectations. Think for example of the so-called "infinite non-periodic decimal fractions" and lawless sequences connected with them that are used to describe real numbers.

The aim of Gödel by constructing L was not to propose a new set theory. His aim was the construction of *inner* models in order to prove the relative consistency of AC and GCH. We have indicated above – when we spoke about the large cardinals – that Gödel himself did not lean towards the restriction of set formation – just the opposite, he leaned towards its extension.

4.4.2 Further remarks and questions

Almost all problems and questions considered by us so far are connected with the actual infinite postulated by the Axiom of Infinity. Without the actual infinite there would be no problems. In fact, the set theory ZF without the Axiom of Infinity is consistent (cf. [115, p. 9]). However, without the infinite mathematics would not go too far. It would be restricted to elementary arithmetic (cf. [23]). Above we have over and over again repeated and stressed that mathematics has created real numbers only thanks to the fact that it has accepted the actual infinite and made it an object of mathematics. But the question is: do we really need the infinite in mathematics to such an extent in which set theory makes it possible?

Do we need for *applied mathematics* a mathematics in which the actual infinite sets of *arbitrary magnitude* are assumed? – We disregard the problem that the concept of an applied mathematics is fuzzy and clouded. – The widespread hypothesis, though not verified, says that the answer to this question is negative. Already Hermann Weyl has shown in his monograph *Das Kontinuum* [365] in 1918 that large parts of the classical mathematical analysis can be developed in a conservative extension of the Peano Arithmetic (PA). New investigations due mainly to Solomon Feferman and Gaisi Takeuti led to the construction of a conservative extension of the arithmetic PA in which the whole classical and modern applied analysis can be formulated. In particular one can formulate there the theory of measurable sets and functions, the general theory of measure and integral as well as fragments of functional analysis. Further support of this hypothesis is provided by results of the so-called reverse mathematics (cf. Section 2.22). It has been shown there that there are weak theories – weaker than theories of Feferman and Takeuti – being conservative extensions of Skolem's Primitive Recursive Arithmetic (PRA) in which important and large parts of analysis and algebra can be reconstructed.

Should the actual infinite be restricted to what is practically necessary? Note that in the modern mathematics there are many domains being far from applications but essentially based on transfinite set theory. The rejection and cancelation of such set theory would mean an impoverishment of mathematics. The infinite is and remains a fundamental object of investigation of mathematics.

There are attempts to look for the foundations of mathematics not in set theory but in category theory developed by Samuel Eilenberg (1913–1998) and Saunders MacLane (1909–2005) in the 1940s. When based on this theory mathematics becomes a knowledge of categories and functors. A category is an abstract object consisting of a class of items and of morphisms between those items that can be connected. Functors are mappings between categories. E. Kleinert discusses in [210, pp. 65 ff.] the complementary relations between concepts of set theory and of category theory. The approach based on the category theory though technically useful and effective in many domains of mathematics (e.g., in topology) does not seem to be elementary and to correspond to the mathematical intuition to such a degree as today's set theory. Hence its meaning is still not the same as the meaning of set theory. Set theory is to a large extent the fundamental mathematical theory providing foundations of the whole mathematics.

4.5 Final remarks

In retrospect we would like to give a survey of what the axioms of both set theories do in fact state. The axioms say how sets in fact are and how we handle them. Since set theory constitutes the background of mathematics, the whole mathematical thinking is affected by set-theoretical principles. We think in sets – and rarely realize it. We would like to try to consider mathematical thinking being stamped by set theory.

Let us begin by the ontological question discussed at the very beginning. The general question concerns the nature and way of existence of mathematical objects. Above there have been remarks that – since sets are the fundament of all mathematical objects – the ontological questions can be reduced to the question concerning sets.

What are sets?

In Subsection 4.2.2 devoted to the problem of universals we have presented the most important conceptions concerning the way of existence of mathematical concepts applied to sets. After giving an insight into the axiomatic of set theory and its problems we ask the question about the nature of sets.

What are sets? The first conception is that given by Cantor in his "definition of a set". Sets consists of objects. From single objects, from single wholes new wholes will be built. This is an activity and achievement of thinking being represented by the concept of a set. So sets are themselves abstract objects of thinking. Their nature is not independent of objects collected by them into wholes even if their existence is autonomous. There are many conceptions concerning mathematical objects – as has been shown in Chapter 2. They should be taken into account when the philosophical question about sets is asked.

The situation is however strange – and typical for questions about the ultimate reasons and foundations. In fact, just sets are objects that should give us information on what mathematical objects are. We expected from set theory a general solution of the ontological problem. So philosophically we are in a sort of an inescapable – but informative – vicious circle.

This circle is clearly seen in the consequent *set*-theoretical procedure, namely in considering from the very beginning sets only as sets of sets. At first glance this gives not much in attempts of explaining the concept of a set. However, we shall point to principles according to which sets are built of sets and the interrelations between them are arranged. The question "What are sets" becomes the question "*How* are sets"? This is the insight that determined for a long time set theory as well as the very mathematics. Only axioms describing interrelations between sets can provide us information on the nature of sets. Some remarks have been made already when axioms of ZF were presented – they will now be partially collected and extended. Nobody can say *what* sets are.

Comments on set-theoretical axioms

Contrary to the case of the concept of number, in the case of the concept of a set one cannot refer to the old philosophical tradition. One finds there no information about sets. Sets were since the antiquity till the modern times mathematically meta-linguistic companions and there was no philosophical reflection on them. The concept of a finite set has been subordinated to a number.

The concept of a set entered temporarily the mathematical consciousness only in connection with the infinity and its paradoxes and disappeared again after the rejection of the actual infinite. It was unimaginable to treat the continuum as a set. Even in the context of the infinity sets were strongly connected with the concept of a number and subordinated to it. The impossibility to imagine an infinite number was one of the arguments against infinite sets. The concept of a set entered the consciousness of mathematicians only in the 19th century – again and now definitely in connection with the infinity. Problems connected with the concept of a set are today first of all problems concerning infinite sets. Only infinite sets required a new mathematical discipline – set theory. Its axioms characterize the nature of sets. Compare in particular the handbooks [92, 102, 117] where the concept of a set is discussed in detail.

The Axiom of Extensionality says that sets are characterized by their extensions. Those extensions can exceed – this is a consequence of the Axiom of Infinity – any imagination. Countable sets are treated as being clear and perspicuous – especially when a listing of their elements is given. However, there are problems with uncountable sets. It is not known how many levels higher than the countable sets they are. This concerns also the real numbers that are so fundamental for every mathematics. For a normal mathematician the orientation in the higher cardinalities is not clear. Recall for example inaccessible singular cardinals mentioned above. Even in the case of the universe V of sets the set theory provides no help. The universe V is in the genuine sense of the word *indescribably large*. There is no property that could characterize this universe. Every property is by the reflection principle satisfied already in a particular level V_α.

The size of sets is restricted by the Axiom Schema of Comprehension. It makes it possible to bound the domain of sets or to distinguish between sets and not-sets. A result of this is that after the old antinomies eliminated by this axiom schema no new antinomies appeared so far. However, the Power Set Axiom unfolds in the domain of infinite sets an inconceivable effect. It acts by transfinite exponents on the size of sets. Already the first step into the infinity leads to the problems connected with the Continuum Hypothesis.

The Replacement Axiom Schema allows the development of new transfinite dimensions. It was just the background of developing it as a principle of building new sets. The point is to secure the projection of ω beyond ω. The name "replacement" does not express its whole meaning. It is a question of replacing and *transferring* given and known sets and structures into unknown transfinite domains. It can be said that the Replacement Axiom secures in the universe of sets our ideas of sets being represented by given and known sets and structures. Reasons for this axiom are not necessarily expectations coming directly from our practical, finite experience with sets. Replacement and transfer are in the case of the finite self-evident to such a degree that they are unrecognized as principles. The reasons are in fact theoretical needs and requirements coming from the extension of sets into "the uncertain", into the infinite. Perhaps this can also explain why the necessity of this axiom has been recognized so late (1922).

The Foundation Axiom is in fact not an axiom securing the formation of new sets – it concerns the reasonable formation of sets and of the universe of sets. It meets concerns of a natural *finite* expectation and reflects our idea of the complexity of building sets that should always *begin* by – relatively – simple prime objects and further proceed to complex elements. The foundation requires the breakup of an *infinite* regress of the membership being included in the very concept of a set, namely to be a set of sets, hence a set of sets of sets, etc.

In the best way, namely explicitly, the hierarchical structure of the "reasonable" universe of sets is described by the levels V_α – the first levels of this hierarchy have been described above. The requirement of the hierarchical structure of the universe of sets is equivalent to the Foundation Axiom.

The Axiom of Choice plays an exposed role among the axioms of set theory – above it was discussed in detail. Now we would like to explicate its special role slightly further.

The selected sets that arise by the Axiom of Choice throw another liberal light on the concept of a set in set theories with AC. Neither the elements that are chosen are definite nor the very set arising by an infinite choice is well defined. No law determines its creation. The selected *set* is claimed and is determined because the Axiom of Extensionality says so.

One of the properties of the Axiom of Choice is that in the finite it does not appear in fact as a special principle of building sets. In a set theory without the Axiom of Infinity (cf. [23]) it trivially holds. It turns to be a principle only when sets, i.e., families of sets are infinite – which was observed first by Peano. The forming of a selected set is hardly noticeable in the finite case because it is not identifiable by a construction rule. The non-constructive character of the choice becomes a great unsolvable problem in the case of the infinite. Example: the principle of well-ordering is obvious in the finite case but the well-ordering of the reals being a consequence of the Axiom of Choice that has for the first time kindled the controversy concerning this axiom (cf. [382]) acted as a long lasting provocation (cf. [323]). The waves have been calmed because one is nowadays careful in applying AC and AC turned out to be useful and even necessary in many branches of mathematics.

It seems to be consequent to currently adjust this principle – hardly noticeable in the finite case and simultaneously indispensable – at the moment when infinite sets are accepted. In fact, the acceptance of this axiom grows in mathematics.

We have seen that the search for new axioms of set theory that could help to solve in one or another way problems connected with the Axiom of Choice or the Continuum Hypothesis lead to interesting results, but so far they were not very successful. The discussion of the new axioms is open. Even worse: generally there is no *broad agreement* concerning the question which farther properties should be linked with the concept of a set and in which way the so far existing set theories should be reformed or completed.

To think mathematically

Sets are determined – as the Axiom of Extensionality states – by their elements. They are as registers of their elements, hence static. If sets form the foundation of mathematical concepts then in principle everything in mathematics stops and stands still. First of all, remember the old intuitive *flowing* continuum which changed into a *stationary* set of single elements identified with the set $\mathbb{R}$. We spoke about it exhaustively in Section 3.3.

To formulate something or to make it precise generally means to reduce it to sets. This influences the thinking. To think mathematically means always to think set-theoretically. Although the set-theoretical thinking is determined not only by axioms we have just commented on, but also by the fact of reducing everything that is mathematical to a single relation, the *membership relation*. It is *a priori* for every element uniquely *registered* in sets. Hence to think mathematically and set-theoretically means insofar to think in a bookkeeping way. It is very precise. Mathematics becomes – figuratively speaking – a "filing cabinet". However, it provides valuable and useful insights into "documents" via mathematical concepts, proofs, propositions and instruments.

A fundamental certificate of our mathematical thinking are natural numbers. They have early been treated by mathematicians as sets, namely as finite cardinals and also today are often associated with this idea. Long ago – since Cantor, Frege and Russell – this was borrowed from other disciplines and dominated in empiricism (cf. Section 2.24.2). Still more serious and hardly conscious anymore is the characterization of natural numbers via set-theoretical Peano axioms (cf. Subsection 5.3.1) in which sets play an important role. In the induction axiom the quantifiers refer to all subsets of the set of natural numbers as if they were all given.

Our structural thinking at general begins with sets that form the basis of structures. Structures are not built anymore. The elements of structures are taken as being ready in given sets. They are not constructed anymore and we do not build structures on them in contrast to the intuitional approach in pre-set-theoretical epochs. Today we are *talking* about elements.

In fact, ways of mathematical thinking are abstract. They reduce and deform the reality, processes and continua. They are destructive – however they produce in this way something effectively *constructive*. What has been lost is mathematically set-theoretically *reconstructed*. What was before unrecognized and not understood becomes in the reconstruction visible in detail. Processes and movements will be grasped as functions, while functions and relations are sets with a variety of properties, and continuity is precisely described. Those are the basic elements of the mathematical turn of the 19th century and main powers of the unprecedented development of mathematics.

The mathematical type of thinking can be seen with reservation and criticized as being only abstract, formal and far from the reality. However, the fact that mathematical methods and propositions have an everyday real meaning is reality. This is universally shown by highly visible and unmanageable applications. Mathematical and set-theoretical thinking opens new approaches. It allows for example – to turn to the natural numbers – deep insights into the complexity of the "real" concept of number that arises by the reconstruction of what has been set-theoretically abstracted (cf. [20]).

The mathematical thinking begins nowadays, as always, with natural numbers – through them it is (by a quasi real standard model) strongly connected with the reality. Because of this, one can say, sets have been "invented" in order to understand numbers. This is clearly indicated by Dedekind's way of thinking in his works about natural and real numbers (cf. [85, 87]).

One can observe a striking turn in the development of set theory – a turn to natural numbers whose foundation were just sets. In fact, natural numbers form in return – closely viewed – the foundation of sets in a certain sense. They provide – in a set-theoretic garment – elements of set-theoretical hierarchy and shed some light at the question about the nature of sets. On the other hand sets allow sophisticated insights into numbers and the concept of number which would not exist without sets.

Infinity

"Definitions of a set" (cf. Section 2.14) were formulated by Cantor in order to describe and to justify the forming of *infinite* sets. The implicitly formulated principles included in those definitions originated from ideas and the experience with finite sets. All those principles lead in a system ZF of set theory *without* infinite sets to a consistent system. Hence axioms of ZF without the Axiom of Infinity describe in an appropriate way our finite concept of a set. Problems arise only when a step towards the actual infinity is made. In the previous sections and chapters we got to know reasons that led towards the actual infinity. The step towards the infinity provided the foundation and produced the clarity, freedom and range in which nowadays mathematics is developed. However, the consistency of ZF or NBG set theories – as was mentioned several times – is in principle not verifiable. So infinity is what it always was, namely *transcendent* – also for a rational mathematical thinking. The infinite remains for set theory, for the theory of the infinite, mysterious.

5 Axiomatic approach and logic

Logic is the morality of speech and thought.[1]

Jan Łukasiewicz

Logic is the hygiene the mathematician practices to keep his ideas healthy and strong.

Hermann Weyl

In Chapter 4 we spoke over and over again about set theory as the foundation of mathematics. It provides the way of speaking and concepts to which principally all concepts of all theories can be reduced. At the same time a chosen set theory is itself a theory. Its construction – as the construction of any other theory – follows the principles of logic. Its concepts and propositions are logically ordered. Set theory delivers – to speak figuratively – raw material for mathematical concepts and propositions. And mathematical logic constructs a network of rules that link mathematical theorems and notions. It is formal, i.e., it is a form – and obviously itself a mathematical theory. Logic provides mathematics with no concepts – with one exception, logic itself.

In general "logic" is a broad concept. For an ordinary person for example $2+2 = 4$ is "logical". For her or him logical means often something like arithmetical or systematic. When a mathematician nowadays speaks about logic then he thinks about something else than a philosopher. Philosophically logic is a doctrine about correct thinking, about concepts and judgements. For a mathematician logic is mathematical logic that describes the formal foundations of mathematical speaking and proving and that investigates mathematical theories.

Mathematics – from the point of view of mathematical logic – is a system of theories, and theories are founded on axioms. From that follows the close connection of axiomatic and logic in today's mathematics. Both, axiomatic base and logic are in the everyday mathematics not necessarily visible, they are nevertheless always in the background. The fact that mathematics is or can be axiomatic-logically ordered is nowadays the basic view and position of mathematicians.

However, it was not always the case – it has been developed only in the 20th century. To see this "new" position as something special we shall look briefly at the history of logic and mathematics that were developed in completely different manners and on different paths. Logic in mathematics was for a long time regarded only as an

1 Logika jest moralnością mowy i myśli.

https://doi.org/10.1515/9783110468335-006

implicit foundation forming the basis of thinking and speaking. Partially it was as syllogistic a philosophical discipline with a epistemological, ontological and linguistical orientation. Only in the 19th century it has been developed as the mathematical logic. This together with set theory and the new axiomatic approach – that have been simultaneously but at the beginning separately developed – led to the situation that mathematics took the task of its foundations in its own hands. Finally it escaped as an independent discipline from the shadows of philosophy.

First we shall give an insight into the character, range and methods of mathematical logic in order to provide a base for understanding the following sections. Such an insight is necessary to understand the big step from the old to the new logic, from the old to the new axiomatic approach and to make clear the results of logic described at the end of the chapter.

5.1 Some elements of mathematical logic

The first task of mathematical logic is the mathematical reflection on mathematics. Logic is like a mirror that reflects what mathematicians are doing. To be able to do this logic needs its own instruments and concepts that can grasp mathematics.

We sketch here briefly the very beginnings of mathematical logic to get an idea of its character and position towards concrete mathematics. We need some basic concepts of logic to refer to them in what follows. We choose a precise example of a logical calculus. As a real introduction we recommend the handbook [103] (we use the approach and notation applied in it) as well as [180]. One finds there also proofs of appropriate theorems. We recommend also the new book [180] by D. W. Hoffmann where the *limits of mathematics* are discussed – many things we only mention are clearly presented there.

Logical instruments are the first thing that should be considered.

5.1.1 Syntax

Logic begins by an analysis of a concrete mathematics. The first problem is the specific mathematical way of speaking. To be able to analyze mathematical expressions a formal **alphabet** of signs is needed – it is necessary to reconstruct and to order mathematical expressions.

Note that such an analysis was successfully done for the first time by Gottlob Frege in his *Begriffschrift* [122] in 1879. Many elements of his analysis – though not directly in Frege's but rather in Peano's symbolism – have for a long time been incorporated into the everyday mathematical way of writing and speaking. Frege was, for example, the first who explicitly distinguished between constants and variables – which today is treated as natural and obvious.

We limit ourselves to the logical *first-order language* – we describe its **syntax**, i.e., the way in which its expressions are constructed as well as its grammar. What "first-order" means will be explained soon.

Definition. Signs of the **alphabet** are the following:

(a)	variables	$x_1, x_2, \ldots, y_1, y_2, \ldots, \ldots, v_1, v_2, \ldots$	(with or without indices)
(b)	connectives	$\neg, \wedge, \vee, \rightarrow, \leftrightarrow$	"not", "and", "or", "if, then", "if and only if"
(c)	quantifiers	$\forall, \exists$	"for all", "there exists"
(d)	=	equality	
(e)	parentheses	), (,], [	
(f)	constants	$k, e, a, \ldots, k_0, k_1, k_2, \ldots$	
	predicate symbols	$P, Q, \ldots, P_1, Q_1, \ldots$	
	function symbols	$F, G, H, \ldots, F_1, F_2, \ldots$	n-ary

Instead of using variables with indices one often writes simply $x, y, v, \ldots$.

Note that this alphabet is rather small when compared to the variety of mathematical expressions in various mathematical theories that can be formally reproduced by it.

In what follows we denote by $\mathcal{S}$ a **set of symbols** from the group (f). It can be empty. An example of $\mathcal{S}$ can be $\{F, k\}$ where F is a symbol of a binary function and k is a constant. They can represent completely different mathematical objects, for example the group operation $*$ and the neutral element e of a group, a metric d and the number 0 or the union of sets and the empty set $\emptyset$. One can see here the function performed by logical symbols that can in an abstract and general, "formal", way represent all possible mathematical objects. Below we shall discuss the connection between formal logical elements and definite mathematical objects.

Logical or formal **terms** are variables of the alphabet, constants from $\mathcal{S}$ or expressions composed of function symbols, variables and constants. They are defined by structural induction – we omit here the precise definition and content ourselves with an example:

$H(F(k, x))$ is a term whereas $F(H, k)$ is not.

Since terms are built by using symbols from $\mathcal{S}$, they are said to be $\mathcal{S}$-**terms**.

Let us outline how **formulas** are defined. It is done inductively in the following way.

Definition. $\mathcal{S}$-**formulas** are constructed according to the following rules:
- If t_1, t_2 are terms then $t_1 = t_2$ is an $\mathcal{S}$-formula.
- If t is a term and P is a 1-digit predicate then Pt is an $\mathcal{S}$-formula.

- If $t_1, t_2, \ldots, t_n$ are terms and R is an n-ary predicate then $Rt_1t_2 \ldots t_n$ is an $\mathcal{S}$-formula.
- If φ, ψ are $\mathcal{S}$-formulas then $\neg\varphi$, $\varphi \wedge \psi$, $\varphi \vee \psi$, $\varphi \rightarrow \psi$, $\varphi \leftrightarrow \psi$ are $\mathcal{S}$-formulas.
- If φ is an $\mathcal{S}$-formula then $\forall x\varphi$ and $\exists x\varphi$ are $\mathcal{S}$-formulas.

A variable is said to be **bound** when it is in the range (marked by parentheses) of a quantifier. Otherwise it is called **free. Sentences** are expressions without free variables.

This is the result of a logical analysis of the language of mathematics. One is dealing here with logical and purely syntactic entities, with formal terms and formulas, that are constructed in the indicated way. They have no meaning.

A remark concerning connectives (b) and quantifiers (c) should be made. In explanations given above like “not”, “and” etc. we already said something that will follow below. In fact, connectives and quantifiers are pure signs without a meaning. Their explanations will become clear only when one moves from the domain of formal expressions to the mathematical reality and assigns to the symbols $\neg$, $\wedge$, $\vee$, $\rightarrow$, $\leftrightarrow$ the meaning “not”, “and”, “or”, “if, then” and “if and only if”, resp., and to the symbols $\forall$, $\exists$ the meaning “for all” and “there exists”, resp. Their meaning is determined *extensionally* – for connectives by appropriate truth-matrices and for quantifiers by *elements* of sets they relate to. Note that the *formal* symbols for quantifiers in the formal alphabet have been used for a long time in the *everyday* mathematical language. The same holds for the connectives – partially in a slightly changed shape.

The determination of the meaning of quantifiers makes also clear what corresponds to formal variables of the alphabet: only *elements*, “individuals” of given sets and *not* subsets of them should correspond to the variables. In this way our language becomes a **first-order language** in which quantifiers can bound only individual variables. One speaks here also about the predicate logic because on the formal site the analysis of mathematical expressions leads to relations, i.e., their formal counterparts called **predicates** – which play the main role in this language. In a **second-order language** quantifiers over predicates or symbols of sets can be used.

Our *standpoint* is now the purely syntactic base of the first-order language. From this we try to consider “mathematical reality” in which we want to *interpret* formal terms, formulas and sentences. So we describe now *semantics* of the first-order language.

5.1.2 Semantics

A mathematical structure consists of a set A – or many sets – and some components. For example: the successor structure of natural numbers $\langle \mathbb{N}, \nu, 1 \rangle$, consists of the set of natural numbers $\mathbb{N}$, a mapping ν and an element 1; an algebraic structure $\langle G, *, e \rangle$ consists of the set G, the binary operation $*$ and the neutral element e.

If on the syntactic side a set of symbols $\mathcal{S}$ is given then one can look an the mathematical side for an appropriate structure with a set A and definite components. Appropriate means here: to any symbol from $\mathcal{S}$ is assigned exactly one component on a set A being the **universe** of the structure. More exactly: any n-ary function symbol of $\mathcal{S}$ is interpreted as an n-ary function on A, every n-ary predicate of $\mathcal{S}$ as an n-ary relation on A, every constant symbol as a fixed element of A, etc.

For example: Let $\mathcal{S} = \{F, k\}$ where F is a binary function symbol and k is a constant symbol. On the mathematical side there is an algebraic structure $\langle G, *, e\rangle$ given with the universe G. The assignment is the mapping α on $\mathcal{S}$ such that $\alpha(F) = *$ and $\alpha(k) = e$. The pair $\mathcal{A} = (A, \alpha)$ consisting of a set and a mapping of symbols, in our example $\mathcal{G} = (G, \alpha)$, is said to be an $\mathcal{S}$-**structure**.

In Fig. 5.1 we depict in a schematic way the mutual transition between signs, terms and formulas in the everyday mathematics and logic: the *logical analysis* of mathematical expressions leads to formal symbols, terms and formulas whose interpretation turns back to mathematical structures.

The contraposition of mathematical structures and their formal analog that is given by the concept of $\mathcal{S}$-structures will become more clear when we assume that $\langle G, *, e\rangle$ is a group and formulate the axioms of groups. In the spare version with the right neutral element e and the right inverse the axioms look like this:

(g_1) $\forall a \forall b \forall c((a * b) * c = a * (b * c))$.

(g_2) $\forall a(a * e = a)$.

(g_3) $\forall a \exists b(a * b = e)$.

These are the axioms of the group $\langle G, *, e\rangle$ in which one uses quantifiers as in the formal expressions. Using symbols F and k they can be written formally in the following way:

φ_1 $\forall x \forall y \forall z(F(F(x, y), z) = F(x, F(y, z)))$.

φ_2 $\forall x(F(x, k) = x)$.

φ_3 $\forall x \exists y(F(x, y) = k)$.

The move from formal axioms to axioms of the concrete group is called an **interpretation.** How symbols are interpreted is given by the $\mathcal{S}$-structure $\mathcal{G} = (G, \alpha)$. It has been already seen that the meaning of quantifiers is connected with elements of G. The interpretation of a single variable is given by a **valuation**, i.e., by a mapping β of the set of variables into the set G. Hence an **interpretation** I is determined by the set G, the mapping α of symbols of $\mathcal{S}$ and the valuation β. Briefly: An interpretation I is a pair $(\mathcal{G}, \beta)$.

For example: If $\langle G, *, e\rangle$ is the group $\langle \mathbb{Q}, +, 0\rangle$ then φ_3, i.e., $\forall x \exists y(F(x, y) = k)$ is interpreted as "for all a there is b such that $a+b = 0$". The valuation β plays no role here since all variables of φ_3 are bounded. We say that φ_3 is **satisfied** in the $\mathcal{S}$-structure $\mathcal{G}$ where, in our example, $G = \mathbb{Q}$.

If in a formula ψ a free variable, say z, occurs then it should be known what the value of z is by the valuation β. For example: If $\beta(z) = 2$ then the formula ψ of the form $\forall x \exists y(F(x, y) = z)$ means "For all a there is b such that $a + b = 2$". In both examples the

Logic

Formal alphabet	Terms	Formulas	Logical sentences
variables $x, y, z, \dots,$ $x_1, x_2, \dots$ logical connectives $\neg, \wedge, \vee, \rightarrow, \leftrightarrow$ quantifiers $\forall, \exists$ equality symbol $=$ parentheses), (;], [*symbols* for: constants $k, e, k_1, k_2, \dots,$ $0, 1, \dots$ functions $F, G, H, F_1, F_2, \dots$ predicates $P, Q, P_1, P_2, \dots$	variables $x, y, z, \dots,$ $x_1, x_2, \dots$ constants $k, e, k_1, k_2, \dots,$ $0, 1, \dots$ $F(x), G(y, k),$ $H(x_1, x_2, \dots, x_n)$ If $t_1, t_2, \dots$ are terms then also $F(t_1), G(t_1, t_2),$ $G(t_1, F(t_2)),$ $H(t_1, t_2, \dots, t_n)$ etc. are terms	$x = k, t_1 = t_2$ $P(t_1), R(F(x), t_1),$ $S(t_1, t_2, \dots, t_n)$ $P(F(t_1)) \rightarrow$ $(R(t_1, t_2) \vee t_1 = t_2)$ $\forall x\, (R(x, y))$ $\exists x(R(x, y)$ $\rightarrow P(y))$	formulas in which no free variables appear.

↑ *Analysis* / *Formalization* ↓ *Interpretation* / *Semantics*

Structures

e. g. arithmetical structures like $\langle \mathbb{N}, +, \cdot \rangle$, $\langle \mathbb{R}, +, < \rangle, \dots$

E.g. arithm. symbols	Arithm. terms	Arithm. formulas	Arithm. sentences
special variables: $n, m, k, l, \dots$ in $\mathbb{N}$ $r, s, t, x, y, \dots$ in $\mathbb{R}$ prop. connectives: non, and, or, $\Rightarrow$ quantifiers: $\forall, \exists, \exists_1,$ for all, exists (exactly one) ... constants: $0, 1, e, \pi, \sqrt{2}$ etc. functions, operations: $f, g, v, \ln, \sin, \dots,$ $+, \cdot, -, :, \dots$ relations: $\in, \leq, \mid, \equiv \bmod m$	variables: $n, m, x, y, \dots$ constants: $0, 1, e, \pi,$ $v(n), \sin(x), x \cdot y,$ $2 + x, \frac{2}{x}, \dots$ $\frac{x^2+2x+1}{x-2}, \sin(\frac{2}{x}),$ $\ln(\cos(x)), \dots$	$x = 2, x + 1 = \frac{1}{x},$ $2^p - 1$ is prime number, $a \mid b, 5 \equiv -9 \bmod 4$ $x_1 > x_2 \Rightarrow$ $f(x_1) < f(x_2),$ p prime number $\Rightarrow$ $2^p - 1$ is prime, $\forall x \in \mathbb{R}(f(x, y) > 0)$	true arithmetical formulas

Mathematics and mathematical practice

Fig. 5.1. Analysis and interpretation – vis-á-vis of mathematical practice and logic.

interpretation leads to a true statement. One says: the interpretation I is a **model** of ψ, or ψ is **satisfied**, or ψ **holds** by the interpretation I, and we write

$$I \models \psi.$$

The same symbol is used for the relation of being a **semantic consequence**. Let us consider the set of sentences

$$\Phi_G = \{\varphi_1, \varphi_2, \varphi_3\}.$$

Here Φ_G is the set of formal axioms of the group theory. The formal counterpart of the existence of the left inverse is

$$\varphi : \forall x \exists y (F(y, x) = k).$$

The existence of the left inverse "follows" – as we used to say – from the axioms of groups given above and one thinks here about a proof leading from the axioms to the existence of the left inverse. *However:* We are now on the formal side. We deal with Φ_G and φ and "φ **follows** from Φ_G" or "φ is a **semantic consequence** of Φ_G" (briefly: $\Phi_G \models \varphi$) means:

> every interpretation satisfying Φ_G satisfies also φ.

What the connections of the semantic consequence and the concept of a proof are will be explained in the following subsection.

Other important semantic concepts are based on the concept of a semantic consequence. A formula φ is said to be a **tautology**, if the following holds:

$$\models \varphi \quad \text{if and only if} \quad \emptyset \models \varphi.$$

This means that φ is satisfied by an arbitrary interpretation, for example $\forall x (x = x)$.

A formula φ (a set of formulas Φ) is said to be **satisfiable** if and only if

> there exists an interpretation satisfying φ (all formulas of Φ).

Formulas φ and ψ are said to be **logically equivalent** if and only if

$$\varphi \models \psi \quad \text{and} \quad \psi \models \varphi.$$

Further semantic notions will be defined in the next sections.

5.1.3 Calculus

To the semantic concept of a consequence corresponds on the syntactic side the concept of a **(formal) proof.** It describes in a formal way what in the everyday practice is called among mathematicians "proof" by reducing the steps of a proof to elementary logical rules. We want to write logical, *formal rules* syntactically as a transition from one sequence of signs to another, more exactly from a **sequent** of formulas to another sequent of formulas.

For example: Let a mathematical statement be given. To it corresponds a formula φ. The statement holds under the assumption of other statements whose formal forms are

$$\varphi_1, \varphi_2, \varphi_3, \dots, \varphi_n.$$

This relation of assumptions and the statement is described syntactically as a sequent

$$\varphi_1, \varphi_2, \varphi_3, \dots, \varphi_n \quad \varphi.$$

Formulas $\varphi_1, \varphi_2, \varphi_3, \dots, \varphi_n$ are called *antecedent* and φ *succedent.*

For example a proof by *reduction ad absurdum* in which a negation of a formula, formally $\neg\varphi$, is assumed, looks syntactically as a sequence of the sequents as follows:

$$\begin{array}{ll} \varphi_1, \varphi_2, \varphi_3, \dots, \varphi_n, \neg\varphi & \psi \\ \varphi_1, \varphi_2, \varphi_3, \dots, \varphi_n, \neg\varphi & \neg\psi \\ \hline \varphi_1, \varphi_2, \varphi_3, \dots, \varphi_n & \varphi \end{array}$$

The horizontal line indicates the transition from the sequents above to the sequent below. Such a transition can be a rule in the set of basic rules that are postulated in order to fix elementary transitions from sequents to sequents. Using them one can obtain further formal transitions formally reflecting the commonly used steps of proofs. In this way a **sequent calculus** $\mathbb{S}$ arises in which concrete proofs can be simulated in a formal and detailed way as deductions by applying rules of it.

We write

$$\varphi_1, \varphi_2, \varphi_3, \dots, \varphi_n \vdash \varphi$$

if φ is **provable** in the calculus. If Φ is a set of formulas then φ is called **formally provable** or **derivable** from Φ, briefly

$$\Phi \vdash \varphi,$$

if there are formulas $\varphi_1, \varphi_2, \varphi_3, \dots, \varphi_n$ in Φ such that $\varphi_1, \varphi_2, \varphi_3, \dots, \varphi_n \vdash \varphi$. If Φ is a system of axioms and $\Phi \vdash \varphi$ then φ is called a **theorem.**

To see how such sequent rules look let us give some more rules. For more clarity we write Δ instead of $\varphi_1, \varphi_2, \varphi_3, \dots, \varphi_n$. Inference rules often used in proofs are, for example, the Modus Ponens or the contraposition. Formally they look like this:

$$\begin{array}{ll} \Delta & \varphi \to \psi \\ \Delta & \varphi \\ \hline \Delta & \psi \end{array} \qquad\qquad \begin{array}{lll} \Delta & \varphi & \psi \\ \hline \Delta & \neg\psi & \neg\varphi \end{array}$$

The following rule that is usually hardly noticed is formalized by the *quantifier rule*:

$$\frac{\Delta \quad \varphi(x/t)}{\Delta \quad \exists x\varphi}$$

It describes formally the following situation: if an example t for the free variable x in φ (under assumptions in Δ) has been found, this is symbolically given as $\varphi(x/t)$, then one can state that there exists an x fulfilling φ. In [103] nine basic rules are given from which all other rules, for example the ones indicated above, can be deduced.

A sequent $\varphi_1, \varphi_2, \varphi_3, \dots, \varphi_n \quad \varphi$ is said to be **sound** if and only if it reflects the semantic consequence, i.e., if $\varphi_1, \varphi_2, \varphi_3, \dots, \varphi_n \models \varphi$ holds. All examples of sequent rules are chosen in such a way that they are sound. Hence one gets the following **Soundness Theorem.**

Theorem 5.1.1. *For all Φ and φ the following assertion holds: if φ is provable from Φ then φ is a semantic consequence of Φ. Briefly:*

$$\textit{If } \Phi \vdash \varphi \textit{ then } \Phi \models \varphi.$$

Kurt Gödel proved in 1929 the converse, called the **Completeness Theorem.**

Theorem 5.1.2 (Gödel). *For all Φ and φ the following assertion holds: if φ is the semantic consequence of Φ then φ is provable from Φ. Briefly:*

$$\textit{If } \Phi \models \varphi \textit{ then } \Phi \vdash \varphi.$$

We close this section by three theorems having far-reaching methodological consequences and forming a background of various logical investigations concerning arithmetic. We write about them in Sections 5.3 and 5.4. Note also that a cardinality of a structure is the cardinality of its universe.

Theorem 5.1.3 (Löwenheim and Skolem). *A finite or countable set of formulas that is satisfiable is satisfiable in a countable structure.*

Theorem 5.1.4 (Tarski). *A formula that is satisfiable in structures of any finite cardinality is satisfiable in an infinite structure.*

Theorem 5.1.5 (Löwenheim, Skolem and Tarski). *If a set of formulas is satisfiable in an infinite structure and κ is an infinite cardinal greater than or equal to the cardinality of the given set of formulas then the set of formulas is satisfiable in a structure of cardinalit κ.*

5.2 Historical remarks

In Chapter 2 we wrote about the beginnings of the axiomatic approach in mathematics by Euclid and about the philosophical sources of it in the works of Plato and Aristotle.

Euclid's approach determined the mathematics till the 19th century. In the 17th and at the beginning of the 18th century it was the explicit or implicit, sometimes remote, background of the mathematical work. However, mathematics has been developed in the 18th century, especially the analysis, in such a dramatic and vague way that it overburdened Euclid's foundations. New foundations, a new framework and a new method were needed. The new foundations and the new framework have been provided by set theory (we wrote about it in Chapter 4) and by the new logic whose first elements and basic concepts were just presented above. The new method is the axiomatic approach in the form commonly used today. One of the consequences of the fact that mathematics took its foundations in its own hand is the fact that it has been definitely detached from the philosophy – earlier it has been in principle a subdiscipline of the latter.

We make only some indications concerning the history of logic that began separately from mathematics and only in the 19th century joined mathematics. Then some words about the development of the new axiomatic approach in the 19th century will be said. Some names and dates are already known from the previous chapters.

5.2.1 From the history of logic

The early beginnings of a philosophical logic can be probably found in the fifth century BC by sophists who taught – for practical and rhetoric reasons – the correct or rather more skillful ways of arguing and speaking. The first explicit testimony of a logic is the syllogistic of *Aristotle* one hundred years earlier. Aristotle explains that a syllogism is "a discourse in which certain (specific) things having been supposed, something different from the things supposed results of necessity because these things are so" [5, 24b18-20]. In the Greek word "syllogismos" (συλλογισμός) the meaning "calculate" is heard that already here at the beginning points to Leibniz's conception and to the algebra of logic of the 19th century.

Syllogisms are logical conclusions of a special form. They connect categorical propositions, i.e., expressions connecting a subject and a predicate. Syllogisms conclude from two categorical propositions, "major premise" and "minor premise" to a third categorical proposition, the "conclusion". The best known example is the following: All humans are mortal (major premise), Socrates is a human (minor premise), Socrates is mortal (conclusion). Here "mortal" is "major term", "Socrates" is "minor term" (subject). Syllogistic is a logic of *names* (individual and general ones) in which in premises and conclusions names (and their extents) are connected. In syllogisms one sees for the first time the idea that the truth of a proposition is formally caused and can be formally determined, that one should look for truth not only in the meaning of names and notions but also in the *form* of propositions in which they occur. Logic fulfilling such an idea has been called *formal logic* in the history of the philosophy.

In the third century BC one finds first attempts of a propositional logic by the Dialectician Philo of Megara (around the turn of the fourth and third century BC) and by Chrysippus of Soli (ca. 281 – ca. 208 BC). The aim of their logic was the clarification of the formal structures of the language. To Philo is ascribed the first paraphrase of truth matrices – the implication (if, then). Chrysippus distinguished between object, denotation and meaning. He gave further formulations for logical connectives of negation, conjunction (and), disjunction (either, or) and implication (if, then) and justified their validity by their truth values. For example: "Either α' or β'. α'. Hence not β'" [183, fragment 1131]. The symbols α' and β' are Greek symbols for "firstly" and "secondly", respectively. Conclusions of these type have been described as "hypothetical syllogisms".

In scholasticism the syllogistic of Aristotle has been readopted. It became the core of logic till the 19th century and has been gradually changed. The main subject of syllogistic in the Middle Ages is the investigation of properties of categorical propositions and of relations between propositions as well as between syllogisms themselves. At that time arose the traditional notation for categorical propositions and for syllogisms. Propositions were distinguished according to their "quantity", i.e., according to the fact whether they are universal (e.g., "all S is P") or particular (e.g., "some S are not P") as well as according to their quality, i.e., whether they are affirmative or negative. One wrote:

SaP	for	all S are P,
SeP	for	no S is P,
Sip	for	some S are P,
SoP	for	some S are not P.

"S" stands here for a "subject", "P" for "predicate". The letters a, e, i, o come from the words "affirmo" and "nego": the first vowel "a" from "affirmo" for the universal, affirmative case, "e" from "nego" for the universal negative case, and the second vowels for particular cases. One can recognize here an early stage of logical symbolism that however has not been developed further. Symbols were simply abbreviations here.

A particular phenomenon of the Middle Ages was the poet, theologian and philosopher *Ramon Llull* (ca. 1232 – ca. 1316); Anglicised Raymond Lully, Raymond Lull; in Latin Raimundus or Raymundus Lullus or Lullius. He is the author of extensive works. Most influential were his works *Ars magna* and *Ars brevis*. Their higher aim was to develop methods that in debates would help to irrefutably persuade irreligious persons or persons of different faith. Llull started from principles expressing properties of God such as "benevolence", "omnipotence", "truth" and from basic concepts like "beginning", "aim", "activity" and "difference", and tried to obtain true conclusions from them. He assigned letters to concepts and sketched a system of a symbolic language. Logic was for him an art of distinguishing the true from the false. He developed a mechanical method of combining concepts and judgements and obtaining in a mechanical way conclusions.

Llull constructed a "logical machine" consisting of seven disks having a common center. On each disk were put words designating concepts – e.g., "a man", "knowledge", "truth" – and logical operations like "difference", "agreement", "contradiction" and "equality". By turning the disks around the common center one got connections between concepts corresponding to partial forms of deduction of syllogistic principles. Many scientists see nowadays in Llull – whose ideas were initially criticized or even rejected – the early precursor and inventor of symbolic logic and of the automatization of reasoning. The ideas of Llull were in fact singular – there was for example even in algebra no comparable symbolism – and they had for a long time no influence on the development of logic.

Leibniz, three and a half centuries later, studied works of Llull. His ideas of *characteristica universalis*, of *ars combinatoria* and *mathesis universalis* were developed under the influence of Llull. Leibniz's ideas and this important stage of the history of logic has been described in detail in Chapter 2. Leibniz did not succeed in realizing the project of an algebraization of argumentation, in developing something like an "algebra universalis". He was engaged in other more realistic projects and this project was in his setting too extensive and in principle not realizable. The essential reasons of his failure – as well as of a partial realization – was the fact that he assumed an absolute, intensional (and not extensional) treatment of concepts and logical forms and did not treat them as conventions. The conception of formalization was not really elaborated.

The idea of mathematization of thinking processes and of the knowledge appeared in the 17th century over and over again. Recall here *René Descartes* and his *mathesis universalis* as well as *Blaise Pascal*'s idea of mathematics as a measure and standard in the world of reason – the latter had, according to Pascal, no meaning for the world of "heart", of ethic and for the domain of the infinite (cf. Section 2.8). Exceeding the idea of mathematization Leibniz formulated an idea of a mathematical logic.

The Leibniz program – as *H. Scholz* (1884–1956) called it (cf. [319, p. 242]) – has later been considered again and again, for example by C. Wolff (1679–1754). It was a philosophical program that again took leave of symbolization and algebraization. Logic in the philosophy remained conceptual, concerned with content and having an epistemological and ontological orientation. The first form of a formal logic in a strict sense was the algebra of logic developed on the mathematical side.

The beginning of a formal, mathematical logic is due to the English mathematicians *George Boole* (1815–1864) and *Augustus De Morgan* (1806–1871). Their merit was to consider and develop parts of formal logic consequently in an algebraic way. Boole in his works *Mathematical Analysis of Logic* [44] in 1847 and *An Investigation of the Laws of Thought on which Are Founded the Mathematical Theories of Logic and Probabilities* [45] in 1854 partially developed the idea that can be found by Leibniz: to symbolize notions, use formulas of the algebraic type to denote logical interrelations, and to develop an algebra of logic.

Boole recognized the fundamental role of logic for mathematics. In [45] he wrote:

> "On the principle of true classification, we ought no longer to associate Logic and Metaphysics, but Logic and Mathematics. [...] A mental discipline which is afforded by the study of Logic, *as an exact science*, is, in species, the same as that afforded by the study of Analysis." ([45, p. 13].)

What is novel by Boole – beside this completely new conception of logic taken later by Frege – is the description of the essence of a formalism. He formulated for the first time the core of a formal logic: A formalism is, he said, a procedure

> "the validity of which does not depend upon the interpretation of the symbols which are employed, but solely upon the laws of their combination." ([45, p. 3].)

The absolutely new element was the principle according to which transformations of expressions do not depend on the meaning, on the interpretation of symbols but exclusively upon rules of transformation that are independent of any interpretation. Boole stressed explicitly that one and the same formal system can be interpreted in various ways.

Symbolic language and formalism were for Boole not ends in themselves. Next to mathematical meaning he emphasized the philosophical aim:

> "That which renders Logic possible, is the existence in our minds of general notions, – our ability to conceive of a class, and to designate its individual members by a common name. The theory of Logic is thus intimately connected with that of Language. A successful attempt to express logical propositions by symbols, the laws of whose combinations should be founded upon the laws of the mental processes which they represent, would, so far, be a step toward a philosophical language." ([45, pp. 4–5].)

Boole presented in the quoted work a theory that should be an extension and clarification of Aristotle's syllogistic. However, its theoretical importance and domain of applicability turned out to be much more extensive. Today's version of his theory is called "Boolean algebra" in recognition of its founder. Boole's system can be interpreted among others as an algebra of sets or as a propositional calculus.

Concepts and propositions were denoted by Boole by Latin capitals like X and Y. To express the four cases of being "true" and "false" he used small Latin letters x, y as "elective symbols". They should be, as he wrote, symbols denoting objects selected according to a given property. He stressed: "The hypothetical Universe 1, shall comprehend all conceivable cases and conjectures of circumstances. The elective symbol x attached to any subject expressive of such cases shall select those cases in which the

Proposition X is true, and similarly for Y and Z" [45, p. 49]. Today we would say that x and y are classes of elements that are X and Y, respectively. The four cases of being "true" and "false" are now expressed symbolically in the following way:

xy	instead of	X is true and Y is true
$x(1-y)$	instead of	X is true and Y is false
$(1-x)y$	instead of	X is false and Y is true
$(1-x)(1-y)$	instead of	X is false and Y is false

By Boole there are no symbols for the implication (if, then) and negation (not). Both are expressed by complex formulas. The symbol + is used as a symbol of the disjunction (either, or).

The classical categorical propositions can be formulated in Boole's symbolism as follows:

All X is Y	as	$x(1-y)=0$
No X is Y	as	$xy=0$
Some X are Y	as	$xy=v$
Some X are not Y	as	$x(1-y)=v$

The letter v seems to correspond here to the English word "some". It is said in [45, pp. 48 f.] that it stands for a class "indefinite in all respect but one" – namely that it contains a member or members. This convention is unsatisfactory and the unclear application of the symbol v leads to a certain inconsistency in Boole's system.

In the same year as *Mathematical Analysis* (1847) also the book *Formal Logic* by De Morgan appeared. It is said that both books have been displayed on the same day in bookshops. De Morgan developed in his book the syllogistic and some ideas of the algebra of logic and of the theory of relations. He introduced symbols for the converse, complement and composition of relations.

One should say here also some words about Bernard Bolzano (1781–1848) – they should extend the remarks made in Chapter 2. Before Boole, Bolzano formulated in his *Wissenschaftslehre* [41] in 1837 a logical concept of a consequence and attempted to clarify the concept of truth. His definition of consequence was complicated and by today's standards not precise enough. However, it should be noticed that Bolzano had at his disposal no logical system of concepts or symbols that is necessary for the required precision.

The core of his approach to provide a definition of a consequence was the following: Bolzano denotes sentences, more exactly expressions, by the capitals $A, B, C, D, \ldots, M, N, O, \ldots$ and calls them "compatible" (*verträglich*) or "unanimous" (*einstimmig*) if there are "ideas" (*Vorstellungen*) which substituted for the variables $i, j, \ldots$ "change them all in true ones" [41, Section 155a]. Sentences $M, N, O, \ldots$ follow from (are consequences of) $A, B, C, D, \ldots$ if "any collection of ideas such that they make true all $A, B, C, D, \ldots$ when substituted for $i, j, \ldots$ make also true $M, N, O, \ldots$". One can easily see here a relationship with the nowadays commonly used concept of a semantic consequence coming from Tarski (cf. Section 5.1).

The concept of truth refers by Bolzano to the reality. A sentence is true if it "states how it is" (cf. [41, Volume I, Section 25]). This is in fact not new and again philosophical – one can philosophize a lot about reality, expressions and their relation. Tarski discussed in [335] colloquial definitions of truth of this type and challenged their possibility. It should be noticed that Bolzano recognized the necessity of a concept of truth especially for formal languages. A well-grounded definition of truth for formalized languages has been given by Tarski in his great work [335].

It seems that Bolzano was the first mathematician who understood what in fact mathematical logic is. His ideas being formulated in a philosophical way had no influence on the further development of mathematical logic in the next decades. Only 100 years later Tarski [336] formulated in a similar way the definition of the semantic consequence making no reference to the work of Bolzano probably for being unaware of it.

The real turn from formal philosophical logic to mathematical logic is due to *Gottlob Frege* (1848–1925). We wrote in the section "Logicism" of Chapter 2 about Frege and his work. It was just Frege who really overcame the boundaries of Aristotle's logic. He created the first comprehensive formal language. The year 1879 in which his *Begriffsschrift* appeared is treated as the most important date in the history of logic since Aristotle. One finds in this work the first formalized axiomatic system for the propositional calculus and the first complete analysis of sentences, i.e., of quantifiers and their axioms. *Begriffsschrift* together with his *Grundlagen der Arithmetik* (1884) and *Grundgesetze der Arithmetik* (in two volumes, 1893 and 1903) should make logic to be the foundation of mathematics. We wrote that Frege's works failed to cause a response from his contemporaries. The reason was first of all his heavy symbolism introduced and uncompromisingly used by him.

The development of a discipline is essentially connected with the development of an appropriate symbolism. Logic – like mathematics – needs such a symbolism. The lack of such a symbolism can be seen as one of the reasons of the delayed development of logic in the previous centuries. To the creation of a suitable symbolism for logic contributed first of all *Giuseppe Peano* (1858–1932). Many denotations introduced by Peano have been adopted directly or in slightly changed form and are today still used in logic and mathematics.

Peano's aim was to present the whole of mathematics in a consequent way by using an artificial symbolic language. Theorems of a given discipline should be deduced from few postulates. The full form of this idea was his project "Formulario" proclaimed by him in 1892 in the journal *Rivista di matematica*. It was connected with Leibniz's idea of *characteristica universalis*. Peano considered the realization of Leibniz's idea a main task of the mathematical sciences. The aim of the project was to express all known theorems of all mathematical disciplines in the symbolic language. Between

1894 and 1908 appeared five volumes of a special journal *Formulario Mathematico* founded by Peano just to publish the results of the project. The last volume contained 4 200 mathematical theorems. Peano treated *Formulario* as the realization of Leibniz's idea. He wrote:

> After two centuries the "dream" of the inventor of the infinitesimal calculus became a reality. We have accomplished the idea of Leibniz.

An example of a theory formulated in a symbolic language was given by Peano already in 1889 in his famous work *Arithmetices principia nova methodo exposita* [266]. One finds there the first axiom system for the arithmetic of natural numbers. Dedekind's work [87] was even earlier, but he formulated his axioms in a set-theoretical way. At the beginning of his work Peano wrote:

> I have represented by signs all ideas which appear in the foundations of arithmetic. With the help of this representation all sentences are expressed by signs. Those signs belong either to logic or to arithmetic. [...] By this notation every sentence becomes a shape as precise as an equation in algebra. Having sentences written in such a way we can deduce from them other sentences – it is done by a process similar to a looking for a solution of an algebraic equation. This was exactly the aim and the reason for writing this paper.

We give here as an example Peano's attempt to formulate arithmetical axioms in a logical way.

Peano chose ten primitive notions of a purely logical nature (not all of them are in fact necessary) and four arithmetical notions. The logical symbols were the following:

P	for an expression or the class of all expressions,
K]	for a class or the class of all classes,
ε	for "is" or the membership relation,
$\cap$	for the conjunction or the intersection of classes – the meaning of $\cap$ depends on the context,
$\cup$	for the disjunction or the union of classes,
$-$	for the negation or the complement of a class,
$\supset$	for the inclusion or implication (the latter is read as: ... is deduced from ...),
$=$	for the equivalence or equality of classes,
Λ	for the false, the empty set,
$(.\varepsilon)$	as the sign of "inversion"; $(x\varepsilon)a$ is read as "x with the property a".

The arithmetical primitive notions were:

N	for the property of being a number or the class of natural numbers,
1	for one,
$a + 1$	for the successor of a,
$=$	for the equality of numbers.

Peano formulated four identity axioms (nowadays they are treated as logical axioms) and the following five arithmetical axioms:

(i) $1\varepsilon\mathrm{N}$,
(ii) $a\varepsilon\mathrm{N}. \supset .a + 1\varepsilon\mathrm{N}$,
(iii) $a, b\varepsilon\mathrm{N}. \supset: a = b. = .a + 1 = b + 1$,
(iv) $a\varepsilon\mathrm{N}. \supset .a + 1 - = 1$,
(v) $k\varepsilon\mathrm{K}.\therefore.1\varepsilon k.\therefore.x\varepsilon\mathrm{N}.x\varepsilon k :\supset_x .x + 1\varepsilon k ::\supset .\mathrm{N} \supset k$.

In today's symbolic they can be written as follows:

(i) $1 \in \mathrm{N}$ (1 is a natural number).
(ii) $a \in \mathrm{N} \to a + 1 \in \mathrm{N}$ (if a is a natural number then $a + 1$ is also a natural number).
(iii) $a, b \in \mathrm{N} \to (a = b \leftrightarrow a + 1 = b + 1)$ (natural numbers are equal if and only if their successors are equal).
(iv) $a \in \mathrm{N} \to (a + 1 \neq 1)$ (a successor is not equal 1).
(v) $k \in \mathrm{K} \wedge 1 \in k \wedge \forall x(x \in \mathrm{N} \wedge x \in k \to x+1 \in k). \to \mathrm{N} \subseteq k$ (if 1 is in k and k is closed under successor then N is a part of k.)

Peano has shown how theorems about natural numbers can be deduced from those axioms. He did not say, however, according to which inference rules theorems are deduced from axioms or other theorems. His deductions were simply sequences of formulas such that one followed from another one in the *intuitive* sense.

Another difficulty of Peano's approach was the problem of clearly distinguishing between logical and set-theoretical notions. One sees that besides logical notions Peano used also set-theoretical notions such as the membership of a class. It was also difficult to separate linguistic levels. For example the metaconcept "expression" – in symbols P – for Peano belonged to an object logico-arithmetical language.

To the distribution of Peano's symbolism and to the acceptance of it among logicians and mathematicians contributed in an essential way the English mathematician, logician and philosopher *Bertrand Russell* (1872–1970). He adopted it with only slight changes in his monumental work *Principia Mathematica* [370] written together with the English–American philosopher and mathematician *Alfred North Whitehead* (1861–1947). We wrote extensively about Russell and the project *Principia Mathematica* in Section 2.19 about logicism. *Principia* played for a long time the role of a starting and reference point of the modern mathematical logic.

Let us summarize the important moments in the development of mathematical logic: In the first half of the 19th century in works of Boole, Bolzano and De Morgan the development of mathematics and the development of logic were for the first time interconnected. Boole clearly recognized the meaning and significance of logic for mathematics. Fundamental elements of logic which so far have been completely embedded into the philosophy, have now been grasped mathematically and symbolized in an algebraic way – they formed a new mathematical discipline, namely the algebra of logic. By Frege and Peano the elements of logic appeared in different ways in the foundation of mathematics. The aim of Peano was to develop a symbolism in which the whole of mathematics could be expressed and mathematical theorems could be ordered by the axiomatic-deductive method. Boole anticipated in sentences quoted above what has been the aim of Frege: the domain that arises as a result of the study of logic should be the entire mathematics. By founding arithmetic on logic Frege attempted to make logic the foundation of analysis and mathematics. In Chapter 2 we have told about Frege's attempts, about the tragedy caused by the discovery of Russell's antinomy and have seen how Russell and Whitehead have continued Frege's project till its realization. We noticed that logic alone – without artificial tricks – cannot form the foundation of mathematics. Not logic alone but logic and set theory nowadays form the foundations of mathematics.

5.2.2 On the history of the axiomatic approach

It has been indicated above that axioms in new areas and in the formal form came to the fore for instance in the works of Frege and Peano. The new view on axioms and their role, that became evident in full clarity only thanks to Hilbert, has been developed in the 19th century parallel to the development of mathematical logic.

The axiomatic approach has been treated as the adequate method for mathematics since Plato, Aristotle and Euclid (cf. Chapter 2). Axioms are elementary statements put at the beginning of a theory. They are considered and accepted as intuitively plausible and assumed without proofs as valid. The best translation of the Greek word "*ἀξίομα*" (axioma) – literally "claim", "demand" – is in the context of mathematics the word "fundamental statement". To formulate axioms one uses primitive notions that nowadays are put undefined at the very beginning of a theory. Axioms can be treated as implicit definitions of primitive notions. Other notions of a theory are defined in terms of primitive notions, and theorems are deduced from axioms.

Euclid and the subsequent mathematicians until the 19th century often tried to define the primitive notions. Such definitions were necessarily of a descriptive philosophical nature. They should show that axioms are connected with given real or abstract objects, or in the *Elements* of Euclid with elements of geometry. Euclid's axioms were *determined with regard to contents*. Euclid's list of axioms for geometry was incomplete and his proofs were – according to our standards nowadays – not always correct. He often referred to non-identified or not explicitly given intuitive assumptions. Nevertheless, Euclid's *Elements* are treated as an ideal approach to mathematics and an ideal way of presenting it – they provided a paradigm for systematization of mathematical knowledge. The very concepts of a proof and of the consequence were understood intuitively by Euclid. A result of this was the fact that in argumentations of *Elements* beside intuitive implicit assumptions there are also gaps and weaknesses. However, Euclid's axiomatic approach despite some deficits was in fact an epoch-making step for the development of the methodology of mathematics, in particular for the systematization of mathematics.

Mathematics in *Elements* as well as in later mathematical works was – from today's point of view – quasi-axiomatic. Its primitive notions made it to be based on philosophy.

However, Euclid's axiomatic approach remained till the 19th century a natural foundation of mathematics and as a "geometrical method" (*more geometrico*) played the role of a pattern of any strict scientific procedure at general – even in philosophy (cf. [62]). It was impossible, even unimaginable to make any objections to it. At the end of the 18th century, by Kant mathematics gained as a domain of "synthetic judgements a priori" additionally philosophical meaning and the arithmetical and geometrical primitive notions were given in his transcendental philosophy an epistemological status. All those seemed to connect mathematics in a new way to philosophy. However, the further development went along other paths. Mathematics began to free itself from philosophy by its own powers.

To the philosophically important period determined by Kant and being today still present in the scientific thinking, belong mathematical works by the great *Carl Friedrich Gauß* (1777–1855). He began to change the mathematical way of thinking (cf. Section 2.13). He was in fact the first who had the courage to think in a non-Euclidean way. Still Euler advocated to hold on to the parallel postulate even if one is unable to prove it because – as he wrote in a letter from 1735 – "from the opposite hypothesis many contradictions arise" (cf. [243, p. 180]). Gauß and a bit later Bolyai and Lobachevsky succeeded to describe mathematically in a consistent way what seemed to be the inconsistency, i.e., the non-Euclidean relationships, hence geometrical "circumstances" different from the Euclidean ones. All Euclidean axioms except the parallel postulate remained valid. This was not philosophically comprehensible anymore at the time when geometry was connected with the reality and described its structure understood in an Euclidean way. Euclidean axioms lost their claim to be absolute. Simultaneously, mathematics made a step to leave the domain of philosophy whose subdiscipline it was by its ontological ties.

The further development of geometry went from Euclidean geometry and its axiomatic to the projective geometry of Jean-Victor Poncelet (1822), and vectors and n-dimensional extensions (1844) used by Hermann Grassmann as a new foundation of geometry (Grassmann went even so far that he was not willing to include Euclidean geometry into mathematics). The mathematical meaning of Grassmann's works was realized much later (1869) by Hermann Hankel and Felix Klein.

In the first half of the 19th century there were besides geometry also other important circumstances that led to a new understanding of axioms. We mean here Gauss who by his geometrical interpretation of complex numbers contributed in an essential way to the change of the status of them – they became the status of quasi real objects and have been accepted as "normal" numbers. Also his fundamental theorem of algebra, proved in his doctoral dissertation, increased the value of complex numbers. Numbers have been developed step by step to a domain of numbers in which there was no natural ordering that so far has been necessarily ascribed to them. The next step was the interpretation and founding of complex numbers as pairs of reals due to *Hamilton* as well as his invention of quaternions – a domain which includes real and complex numbers. In the case of quaternions there disappeared another property treated so far as a condition and an unexpressed axiom when speaking about numbers, i.e., the commutativity of multiplication. One started to treat arithmetical properties in a free way and as available. The concept of an arithmetical operation evolved into the concept of a general algebraic operation (cf. for example Grassmann).

A new concept played an essential role in a free interpretation of axioms – namely groups. They played an important role in the theory introduced by *Évariste Galois* (1811–1832) to answer the question about the solvability of equations. They paved the way for a new algebra – far from Euclidean axioms. Groups appear in quite different contexts as groups of substitutions (automorphisms) by Galois, in arithmetic or geometry. Groups began to get free from those connections and to become independent mathematical objects. Arthur Cayley (1821–1895) formulated in 1854 the abstract concept of a group. That year is treated as the beginning of the theory of groups (cf. [380]). Abstract groups and their properties expressed as axioms of groups were made to be interpreted in various ways by their different origins.

In the second half of the 19th century there are further steps towards the new axiomatic approach. An important step was done by Dedekind. Before set theory has been introduced Dedekind formulated axioms of natural numbers on the base of set-theoretical concepts. His famous work *Was sind und was sollen die Zahlen?* [87] comes from the period of 1872–1878 and before as he wrote in the preface. We wrote in Chapter 2 about Dedekind and his work in detail. It was in fact a unique undertaking to leave the domain of familiar numbers and to attempt to found them on a new base that seemed then dim. So far the natural numbers being a base of all mathematics of numbers were given intuitively. There were no arithmetical axioms neither by Euclid nor by later authors. Dedekind wrote about his risky enterprise:

> But I feel conscious that many a reader will scarcely recognize in the shadowy form which I bring him his numbers which all his life have accompanied him as faithful and familiar friends. ([87, p. IV].)

Dedekind wrote about the necessity of justifying mathematics what became a sign of the times:

> In science nothing capable of proof ought to be accepted without proof. ([87, p. III].)

He refers in a footnote to the "most new presentations" concerning this "simplest science" which can be found in works of Schröder, Kronecker and Helmholtz – this "obvious requirement" by "no means can be seen as fulfilled" there.

Let us turn back again to geometry which played an important role in the development of the new approach to axioms. *Moritz Pasch* (1843–1930) realized that the Euclidean axioms were inconsequent, little systematic and full of gaps bypassed by intuition. He attempted to reform the Euclidean axioms (as some others before him) and formulated three groups of axioms: axioms of incidence, axioms of order and axioms of congruence. His aim was stated again in an Euclidean way of thinking:

> "The basic principles (*Grundsätze*) should comprise in a *complete* [authors' emphasis] way the empiric material handled by mathematics so that after formulating them there is no need to turn back to sensory perception."

The source of axioms is for Pasch still the reality but his aim is to get free from it, from any "empirical material" and every "sensory perception". What is left then is a formal framework given by axioms. And this was resumed by Hilbert.

David Hilbert (1862–1943) referred in his foundations of geometry to Pasch and took his groups of axioms. However, his conception of an axiomatic approach is explicitly from the very beginning new. Axioms do not refer to any special objects – only interrelations between given elements are important. Objects are undetermined. The formulation of the "Explanation" (*Erklärung*) in *Grundlagen der Geometrie* [165, Chapter 1, Section 1] has many times been copied but unrivalled:

> "Let us consider three distinct systems of things. The things composing the first system, we will call points and designate them by the letters $A, B, C, \ldots$; those of the second, we will call straight lines and designate them by the letters $a, b, c, \ldots$; and those of the third system, we will call planes and designate them by the Greek letters $\alpha, \beta, \gamma, \ldots$" etc.

The difference to previous concepts of axiom systems is concentrated here on one word: *call* (*nennen*). Hence things of the first system *are* not points anymore. Hilbert's dictum about things the axioms speak about became famous. It cannot be missing here: Instead of "point", "straight line" and "plane" one can say also "beer mug", "table" and "chair". Important are only the interrelations between them described by axioms.

In a letter to Gottlob Frege from December 29, 1899, he wrote (cf. [125, p. 67]):

> Yes, it is evident that one can treat such theory only as a network or schema of concepts besides their necessary interrelations, and to think of basic elements as being any objects. If I think of my points as being any system of objects, for example the system: love, law, chimney-sweep [...], and I treat my axioms as [expressing] interconnections between those objects, then my theorems, e.g., the theorem of Pythagoras, hold also for those objects. In other words: any such theory can always be applied to infinitely many systems of basic elements.

This is the new, purely formal view on the axiomatic approach – it was logically described in Section 5.1. However, it should be noticed that by Hilbert the formal frames were motivated as regards content, i.e., he considered theories together with suitable nonempty domains (*Bereiche*) which indicated the range of the individual variables of the theory and the interpretation of the nonlogical symbols. Hilbert, as a mathematician and having little interest for philosophical questions, was not interested in establishing precisely the ontological status of mathematical objects. It can be even said that the sense of his program of formalization was calling on people to turn their mathematical and philosophical attention away from the problem of the objects of mathematical theories and turn it toward a critical examination of methods and assertions of theories and to clarify their consistency and completeness. Hilbert's program has been described in detail in Section 2.22.

The new formal approach, which was so early and clearly characterized firstly by Hilbert, determines today's views on axioms and the axiomatic approach of every mathematician. However, it needed some time to become accepted. For example for Felix Klein (1849–1925), Hilbert's colleague from Göttingen, axioms were still in 1909 "evident truths" and axioms of geometry were "not arbitrary but reasonable sentences" [209, Volume II, p. 202]. Obviously, he did not comprehend the sense of the new axiomatic approach. H. Meschkowski ascribes him in [240] the following dictum: "If a mathematician has no ideas then he practises axiomatic approach".

How difficult it was to free oneself from the old ties and old way of thinking was shown in another context by quotations from Hankel concerning the conflict between magnitudes and real numbers (cf. Chapter 3). Also other examples show how slowly and hesitantly the mathematical thinking has been chosen. Even Peano and Frege who in an essential way contributed to the development of mathematical logic and axiomatic foundations of mathematics, were in a sense imprisoned by the contentual thinking. Above we stressed Peano's axiomatic approach to natural numbers and Frege's axioms having a formal character to an extent that has never appeared before. However, Peano connected symbols used by him with thoughts. He wrote in 1896: "In this way a one-to-one correspondence between thoughts and symbols is established, a correspondence that cannot be found in the colloquial language". This is still under the influence of Leibniz. In fact, by such an approach symbols of, for example, arithmetic are not free interpretable but notionally and "unequivocally" connected with numbers.

Frege expressed his traditional conception of axioms very clearly in a letter to Hilbert in the following way:

> "Axioms are according to me statements that are true but not proved because our cognition of them comes from a source of cognition different from logic, which is called perception of space. The verity of the axioms implies by itself that they are consistent. This requires no proof."

It is difficult to interpret this in the case of his axioms for, e.g., propositional calculus. It can be assumed that by Frege they tell solely about linguistic statements. However, the source of the validity is uncertain. The axioms seem to be natural as laws of pure thinking and self-evident as a matter of thinking.

Frege says here clearly that by such an intuitive and with regard to content motivated approach to axioms there is no place for the idea of an interpretation of symbols and for the question on the mutual consistency of axioms.

Hilbert set a high value on the new axiomatic method and on the mathematics which created it and stressed it. He regarded the axiomatic approach to mathematics as a universal pattern for any science. In the work *Axiomatisches Denken* [171, p. 405] he wrote:

> When we assemble the facts of a definite, more-or-less comprehensive field of knowledge, we soon notice that these facts are capable of being ordered. This ordering always comes about with the help of a certain *framework of concepts* (*Fachwerk von Begriffen*) in the following way: each individual object of the field of knowledge corresponds to a concept of this framework and every fact within the field of knowledge to a logical relation between concepts. The framework of concepts is nothing other than the *theory* of the field of knowledge.

In the same work he also wrote:

> I believe: anything at all that can be the subject of scientific thought becomes dependent on the axiomatic method, and thereby indirectly on mathematics, as soon as it is ripe for the formation of a theory.

A bit later he wrote:

> [...] we become ever more conscious of the unity of our knowledge. In the sign of the axiomatic method, mathematics is summoned to a leading role in science.

We are nowadays far from the unity of sciences. One cannot see serious applications of the axiomatic method beyond mathematics, theoretical physics and computer science. However, the axiomatic method has surely contributed to the "unity of knowledge" in mathematics. The interpretability of axioms establishes connections between theories. The unity is established by the common foundation provided by set theory and logic.

5.3 Logical axioms and theories

What is really new in the new axiomatic approach? Hilbert's words quoted above describe this accurately and indicate the difference with respect to the old Euclidean approach. The starting point of the old approach was the somehow given evidence of axioms and the intuitive clarity of primitive notions described philosophically in definitions. The axioms were *ontologically* founded. The novelty of the new approach consists in the free interpretability of axioms and in the undefinability of primitive notions. *Axioms themselves* define the primitive notions, they provide implicit definitions of them. What this new characteristics mathematically means becomes clear in its full meaning only when one considers the logical procedure behind it. We have become acquainted with it in outlines in Section 5.1 and will carry on with it in this section. The new axiomatic approach opened mathematics to logical analysis.

In Section 5.1 we have given as an example the formal group theory. It differs essentially from the usually given theory of groups. We have stressed above the differences but we gave the usual axioms of groups in such a way that one can see the transition to the formal version. In the research practice one begins with a *set* G and an operation $*$ on G and then formulates the axioms of a group. This means that the axioms are usually formulated set-theoretically and the *axioms of set theory* are assumed at this moment and in everything that follows.

In the purely logical axiomatic approach one proceeds in another way: theories are presented as *self-contained*. Their primitive notions are explicitly given and reproduced by formal symbols. Next, axioms are formulated in a formal language using those symbols. There are usually no symbols for sets. Only by interpreting axioms one needs sets when structures are considered – in fact the universe in which the variables will be interpreted is needed. One should add that there are however formalizations in which symbols and some expressions are interpreted set-theoretically. We shall make some remarks about both forms below.

We were talking all the time about theories. But what are **theories**? Also this concept has been made precise by logic. Briefly one can say: a theory is a collection of sentences fulfilled/satisfied in a structure. Generally: A theory T is a set of sentences closed under consequence.

More exactly: Consider a mathematical structure, e.g., a given group $\langle G, *, e\rangle$ as in Section 5.1. To $*$ and e of the structure correspond symbols F and k. Then $\mathcal{G} = (G, \alpha)$, hence the universe G and the interpretation α of symbols F and k by $*$ and e, respectively, is said to be a **structure** in the logical sense with the set of symbols $\mathcal{S} = \{F, k\}$ (cf. Section 5.1). If φ is a sentence and

$$\varphi \text{ is } \textbf{true} \text{ in } \mathcal{G}, \text{ briefly: } \mathcal{G} \models \varphi$$

then $\mathcal{G}$ is called a **model** of φ. The **theory** of $\mathcal{G}$, denoted by $\mathrm{Th}(\mathcal{G})$, is the set of all sentences satisfied in the structure $\mathcal{G}$. We write

$$\mathrm{Th}(\mathcal{G}) = \{\varphi \mid \varphi \text{ is a sentence and } \mathcal{G} \models \varphi\}.$$

This theory is **complete** since by definition for any sentence φ either φ or $\neg\varphi$ belongs to $\mathrm{Th}(\mathcal{G})$.

Generally: If $\mathcal{S}$ is a set of symbols then a set T of $\mathcal{S}$-sentences is said to be a **theory** if any consequence of T belongs to T. Hence for example

$$\Phi_G^{\models} = \text{the set of all sentences being consequences of (following from) axioms of } \Phi_G$$

is a theory, more exactly the group theory.

An important question arises when for example Peano arithmetic, hence all sentences *following* from arithmetical axioms is compared with the set of sentences *true* in the structure of natural numbers, i.e., with the theory of this structure. This generates the question about the axiomatizability of the elementary arithmetic – it will be considered in Section 5.4.1.

A theory T and the set of its axioms are called **consistent** if and only if no contradiction, i.e., a formula φ and its negation $\neg\varphi$, can be deduced from it. A theory is consistent if and only if it has a model.

5.3.1 Peano arithmetic

We would like to present in detail the purely logical and axiomatic approach and give at the beginning the fundamental example of the arithmetic of natural numbers. Since the logical axioms of arithmetic have been formulated for the first time by Peano, it is called **Peano arithmetic** and briefly denoted **PA.** Peano's historical attempt to formulate arithmetical axioms in which still set-theoretical elements were used has been described in Section 5.2.

Set-theoretical vs. logical

To make the difference clear between the usual set-theoretical formulation we first give arithmetical axioms in the usual, i.e., set-theoretical formulation. Those axioms were first formulated by Dedekind, however they are wrongly called "Peano axioms".

In a set-theoretical setting one needs only:

$$\text{a set } N, \text{ a function } \nu \text{ on } N, \text{ an element } 1 \in N.$$

We denote the following arithmetical axioms here (to distinguish them from the Peano axioms) by (D1), (D2) and (D3):

(D1) $\forall n \in N(\nu(n) \neq 1)$.
(D2) $\forall m, n \in N(m \neq n \Rightarrow \nu(m) \neq \nu(n))$.
(D3) If $A \subseteq N$ and
  (i) $1 \in A$,
  (ii) $\forall n(n \in A \Rightarrow \nu(n) \in A)$,
  then $A = N$.

The function v describes the successor function of counting that begins with 1. If the axioms are satisfied then the elements of N are called “natural numbers”. More exactly: N is called a model of “natural numbers” that are thought to be given in some way really (or as sets). The essential difference between the set-theoretical and the logical formulation concerns the “induction axiom” (D3) given below as (A7).

In the logical approach arithmetical axioms for natural numbers are formulated in the language of the first-order predicate calculus – we wrote about it in Section 5.1. Only logical symbols representing arithmetical primitive notions are used. The primitive notions are equality, zero, successor, addition and multiplication. They are denoted by the following symbols:

- = for equality,
- the individual constant 0,
- the unary function symbol S,
- the binary function symbols + and ·

The symbol S, coming from “successor”, symbolizes the operation of successor. As we see, symbols usually applied in mathematical practice are used as logical symbols now. This fact suggests the interpretation of those symbols and makes the axioms easier to understand. However, it should be stressed that despite of their shape the symbols can be arbitrarily interpreted.

The arithmetical axioms are the following:

(A1) $S(x) = S(y) \to x = y$,
(A2) $\neg(0 = S(x))$,
(A3) $x + 0 = x$,
(A4) $x + S(y) = S(x + y)$,
(A5) $x \cdot 0 = 0$,
(A6) $x \cdot S(y) = x \cdot y + x$,
(A7) $[\varphi(0) \wedge \forall x(\varphi(x) \to \varphi(S(x)))] \to \forall x \varphi(x)$.

If the symbols are interpreted arithmetically in the usual way then (A1) means that the successor function is injective (i.e., one-to-one), (A2) says that 0 is not a successor of any number, (A3) and (A4) characterize addition, (A5) and (A6) characterize multiplication and (A7) is a schema of countably many axioms describing the induction for countably many arithmetical expressions φ. This is the logical counterpart of the set-theoretical induction axiom (D3). The set of axioms (A1)–(A7) is denoted – similarly to the whole Peano arithmetic – by **PA**.

One can see that the set-theoretical means and the axioms (D1)–(D3) seem to be essentially more economical. This is made possible by the powerful set theory laying at the foundation – it gives further instruments and enables the definition of arithmetical operations in terms of the successor function only.

The set-theoretical embedding of arithmetic is “comfortable”. However, sets in the characterization of arithmetic are in a certain sense “foreign bodies”. This is clearly visible for the subsets in the induction axiom (D3). Sets of numbers are in fact arithmetical abstract objects that are only *developed* in arithmetic. However, in the induction axiom they are foremost sets and play an essential role in the set-theoretical characterization of arithmetic. There one speaks about *all subsets* of N, hence about the power set of N, and this set is problematic since its cardinality is mathematically not uniquely determined (cf. Continuum Hypothesis). This is in fact an ultralarge problematic background of this approach.

In a logical setting, arithmetic is free of any foreign influences. Instead of subsets one speaks here only about arithmetical formulas that are purely arithmetical. The set-theoretical induction axiom (D3) becomes now the logical axiom schema (A7) because quantifying over formulas is not allowed. The formulation of the Peano axioms in the language of the first-order predicate calculus is a quasi "natural" description of the structure of natural numbers. However, it is laborious – and static. In fact, it registers the arithmetical interrelations, isolates them and masks the arithmetical connections as for example between addition and multiplication.

We shall see that PA is not strong enough to grasp the arithmetic of natural numbers in a "unique way", i.e., by demarcating it from other structures. PA has non-isomorphic models. The situation with the set-theoretical arithmetical axioms characterizing natural numbers is different – they determine models uniquely up to isomorphism. The logical setting of PA is of smaller practical meaning, however its principal meaning is high.

Standard model and satisfaction

Logical investigation of natural numbers begins in fact with a *philosophical assumption.* Its aim is to deduce properties of natural numbers from some properties of them treated as "fundamental". One starts from a fixed conception of natural numbers and treats them as given in a certain sense – by intuition, as sets, by a set-theoretical description, etc. So did Peano when he formulated the axioms and so it is done involuntarily today when one reads the axioms of PA.

Those in some way "given" natural numbers together with "concrete" operations on them form the structure of natural numbers:

$$\mathcal{N}_0 = \langle \mathbf{N}, 0, S, +, \cdot \rangle,$$

where $\mathbf{N}$ is the set of "given" natural numbers, S is the symbol for successor that simulates the counting with natural numbers, 0 is the symbol for the "real" zero, and + and $\cdot$ are symbols for the "concrete" operations "plus" and "times", respectively. Note that the same symbols are used here and in the formal language for arithmetic. This indicates how the idea of the "standard model" is present in logic. However, the formal and the concrete arithmetical level should be accurately distinguished.

Formal arithmetical expressions and sentences φ in PA are initially nothing more than sequences of symbols. When an expression φ can be *deduced* in PA by the inference rule of the predicate calculus, i.e., when

$$\mathrm{PA} \vdash \varphi,$$

then φ is said to be an arithmetical **theorem**. Arithmetical sentences can be *interpreted* in $\mathcal{N}_0$ – and possibly in other structures. Then symbols occurring in φ get a concrete *meaning* and φ becomes a proposition. All axioms and theorems will be **satisfied** in $\mathcal{N}_0$ by such an *interpretation*. This means that

$$\mathcal{N}_0 \vDash \mathrm{PA},$$

i.e.,

$$\mathcal{N}_0 \vDash \varphi \quad \text{for any } \varphi \text{ in PA}.$$

5.3.2 On the axioms for real numbers

We provide also two sets of axioms for real numbers – a set-theoretical one and a logical one. Set-theoretical axioms will be compared with logical ones in order to indicate the specific aspect of the logical approach. To express the property of completeness one can use various versions of axioms. We shall use the idea of the upper bound.

What should be done? One wants to describe reals as an ordered field. It is well known how to do this – the set-theoretical description is analogous to the logical one. It is just the axiom of completeness – independently of the way of formulating it – which causes problems. In the first-order logic it must be formulated in another way as would be expected – and in the case of reals it especially causes difficulties. This situation is different from the situation with the axiom of induction in the case of natural numbers. In fact, it seems to be impossible to speak about completeness without using sets that should have upper bounds. The situation is here different as it was for natural numbers because sets belong directly to the characterization of reals and as "arithmetically external elements" they seem to be irreplaceable. So we shall try to grasp the completeness via formulas by paraphrasing arithmetically the situation of an upper bound for any formula. It is a bit laborious but not difficult. On the other hand the fact that sets cannot be used turns out to be serious in other respects.

Set-theoretical vs. logical

First we give set-theoretical axioms – one is thinking about them doing mathematics in the usual way. To grasp set-theoretically the arithmetic of reals one needs

a set R, elements $0, 1 \in R$, operations $+$ and $\cdot$, and a relation $<$ on R.

The set-theoretical axioms describing R as an ordered field are the following:

$$
\begin{aligned}
&0 \neq 1,\\
&\forall x, y, z \in R[(x+y)+z = x+(y+z)],\\
&\forall x, y \in R(x+y = y+x),\\
&\forall x \in R(x+0 = x),\\
&\forall x, y, z \in R[(x\cdot y)\cdot z = x\cdot(y\cdot z)],\\
&\forall x, y \in R(x\cdot y = y\cdot x),\\
&\forall x \in R(x\cdot 1 = x),\\
&\forall x, y, z \in R[x\cdot(y+z) = (x\cdot y)+(x\cdot z)],\\
&\forall x, y \in R\exists z \in R(x = y+z),\\
&\forall x, y \in R[y \neq 0 \to \exists z \in R(x = y\cdot z)],\\
&\forall x, y, z \in R[y < z \to x+y < x+z],\\
&\forall x \in R\,\forall y \in R\,\forall z \in R[0 < x \wedge y < z \to (x\cdot y < x\cdot z)],\\
&\forall x \in R\neg(x < x),\\
&\forall x, y, z \in R[x < y \wedge y < z \to x < z],\\
&\forall x, y \in R[x < y \vee x = y \vee y < x].
\end{aligned}
$$

The set-theoretical completeness axiom is

$$
\begin{aligned}
&\forall X \subseteq R(\neg X = \emptyset \wedge \exists b\forall x(x \in X \to x \leq b)\\
&\qquad \to \exists g(\forall x(x \in X \to x \leq g) \wedge (\forall c(\forall x(x \in X \to x \leq c)) \to g \leq c))).
\end{aligned}
$$

Algebraic axioms of fields, which we now want to formulate using methods of the first-order logic, look almost like their set-theoretical counterparts. The difference is that now one cannot refer to any set. The following nonlogical symbols are needed:

constants 0, 1, two binary function symbols $+, \cdot$ and a binary predicate $<$.

The nonlogical axioms are the following:

$$
\begin{aligned}
&0 \neq 1,\\
&\forall x\forall y\forall z[(x+y)+z = x+(y+z)],\\
&\forall x\forall y(x+y = y+x),\\
&\forall x(x+0 = x),\\
&\forall x\forall y\forall z[(x\cdot y)\cdot z = x\cdot(y\cdot z)],\\
&\forall x\forall y(x\cdot y = y\cdot x),\\
&\forall x(x\cdot 1 = x),
\end{aligned}
$$

$$\begin{aligned}
&\forall x \forall y \forall z[x \cdot (y+z) = (x \cdot y) + (x \cdot z)],\\
&\forall x \forall y \exists z(x = y + z),\\
&\forall x \forall y(y \neq 0 \rightarrow \exists z(x = y \cdot z)),\\
&\forall x \forall y \forall z(y < z \rightarrow x + y < x + z),\\
&\forall x \forall y \forall z[0 < x \wedge y < z \rightarrow (x \cdot y < x \cdot z)],\\
&\forall x \neg(x < x),\\
&\forall x \forall y \forall z(x < y \wedge y < z \rightarrow x < z),\\
&\forall x \forall y(x < y \vee x = y \vee y < x).
\end{aligned}$$

The basic difference between the set-theoretical and the logical-arithmetical axioms consists in the formulation of the completeness axiom. In the first-order language it will become a schema when one uses formulas instead of sets:

$$\begin{aligned}
&\exists x \varphi(x) \wedge \exists b \forall x(\varphi(x) \rightarrow x \leq b)\\
&\quad \rightarrow \exists g(\forall x(\varphi(x) \rightarrow x \leq g) \wedge \forall c(\forall x(\varphi(x) \rightarrow x \leq c)) \rightarrow g \leq c).
\end{aligned}$$

In this way one has set the real numbers free of set-theoretical external elements and gave the purely arithmetical (and order-theoretical) description. And again the purely logical-arithmetical axioms – in contrast to the set-theoretical ones – do not characterize the real numbers in a unique way. Even more: there are no first-order arithmetical axioms that could do this. Arithmetical first-order axioms allow nonstandard models.

Models

The standard model of real numbers to which everything logically refers is denoted by

$$\mathcal{R} = \langle \mathbf{R}, 0, 1, +, \cdot, < \rangle.$$

It is sometimes presented geometrically on the number line or set-theoretically according to one of the constructions of the reals. All purely arithmetical axioms given above are satisfied in $\mathcal{R}$. However, they are satisfied also for example in the structure

$$^*\mathcal{R} = \langle ^*\mathbf{R}, 0, 1, +, \cdot, < \rangle,$$

i.e., in the structure of hyperreal numbers – we wrote about them in Section 3.3.8; they will also be described in Chapter 6. Thus $^*\mathcal{R}$ is “elementarily equivalent” to $\mathcal{R}$. This means that all formulas without free variables formulated in the first-order language and satisfied in $\mathcal{R}$ are also satisfied in $^*\mathcal{R}$ and vice versa. Hence $\mathcal{R}$ and $^*\mathcal{R}$ cannot be “internally” distinguished in the first-order logic. This can be done only when one looks at both structures “from the outside”, i.e., when non-first-order means, e.g., set-theoretical ones, are used. At this “external” level $\mathcal{R}$ and $^*\mathcal{R}$ are not isomorphic: in fact, $\mathcal{R}$ is archimedean and $^*\mathcal{R}$ is not. More about this can be found in Chapter 6.

Remarks

Axioms for real numbers not using sets are uninteresting from the point of view of mathematical practice. Practice means here mathematical analysis based on reals. It is a matter not only of the arithmetic but especially of sets. The real numbers should represent the linear continuum and this is nowadays regarded as a set. We wrote in detail about this in Section 3.3. The real numbers should make just *sets* available: subsets, intervals, sequences. They cannot be seen in the logical axioms given above. Many things should be added here to make it a suitable background for mathematical work.

Such suitable sets of axioms for real numbers were given by Hilbert [166] and Tarski [338]. They belong to the second type of the logical axiomatic approach – there one uses symbols for primitive notions of the considered theory as well as symbols for primitive notions of set theory. Hilbert’s axioms are augmented by non-arithmetical set-theoretical elements that we just wanted to remove from a logical axiomatic approach. They are so strong that they enable to formulate the completeness axiom in a set-theoretical way and to characterize real numbers up to isomorphism.

Purely logico-arithmetical axioms of the first order – being uninteresting and not sufficient from the point of view of applications – are, similar to the case of elementary arithmetic, rather of a foundational interest. This is the case when one considers for example problems of limits of formalization, hence philosophical aspects of logic. However, note that logico-arithmetical axioms of the first order lead on to nonstandard Models, to hyperreal numbers which are of special mathematical interest. We speak about this in Chapter 6.

Problems of symbolization – in the context of epistemological and logical investigations – as well as the content-formal and symbolic-formal possibilities and limits of the logical approach to structures of numbers are considered in the book [384] by R. Ziegler.

5.4 On the arithmetic of natural numbers

Consider the axiom system PA of the Peano arithmetic described above. Two aspects can be distinguished:

(i) Syntactic aspect: syntactic properties of PA are examined, hence properties referring only to symbols and expressions of the language of PA and disregarding any interpretations, i.e., possible meaning of expressions in some models. Here are first of all questions concerning provability, completeness, decidability.

(ii) Semantic aspect: this concerns interpretations of considered systems and, for example, the question of categoricity – whether all models are isomorphic or there are "nonstandard models".

5.4.1 Syntactical aspect

With any system of axioms Φ at least two expectations are connected:

(1) Axioms should be chosen in such a way that it is clear which formulas are the axioms of Φ and which sentences should be deduced from them. This means that there should be given a simple, for example programmable/algorithmic, procedure by which for any expression φ it could be decided in a finite number of steps whether φ belongs to the set of axioms Φ or not. Then it is said that Φ is a **recursive** set of axioms and that the theory Th(Φ), i.e., the set of all sentences that can be deduced from Φ, is **recursively axiomatizable** or simply **axiomatizable.**

(2) The axiom system Φ and the theory based on it should be **complete**. This means: for any sentence φ in the language of Φ either φ or $\neg\varphi$ can be deduced from Φ.

The axiom system PA of the Peano arithmetic obviously satisfies the first expectation. One hoped for a long time – encouraged by other successes in the domain of the foundations of mathematics – that also the second one will be satisfied. In fact, at the beginning some positive results for fragments of arithmetic have been obtained. Kurt Gödel proved in 1929 the completeness of the first-order predicate calculus (only the quantification over individual variables is allowed here). He showed that for any formula being a consequence there exists a formal proof of it. However, another result of Gödel destroyed any hope of completeness of a system of arithmetic fulfilling the moderate condition (1). He proved the following **First Incompleteness Theorem.**

Theorem 5.4.1 (Gödel, 1931). *Every consistent theory containing* PA *and based on a recursive set of axioms is incomplete, i.e., for any consistent extension of* PA *there are sentences φ such that neither φ nor $\neg\varphi$ are provable in it.*

Consequently, the Peano arithmetic itself – provided it is consistent – is incomplete. Similarly any consistent extension of PA is incomplete. It is said that PA is essentially incomplete.

In the proof of this theorem an arithmetization (sometimes also called Gödelization) of syntax is used. To every formula φ corresponds its Gödel number $\ulcorner\varphi\urcorner$. Using this one can translate expressions concerning the syntax of PA, hence metamathematical expressions, into arithmetical ones (cf. [180, Section 4.2.1] and [250, Chapter 2]). Next one constructs – using the idea of the Liar paradox – a metamathematical sentence ψ stating its own unprovability:

$$\psi \leftrightarrow \psi \text{ is unprovable in } \Phi,$$

where the latter is formulated in the language of arithmetic (which is possible thanks to the arithmetization of syntax). In fact, neither ψ nor $\neg\psi$ can be deduced from Φ. J. Paris, L. Harrington and L. Kirby gave in 1979 and 1982, respectively, examples of undecidable sentences of a mathematical, more exactly combinatorial and number-theoretical content. Compare the book by D. Hoffmann [180,

Section 4.5] as well as his book *Die Gödelschen Unvollständigkeitssätze* [181] where Gödel's proof is presented in detail.

Connected with the problem of completeness of theories is the problem of the decidability. A set Ψ of sentences is said to be **decidable** if and only if it is recursive (see above), i.e., if there is a procedure by which for any φ it can be said in a finite number of steps whether φ belongs to Ψ or not. In the case of a theory T it means that for any formula φ it can be decided in a finite number of steps whether φ is *provable* in T or not.

Connections between decidability and completeness as well as undecidability and incompleteness of theories are given by the following theorems. Note that consistency is always assumed here.

Theorem 5.4.2. *If a theory* T *is recursively axiomatizable and complete then* T *is decidable. If* T *is recursively axiomatizable and undecidable then* T *is incomplete.*

Gödel proved the following theorem.

Theorem 5.4.3. PA *and every extension of* PA *is undecidable. Briefly:* PA is essentially undecidable. *Furthermore, no complete extension of* PA *is axiomatizable.*

By **elementary arithmetic** we mean here a set of all sentences true in the structure $\mathcal{N}_0$, i.e., the theory of natural numbers $\mathrm{Th}(\mathcal{N}_0)$. It is clear that for any sentence φ either φ or $\neg\varphi$ belongs to $\mathrm{Th}(\mathcal{N}_0)$. This means that $\mathrm{Th}(\mathcal{N}_0)$ is complete. Hence:

Theorem 5.4.4. *The elementary arithmetic is not axiomatizable.* PA *and every axiomatizable extension of* PA *is a proper part of the elementary arithmetic.*

Let us quote some general theorems on arithmetical theories.

Theorem 5.4.5. *Every arithmetical theory T in the language of PA whose axioms are true in $\mathcal{N}_0$ (or in any other model of* PA*) is undecidable. In particular every such subtheory T of* PA *is undecidable.*

The situation changes when the language is smaller than the full language of PA and only a part of it. Consider subtheories T_S or T_+ of PA whose nonlogical symbols are only $0, S$ or $0, S, +$, respectively. The theory T_S is based on the following nonlogical axioms:

- $0 \neq S(x)$,
- $S(x) = S(y) \rightarrow x = y$,
- the axiom schema of induction for formulas φ of the language of T_S, i.e., formulas in which only $0, S$ as nonlogical can occur.

The theory T_+ is based on the above axioms and on the following ones:

(i) $x + 0 = x$,
(ii) $x + S(y) = S(x + y)$,
(iii) axiom scheme of induction for formulas φ of the language of T_+, i.e., formulas in which only $0, S, +$ as nonlogical symbols can occur.

For those theories the following theorems hold (in contrast to PA).

Theorem 5.4.6 (Herbrand, 1928). *The theory T_S is complete and decidable.*

Theorem 5.4.7 (Presburger, 1930). *The theory* T_+ *is complete and decidable.*

Since $\langle \mathbf{N}, 0, S\rangle$ is a model of T_S and $\langle \mathbf{N}, 0, S, +\rangle$ is a model of T_+, for any sentence φ in the language of T_S or T+, respectively, holds,

$$\langle \mathbf{N}, 0, S\rangle \models \varphi \quad \text{if and only if} \quad T_S \vdash \varphi,$$
$$\langle \mathbf{N}, 0, S, +\rangle \models \varphi \quad \text{if and only if} \quad T_+ \vdash \varphi.$$

This means that both successor and addition can be characterized syntactically and semantically.

Let $T_\cdot$ be the semantically defined theory of multiplication, i.e.,

$$T_\cdot = \mathrm{Th}(\langle \mathbf{N}, 0, S, \cdot\rangle).$$

We now have the following theorem.

Theorem 5.4.8 (Skolem, 1930). *The theory* $T_\cdot$ *is decidable.*

A syntactical characterization of $T_\cdot$ has been given only 1981 by Cegielski. Comparing the theorems given above with the First Incompleteness Theorem one comes to the following conclusion concerning connections between addition and multiplication.

Theorem 5.4.9. *The theory of addition and the theory of multiplication are decidable and complete. However, the theory PA of addition and multiplication is undecidable and incomplete. Hence addition cannot be defined by S and · as well as multiplication cannot be defined by S and +.*

Note that the latter statements concerning undefinability do not hold when set-theoretical means – not available in PA and its subtheories – are allowed.

5.4.2 Semantical aspect

From the semantical point of view one expects first of all that an axiom system for arithmetic will characterize the set of natural numbers in a unique and complete way, i.e., that the structure

$$\mathcal{N}_0 = \langle \mathbf{N}, 0, S, +, \cdot\rangle$$

will be the unique – up to isomorphism – model of this axiom system. However, this expectation of a categorical axiom system for arithmetic cannot be fulfilled in the first-order logic – in fact there are "plenty of" **nonstandard models** of "various type".

The reason is that the Peano arithmetic has a model $\mathcal{N}_0 = \langle \mathbf{N}, 0, S, +, \cdot\rangle$ and that it is infinite. By theorems of model theory (cf. Section 5.1 – theorems of Löwenheim, Skolem and Tarski) there are models of any infinite cardinality: For any uncountable cardinal κ there exists a model

$$\mathcal{M} = \langle M, o, s, \oplus, \otimes, \rangle$$

of PA such that

- the cardinality of M is κ,
- $o \in M$,
- $s : M \to M$,
- $\oplus, \otimes : M \times M \to M$,

and the axioms of PA are true in $\mathcal{M}$:

$$\mathcal{M} \models \text{PA}.$$

Except those uncountable models there are also "plenty of" countable models that are essentially different from the standard model $\mathcal{N}_0$.

Theorem 5.4.10 (Skolem, 1934). *There are countable nonstandard models of arithmetic, i.e., countable models that are not isomorphic to* $\mathcal{N}_0$. *More exactly: There are countable nonstandard models of PA and of the whole elementary arithmetic* $\text{Th}(\mathcal{N}_0)$.

A characteristic feature of nonstandard models is the fact that there are nonstandard elements a in them such that

$$a > 0, \quad a > S(0), \quad a > SS(0), \quad \ldots, \quad a > \underbrace{S\ldots S}_{n}(0), \quad \ldots$$

Hence they are "infinitely large numbers". The order < between elements is defined by addition in the following way:

$$x < y \leftrightarrow \exists z(z \neq 0 \wedge x + z = y).$$

One can show that the order in any countable nonstandard model $\mathcal{M}$ is of the type

$$\omega + (\omega^* + \omega) \cdot \eta,$$

where ω is the order-type of the set of natural numbers $\mathbf{N}$, ω^* is the order-type of the set of negative integers and η is the order-type of rational numbers. This means that a typical countable nonstandard model can be imagined as:

$$1, 2, 3, \ldots \quad \ldots, a_{-2}, a_{-1}, a, a_1, \ldots \quad \ldots, b_{-2}, b_{-1}, b, b_1, \ldots \quad \ldots, c_{-2}, c_{-1}, c, c_1, \ldots.$$

At the beginning there are natural numbers as usual. Then follow nonstandard numbers in sections isomorphic to $\mathbb{Z}$. There are countably many such sections and they are ordered densely as the rational numbers: between every two such sections there is another isomorphic section.

By the theorem of Skolem there are countable structures

$$\mathcal{M} = \langle M, o, s, \oplus, \otimes \rangle$$

such that for any arithmetical sentence φ one has

$$\mathcal{N}_0 \models \varphi \quad \text{if and only if} \quad \mathcal{M} \models \varphi.$$

Then $\mathcal{M}$ and $\mathcal{N}_0$ are **elementarily equivalent** and one writes

$$\mathcal{M} \equiv \mathcal{N}_0.$$

On the other hand they are not isomorphic:

$$\mathcal{N}_0 \not\cong \mathcal{M}.$$

Even a stronger theorem – being a consequence of Gödel's Incompleteness and Completeness theorems – holds.

Theorem 5.4.11. *There are* $2^{\aleph_0}$ *countable non-elementarily equivalent models of PA.*

By $\aleph_0$ the cardinality of countable sets is denoted, and $2^{\aleph_0}$ means "uncountably many". For any two such models $\mathcal{M}_1$ and $\mathcal{M}_2$ one has

$$\mathcal{M}_1 \not\equiv \mathcal{M}_2,$$

i.e., there exists a sentence φ such that $\mathcal{M}_1 \models \varphi$ and $\mathcal{M}_2 \models \neg\varphi$. Hence

$$\mathcal{M}_1 \not\cong \mathcal{M}_2.$$

Consequently, there exists a rich variety of countable models of PA. The aim to characterize axiomatically natural numbers in a unique way cannot be achieved by far in the first-order logic.

On the other hand in **second-order logic** a categorical characterization of natural numbers can be given.

Consider the arithmetic of second order A_2^-. It is formulated in a language with two sorts of variables:

- $x_1, x_2, x_3, \ldots$ (individual variables),
- $X_1, X_2, X_3, \ldots$ (set variables).

Its nonlogical symbols are $S, 0, +, \cdot, \in$. The nonlogical axioms of A_2^- are the following:

- axioms of PA without the axiom scheme of induction,
- extensionality axiom:
$$\forall x(x \in X \leftrightarrow x \in Y) \rightarrow X = Y,$$
- induction axiom:
$$0 \in X \wedge \forall x(x \in X \rightarrow S(x) \in X) \rightarrow \forall x(x \in X),$$
- comprehension axiom:
$$\exists X \forall x(x \in X \leftrightarrow \varphi(x, \ldots)),$$

where φ is any formula of the second-order language in which X is not free. The dots "..." denote free individual or set variables occurring in φ. Hence the comprehension axiom is in fact an axiom scheme. Instead of the axiom scheme of induction in the first-order language one has here a single second-order axiom of induction.

PA is of course a subtheory of the theory A_2^-, i.e., PA $\subseteq A_2^-$. Models of A_2^- can be obtained in the following way: Let $\langle M, 0, S, +, \cdot \rangle$ be a model of PA, i.e.,

$$\langle M, 0, S, +, \cdot \rangle \models \text{PA}.$$

Let $\mathcal{A} = \langle \mathcal{X}, M, 0, S, +, \cdot, \varepsilon \rangle$, where $\mathcal{X}$ is a set of subsets of M and ε is an interpretation of the membership relation. It is said that $\mathcal{A}$ is a model of A_2^- if and only if all axioms of A_2^- are true in $\mathcal{A}$. Then one writes

$$\mathcal{A} = \langle \mathcal{X}, M, 0, S, +, \cdot, \varepsilon \rangle \models A_2^-.$$

If $\mathcal{X} = \mathcal{P}(M)$ and ε is interpreted as the set-theoretical membership relation $\in$ then the structure

$$\mathcal{A} = \langle \mathcal{X}, M, 0, S, +, \cdot, \in \rangle$$

is said to be standard with respect to set-theoretical notions. In such model the comprehension axiom is always satisfied. In particular,

$$\langle \mathcal{P}(\mathbf{N}), \mathbf{N}, 0, S, +, \cdot, \in \rangle \models A_2^-.$$

Theorem 5.4.12. *Let*

$$\mathcal{A}_1 = \langle \mathfrak{P}(M_1), M_1, o, s, \oplus, \otimes, \in \rangle,$$
$$\mathcal{A}_2 = \langle \mathfrak{P}(M_2), M_2, o', s', \oplus', \otimes', \in \rangle$$

be models of A_2^- and let M_1 and M_2 be of the same cardinality. Then

$$\mathcal{A}_1 \simeq \mathcal{A}_2.$$

If additionally M_1 and M_2 are countable then

$$\mathcal{A}_1 \simeq \mathcal{A}_2 \simeq \langle \mathfrak{P}(\mathbf{N}), \mathbf{N}, 0, S, +, \cdot, \in \rangle.$$

It is important that in all structures in the theorem the universe for set-variables is always the whole power set of the first-order universe.

So arithmetic can be described categorically in an axiomatic way in the second-order logic. Primitive notions can be even restricted to $0, S, \in$. Denote by A_2^* the subtheory of A_2^- obtained by restricting the nonlogical symbols to $0, S, \in$. The latter theorem holds also for models of A_2^*.

This theorem can possibly be an explanation why Peano in his historical characterization of natural numbers used not only first-order axioms but also set-theoretical notions.

Some further theorems indicate limits of the axiomatic method and of mathematical theories – for example theorems of Tarski and of Löwenheim and Skolem (cf. Section 5.1). Tarski proved also the **undefinability of the concept of truth** in a theory [335].

Tarski uses in his theorem the classical approach to the concept of truth of an expression. For example when is the sentence "Snow is white" true? It is true if and only if snow is white. Generally: A formula φ in the language of a theory T is true if and only if φ is fulfilled. Is there a formula Φ of the language of the theory T expressing the truth of formulas φ? Tarski proved that for any theory T – satisfying some natural conditions – there is no such formula Φ. More exactly:

> There exists *no* formula Φ such that for any formula φ of the language of T one has
>
> $$\Phi(\varphi) \leftrightarrow \varphi.$$

In particular the concept of truth for arithmetical formulas is not definable in the arithmetic of natural numbers.

It should be added and emphasized that Tarski gave in the quoted work a **definition of the concept of truth** – it was given in a (formalized) **metalanguage** that is stronger than the object language of the theory T. The whole problem and its explanations are complicated. It is a question of the concept of truth restricted only to *formalized* languages. Take as an example the arithmetical language of the theory PA. Let arithmetical symbols be interpreted in a model $\mathcal{M}$. Let φ be an arithmetical formula with free variables. Then

> φ *is true in the model* $\mathcal{M}$ if and only if φ is satisfied by all valuations of free variables.

Tarski expressed this formulation (belonging to the metalanguage!) in the case of PA by an expression Φ of a formalized metalanguage.

Skolem's Theorem 5.4.10 states that any systems of axioms for the arithmetic of natural numbers has a nonstandard model, i.e., models being different than the usual model consisting of the natural numbers $0, 1, 2, 3, \ldots,$ with operations $+$ and $\cdot$. Consequently, even the simple natural numbers cannot be uniquely characterized by the axiomatic-deductive method. Any attempt to describe axiomatically the domain of natural numbers will fail. Any attempt admits "unwanted" unintended interpretations. The same holds for other theories, in fact for any consistent theory axiomatizable in the first-order logic. The reason is not the choice of axioms. It is in the nature of the axiomatic method itself.

There are other theorems indicating the limits of the axiomatic method. Gödel announced a theorem called today the **Second Incompleteness Theorem.** It states that no consistent theory T containing the Peano arithmetic PA can prove its own consistency. This means that in a proof of the consistency of such a theory stronger methods are necessary, i.e., methods that are stronger than those available in T. In particular the consistency of the arithmetic PA cannot be proved in PA itself, i.e., by arithmetical methods only. One should use for example the transfinite induction up to a higher ordinal – as Gentzen has shown.

5.5 Truth and provability

Throughout this book we are permanently talking about truth as well as about true expressions and sentences. Above we told about the theoretical undefinability and about a definition of the concept of truth that can be formulated in a metalanguage. In the historical development of mathematics the concept of truth was always directly connected with the concept of provability.

Truth was understood in the history of philosophy and of mathematics in various ways. It was considered as an idea from and in a higher world, as a property of divine creators, an *a priori* property of sentences, as a correspondence between thinking and mind on the one hand and the reality on the other, or as being deducible from given assumptions, hence as provability. In the case of mathematics the most important were both latter characterizations. They were the base of the high reputation of mathematics being a domain of true statements.

Since Euclid mathematics begins with axioms, which were treated as simple, primitive and undisputable truths because they corresponded to the reality. From them emerged by deduction the mathematical world of true statements. Since the axioms were accepted as truth, mathematical truth equaled provability. Everything that can be proved is true, and vice versa, any mathematical truth can be proved. One was convinced of this. What is a proof was in fact a matter of an agreement among mathematicians (as today still is in the usual mathematical research practice).

Such conception prevailed till the 20th century and was in the background of the approach whose aim was to specify the concepts of a proof at the beginning of the 20th century (cf. Section 5.1). The way to the distinction between truth and provability was hard and long. The process of the development of the consciousness of the difference between truth and provability was philosophically and psychologically difficult and in a certain sense full of dramatic events. The crucial point in this development was the First Incompleteness Theorem of Gödel in the year 1931.

5.5.1 Formal truth

Prior came Hilbert's program that induced, among others, Gödels investigations. Prior to Hilbert's program came his work *Grundlagen der Geometrie* in which the step towards a new axiomatic approach was made – it was essentially different than the old Euclidean one. We wrote about it in Section 5.2.2.

The aims of the new axiomatic approach were indicated by Hilbert in a letter to Gottlob Frege from December 29, 1899. He explained there why the geometry should be axiomatized. He wrote [125, p. 67]:

> "I was forced to construct my systems of axioms by a necessity: I wanted to have a possibility to understand those geometrical propositions which in my opinion are the most important results of geometrical researches: that the Parallel Axiom is not a consequence of other axioms, and similarly for the archimedean one, ..."

Hilbert left in his geometrical axioms the reality – though they seemed to be close to the Euclidean ones. It can be said that he resigned from the old philosophical truth. Axioms were now only formal statements without real meaning.

Hilbert treated mathematical theories as "formal systems". He considered the axiomatic method as a method that enables to order systematically any sufficiently developed domain. The essence and aim of an axiomatic theory was for Hilbert to determine a position of single statements in a given system and to clarify the interrelations between them. One of the results of investigations carried out in the frameworks of Hilbert's program was the fact that the so far rather informal concept of a proof has been replaced by the concept of a formal proof and of the consequence and in this way specified (cf. Sections 2.20 and 5.1). Truth seemed to be mathematically reducible to provability and consistency of formal systems. The historical concept of truth seemed to be mathematically clarified.

Gödel's incompleteness theorems challenged this reduction. The way to them led via Hilbert and Hilbert's program.

Hilbert's aim was to found mathematical knowledge and mathematical truth solely by *syntactic means*. He neglected the semantical side. To do this all mathematical theories (and finally the whole mathematics) should be formalized in order to investigate them as systems of symbols abstracting from any content.

Such formal axiomatic systems should satisfy three conditions: they should be complete and consistent and their axioms should be mutually independent. Consistency was for Hilbert a – rather the only – criterion of mathematical truth and of the existence of mathematical objects in theories. He wrote the following in the letter to Frege quoted above (cf. [125, p. 66]).

> "If the arbitrarily given axioms do not contradict one another with all their consequences then they are true and the things defined by the axioms do exist."

The existence and the truth are given here for the first time a formal foundation. Their justification comes from the formal level and is transferred to mathematical structures via – in principle arbitrary – interpretations of nonlogical symbols and variables occurring in axioms.

It was presumed in Hilbert's approach that theories are categorical, i.e., their models are determined by axioms in a unique way up to isomorphism. This requirement is strongly connected with the assumption of the *completeness* of theories. And this turn out to be important for our considerations. In fact, completeness should play an essential role in Hilbert's way to the truth.

5.5.2 Completeness and truth

We wrote about completeness in Section 5.3. Where did the concept of completeness come from and how has it been developed?

Recall that in *Grundlagen der Geometrie* Hilbert postulated explicitly completeness as one of the axioms (more precisely it was not present in the first edition of *Grundlagen*, but was included first in the French translation and then in the second edition of 1903). The axiom V2 stated:

> "Elements of geometry (i.e., points, lines and planes) form a system that does not admit any extension provided all the mentioned axioms are preserved."

In later editions of *Grundlagen* beginning with the seventh edition from 1930 Hilbert replaced this axiom by the axiom of linear completeness stating that:

> "Points of a line form a system which admits no extension provided the linear order of the line (Theorem 6), the first congruence axiom and archimedean axioms (i.e., axioms I1–2, II, III1, V1) are preserved."

In Hilbert's lecture [169] at the Third International Congress of Mathematicians held in Heidelberg in 1904 one finds such an axiom of the real numbers. Later Hilbert understood completeness as an inner property of a system – as it is understood nowadays.

In his lectures *Logische Principien des mathematischen Denkens* from 1905 Hilbert characterized completeness as the demand that the axioms suffice to prove all "facts" of the theory in question. In [168] the following words can be found:

> "We will have to demand that all other facts (*Thatsachen*) of the given field are consequences of the axioms."

Hilbert's conviction as to the solvability of every mathematical problem – expressed for example in his 1900 Paris lecture [167] and repeated in his opening address "Naturerkennen und Logik" [174] before the Society of German Scientists and Physicians in Königsberg in September 1930 – can be treated as informal reflection of his belief in completeness of axiomatic theories. In his Paris lecture he said full of optimism:

> "The conviction of the solvability of any mathematical problem is for us a strong motive in this work; we hear the whole time the call: *There is a problem, look for a solution. You can find it by a pure thinking; there is no Ignorabimus in the mathematics.*" ([167, p. 298].)

He repeated this in the paper "Naturerkennen und Logik" [174]. In his speach at the local radio station in Königsberg in September 1930 he formulated this in the following way:

> "For the mathematician there is no Ignorabimus, and, in my opinion, not at all for natural science either. [...] The true reason why [no one] has succeeded in finding an unsolvable problem is, in my opinion, that there is *no* unsolvable problem. In contrast to the foolish Ignorabimus, our credo avers: We must know. We shall know."

Turn back to the central concept of completeness. In Hilbert's lecture in August 1900 in Paris this concept appears in the following form:

> "When we are engaged in investigating the foundations of a science, we must set up a system of axioms which contains an exact and complete description of the relations subsisting between the elementary ideas of that science. The axioms so set up are at the same time the definitions of those elementary ideas; and no statement within the realm of the science whose foundations we are testing is held to be correct unless it can be derived from those axioms by means of a finite number of logical steps." ([167, pp. 299–300].)

The phrases "exact and complete description" (*genaue und vollständige Beschreibung*) can be understood in such a way that axioms of a system allow to decide the truth or falsity of every statement of the given theory. All axiomatic systems considered by Hilbert himself were complete – more exactly: categorical, i.e., models of given theories were isomorphic. However, they always contained axioms formulated in the second-order logic. Indeed his axiomatization of geometry from *Grundlagen* as well as his axiomatization of arithmetic published in 1900 were second-order. Each of those systems had a second-order archimedean axiom and both had a metatheoretical "completeness axiom" stating that the structure under consideration was maximal with respect to the remaining axioms.[2]

The demand discussed here would imply that a system of axioms complete in this sense is possible only for sufficiently advanced theories. On the other hand Hilbert called for complete systems of axioms also for theories being developed. In "Mathematische Probleme" [167, p. 295] he wrote:

> "[...] wherever mathematical concepts emerge from epistemological considerations or from geometry or from theories of science, mathematics acquires the task of investigating the principles lying at the basis of these concepts and defining [...] these through a simple and complete system of axioms."

One should also add here that Hilbert admitted the possibility that a mathematical problem may have a negative solution, i.e., that one can show the impossibility of a positive solution on the basis of a considered axiom system. Again in "Mathematische Probleme" [167, p. 297] he wrote:

> Occasionally it happens that we seek the solution under insufficient hypotheses or in an incorrect sense, and for this reason we do not succeed. The problem then arises: to show the impossibility of the solution under the given hypotheses, or in the sense contemplated [...] and we perceive that old and difficult problems [...] have finally found fully satisfactory and rigorous solutions, although in another sense than that originally intended. It is probably *this* important fact along with other philosophical reasons that give rise to the conviction [...] that every definite mathematical problem must necessarily be susceptible of an exact settlement, either in the form of an actual answer to the question asked, or by a proof of the impossibility of its solution and therewith the necessary failure of all attempts.

5.5.3 Syntactic reduction of truth

In Hilbert's lectures [170] from 1917–1918 one finds completeness in the sense of maximal consistency, i.e., a system is complete if and only if for any non-derivable sentence that it is added to the system the system becomes inconsistent:

$$\forall \varphi (T \text{ non} \vdash \varphi \rightarrow T \cup \{\varphi\} \text{ inconsistent}).$$

2 Note that this axiom was not even properly a second-order axiom.

He stated there [170, p. 152]:

> "Let us now turn to the question of *completeness*. We want to call the system of axioms under consideration complete if we always obtain an inconsistent system of axioms by adding a formula which is so far not derivable to the system of basic formulas."

In [178] Hilbert wrote:

> "In the stronger sense of the word, completeness means that the addition of a previously unprovable formula to the axiom system always yields a contradiction."

In his lecture at the International Congress of Mathematicians in Bologna in 1928 Hilbert distinguished two aspects of completeness: (a) completeness for the first-order predicate calculus – completeness with respect to validity in all interpretations, hence the semantic completeness, and (b) completeness for a system of elementary number theory – formal completeness, in the sense of maximal consistency, i.e., so-called Post-completeness, hence the syntactical completeness (cf. [175]).

Syntactical completeness seems to replace the semantical completeness as well as the concepts of satisfaction and truth. The reasons of emphasizing here the syntactic and finite methods together with the requirement of the completeness were described later by Gödel in the following way:

> "[...] formalists considered formal demonstrability to be an *analysis* of the concept of mathematical truth and, therefore were of course not in a position to *distinguish* the two."
> ([360, p. 9].)

Indeed, the informal concept of truth was not commonly accepted as a definite mathematical notion at that time – it has been done under the influence of formalistic mathematics. Only in 1933 Tarski gave the precise definition of the concept of truth – it has been explained above in Section 5.4. However, there were fierce logico-philosophical discussions concerning the acceptance of it.

Gödel wrote in a crossed-out passage of a draft of his reply to a letter of the student Yossef Balas (cf. [362, pp. 84–85]):

> "[...] a concept of objective mathematical truth as opposed to demonstrability was viewed with greatest suspicion and widely rejected as meaningless."

This opinion of Gödel is confirmed by remarks of Rudolf Carnap. When he invited A. Tarski to speak on the concept of truth at the September 1935 International Congress for Scientific Philosophy he wrote the following in his diary

> "Tarski was very sceptical. He thought that most philosophers, even those working in modern logic, would be not only indifferent, but hostile to the explication of the concept of truth."

And indeed at the Congress "[...] there was vehement opposition even on the side of our philosophical friends" [72, pp. 61–62].

All these explains in some sense why Hilbert and formalists who determined at that time logic and philosophical discussions preferred to deal in their metamathematics solely with the forms of the formulas, using only finitary reasoning which were considered to be save – contrary to semantical reasonings which were non-finitary and consequently not save.

David Hilbert and Wilhelm Ackermann formulated in 1928 the problem of completeness of logic explicitly in their book *Grundzüge der theoretischen Logik* [178] in the following way:

> "The completeness of an axiom system may be defined in two ways. First, it may be taken to mean that all the true formulas of a certain domain which is characterized by content can be proved from the set of axioms. However, the concept of completeness may also be more strictly formulated, so that an axiom system is termed complete only if a contradiction always arises when there is added to the axioms a formula not previously provable from them."

So the second way of understanding is here the maximal consistency – we wrote about it above. The problem was solved by Gödel in 1929 in his doctoral dissertation where he showed that the first-order logic is complete, i.e., every true statement can be derived from the axioms. Moreover, he proved that, in the first-order logic, every consistent axiom system has a model (cf. Section 5.1.3). Gödel wrote in [135] that by *completeness* he meant that

> "every valid formula expressible in the restricted functional calculus [...] can be derived from the axioms by means of a finite sequence of formal inferences."

And he added that this "can easily be seen to be equivalent to the following: Every consistent axiom system consisting of only *Zählaussagen* [counting expressions – authors' remark] has a realization" as well as that "[t]he latter formulation seems also to be of some interest in itself, since the solution of this question represents in a certain sense a theoretical completion of the usual method for proving consistency" [135, p. 61].

Gödel's Completeness Theorem (1929) seems to confirm the formalistic view that truth is the same as provability. It shows that the logical methods admitted by the concept of provability are appropriate and sufficient, that they reflect mathematical proofs in a proper way and secure the truth of mathematical statements.

Notice that the proof of the completeness theorem assumes the concept of satisfaction in a structure (cf. Section 5.1.2). "Satisfaction" there means "validity" or "truth". However, this concept has not been discussed either in Gödel's doctoral dissertation (1929) or in the published version of it (1930). In fact, "truth" has been reduced in a way to the concept of "truth". There was in fact a long tradition of use of the informal notion of satisfiability – compare works of Löwenheim, Skolem and others.

5.5.4 Truth is unequal to provability

Some months later happened the decisive thing. In his famous paper from 1931 Gödel proved that arithmetic of natural numbers and all richer theories are *essentially incomplete* (provided they are consistent) (cf. Section 5.4.1). There exist true arithmetical sentences that on the base of axioms of arithmetic can be neither proved nor rejected. The incompleteness theorems (cf. Section 5.4.1) showed that the reduction of truth to provability is not possible. Hence provability *is not equal* to truth.

How did Gödel come to the idea to prove such a theorem? He wrote himself on his discovery in a draft of a reply to a letter dated May 27, 1970, from Yossef Balas, then a student at the University of Northern Iowa (cf. [362, pp. 84–85]). Gödel indicated there that it was precisely his recognition of the contrast between the formal definability of provability and the formal undefinability of truth that led him to his discovery of incompleteness. There one finds also the following statement:

> "[...] long before, I had found the *correct* solution of the semantic paradoxes in the fact that truth in a language cannot be defined in itself."

A well-known example of a semantical paradox is the Liar paradox. Gödel used an intuitive not precise concept of truth. It should be however stressed that he was convinced of the mathematical significance and objectivity of this concept. Did he see the necessity of an analysis of it? The answer seems to be positive. In a letter to Rudolf Carnap from September 11, 1932, he announced [214]:

> "On the base of ideas I will give in the second part of my paper the definition of "true" and I am of the opinion that this cannot be done in another way and that the higher induction calculus cannot be semantically [i.e., at that time "syntactically"] grasped."

The concept of truth was defined only in 1933 by Tarski – in a formalized metalanguage – and it was precisely proved that this concept is undefinable *in* a formal language itself. This has been explained at the end of Section 5.4.

Gödel even avoided the terms "true" and "truth" as well as the very concept of being true – he used the terms "richtige Formel" (correct formula) and "inhaltlich richtige Formel" (correct formula with regard to content) and never the term "wahre Formel" (true formula). What were the reasons of this? An answer can be found in a crossed-out passage of a draft of Gödel's reply to a letter of the student Yossef Balas (already mentioned above). Gödel wrote there:

> "However, in consequence of the philosophical prejudices of our times 1. nobody was looking for a relative consistency proof because [it] was considered axiomatic that a consistency proof must be finitary in order to make sense, 2. a concept of objective mathematical truth as opposed to demonstrability was viewed with greatest suspicion and widely rejected as meaningless."

It can be supposed that – as Feferman formulated it (cf. [113]) –

> "[...] Gödel feared that work assuming such a concept [i.e., the concept of mathematical truth — authors' remark] would be rejected by the foundational establishment, dominated as it was by Hilbert's ideas. Thus he sought to extract results from it which would make perfectly good sense even to those who eschewed all non-finitary methods in mathematics."

It should be noticed that Gödel's own philosophy of mathematics was in fact platonist. He was convinced that:

> "It was the anti-Platonic prejudice which prevented people from getting my results. This fact is a clear proof that the prejudice is a mistake." (Cf. [363, p. 83].)

The discovery of the phenomena of incompleteness and undefinability of the concept of truth has shown that the definable concept of formal provability cannot be treated as the explanation of the concept of mathematical truth. It has been also shown that Hilbert's program to found and to justify the mathematics by finitistic methods had to fail.

Another philosophically important fact is that Gödel was in fact "a rationalistic optimist" – to use Hao Wang's term. He believed that mathematics is a system of true statements that is complete in another very general sense:

> "Every precisely formulated yes-or-no question in mathematics must have a clear-cut answer." (Cf. [145, pp. 378–379].)

It resembles Hilbert's optimism we saw above. Gödel rejected however – in the light of his Incompleteness Theorem – the idea that the basis of these truths is their derivability from axioms. In his Gibbs lecture of 1951 Gödel distinguishes between the system of all true mathematical propositions from that of all demonstrable mathematical propositions, calling them, respectively, mathematics in the objective and subjective sense. He claimed also that it is objective mathematics that no axiom system can fully comprise.

Truth in mathematics cannot be – as Gödel's incompleteness theorems showed – understood as provability, hence by syntactic finitary methods. Truth can be only approximated by such methods. Consequently, Hilbert's point of view should be extended to achieve the accordance between truth and provability. By which – obviously non-finitary – methods can this be done?

5.5.5 Search for the way out

Hilbert in his lecture in Hamburg in December 1930 (cf. [176]) proposed to admit a new rule of inference to be able to realize his program. This rule is similar to the ω-rule, but it has a rather informal character and a system obtained by admitting it would be semi-formal. In fact, Hilbert proposed that

> "whenever $A(z)$ is a quantifier-free formula for which it can be shown (finitarily) that $A(z)$ is a correct (*richtige*) numerical formula for each particular numerical instance z, then its universal generalization $\forall x A(x)$ may be taken as a new premise (*Ausgangsformel*) in all further proofs." ([176, p. 194].)

In the preface to the first volume of Hilbert and Bernays' monograph *Grundlagen der Mathematik* they wrote:

> "[...] the occasionally held opinion, that from the results of Gödel follows the non-executability of my Proof Theory, is shown to be erroneous. This result shows indeed only that for more advanced consistency proofs one must use the finite standpoint in a deeper way than is necessary for the consideration of elementary formalism."

What does it mean "to use the finite standpoint in a deeper way"? We see how Hilbert tried to defend his program – in fact by such curious and ambiguous arguments.

Gödel pointed in many places that new axioms are needed to settle both undecidable arithmetical and set-theoretic propositions. In the footnote 48^a (evidently an afterthought) to [136] he wrote:

> "As will be shown in Part II of this paper, the true reason for the incompleteness inherent in all formal systems of mathematics is that the formation of ever higher types can be continued into the transfinite [...] while in any formal system at most denumerably many of them are available. For it can be shown that the undecidable propositions constructed here become decidable whenever appropriate higher types are added (for example, the type ω to the system P). An analogous situation prevails for the axiom system of set theory."

In handwritten notes in English from the 1930s, evidently for a lecture, one finds the following words of Gödel [147, p. 164]:

> "[...] number-theoretic questions which are undecidable in a given formalism are always decidable by evident inferences not expressible in the given formalism. As to the evidence of these new inferences, they turn out to be exactly as evident as those of the given formalism. So the result is rather that it is not possible to formalize mathematical evidence even in the domain of number theory, but the conviction about which Hilbert speaks [i.e., the conviction of the solvability of every well-formulated mathematical problem – authors' remark] remains entirely untouched. Another way of putting the result is this: It is not possible to mechanize mathematical reasoning, i.e., it will never be possible to replace the mathematician by a machine, even if you confine yourself to number-theoretic problems."

In 1933 Gödel wrote [140, p. 35]:

> "[...] there are number-theoretic problems that cannot be solved with number-theoretic, but only with analytic or, respectively, set-theoretic methods."

In a similar way he wrote in [146, p. 48]:

> "[...] there are arithmetic propositions which cannot be proved even by analysis but only by methods involving extremely large infinite cardinals and similar things."

In [143, p. 151] Gödel called for an effort to use progressively more powerful transfinite theories to derive new arithmetical theorems. Rudolf Carnap formulated this in the following way (cf. [71, p. 274]):

> "[...] *all that is mathematical can be formalized; yet the whole of mathematics cannot be grasped by one system* but an infinite series of still richer and richer languages is necessary."

Similar ideas can be found by Alan Turing in [347]. He proposed to construct a transfinite sequence of still stronger and stronger systems of logic.

Also Zermelo proposed to allow infinitary methods to overcome restrictions revealed by Gödel. According to Zermelo the existence of undecidable propositions was a consequence of the restriction of the notion of proof to finitistic methods (he spoke here about "finitistic prejudice"). This situation could be changed if one used a more general "scheme" of proof. Zermelo had here an infinitary logic in mind, in which there were infinitely long sentences and rules of inference with infinitely many premises. In such a logic, he insisted, "*all* propositions are decidable!" He thought of quantifiers as infinitary conjunctions or disjunctions of unrestricted cardinality and conceived of proofs not as formal deductions from given axioms but as metamathematical determinations of the truth or falsity of a proposition. Thus syntactic considerations played no role in his thinking.[3]

5.5.6 Concluding remarks

It has been shown above that the development of the consciousness of the difference between truth and provability was in fact long and hard. An essential role played here Gödel's incompleteness theorems. They have shown – together with Tarski's theorem on the undefinability of the concept of truth – that syntactical methods of formal logic are restricted and that semantics cannot be syntactically grasped and replaced by syntactic methods.

Note that the very distinction between provability and truth in mathematics presupposes some philosophical assumptions. In fact, for pure formalists and for orthodox intuitionists there exists no "truth *versus* proof" problem. For them a mathematical statement is true just in case it is provable, and proofs are syntactic or mental constructions of our own making. They are thinking in a formal or cognitive world, always closed in itself. In the case of a platonist (realist) philosophy of mathematics – in which concepts possess their own reality – the situation is different. One can say that the platonist approach to mathematics enabled Gödel to state the problem and to be able to distinguish between proof and truth, between syntax and semantics. Note that, as indicated above, Hilbert was not interested in philosophical questions and did not consider them. Maybe he tried to ignore the difference between truth and provability in his formalistic assumptions. Maybe he did not recognize it.

5.6 Final conclusions

In mathematical logic – where one abstracts from axioms of concrete mathematical structures – free interpretability of axioms becomes one of the main features. Axioms are considered in a purely formal way. At this level logic faces mathematics. In logic several concepts used so far in an intuitive and not strict way are specified. Let us mention some central concepts like the concept of language, proof, interpretation, consequence, model, theory and truth. Mathematical logic made it possible to distinguish clearly object language and metalanguage, which has been and will be impossible

3 Note that the time was not yet ripe for such an infinitary logic. Systems of such a logic, though in a more restricted form than demanded by Zermelo, and without escaping incompleteness, were constructed in the mid-fifties in works of Henkin, Karp and Tarski.

without using tools of logic. Logic used this distinction artificially – for instance in the proofs of the incompleteness theorems.

5.6.1 Logic as the background of mathematics

Results of mathematical logic – in particular the completeness theorem (cf. Theorem 5.1.2) – indicate that *the axiomatic method of the first-order logic* provides methods and tools that enable adequate presentation and specifying the mathematical knowledge. Second-order logic has from that point of view some deficits: the completeness theorem does not hold in this case. There exists no correct system of inference rules that would be complete, i.e., that would make it possible to prove formally any semantic consequence. The axiomatic method in the framework of first-order logic establishes today the paradigmatic core with respect to mathematics – even if it does not fulfill all expectations.

In previous sections some theorems of mathematical logic have been presented. They are not only theorems of the domain of mathematical logic but have general meaning and are of philosophical interest. It is the matter of significance of the axiomatic method, of limits of mathematical theories and consequently of the mathematical method in general. All theorems mentioned above indicate a certain "weakness" of the axiomatic method.

From the syntactical point of view it is impossible to give a complete and decidable axiom system of natural numbers. Additionally in logical investigations of arithmetic it must be assumed that axiom systems like PA are consistent – this is in fact a natural but necessary assumption. The consistency of such systems – as well as of all axiom systems extending PA – cannot be proved in those systems themselves. This is stated by Gödel's Second Incompleteness Theorem. The incompleteness can neither be eliminated in second-order predicate calculus nor by using set-theoretical means.

From a semantical point of view it is unsatisfactory that first-order axiom systems for natural numbers are not categorical. The structure of natural numbers cannot be adequately characterized by first-order axioms. There exist nonstandard models of various types. Only by applying some set-theoretical concepts or by embedding arithmetic in set theory – as is usually done – a categorical characterization can be obtained. Theorems mentioned in Section 5.4 show that non-arithmetical means, i.e., set-theoretical or equivalent ones, are indispensable to get a categorical system.

One can consider appropriate set theories given as axiomatic theories in a similar way as arithmetic. Set theory like ZF can then be used to describe the mathematical universe. All objects there are of the same type – namely sets – and Peano axioms can be formulated in a set-theoretical manner like it was the case in the works of Dedekind, without distinguishing between individuals and sets and not being forced to leave first-order logic (cf. [102, pp. 133 ff.]). Consider a fixed model of ZF – a categorical characterization of natural numbers can be simulated in it. This means that the first-order logic via set theory provides a paradigmatic framework also for the arithmetic of natural numbers.

In the framework of set theory, arithmetic – beside the set-theoretical symbol of membership – has two symbols: S for the successor function and 0 for null. It is well known how further operations can be defined. On the base of the arithmetic of natural numbers one can construct – using some transfinite means – the arithmetic of real numbers as well as higher arithmetics.

Set theory – based for example on the ZF axioms – is not free of problems indicated in the case of arithmetic. It includes arithmetic and consequently – by Gödel's Incompleteness Theorem – is itself incomplete. We often pointed to this fact in previous chapters. The Second Incompleteness Theorem states that the consistency of the theory ZF cannot be proved in ZF itself. Since set theory comprises in a certain sense the whole of mathematics – it comprises all other mathematical theories – we do not know and will never know whether mathematics is consistent. The consistency of mathematics cannot be proved by mathematical means.

5.6.2 Consequences for the shape of mathematics

It can be said that the year 1879 marks the beginning of mathematical logic. In that year *Begriffsschrift* by Gottlob Frege has been published (cf. [122]). Frege's aim was to present logic as the foundation of elementary arithmetic and to make it in this way the foundation of the whole mathematics. This project has finally been realized by Russell and Whitehead in their monumental *Principia Mathematica*. However, some set-theoretical means were needed here. Hence logic *and* set theory became the foundation of mathematics. Foundations of mathematics passed from the philosophy into the hands of mathematics. Mathematics has finally been delivered from philosophy whose sub-domain it in principle was so far due to its philosophical sources.

A bit later Peano started his project "Formulario" – its aim was to express all of mathematics in a symbolic language. Some symbols and notations introduced by him have been adopted with small changes in the language of mathematics. In Peano's project axiomatic organization of propositions of particular domains has been considered. An example is provided in his work *Arithmetices principia nova methodo exposita* [266] from 1889 in which the Peano axioms were formulated. Symbolic language together with the axiomatic approach prepared logic the access to mathematics. By logical analysis of mathematical theories and by their formal representation, mathematical logic became a mirror and mathematical consciousness of mathematics – to express it figuratively.

Set theory and logic as well as – in the latter – the formal axiomatic approach initiated in the 19th century contributed to the new paradigm of the present-day methodology of mathematics. What is characteristic for this paradigm? The following features can be mentioned (cf. [13]).

(i) Set theory is the fundamental domain of the whole mathematics in two senses: First, set-theoretical means are used in every mathematical discipline, i.e., basic concepts of set theory belong to a language of any theory. At least some set-theoretical axioms are among axioms of any theory.[4] Second, set theory forms – together with logic – the fundament of mathematics. This means that all mathematical notions can be defined by primitive notions of set theory and all theorems of mathematics can be deduced from axioms of set theory.
(ii) Languages of mathematical theories are strictly separated from the natural colloquial language. They are artificial languages and the meaning of their terms is described exclusively by axioms.
(iii) In theories some primitive concepts are distinguished and all other notions are defined in terms of them according to precise rules of defining notions.
(iv) Mathematical theories have been axiomatized.
(v) There is a precise and strict distinction between a mathematical theory and its language on the one hand and metatheory and its metalanguage on the other (the distinction was explicitly made by A. Tarski).
(vi) The concept of a proof and the concept of consequence are fundamental both for the methodology of mathematics and for the foundations of mathematics. They have been specified by means of mathematical logic.

The concepts of a proof and of consequence have always been used by mathematicians. However, till the end of the 19th century they were understood intuitively and consequently were unclear. An exception was Bolzano. However, he did not have at his disposal necessary logical means we mentioned above to be able to really clarify the concept of consequence. The logical concept of semantical consequence has been defined only by Alfred Tarski in 1935, hence almost 60 years after the initiation of mathematical logic by Frege. The concept of a proof has been clarified in the framework of the development of logic. Let us mention G. Frege, B. Russell, A. N. Whitehead, D. Hilbert, P. Bernays, W. Ackermann, S. Jaśkowski and G. Gentzen who made here the significant contribution. The first precise definition of a proof has been given in the book *Grundzüge der theoretischen Logik* [178] by Hilbert and Ackermann in 1928.

Practically all mathematical theories have been axiomatized. This does not mean however that axioms are in a unique way fixed. We want to stress only *the principle* that in mathematical theories all theorems should be logically deduced from axioms and only from axioms without reference to any principles that does not come from axioms or from logic.

4 For example in the case of the arithmetic of the reals (cf. Section 5.3.2), various geometries, mathematical analysis and algebra. Of course there are theories that contain no set-theoretical components. However, they are usually far from mathematical research practice and are considered only in foundational investigations.

The axiomatic approach was always the method of mathematics. It was *the* mathematical method of presenting and developing theories and it does not seem that this will be changed. Just logic and the axiomatic approach led us to the limits of mathematics. It is the merit of mathematical logic to indicate those limits – not only for mathematics but for the theoretical approach in all sciences. In fact, any genuine theory – to use the words of Hilbert (see above) – orders its notions and propositions in an axiomatic way. If this is true then the significance and meaning of logico-mathematical results concerning consistency, incompleteness, axiomatizability, definability, decidability and truth go beyond mathematics.

6 Thinking and calculating infinitesimally – First nonstandard steps

One should know
that a line is not composed of points.[1]

Gottfried Wilhelm Leibniz

6.1 Preliminary remark

Does there "really" exist the infinitely small, the infinitesimal? This question is a philosophical one like "does there exist the infinitely large" in the reality, e.g., infinite sets? We think the answer to both questions is "no". In Section 3.2.1 in connection with the second question it has been said that *potentially* the infinite seems to be necessary, in particular in thinking. It arises in the process of counting "1,2,3, ...". In the 19th century the infinity has been mathematically made *actual* when it was claimed axiomatically, i.e., by *putting* curly brackets on the left- and right-hand side: $\{1, 2, 3, \ldots\}$.

In the 17th century, almost 200 years earlier, the infinitesimal, the infinitely small, has been actual – it has accompanied for a long time the mathematical thinking and by Archimedes and Cavalieri it already was nearly actual. This development has been described in Section 3.3. However, we are of the opinion that – till their appearance by Leibniz and Newton – infinitesimals had the heuristic character, the character of an intellectual game. Both those epoch-making scientists made in the 17th century the actual infinitesimal to be a mathematical instrument – and this had an enormous impact.

Are there moments when the infinitesimal, the infinitely small, appears potentially and leads to the actual?

Motives

Nowadays the process $1, \frac{1}{2}, \frac{1}{3}, \frac{1}{4}, \ldots$ is treated without further ado simply as actually infinite, in particular as a sequence $(\frac{1}{n})$. However, mathematically one usually shields oneself against an infinitely small number coming after $(\frac{1}{n})$. One follows the experience of measuring, persists to be archimedean and rejects the old calculus of infinitesimals. In dealing with infinitely large numbers one was a bit more generous and used for example without any doubt the symbol ∞. However, one does not calculate with the

1 Man muss wissen, dass eine Linie nicht aus Punkten zusammengesetzt ist.

https://doi.org/10.1515/9783110468335-007

infinite – at most set-theoretically and marginally with infinite cardinals and ordinals in order to develop an "exotic" arithmetic.

It seems that infinitesimals historically appeared in a way similar to actually infinite sets – the latter do not come from the arithmetic but have been invented and enforced to control the continuum. In fact, it was just the continuum that quasi provoked the infinitesimals to be controlled. The long and changeful way to that has been sketched in Sections 3.3.2–3.3.8. However, the ideas that led to infinitesimals became alien since one thinks in terms of limits. This is indicated by quotations in Section 3.3.7.

Nowadays there exists at least one – nearly popular – place where there are openly different opinions – for or against the infinitesimals. Even keenest minds get into the danger of splitting when there appears the question: Is 0.999 . . . smaller than or equal to 1? Note that even the limit-"party" crumbles among full-grown mathematicians. This question has been mentioned in Section 3.3 and will now be considered in detail. It clearly indicates the infinitely small, prepares the actual infinitesimal and makes it plausible. Being infinitely small – and for the ones "knowing" simply stupid – the question about 0.999 . . . spreads into dimensions mentioned in Section 3.3. It finally forces that more should be said about the mathematical foundations of the infinitesimals and gives reason to make the very first steps towards the calculus of infinitesimals and to notice the specific features of the infinitesimal way of thinking.

6.2 The question about 0.999 . . .

This question is often asked in classes and the answer usually given is: "0.999 . . . < 1!". It is impossible to prevent this answer neither by genuine arguments nor by concrete proofs. We ourselves were taught that 0.999 . . . = 1. We know this. However, there seem to remain some doubts.

We would like to consider how this is possible and take into account results of a study [14].

6.2.1 Empirical approach

Ludwig Bauer researched in detail and thoughtfully into the problem of this section. The results of the inquiry among 256 pupils of a secondary school in Bavaria (classes 7 till 12) and 50 students of mathematics (after the third semester) are the following:

	a) 0.999 . . . < 1	b) 0.999 . . . = 1	Abstention	Invalid
Secondary school	72.2 %	31.6 %	3.1 %	4.3 %
Undergraduates	50 %	50 %	0 %	0 %

There was no difference between justifications given by students who *finished* courses in Analysis I and II and pupils of the secondary school. The reasons for answers given by pupils will be given below.

Results of our own experiment with 51 students were surprisingly the following:

	a) 0.999 . . . < 1	b) 0.999 . . . = 1	Abstention	Invalid
Undergraduates	90.2 %	9.8 %	0 %	0 %

In two random samples of 17 teachers and lecturers 47.1 % chose the answer 0.999 . . . < 1. Even teachers are not sure when asked "privately". In teaching they should be sure of course.

So far the data. They show that in spite of all mathematics the majority of pupils do not learn or do not want to learn what they should learn. How is this possible?

6.2.2 The problem

The problem is

$$0.99999999999999999999999999999999999999999999999999999\ldots$$

By writing usually only few nines, by the harmless "etc." or the dots ". . . ", by saying "0 comma period 9" and the expression "$0.\overline{9}$", the inconceivable infinity that follows is belittled. The infinite seems to be grasped by the way of writing and speaking. It will be seen that there is some danger in the case of 0.999 . . . even though it is a "modest" fraction among the infinite decimal ones. There are more problems with "infinite non-periodic decimal fractions" (cf. Section 1.7).

When such an entity consisting of 0 and an infinite sequence of nines are compared with the number 1 then there appears the problem of the infinity. What will usually be done in such case? One follows the nines, compares 0.9, then 0.99, then in a distant probably 0.999999999999999999 etc. with the number 1. In fact, one sets out the way through the infinite sequence of such stations and everybody is of the opinion that

– the infinite sequence (0.9, 0.99, 0.999, . . .) strives to 1.

Nobody has any doubt. And now comes the dissent. For reasons we will try to explain some people say:

(a) "It *never* catches 1!"

Others claim:

(b) "Sure!"

First consider the second opinion (b).

0.999 . . . = 1. Why so?

Let us quote some answers from a list of collected comments of 29 pupils (cf. [14, p. 100]) which in our opinion are characteristic and important.

0.999 . . . = 1. Justifications

- “We learned this.”

This answer – in various forms – appears very often.

- “Because it is already almost 1.”
- “0.999 . . . = 1 because the number that is missing is infinitely small.”

Both sound unconvincing and contradict in fact the decision of the equality. However, the reference to the infinitely small is striking.

The mysticism of the infinitely large appears in the following argumentations:

- “Since 0.999 . . . goes to the infinite and approaches 1, one is able to say that 0.999 . . . = 1.”
- “Eventually, 0.999 . . . will become 1 after umpteen 0.999 . . . 999.”

“One can say” does not sound quite convincing. “Say” is something else than “is”. Interesting is *the domain between the large finite (“umpteen”) and the infinite* appearing in the second comment above.

Among statements of pupils there are three proofs that are more or less carried out using the equalities $\frac{1}{3} = 0.333\ldots$ or $\frac{1}{9} = 0.111\ldots$. Now we consider well-known and convincing proofs.

0.999 . . . = 1. Proofs

Proof 1. The first proof starts with $\frac{1}{3} = 0.333\ldots$:

$$\begin{array}{rcl} \frac{1}{3} & = & 0.333\ldots \\ 3 \cdot \frac{1}{3} & = & 0.999\ldots \\ \hline 1 & = & 0.999\ldots \end{array}$$

Proof 2.

$$\begin{array}{lcl} 1 + 0.999\ldots & = & 1.999\ldots \\ 1.999\ldots : 2 & = & 0.999\ldots \end{array}$$

The result is: The arithmetical mean of 0.999 . . . and 1 is 0.999 . . . . But this is impossible. Hence,

> “[. . .] between 0.999 . . . and 1 [there is] no number. Hence 0.999 . . . and 1 represent the same number value.”

That was the answer of a mathematician in *Mitteilungen der DMV* 2003/2 (cf. [26, p. 89]) to the question of a pupil.

Proof 3. The third proof is algebraic and uses rules for calculations with decimal fractions:

$$\begin{array}{rrcl} & 10 \cdot & 0.999\ldots & = 9.999\ldots \\ - & 1 \cdot & 0.999\ldots & = 0.999\ldots \\ \hline & 9 \cdot & 0.999\ldots & = 9 \\ & & 0.999\ldots & = 1 \end{array}$$

Proof 4. The fourth proof comes from Analysis I:

$$0.9 + 0.09 + 0.009 + \cdots = \sum_{i=1}^{\infty} \frac{9}{10^i} = 1$$

because for the sequence (s_n) of partial sums $s_1 = 0.9$, $s_2 = 0.99$, $s_3 = 0.999$ etc. it holds

$$\forall \varepsilon > 0\, \exists N\, \forall n > N(|1 - s_n| < \varepsilon).$$

Looking at the above proofs it seems to be incomprehensible how anyone obtains an answer different from $0.999\ldots = 1$. Those proofs are known to pupils and certainly they are known to students and teachers. In classes all state and justify that $0.999\ldots = 1$. However, the majority does not believe it.

0.999 . . . = 1. Problems

Is there any criticism towards the above proofs? In the first three of them the notation and calculating rules were transferred without any concern and reflection from the domain of *finite* decimal fractions to the *infinite* decimal fractions. This problem can be easily seen in the case of Proof 3.

Infinite calculations

Let us first calculate somewhat differently by going back to the original meaning of the infinite sum of $0.999\ldots$ and $9,999\ldots$.

$$9.999\ldots - 0.999\ldots =$$

$$\begin{array}{rlclclcl} & (9 & + & 0.9 & + & 0.09 & + & \cdots) \\ - & (0.9 & + & 0.09 & + & 0.009 & + & \cdots) \\ \hline & 8.1 & + & 0.81 & + & 0.081 & + & \cdots \end{array}$$

The provisional result is an infinite summation. The auxiliary calculation gives – on the infinite way to the final result – partial results one after another, i.e., the following sequence of partial sums:

$$(8.91, 8.991, 8.9991, \ldots).$$

What will be the final result when this infinite sequence is taken into account? If the final result is written as usual as infinite decimal fraction then it is of course

$$8.999\ldots.$$

Notice that the position 1 at the end of every provisional result is ignored. It vanishes beyond the infinite. Such phenomenon does not occur by finite decimal fractions.

It becomes clear that calculating with an infinite number of positions is something special and perhaps even highly questionable. In particular in the case of periodic decimal fractions one dares to do such *infinite calculations*. Actually proofs are needed in order to justify calculations like those used in Proof 4 in which the limit argumentation is applied.

If the new result given above is accepted then Proof 3 can be presented in the following way.

New Proof 3.

$$\begin{array}{rrcl} & 10 \cdot 0.999\ldots & = & 9,999\ldots \\ - & 1 \cdot 0.999\ldots & = & 0.999\ldots \\ \hline & 9 \cdot 0.999\ldots & = & 8.999\ldots \\ & 0.999\ldots & = & 0.999\ldots \end{array}$$

Nothing has been gained.

The first phenomenon of the vanishing of a position in the infinite at the end of a partial sum can also be seen in the second proof.

Is $0.333\ldots = \frac{1}{3}$?

The first proof reduces the statement "$0.999\ldots = 1$" to $0.333\ldots = \frac{1}{3}$. To ask whether "$0.999\ldots = 1$" means to call in question whether $0.333\ldots = \frac{1}{3}$. Can this be done, is it allowed? We shall see. A small inquiry at a seminar gave however a better result for $0.333\ldots = \frac{1}{3}$ than for $0.999\ldots = 1$. Teachers voted even "better".

	a) $0.333\ldots < \frac{1}{3}$	b) $0.333\ldots = \frac{1}{3}$	Abstention	Invalid
Undergraduates	23.5 %	76.5 %	0 %	0 %

In fact, behind $0.333\ldots$ there is also a hidden infinite calculation. The division of 1 by 3 that goes always further and further – what is marked here by dots – is practically experienced, which is not the case for $0.999\ldots$. Therefore, the result $0.333\ldots = \frac{1}{3}$ is more accepted. After all, this division is not *fully* calculated. The problem of infinite calculations is that they are never fully done. This can perhaps explain why sometimes answers like "$0.333\ldots < \frac{1}{3}$" are given.

Will the limit value be achieved?

We will not doubt Proof 4. Beside rather technical problems in writing and calculating with the infinite there appears the old and great problem with the concept of the limit value – it is again and seriously connected with the infinite.

A problem is first of all – independently of the phenomenon of the infinite – that putting "$0.999\ldots = 1$" requires indirectly the treatment of a sequence as a number. In fact, when one wants to compare 0.999 . . . with 1 then – as we saw above – there appears the sequence of partial sums $(0.9, 0.99, 0.999, \ldots)$ that is 0.999 . . . and that should be equal to 1. Though the first question of the inquiry was about the *convergence* towards 1 and the sequence $(1, 1, 1, \ldots)$ not only converges to 1 but *is* 1, many of the students, in fact 43.1 % (and about 82 % of lecturers) rejected the statement "the sequence is equal to 1":

	$(1, 1, 1, \ldots) \neq 1$	$(1, 1, 1, \ldots) = 1$	Abstention	Invalid
Undergraduates	43.1 %	56.9 %	0 %	0 %

"Sequence = number" is the beginning of the construction of the real numbers – it then becomes "class of sequences = number". Pupils and students do not know such things. On the other hand "sequence = number" contradicts in a way the mathematical convention according to which n-tuples of 1s, here sequences consisting of 1s should be distinguished from 1 as members of a sequence.

The concept "limit value" leaves it open whether this value will be reached or not. Since in the case of 0.999 . . . the infinite sequence of partial sums and the sequences of nines cannot be passed through, the natural intuition says that the limit value will not be reached and therefore $0.999\ldots < 1$. The argumentation using "$\forall\varepsilon > 0 \exists N \forall n > N$" does not say any more but only that 1 is the inaccessible limit value.

However, mathematics has put $0.999\ldots = 1$ as if the infinite sequence were finished by 1. But the intuition is not impressed by it – as was seen in the inquiry. It is stronger. Infinite sequences are intuitively treated as *potentially infinite*, as open and in principle not "lockable". Since the argumentation using "$\forall\varepsilon > 0 \exists N \forall n > N$" does not lead up to the limit value, it is not possible to teach that "$0.999\ldots = 1$". The concept of the limit value has been removed from some syllabuses. There is a danger that it completely disappears from schools.

The problem of the actual infinity in sequences

The requirement of the actual infinity in treating infinite sets like finite ones means in the case of the infinite sequence of nines in 0.999 . . . and of the sequence of partial sums (0.9, 0.99, 0.999, ...) that they should be treated as closed. However, as was said, this is resisted by intuition. There is a general problem of treating sequences as being closed.

However, for some sequences, especially for *one* particular sequence this is not the case. It is common practice accepted by all, also by pupils, that one writes $\mathbb{N} = \{1, 2, 3, 4, \dots\}$ and closes by "}" the never ending sequence $1, 2, 3, \dots$. The "set" of natural numbers seems to be unproblematic. Why is it not transferred to the case of an arbitrary sequence?

We try to understand this in the following way: The natural numbers do appear all at once by the simple law of counting – as it was described by Cantor in one of his "definitions of a set". Building a set, one abstracts from the order of elements – however, here the building of $\mathbb{N}$ is visible in a "natural way" thanks to its elements, the natural numbers.

In the case of infinite sequences of real numbers, in the center of attention is the *order* of sequence members. Members of the sequence are seen as one coming after the other – not always according to a simple law. To treat such an infinite process as closed and finished seems to be impossible for a normal way of thinking. Sets abstract from the order of its members. Therefore building an infinite set of sequence members is complicated for the character of *sequence* gets lost. Elements of such a set that earlier formed a sequence cannot represent a succession of elements – as natural numbers do.

The law according to which nines are ordered in $0.999\dots$ is absurdly simple, however to treat the sequence of nines as closed would make it to be the set of indistinguishable nines, like $\{9\}$. Nothing is left of the succession of nines that were on the positions of the sequence.

$0.999\dots < 1$. Why?

We again quote from the list of justifications in [14, p. 99] given by pupils and from 40 answers we choose some being exemplary and important.

$0.999\dots < 1$. Justifications

- "Mathematically is $0.999\dots = \frac{9}{9} = 1$. However, since one can put infinitely many nines one after another, this number is always smaller than 1.

This is a beautiful example of the case that mathematical knowledge is irrelevant when it contradicts the intuition. This is peculiar to mathematics. Mathematics is another world that has little to do with the reality.

- "The period is infinite but 1 will never be reached."
- "The number $0.000\,000\,000\,00\dots 1$ lacks to the number $0.099\,999\,999\dots$ to be equal to 1."

The number 1 is only the limit value. It is spoken very often of “little”, “unimaginably close”, “small difference”. What is surprising is that a difference is given (cf. [379, pp. 89 ff.]). It is infinitely small.

- “Zero point something is always smaller than 1.”
- “0.999 . . . is not a ready whole. 1 is a whole.”

This is convincing for everybody. Mathematicians are voiceless or should go far back in their explanations – however without results.

0.999 . . . < 1. Proofs?

Let us try to give a proof that looks like a mathematical one. Therefore, we again ask ourselves what it in fact means to compare 0.999 . . . with 1. We answer as it was done above:

- This means to compare 0.9 with 1, then 0.99 with 1, next 0.999 with 1 etc.,

hence to put an infinite sequence of comparisons. Formally this can be formulated as follows: to compare 0.999 . . . with 1 means to compare the sequences

$$(0.9, 0.99, 0.999, \ldots) \quad \text{and} \quad (1, 1, 1, \ldots).$$

Since

$$0.9 < 1, \quad 0.99 < 1, \quad 0.999 < 1, \ldots,$$

every member of the first sequence is smaller than the appropriate member of the second sequence and consequently in a natural way

$$(0.9, 0.99, 0.999, \ldots) < (1, 1, 1, \ldots).$$

The first sequence of partial sums is 0.999 . . . , and $(1, 1, 1, \ldots)$ is a special way of writing 1.

- Consequently, it holds 0.999 . . . < 1.

This method works also for 0.333 . . . : To compare 0.333 . . . with $\frac{1}{3}$ means to compare first 0.3 with $\frac{1}{3}$, then 0.33 with $\frac{1}{3}$ etc., hence to compare $(0.3, 0.33, 0.333, \ldots)$ with $(\frac{1}{3}, \frac{1}{3}, \frac{1}{3}, \ldots)$.

Since $0.3 < \frac{1}{3}$, $0.33 < \frac{1}{3}$, $0.333 < \frac{1}{3}$ etc. it holds

$$(0.3, 0.33, 0.333, \ldots) < \left(\frac{1}{3}, \frac{1}{3}, \frac{1}{3}, \ldots\right).$$

- So is 0.333 . . . < $\frac{1}{3}$?

Is this an imagination? No, it is mathematics since that time when C. Schmieden and D. Laugwitz in 1958 and A. Robinson in 1961 developed it. It dates back to Leibniz (1646–1716).

First remark on the infinitesimal mathematics

Laugwitz starts in [226, Chapter 2] from the domain of *all* sequences of reals. This domain is an ordered ring. From that ring Laugwitz constructs an ordered field of hyperreal numbers $^*\mathbb{R}$. Hyperreal numbers are classes of sequences of reals – as reals are classes of sequences of rational numbers (instead of classes in what follows we write only representatives of them). In the field $^*\mathbb{R}$ the real numbers $\mathbb{R}$ are included. The real numbers are represented by constant sequences $(r, r, r, \ldots)$ – we simply write r. Important is here the definition of the order.

Definition 6.2.1. $(a_1, a_2, a_3, \ldots) < (b_1, b_2, b_3, \ldots)$ if and only if $a_i < b_i$ for all or *almost all i.*

What "*almost all*" means here is in fact complicated. However, always when $a_i < b_i$ *for all but a finite number* (cf. [226, p. 97]) of members of the sequence then

$$(a_1, a_2, a_3, \ldots) < (b_1, b_2, b_3, \ldots).$$

Hence in the field $^*\mathbb{R}$ it always holds: $0.333\ldots < \frac{1}{3}$ and $0.999\ldots < 1$, exactly as it has been described above.

In such a way our goal has been obtained. We can answer now the question concerning $0.999\ldots$.

6.2.3 Answer

"Is $0.999\ldots < 1$" **or** $0.999\ldots = 1$"? The answer is:
- "Both!"

"$0.999\ldots < 1$ **and** $0.999\ldots = 1$."

One can say:
- In mathematics *with limit values* one has $0.999\ldots = 1$.
- In the *infinitesimal* mathematics one has $0.999\ldots < 1$.

We would like to indicate some consequences of choosing this or that answer and applying this or that mathematics. Mathematics with limit values is nowadays standard and has been developed successfully for 140 years. It is the "standard mathematics". It is everyday mathematics and there is no need to discuss its consequences. Some problems met in teaching connected with it have been indicated above. Strange is only the fact that nowadays in schools where problems are connected with limit values, limit-value mathematics is taught to a large extent without limits, that is, without the fundament.

We will soon make a remark concerning the other one, namely the infinitesimal mathematics. But first we ask: why it is called this way. "Infinitesimal" means "infinitely small" and the following story indicates where those infinitely small objects can be found.

Anecdote

A pupil of the sixth class, her name is Lina, has problems with $0.999\ldots = 1$. She is writing to *the* authority, namely to Deutsche Mathematiker-Vereinigung (DMV) [Association of German Mathematicians]. There the answer should be known. In *Mitteilungen der DMV* (reprinted in the book [26, pp. 88–91])[2] one finds her letter:

> "Dear mathematicians,
> I am in the 6th class and we are learning now about periodic decimal fractions. We have learned that $\frac{1}{9} = 0.111\ldots$, $\frac{3}{9} = 0.333\ldots$ etc. But what is $0.999\ldots$? Our teacher said that it was $\frac{9}{9}$. But this can not be so. It would then be 1 and yet $0.999\ldots$ is infinitely smaller than 1. Does there exist $0.999\ldots$ at all? However, a number which I can imagine must even exist. How can one get $0.999\ldots$? I would be glad to receive an answer from you.
> *Lina*"

A friendly mathematician has read the letter of Lina to *Mitteilungen der DMV* and he is of the opinion that she should be given an answer. However, it is disappointing. At the very beginning it is written:

> "Your teacher is indeed right, $0.999\ldots$ is really equal to 1."

Further it is written about notation for fractions and at length about *standard* calculations of the arithmetical means. He comes to the conclusion that the arithmetical mean of $0.999\ldots$ and 1 cannot be $0.999\ldots$, unless both numbers have "the same number value". At the very end he responds to Lina's "infinitely small" and gives her words of "comfort [...] if it is necessary":

> "Later you will learn that an infinitely small is equal to zero."

Now we want to know just this.

Second infinitesimal remark: some calculations in $^*\mathbb{R}$

Our candidate for "infinitely small" is – like for Lina – the difference α between 1 and $0.999\ldots$. In the domain of real numbers – when one thinks in the mathematics with limit values – both have "the same number value". In the domain of hyperreals they do not. We are sure that the difference should be infinitely small. This must be shown.

---

2 This book is usually given as a reward to high-school graduates for particularly high achievements in the final secondary-school examinations in mathematics.

Determine $\alpha = 1 - 0.999\ldots$!

The term "1–0.999 . . ." means "(1, 1, 1, . . .)–(0.9, 0.99, 0.999, . . .)". One calculates:

$$\begin{array}{rlllll} & (1.0, & 1.00, & 1.000, & 1.0000, & \ldots) \\ - & (0.9, & 0.99, & 0.999, & 0.9999, & \ldots) \\ \hline & (0.1, & 0.01, & 0.001, & 0.0001, & \ldots) \end{array}$$

This is the sequence

$$\left(\frac{1}{10}, \frac{1}{100}, \frac{1}{1000}, \frac{1}{10000}, \ldots\right).$$

Hence the result is

$$\alpha = 1 - 0.999\ldots = (0.1, 0.01, 0.001, \ldots) = \left(\frac{1}{10}, \frac{1}{10^2}, \frac{1}{10^3}, \ldots\right).$$

It can be immediately seen that one cannot write α as an ordinary decimal fraction. It is a new, not real but hyperreal number.

How large is α?

If α is "infinitely small" then it is surely not null: $\alpha > 0$ when

$$(0.1, 0.01, 0.001, \ldots) > (0, 0, 0, \ldots).$$

In fact, every position of the first sequence is greater that every position of the second one.

Lina's mathematical penfriend is not right if we now show that α is smaller than every negative power of ten, hence smaller than every real number and consequently *infinitely small.* One has

$$\alpha < \frac{1}{10} \quad \text{for} \quad \left(\frac{1}{10}, \frac{1}{10^2}, \frac{1}{10^3}, \ldots\right) < \left(\frac{1}{10}, \frac{1}{10}, \frac{1}{10}, \ldots\right).$$

In fact, all members – except the first one – of the first sequence are smaller than the respective members of the second sequence. Moreover,

$$\alpha < \frac{1}{10^2} \quad \text{for} \quad \left(\frac{1}{10}, \frac{1}{10^2}, \frac{1}{10^3}, \ldots\right) < \left(\frac{1}{10^2}, \frac{1}{10^2}, \frac{1}{10^2}, \ldots\right).$$

In fact, all members – except the first two ones – of the first sequence are smaller than the respective members of the second sequence. Furthermore,

$$\alpha < \frac{1}{10^3} \quad \text{for} \quad \left(\frac{1}{10}, \frac{1}{10^2}, \frac{1}{10^3}, \ldots\right) < \left(\frac{1}{10^3}, \frac{1}{10^3}, \frac{1}{10^3}, \ldots\right).$$

All members – except the first three ones – of the first sequence are smaller than the respective members of the second sequence etc. Hence,

$$\alpha < \frac{1}{10^i}$$

for each fixed i and for

$$\left(\frac{1}{10}, \frac{1}{10^2}, \dots \frac{1}{10^i}, \frac{1}{10^{i+1}}, \dots\right) < \left(\frac{1}{10^i}, \frac{1}{10^i}, \frac{1}{10^i}, \dots\right).$$

In fact, all members – except the first i ones – of the first sequence are smaller than the respective members of the second sequence.

As a result we obtain:

– The number α is *infinitely small* and *not* equal to zero.

Further hyperreal numbers

Also

$$\omega = \left(\frac{1}{1}, \frac{1}{2}, \frac{1}{3}, \dots\right)$$

is infinitely small because

$$\omega < \varepsilon \quad \text{for all real } \varepsilon > 0.$$

The reason is similar to the case of α. Also in the domain of the infinitely small magnitudes can be compared. It holds for example

$$\alpha < \omega,$$

which can be immediately seen when the respective sequences are written.

Beside the infinitely small numbers there exist also infinitely large numbers:

$$\Omega = (1, 2, 3, \dots) \quad \text{is infinitely large}$$

because $(1, 2, 3, \dots) > (n, n, n, \dots)$ for every n. Considering now how it would be divided – every member of the dividend sequence by the respective member of the divisor sequence – it would be clear that

$$\omega = \frac{1}{\Omega}.$$

Calculating the arithmetical mean

Let us calculate at last the arithmetical mean of 0.999 . . . and 1 in order to confront it with the above calculations of the mathematician who wanted to convince Lina. We want to calculate

$$\frac{0.999\ldots + 1}{2}.$$

We have

$$\begin{array}{rlllll} & (0.9, & 0.99, & 0.999, & 0.9999, & \dots) \\ + & (1.0, & 1.00, & 1.000, & 1.0000, & \dots) \\ \hline & (1.9, & 1.99, & 1.999, & 1.9999, & \dots) \end{array}$$

Further it holds

$$\frac{0.999\ldots+1}{2} = \frac{(1.9, 1.99, 1.999, \ldots)}{(2, 2, 2, \ldots)} = \left(\frac{1.9}{2}, \frac{1.99}{2}, \frac{1.999}{2}, \ldots\right) = (0.95, 0.995, 0.9995, \ldots).$$

So we have

$$0.999\ldots < (0.95, 0.995, 0.9995, \ldots) < 1$$

and notice that in the domain of hyperreal numbers it can distinguished in a fine way beyond the measurable. Between $0.999\ldots$ and 1 there are infinitely many hyperreal numbers.

At the end notice once again that we were a bit generous in notation. We calculated with sequences and the results are all correct. We should have calculated with classes of sequences. Our sequences written and considered above represent those classes. In the domain of reals one is doing the same when one has to prove that $\sqrt{2} \cdot \sqrt{3} = \sqrt{6}$.

Gaps in teaching?

Officially the hyperreal numbers do not appear in teaching at mathematical faculties of universities. It looks as if they were unknown. Even authors of widely recognized handbooks seem to suppress that there exist infinitely small numbers similarly to the reals. This assumption is suggested by the following quote from a handbook of analysis [25, p. 237]:

> "[...] one should perhaps note that the founding fathers [of analysis – authors' remark] imagined in their work that there exist something like 'infinitely small magnitudes'. You are in a good society when you have problems with this interpretation. You should [...] never use expressions *dy* and *dx* as independent magnitudes."

The differentials dx, dy are the old infinitely small magnitudes by Leibniz.

In the very beautiful book *Kleines Mathematikum – Die 101 wichtigsten Fragen und Antworten zur Mathematik* [33][3] by the known mathematician A. Beutelspacher one finds again the question "is $0.999\ldots = 1$?" as question 61. The quick answer is:

> "[...] in fact $0.999\ldots$ is 1."

3 Small Mathematicum – the 101 Most Important Questions and Answers to Mathematics

Beutelspacher treats the problem of the limit value not so seriously as we do and justifies extensively his claim using convergence of the sequence (0.9, 0.99, 0.999, . . .) in the following way:

> "The distance will be smaller than any number we can imagine."

However, we and Lina have imagined a number – an infinitely small, our α – that gives the distance between 0.999 . . . and 1, is *not zero* and is smaller than any *real* number that can be imagined. It shows that:

> "[. . .] in fact 0.999 . . . is *not equal* to 1."

How does it come to the fact that infinitely small numbers are overlooked or ignored? One of the reasons is that infinitesimals are connected with an assumption concerning the usual treatment of the "continuum", of the geometrical space. A symbol of this treatment is the number line identified with the real numbers. This development has been presented in Section 3.3, in particular in Section 3.3.6.

6.2.4 Final remarks

We have shown how deep this small question whether "(a) $0.999\ldots < 1$" or (b) $0.999\ldots = 1$?" is connected with the foundations of mathematics. It leads to a dichotomy of mathematics and the following answer we give indicates this:

- "It holds $0.999\ldots < 1$ **and** $0.999\ldots = 1$!"

Answer (a) is known nowadays to rather few persons. If one thinks – what is nowadays almost a common practice – that the *unique* answer (b) *should* be given then one thinks that

(b) the real numbers *are* the linear continuum.

We have sketched that this is not mathematically reliable. This is shown by methods that lead to hyperreal numbers. They show even more:

(a) The continuum cannot be grasped mathematically.

As long as one adheres to the unique answer (b) – what nowadays predominantly is the case – and does not take into account the possibility and consequences of hyperreal numbers then mathematics splits into

- $(0.999\ldots < 1)$-mathematics **and** $(0.999\ldots = 1)$-mathematics.

However, this is entirely not necessary – as has been remarked in Section 3.3 and as will be more exactly seen below. This situation is not serious for the mathematical practice, it is curious.

On the other hand serious for the view of mathematics and its relations to the reality is the background that began with the history of 0.999 . . . . There can be seen not only mathematical but also philosophical aspects.

We would like to refer to two small words: "really" and "in fact". Both the mathematician who answered Lina as well as A. Beutelspacher used them:

- "Your teacher is indeed right, 0.999 . . . is **really** equal to 1."
- "0.999 . . . is **in fact** equal to 1."

What is the situation concerning "in fact" and "really"? Our look into the history of mathematics, in particular in Section 3.3, shows an interesting development – and a sudden break, a revolution in mathematical thinking. It can be said:

Mathematics has left the reality in the 1870s.

With the actual infinity introduced by Georg Cantor it has risked a step away from the finite world. Mathematicians learned to look at the world by set-theoretical eyes. They have converted the homogeneous continuum into a set of isolated elements. The invaluable fruits of this revolution were the real numbers and the fundamental insights into properties of the continuum. The reals stand at the beginning of a uniquely story of success.

However, real numbers are a *theory* and *not* the reality. Perhaps just here the difficulties for mathematics are located. We have already indicated this. Mathematics seemingly believed and believes that by real numbers a part of the reality, namely the continuum, has been grasped. It may be difficult to call this belief in question, even to ask the appropriate question.

Up to rational numbers, numbers are closely connected with the reality. In the case of reals there is no such connection; similarly for hyperreals.

- $\mathbb{R}$ and ${}^*\mathbb{R}$ are only two *models* of one and the same continuum.

The following should be said and we want to close the question of 0.999 . . . with these words:

- 0.999 . . . is neither **really** nor **in fact** equal to 1.
- 0.999 . . . is neither **really** nor **in fact** smaller than 1.

But:

- 0.999 . . . is *theoretically* equal to 1

and

- 0.999 . . . is *theoretically* smaller than 1.

6.3 A bit of infinitesimal calculus

After the introduction that made possible the idea of the potentially as well as actually infinitely small we now come closer to infinitesimals in practice.

Mathematical foundations provided mathematics with infinitesimals in the middle of the 20th century – in fact they appeared already earlier and with them began the rise of the mathematical analysis in the 17th century. Infinitely small and infinitely large numbers – the nonstandard numbers – have been added to the real numbers (cf. Section 3.3.4). Logic teaches us that there are nonstandard models of real numbers and how they can be understood (cf. Section 5.3.2). They are constructed in set theory. They are mathematically fundamental and extremely practical. Since the nonstandard numbers are still rather not standard and appear very seldom we want to say more about them.

In Section 3.3.8 some details of nonstandard numbers have been given – they form the foundations of another analysis, of a really infinitesimal *calculus*. Here we shall learn at the beginning how to calculate with nonstandard numbers and introduce first elementary concepts and theorems of analysis in order to show how different here the way of thinking is and how simple the calculations with infinitesimals are that replace the complicated considerations with limit values. At the end we look at a construction of hyperreal numbers and at their logical peculiarity.

6.3.1 Infinitesimal calculations

Assume *naively* that there exist infinitely small and infinitely large numbers. The new infinitely small numbers are smaller than *all* real numbers and the infinitely large numbers are greater than *all* real numbers. We already know vaguely how they are constructed and how one can calculated with them. New numbers arise when infinitely small numbers are added to arbitrary real numbers. The "old" real numbers are called *standard numbers* and the new ones are called *nonstandard numbers*. Together they form the field of the hyperreal numbers $^*\mathbb{R}$ within which we calculate.

One should look closer at algebraic interrelations between the normal reals and the new numbers – here some notations will be introduced. On this elementary base one can intuitively and still in a well-grounded manner speak about differentials, derivatives and integrals – as will later be noticed.

Preliminaries

The new numbers in $^*\mathbb{R}$ will usually be denoted by small Greek letters $\alpha, \beta, \ldots, \kappa, \lambda, \mu, \ldots$. If a number μ is *infinitely large* or, as can be more exactly said *positively infinite* then one writes "$\mu \gg 1$". If α is *infinitely small* or *infinitesimal* then one writes $\alpha \approx 0$. If α and β are *infinitely close* to each other, i.e., $\alpha - \beta \approx 0$ then $\alpha \approx \beta$ is written. For example $0.999\ldots \approx 1$.

Besides $0.999\ldots$ we have learned above further hyperreal, not real numbers:

$$0.333\ldots \approx \frac{1}{3},$$
$$\alpha = 1 - 0.999\ldots \approx 0,$$
$$\omega = \left(\frac{1}{1}, \frac{1}{2}, \frac{1}{3}, \ldots\right) \approx 0,$$
$$\Omega = (1, 2, 3, \ldots) \gg 1.$$

The new situation can be illustrated by using a magnifying glass – with an infinite magnification – and put it on 0 and 1. By an "infinite telescope" we are looking into the infinite distance.

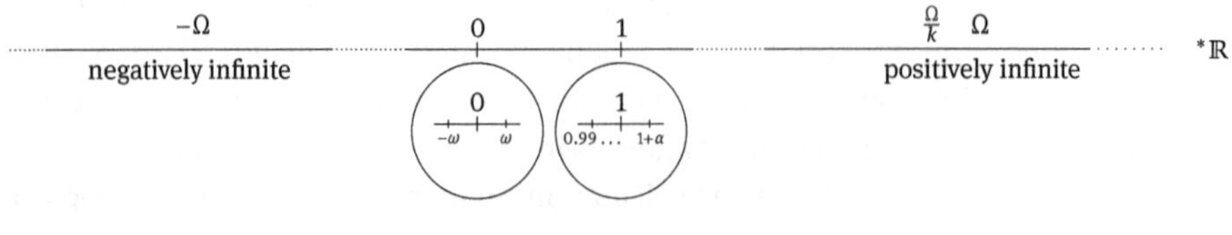

Note that this is not a methodical trick – it is mathematically justified as Karl Kuhlemann proved in [218].

If $\mu \gg 1$, hence μ is *infinitely positive*, then $\mu > r$ for all $r \in \mathbb{R}$. If $\beta \approx 0$ and β is positive then $\beta < r$ for all positive r.

A number y is said to be bounded if there exists a finite natural number n such that $|y| \leq n$. If y is bounded and not infinitesimal then y is called *finite*. The standard part of a bounded number y is the real number r such that $r \approx y$. It will be shown that there exists such an r and that it is uniquely determined.

Some calculation rules

To calculate with hyperreal numbers usually means that for both – old as well as new numbers – all algebraic laws of an ordered field hold. Unusual is only that the archimedean axiom does not hold in $^*\mathbb{R}$ – more exactly it does not hold without some restrictions which will be indicated later. In $\mathbb{R}$ it holds:

(A) For all positive real numbers r there exists a natural number n such that $r < n$.

If $0 < r_1 < r_2$ and both belong to $\mathbb{R}$ then there exists n such that $n > \frac{r_2}{r_1}$ or $n \cdot r_1 > r_2$. In the latter case – provided r_1, r_2 are interpreted as magnitudes or segments – the archimedean axiom is conveyed. Geometrically stated it says: Any given segment r_2 can be exceeded by putting many times the smaller segment r_1. For infinitely small segments – as they were imagined by Leibniz and Euler – it does not hold. For hyperreal numbers, hence in $^*\mathbb{R}$, the analogous property (A) does not hold. For example $\Omega = (1, 2, 3, \ldots) > n = (n, n, n, \ldots)$ for every $n \in \mathbb{N}$.

Let us put together some rules that can easily be proved:

(a) For every infinite number there exists an inverse element being infinitesimal and vice versa.
(b) If $\alpha \approx 0$ and $\beta \approx 0$ then $\alpha + \beta \approx 0$ and $\alpha \cdot \beta \approx 0$.
(c) If $\alpha \approx \beta$ and $\beta \approx \gamma$ then $\alpha \approx \gamma$.
(d) If $\alpha \approx 0$ and γ is bounded then $\alpha \cdot \gamma \approx 0$ and it is infinitesimal.
(e) If $\gamma \gg 1$, $\delta > \gamma$ and η is finite then $\delta \gg 1$, $\gamma + \eta \gg 1$ and $\gamma - \eta \gg 1$.

All those corresponds to our expectations. We want to justify partially the above statements and do this in the case of (a), (b) and (d). So let us say at the beginning something about the uniquely determined standard part of every *bounded* nonstandard number y:

- In any case there exists a standard part s for y: s is the Dedekind cut generated in $\mathbb{Q}$ by y. Necessarily $y - s \approx 0$, and hence $y \approx s$. This cannot hold for a *second* real number t. Indeed, then there would hold $y \approx t$, and hence $s \approx t$, which is impossible because the distance between two real numbers cannot be infinitesimal.
- Hence every nonstandard number y can be represented as $y = s + \alpha$ where s is real and α is infinitesimal.

We now prove (a), (b) and (d), respectively.

(a) In fact, nothing must be shown. This is usual by calculations within a field. An example is already known: ω is the inverse element of Ω. In fact,

$$\omega \cdot \Omega = \left(\frac{1}{1}, \frac{1}{2}, \frac{1}{3}, \ldots\right) \cdot (1, 2, 3, \ldots) = (1, 1, 1, \ldots) = 1$$

What is α^{-1}? It holds

$$\alpha = \left(\frac{1}{10}^{n}\right) = \frac{1}{10}^{\Omega}.$$

Hence $\alpha^{-1} = 10^{\Omega}$.

(b) Let $\alpha \approx 0$ and $\beta \approx 0$ and let both be positive. It has to be proved that $\alpha + \beta \approx 0$ and $\alpha \cdot \beta \approx 0$, hence $\alpha + \beta < r$ and $\alpha \cdot \beta < r$ for all positive real numbers r. So let a number r be given. Both α and β are smaller than $\frac{r}{2}$ and $\sqrt{r}$, hence $\alpha + \beta < r$ and $\alpha \cdot \beta < r$.

(d) Let $\alpha \approx 0$ and y be finite and both positive. It has to be shown that $\alpha \cdot y < r$ for all r. So let r be given and let t be the standard part of y, i.e., $y = t + \beta$ where β is infinitesimal. Then $\alpha < \frac{r}{t}$, hence $\alpha \cdot t < r$ and

$$\alpha \cdot y = \alpha \cdot (t + \beta) = \alpha \cdot t + \alpha \cdot \beta \approx \alpha \cdot t < r.$$

The proofs of properties (c) and (e) are similar.

6.3.2 Continuity, differential quotient, derivative

In the 17th century Leibniz introduced the symbols $dx, dy, \ldots$ for infinitely small magnitudes, segments and numbers – they are still used today; however, their old meaning is usually unknown or, as we saw, sometimes rejected. The terms $dx, dy, \ldots$ are today only symbolic figures and indicate in the standard analysis usually limit processes (treated as closed) of "x-differences", "y-differences", etc. We use them in their original meaning, in particular as infinitely small numbers. For further explanations see [226, Chapter I].

Continuity

Let f be a real function. Consider it as being extended to $^*\mathbb{R}$. It can be precisely described how it can be done. We will come back to this later when we speak about the extension of $\mathbb{R}$ to $^*\mathbb{R}$. Now let us remain at intuition only.

What does it mean intuitively that "f is continuous at x"? Something like: if x is changed a bit then also $f(x)$ changes a bit – and it does not make any jumps. Just this intuitive view can be translated into the language of infinitesimals. There it means: "f is continuous at x" if infinitesimal changes of the argument x generate also infinitesimal changes of the values of f. So one can define:

f is said to be *continuous* if and only if $f(x) \approx f(x + dx)$ for infinitesimal dx.

In the case of real functions this makes the usually applied indirect, complicated ε-δ-definition superfluous.

Differential quotient *and* derivative

Let f be a linear function. The quotient of $\Delta y = f(x_1) - f(x_2)$ of values of the function and $\Delta x = x_1 - x_2$ of the arguments x_1 and x_2 describes the slope of the line g in the picture being the graph of f. This idea is usually transferred to the case of functions f whose graphs are curves and one writes the "differential quotients" $\frac{dy}{dx}$ for the slope of the tangent at the point $(x_0, f(x_0))$. This should recall the difference quotients $\frac{\Delta y}{\Delta x}$ whose denominator and numerator disappear in a limit value process of secants to tangent.

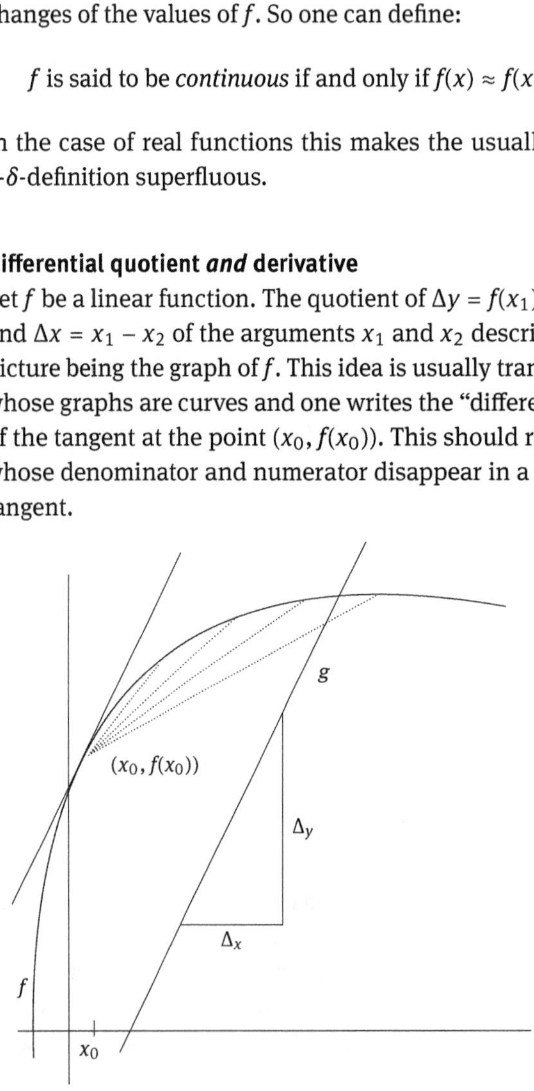

The term $\frac{dy}{dx}$ is usually nothing more than a way of writing and in fact it is no genuine quotient. However, we have infinitely small differences dx, dy at our disposal. Using them one can understand $\frac{dy}{dx}$ as genuine differential *quotients* that give directly the slope of the tangent at the point $(x_0, f(x_0))$ provided it exists. We define:

- If f is a real function and $dy = f(x_0 + dx) - f(x_0)$ for infinitesimals $dx \neq 0$ then $\frac{dy}{dx}$ is said to be the *differential quotient*.

The "differential quotients" should be distinguished from the real derivative. This can be seen at once in the standard example of $f(x) = x^2$. One has

$$\frac{dy}{dx} = \frac{f(x_o + dx) - f(x_o)}{dx} = \frac{(x_o + dx)^2 - x_o^2}{dx} = \frac{x_o^2 + 2x_o dx + dx^2 - x_o^2}{dx} = 2x_o + dx.$$

The standard part $2x_0$ of $\frac{dy}{dx}$ is the derivative. So we extend the definition:

- If f is a real function, $dy = f(x_0 + dx) - f(x_0)$ for infinitesimals $dx \neq 0$ and there exists a real a such that $\frac{dy}{dx} \approx a$ then a is said to be the derivative of f at the point x_0 and f is said to be differentiable at x_0. The derivative a at x_0 is denoted by $f'(x_0)$.

If $\frac{dy}{dx}$ for infinitesimals dx is "strongly" different, i.e., not infinitesimal, then the derivative does not exist. The standard example for this is the function f such that $f(x) = |x|$ at $x_0 = 0$.

Calculating with infinitesimals enables us to set more "conveniently" various *differentiation rules*. Let us limit ourselves to the case of the *chain rule* that is rather delicate and complicated for calculating with the limit values.

Let $y = f(x)$, $z = g(y)$ be differentiable functions and let $z = \varphi(x) = g(f(x))$. Then it holds

$$dy = f(x + dx) - f(x) \quad \text{and} \quad dz = g(y + dy) - g(y).$$

Inserting dy into the last equation and writing $f(x)$ instead of y, one gets

$$dz = g(f(x) + f(x + dx) - f(x)) - g(f(x)) = g(f(x + dx)) - g(f(x)) = \varphi(x + dx) - \varphi(x).$$

It holds

$$\frac{dz}{dy} \approx g'(y) \quad \text{and} \quad \frac{dy}{dx} \approx f'(x).$$

Taking into account the trivial equality $\frac{dz}{dx} = \frac{dz}{dy} \cdot \frac{dy}{dx}$, one gets

$$\varphi'(x) = g'(y) \cdot f'(x) = g'(f(x)) \cdot f'(x).$$

Integral

We introduce the integral only for monotonously growing functions. In the case of monotonously decreasing functions one proceeds analogously. Our approach fully corresponds to the usual one in the case of the Riemann integral. We only show how all can be done without using limit values. In fact, we shall calculate with infinitesimals – with infinitely small numbers – and, which is here new, with the infinite, infinitely large numbers. In particular we consider sums of $\mu \gg 1$, infinitely many, summands.

So let f be a function monotonously growing on the interval $[a, b]$, i.e., for all $x_1 \leq x_2$ in this interval one has $f(x_1) \leq f(x_2)$. Let $\mu \gg 1$. Let the interval $[a, b]$ be partitioned in the following way:

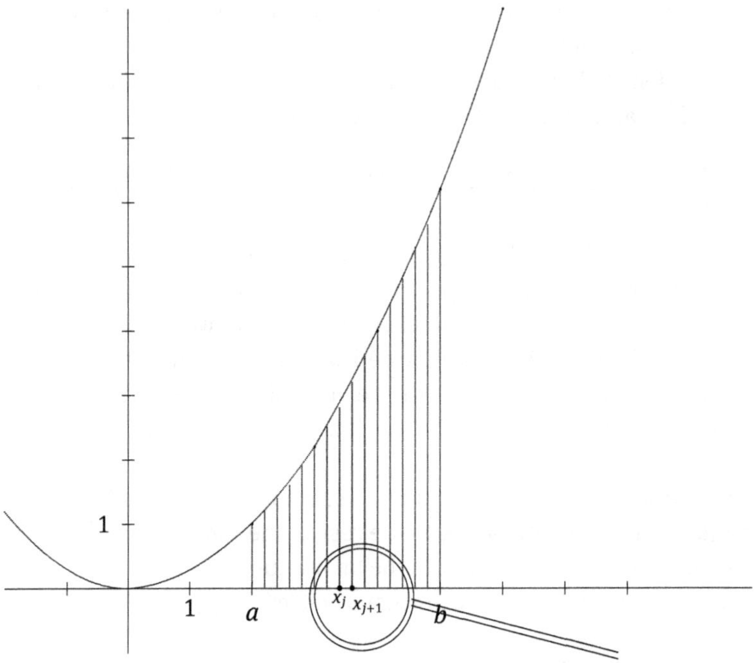

Fig. 6.1. Function f monotonously growing on the infinitesimally partitioned interval $[a, b]$ – with a magnifying glass.

$$a = \xi_0, \xi_1, \xi_2, \ldots, \xi_\mu = b$$

and

$$0 < d\xi_k = \xi_{k+1} - \xi_k \approx 0.$$

Additionally let

$$\xi_k \leq \hat{\xi}_k \leq \xi_{k+1}$$

such that every $\hat{\xi}_k$ is in the interval $[\xi_k, \xi_{k+1}]$. Let the lengths of the partial intervals be at most $\delta \approx 0$. We define the approximation sum S, the lower sum Ls and the upper sum Us of the infinitesimal areas over $\mu \gg 1$ many partial intervals up to the graph of f in the following way:

$$\mathrm{S} = \sum_{k=0}^{\mu} f(\hat{\xi}_k)d\xi_k, \quad \mathrm{Us} = \sum_{k=0}^{\mu} f(\xi_k)d\xi_k, \quad \mathrm{Os} = \sum_{k=0}^{\mu} f(\xi_{k+1})d\xi_k.$$

Since f is monotonously growing it holds

$$f(a)(b-a) \leq \mathrm{Ls} \leq \mathrm{S} \leq \mathrm{Us} \leq f(b)(b-a).$$

Since $\delta \approx 0$ one has

$$\begin{aligned} 0 &\leq \mathrm{Us} - \mathrm{Ls} \\ &= \sum_{k=0}^{\mu} (f(\xi_{k+1}) - f(\xi_k)) d\xi_k \\ &\leq \delta \sum_{k=0}^{\mu} f(\xi_{k+1}) - f(\xi_k) \\ &\leq \delta(f(a) - f(b)) \approx 0, \end{aligned}$$

provided $a, b, f(a), f(b)$ are finite. In this way the sums S, Ls and Us have the same standard part st(S) that gives the value of the definite integral:

$$\int_a^b f(x)\,dx := \mathrm{st}(\mathrm{S}).$$

In the case of continuous not necessarily monotonous functions arbitrary infinite sums are considered and it should be proved that they are infinitely close to each other and consequently have the same standard part – the latter is defined as the integral. This is a bit laborious [226, pp. 41 f.]. However, the general case of continuous functions whose peak-to-valley value is locally bounded can easily be reduced to the case of growing and decreasing functions.

Standard example: the definite integral $\int_a^b x^2\,dx$

Let $a = 0$. We calculate with infinitesimals and with infinite numbers as with the usual numbers. In particular we have the following formula for the sum of squares of natural numbers:

$$\sum_{k=1}^{\mu} k^2 = \frac{\mu(\mu+1)(2\mu+1)}{6}$$

for infinitely large μ.

Let the interval $[0, b]$ be partitioned into infinitely many, say μ, infinitely small partial intervals $[\xi_k, \xi_{k+1}]$, $k = 0, \ldots, \mu$ of the length $\omega \approx 0$. Then $b = \mu \cdot \omega$ and

$$\begin{aligned} \int_0^b x^2\,dx &\approx \sum_{k=1}^{\mu} (k \cdot \omega)^2 \omega \\ &= \omega^3 \cdot \sum_{k=1}^{\mu} k^2 \\ &= \omega^3 \frac{\mu(\mu+1)(2\mu+1)}{6} \end{aligned}$$

$$\begin{aligned}
&= \frac{\omega\mu \cdot \omega(\mu+1) \cdot \omega(2\mu+1)}{6} \\
&= \frac{\omega\mu \cdot (\omega\mu+\omega) \cdot (2\omega\mu+\omega)}{6} \\
&= \frac{b \cdot (b+\omega) \cdot (2b+\omega)}{6} \\
&= \frac{b \cdot (2b^2+3b\omega+\omega^2)}{6} \\
&= \frac{1}{3}b^3 + \frac{1}{2}b^2\omega + \frac{1}{6}b\,\omega^2 \approx \frac{1}{3}b^3
\end{aligned}$$

since ω is infinitely small and b is finite. Since on the left- and the right-hand side there are real numbers one has

$$\int_a^b x^2\,dx = \frac{1}{3}b^3.$$

Elementary theorems

It was and is often emphasized that till the 19th century analysis had no foundations. This impression arises when today's standpoint is applied and from it the history is considered. Our way of looking at history tends to judge by today's standards and rules.

At that time the foundations were different. It was the geometrical intuition and this was not justified – as was shown in Section 3.3. Additionally there came – since Leibniz – infinitesimals that continued Aristotle's intuition and whose principles have been formulated by Leibniz (cf. Section 3.3.4). The intuition and the infinitesimals – eliminated in the 19th century and mathematically considered today as alien – made some issues directly comprehensible and the question about their justification was not put. A good example of the power of a well-founded intuition and of infinitesimals is provided by the Nullstellensatz that is often demonstrated as an achievement of the new mathematics in comparison to the old one.

Nullstellensatz

(f) A function continuous on an interval and having on this interval positive and negative values as arguments has also the value zero.

Nobody doubted this in the 18th century – and also today nobody doubts this. One imagined – and imagines today – a continuous graph that inevitably cuts the x-axis as a line cuts another line. Doubts and the requirement of a justification *arose* only in connection with attempts to comprehend the continuity "non-geometrically", hence "purely", and to take the linear continuum set-theoretically as a set of points without infinitesimals that have then been used intuitively.

While in the standard approach the point of zero in the proof of the Nullstellensatz is "sighted" only as a limit, in the infinitesimal approach it is concretely "captured". It is caught by nesting with infinitesimal "boxes".

Proof of (f). Let f be a continuous real function. Consider a negative value $f(a)$ and a positive value $f(b)$. Consider the interval $[a, b]$ as partitioned into infinitely many infinitesimal intervals $[\xi_\kappa, \xi_{\kappa+1}]$. Let λ be the largest index such that $f(\xi_\lambda) \leq 0$ and let x be a real number such that $\xi_\lambda \approx x \approx \xi_{\lambda+1}$. Then $0 \leq f(\xi_{\lambda+1})$ and by the continuity of f one has $0 \geq f(\xi_\lambda) \approx f(x) \approx f(\xi_{\lambda+1}) \geq 0$. Since x is a real number, $f(x) = 0$. □

Our type of argumentation that will be discussed later is typical for thinking with infinitesimals and for the infinitesimal intuition. In the 18th century one quasi *saw* what we have demonstrated technically.

Maxima and minima

The infinitesimal procedure of infinitesimal partition of intervals will be applied now also in the following theorem.

(g) A function continuous in the interval $[a, b]$ attains there the least and the greatest value.

Proof. Assume that $[a, b]$ is partitioned into infinitely many infinitesimal intervals $[\xi_\kappa, \xi_{\kappa+1}]$. Then there exist the least and the greatest element of $f(\xi_\kappa)$, say $f(\xi_\lambda)$ and $f(\xi_\mu)$, respectively. Let x_l and x_m be real numbers such that $x_l \approx \xi_\lambda$ and $x_m \approx \xi_\mu$. So the real number $f(x_l) \approx f(\xi_\lambda)$ is the minimum and the real number $f(x_m) \approx f(\xi_\mu)$ is the maximum. □

(h) Let $f : [a, b] \to \mathbb{R}$ be differentiable. If x_0 is a local extremum inside the interval $[a, b]$ then $f'(x_0) = 0$.

Proof. Suppose that x_0 is a minimum. Let the interval $[a, b]$ be partitioned into infinitely many infinitesimal intervals $[\xi_\kappa, \xi_{\kappa+1}]$. Choose the greatest κ, say λ, such that $f(\xi_\lambda) \leq f(x_0)$. Then x_0 is in the interval $[\xi_\lambda, \xi_{\lambda+1}]$. So

$$f(\xi_\lambda) \leq f(x_0) \geq f(\xi_{\lambda+1}),$$

and hence

$$0 \leq \frac{f(\xi_\lambda) - f(x_0)}{\xi_\lambda - x_0} \approx f'(x_0) \approx \frac{f(\xi_{\lambda+1}) - f(x_0)}{\xi_{\lambda+1} - x_0} \leq 0.$$

Consequently, $f'(x_0) = 0$. □

One now obtains *Rolle's Theorem.*

(i) If $f(a) = f(b) = 0$ and f is differentiable in $[a, b]$ then there exists an x_0 such that $f'(x_0) = 0$.

In fact, if f is not constant then in the interval $[a, b]$ it attains an extremum.

Mean value theorem

As the continuous graph of f from the above Nullstellensatz cuts further lines parallel to the x-axis, f attains every other value between $f(a)$ and $f(b)$. If f' is continuous then the same holds for f'. In particular this holds for slopes, hence for the values of f' between $(a, f(a))$ and $(b, f(b))$.

(j) Let f be differentiable and let f' be continuous. Then in the interval $[a, b]$ there exists some s such that
$$f'(s) = \frac{f(a) - f(b)}{a - b}.$$

The assumption that f' is continuous is unnecessary. In fact, the situation that $f(a) = f(b)$ for a continuous function in $[a, b]$ can be obtained by a simple manipulation, in particular by adding a linear function. Hence after some reformulations (j) follows by Rolle's Theorem.

If in the mean value theorem one writes the variable x instead of a, h instead of the difference $b - a$ and one multiplies by h, then one obtains a formula in which we can think about h as being infinitesimal, hence for infinitesimal intervals we have the following result:

(k) $f(x + h) - f(x) = f'(x + r \cdot h) \cdot h$ for r such that $0 < r < 1$.

By the mean value theorem it follows immediately:

(l) A function is constant if its derivative is 0.

Both latter statements form essential elements of a simple proof of the main theorem in which infinitesimals are used. We follow the ideas formulated in [226, pp. 42 f.].

Main theorem

Let $F'(x) = f(x)$. We show that
$$\int_a^b f(x)\,dx = F(b) - F(a). \tag{6.1}$$

Proof. Let us partition the interval $[a, b]$ into μ infinitesimal intervals $[\xi_k, \xi_{k+1}]$ of the length h and apply (k). Assertion (k) states that for every of those intervals there exists a $\rho < 1$ such that
$$F(\xi_k + h) - F(\xi_k) = F'(\xi_k + \rho \cdot h) \cdot h.$$

Put $\xi_k^* = \xi_k + \rho \cdot h$. Then
$$\begin{aligned} F(b) - F(a) &= \sum_{k=1}^{\mu} F(\xi_k + h) - F(\xi_k) \\ &= \sum_{k=1}^{\mu} F'(\xi_k + \rho \cdot h) \cdot h \\ &= \sum_{k=1}^{\mu} F'(\xi_k^*) \cdot h \\ &\approx \int_a^b f(x)\,dx. \end{aligned}$$

Since $F(b) - F(a)$ and $\int_a^b f(x)\,dx$ are real, equality (6.1) holds. □

If b is variable and replaced by x, then

$$F(x) = \int_a^x f(x)\,dx + F(a)$$

is a real function such that $F'(x) = f(x)$. If G is such that $G'(x) = f(x)$, then we have $(F - G)' = F' - G' = 0$. By (1) it follows that $F - G$ is a constant C. In this way one has all that is needed to formulate the main theorem.

Theorem 6.3.1 (Main Theorem). *If f is a continuous real function in the interval $[a, b]$ and $F' = f$ for a differentiable function F, then*

$$\int_a^b f(x)\,dx = F(b) - F(a).$$

If G is any function such that $G' = f$, then

$$G(x) = \int_a^x f(t)\,dt + C$$

for $x \in [a, b]$.

An infinitesimal calculating procedure for a natural logarithm

We present now a small application of the above infinitesimal results and continue in this way calculations with infinitesimals.

Consider the monotonically decreasing function $f(x) = \frac{1}{x}$ and define

$$L(b) := \int_1^b \frac{1}{x}\,dx.$$

Let the interval $[1, b]$ be partitioned into infinitely many, say $\mu \gg 1$, infinitely small partial intervals of the length $\alpha \approx 0$. So $b = 1 + \mu \cdot \alpha$ and

$$L(b) = \int_1^b \frac{1}{x}\,dx \approx \sum_{k=0}^{\mu} \frac{1}{1 + k\alpha} \cdot \alpha \tag{6.2}$$

is a presentation of $L(b)$ as an infinite sum.

Let c be finite. Since $1 < b = 1 + \mu \cdot \alpha$, one has $c < b \cdot c = c + \mu \cdot \alpha \cdot c$. But $\alpha \cdot c \approx 0$ is infinitesimal. Let the interval $[c, bc]$ be partitioned into infinitesimal partial intervals of the length αc. Then

$$\int_c^{cb} \frac{1}{x}\,dx \approx \sum_{k=0}^{\mu} \frac{1}{c + k \cdot c\alpha} \cdot \alpha c,$$

and by (6.2) one gets

$$\int_c^{c\cdot b} \frac{1}{x}\,dx \approx \sum_{k=0}^{\mu} \frac{1}{c+k\cdot c\alpha}\cdot \alpha c = \sum_{k=0}^{\mu} \frac{1}{1+k\alpha}\cdot \alpha \approx \int_1^b \frac{1}{x}\,dx = L(b). \tag{6.3}$$

Now

$$\int_1^{c\cdot b} \frac{1}{x}\,dx = L(c\cdot b)$$

can be calculated. Decompose the interval $[1, bc]$ into $[1, c]$ and $[c, bc]$. Decompose further the interval $[a, c]$ into $\lambda \gg 1$ subintervals of the length $\beta \approx 0$ and the interval $[c, bc]$ as above into infinitesimal subintervals of the length αc. Finally, one gets an infinitesimal partition of the interval $[a, bc]$. Let us calculate by applying (6.3):

$$\begin{aligned} L(c\cdot b) &= \int_1^{c\cdot b} \frac{1}{x}\,dx \\ &\approx \sum_{k=0}^{\lambda} \frac{1}{1+k\beta}\cdot\beta + \sum_{k=0}^{\mu} \frac{1}{c+k\cdot c\alpha}\cdot \alpha c \\ &\approx \int_1^{c} \frac{1}{x}\,dx + \int_c^{c\cdot b} \frac{1}{x}\,dx \\ &= L(c) + L(b). \end{aligned}$$

It follows that

$$L(b\cdot c) = L(b) - L(c)$$

for all $b, c \geq 1$. Hence obviously $L(1) = 0$.

Finally, consider the chain rule: we know that $L'(x) = \frac{1}{x}$. Put $\varphi(x) = L(\exp(x))$ and one gets

$$\varphi'(x) = L'(exp(x))\cdot \exp'(x) = \frac{1}{\exp(x)}\cdot \exp(x) = 1.$$

By the Main Theorem one has

$$\varphi(x) = L(\exp(x)) = x + C.$$

But $C = 0$ because $\varphi(0) = L(\exp(0)) = L(1) = 0$. So L is the inverse of exp and $L = \ln$.

6.4 On the construction of hyperreal numbers

When we started our above calculations the situation was similar to the situation of the passage from the rational numbers to the real numbers. This passage is hardly or not at all noticed in school teaching as remarked in Chapter 1, or one forgets what one has been taught. It is calculated with real numbers without actually knowing what they are – they are assumed as given. Rational intervals are for example filled by real numbers and rational functions are considered to be simply extended to real numbers. Something similar has been done above when we went from the real numbers to the hyperreal ones. The idea of a number line helps in both cases.

Now we want to carry out in the case of hyperreal numbers what is seldom done in the case of real numbers, i.e., to indicate the procedure of extending and to discuss problems connected with it. Some remarks on this have been already made in Section 6.2. We consider at the beginning – to get the first insight into the construction – the natural numbers and their extension to nonstandard natural numbers since this is more clear.

6.4.1 Hypernatural numbers

Consider the set of all infinite sequences of natural numbers denoted by (a_n), (b_n), (c_n) etc. Sequences (a_n), (b_n) are identified if they are "finally" identical, i.e., identical for all n from a certain point n_0. It is said that "almost all" members a_n, b_n of the sequences are equal:

$$(a_n) \sim (b_n) :\Leftrightarrow a_n = b_n \text{ for almost all } n.$$

The relation $\sim$ is an equivalence relation. Instead of the abstraction classes $\overline{(a_n)}$ we write for the moment simply their representatives (a_n).

We say (a_n) is smaller than (b_n) if almost all a_n are smaller than b_n, i.e., $(a_n) < (b_n)$ if $a_n < b_n$ for almost all n. Addition and multiplication are defined by addition and multiplication of members of appropriate sequences, e.g., $(a_n) + (b_n) = (a_n + b_n)$. Natural numbers n are represented by constant sequences $(n, n, n, \ldots)$. Instead of $(n, n, n, \ldots)$ one simply writes n.

The first and most natural sequence of natural numbers is the sequence used by counting, i.e.,

$$\Omega := (1, 2, 3, \ldots).$$

More exactly, Ω is represented by $(1, 2, 3, \ldots) = (n)$, $\Omega + 1$ by $(2, 3, 4, \ldots) = (n + 1)$, $2 \cdot \Omega$ by $(2, 4, 6, \ldots) = (2n)$, $2^{\Omega} = (2, 4, 8, 16, \ldots) = (2^n)$, etc. There exists no smallest infinite natural number. For example $\Omega - 3 = (0, 0, 0, 1, 2, 3, \ldots)$.

Of course Ω is – as we saw above – infinitely large. In fact, for all n it holds

$$n = (n, n, n, \ldots) < (1, 2, 3, \ldots) = \Omega$$

since every n is smaller than almost all k in $(1, 2, 3, \ldots)$.

By the indicated construction one gets an arithmetical domain of calculating, a semiring. In order to make it work better the identification of sequences should be extended. We return to this below. At the end we get a linearly ordered semiring without zero divisors. It is denoted by $^*\mathbb{N}$.

The following picture of $^*\mathbb{N}$ fully corresponds to the intuition indicated already in the picture above: first there are the standard natural numbers, then the infinitely large numbers:

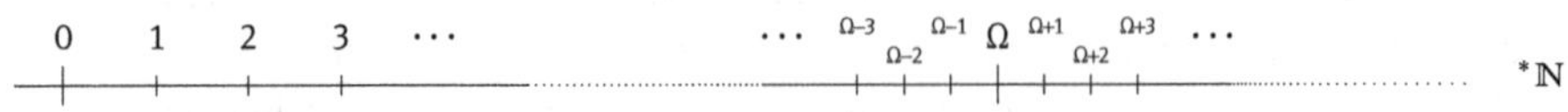

Considering for example unit fractions of infinitely large nonstandard numbers one gets infinitely small numbers. For example the number

$$\omega = \frac{1}{\Omega} = \left(\frac{1}{1}, \frac{1}{2}, \frac{1}{3}, \frac{1}{4}, \dots\right)$$

is infinitely small. In fact, every real number r represented by the constant sequence $(r, r, r, \dots)$ is bigger than ω because almost all $\frac{1}{k}$ are smaller than every given real r.

The construction of hyperreal numbers proceeds according to the same principle – now we describe it more exactly. Note that the next point will set-theoretically be a bit more difficult.

6.4.2 Hyperreal numbers

Now our starting point is the ring of all sequences of real numbers, both convergent and divergent. Denote by $\mathbb{R}^{\mathbb{N}}$ the ring of such sequences. Arithmetical operations are again defined in $\mathbb{R}^{\mathbb{N}}$ by operations on members of appropriate sequences, e.g., $(a_n) \cdot (b_n) = (c_n) :\Leftrightarrow a_n \cdot b_n = c_n$ for all n. Real numbers are represented in this ring by constant sequences. The real numbers can be obtained here in the usual way: The starting point of Cantor's construction of the real numbers is the ring $\mathbb{Q}^{\mathbb{N}}$ of sequences. Now $\mathbb{R}$ is the quotient ring of the ring of fundamental sequences modulo the maximal ideal of zero sequences.

A hardly more complicated construction of Schmieden and Laugwitz from the year 1958 of the – from the outside – non-archimedean field uses the whole ring $\mathbb{R}^{\mathbb{N}}$ of sequences and another, smaller ideal, namely the ideal V of eventually vanishing sequences, i.e., such sequences that only a finite number of their members is not equal to zero.

With sequences in V in particular the set Cof of *cofinal sets* in $\mathbb{N}$ is connected, i.e., of sets M such that $\mathbb{N} \setminus M$ is finite:

$$\mathrm{Cof} = \{M \subseteq \mathbb{N} \mid \mathbb{N} \setminus M \text{ is finite}\}.$$

Hence

$$(c_n) \in V \quad \text{if and only if} \quad \{n \mid c_n = 0\} \in \mathrm{Cof}\,.$$

Sequences (a_n), (b_n) denoted simply by a, b, are equivalent if $a - b \in V$, i.e.,

$$a \sim b \quad \text{if and only if} \quad \{n \mid a_n = b_n\} \in \text{Cof},$$

hence if almost all members of the sequences are identical. The relation $\sim$ is an equivalence relation.

The equivalence classes, hence elements of $\mathbb{R}^{\mathbb{N}}/V$, are denoted by small Greek letters $\alpha, \beta, \gamma, \ldots$. They are represented by sequences $(a_n), (b_n), (c_n), \ldots$. The set of real numbers is again represented by constant sequences (for example 1 by $(1, 1, 1, \ldots)$ as above). For elements $\alpha, \beta \in \mathbb{R}^{\mathbb{N}}/V$ the following relations hold:

$$\begin{aligned}
\alpha = \beta &\quad \text{if and only if} \quad a \sim b, \text{ i.e., } \{n \mid a_n = b_n\} \in \text{Cof},\\
\alpha + \beta = \gamma &\quad \text{if and only if} \quad \{n \mid a_n + b_n = c_n\} \in \text{Cof},\\
\alpha \cdot \beta = \gamma &\quad \text{if and only if} \quad \{n \mid a_n \cdot b_n = c_n\} \in \text{Cof},\\
\alpha \cdot \beta = 0 &\quad \text{if and only if} \quad \{n \mid a_n \cdot b_n = 0\} \in \text{Cof}.
\end{aligned}$$

The example of $\alpha \cdot \beta$ where $a_n = 1 + (-1)^n$ and $b_n = 1 - (-1)^n$ shows that in the ring $\mathbb{R}^{\mathbb{N}}/V$ there are zero divisors. The way to a field of hyperreal numbers is abstract. We just come to this. However, first consider the order.

Let

$$\alpha < \beta :\Leftrightarrow \{n \mid a_n < b_n\} \in \text{Cof}.$$

For example

$$(0.9, 0.99, 0.999, \ldots) < (1, 1, 1, \ldots).$$

The relation $<$ is transitive since Cof is a filter. This means that $<$ is an order relation. Moreover, $<$ is not linear.

What about the power $(-1)^{\Omega}$? This is the class of the representative $((-1)^n)$. It holds neither $(-1)^{\Omega} < 0$ nor $(-1)^{\Omega} = 0$ nor $(-1)^{\Omega} > 0$. This is a counterexample indicating that $<$ is not linear. This defect can be removed when on the base of $\mathbb{R}^{\mathbb{N}}/V$ a field can be obtained. The way to this is abstract and not constructive. The Axiom of Choice has to be assumed.

The starting point of the procedure is the above free filter Cof. With Cof the ideal V is connected. Every better free filter F contains Cof and generates an ideal containing V, say I_F. For sequences a, b it holds

$$a \in I_F \quad \text{if and only if} \quad \{n \mid a_n = 0\} \in F,$$

and a and b are equivalent if and only if $a - b \in I_F$ which in turn is the case if and only if $\{n \mid a_n = b_n\} \in F$.

Zorn's Lemma, hence the Axiom of Choice, implies that in the ordered set of such free filters there exists a maximal free filter U. Now $\mathbb{R}^{\mathbb{N}}/I_U$ is an ordered field, the field of *hyperreal numbers* ${}^*\mathbb{R}$. Its existence follows by the Axiom of Choice. However, one does not know in detail how it looks. It is not uniquely determined. But we know to a large extent how one can calculate in it, namely like in $\mathbb{R}^{\mathbb{N}}/V$. This suffices for the practice.

Because of applying Axiom of Choice in this construction – which in fact is often exaggerated – Laugwitz describes the hyperreal numbers also in an algebraic-logical way as a result of kind of adjoining an infinitely large element Ω (cf. [226, Chapter 2]). He generally starts with a field K, extends it to the "Omega-numbers" ${}^{\Omega}K$ and follows by Leibniz's principle (approximately logically formulated) "the rules for the finite hold also for the infinite" the following:

> "Let $A(\cdot)$ be a formula formulated in a language of K. If there is an $n_0 \in \mathbb{N}$ such that for all $n \geq n_0$ the formula $A(n)$ is true in the theory of K then $A(\Omega)$ should be included as a true sentence into the new theory ${}^{\Omega}K$." ([226, p. 88].)

Also here the result is not unique. However, this quasi-algebraic construction is interesting since in it the usual algebraic adjoining is used and this new type of adjoining is repeatable, as can immediately be seen. Also the construction with the ring of sequences is of course repeatable.

6.4.3 How will ${}^*\mathbb{R}$ become a model of $\mathbb{R}$?

The field $\mathbb{R}$ is archimedean and ${}^*\mathbb{R}$ is not. Ask – slightly freely formulated: How can ${}^*\mathbb{R}$ become a model of $\mathbb{R}$? The fact that ${}^*\mathbb{N}$ is a model of $\mathbb{N}$ has been "secretly" used above in the calculations. And again the answer can only be roughly described.

Briefly: Look at ${}^*\mathbb{R}$ with "purely arithmetical eyes". One sees the so-called "*internal*" sets, relations and functions and all is as in $\mathbb{R}$. What does "purely arithmetical" mean, what does "internal" mean? For example, an internal set arises when in properties describing its elements only the symbols $+, \cdot, <, =, 0, 1$ are used. Axioms of the real numbers are formulated by those symbols. The completeness axiom is now an axiom scheme in which only formulas φ can appear that are formulated by those symbols (cf. Section 5.3.2). A. Robinson – and this was a part of his epochal discoveries – noticed that fields like ${}^*\mathbb{R}$ are models of arithmetically formulated axioms of the real numbers $\mathbb{R}$. The standard model is $\mathbb{R}$, and ${}^*\mathbb{R}$ is a nonstandard one. "Model" means here that all formulas true in $\mathbb{R}$ hold also *in* ${}^*\mathbb{R}$.

One looks at ${}^*\mathbb{R}$ "from the outside" when symbols like $\approx, \gg$ introduced above are used or when expressions like "finite", "infinite", "infinitesimal", "real" and "nonstandard" are applied. *External* and not internal are for example the set of finite real numbers and the set of natural numbers, hence $\mathbb{R}$ and $\mathbb{N}$, as well as the set of "infinitely large" natural numbers such as $\ldots, \Omega - 1, \Omega, \Omega + 1, \ldots$.

Another example that seems at first glance to be paradoxical: We know that $0.999\ldots$ as a sequence of partial sums – more exactly as a class of sequences – is an element of ${}^*\mathbb{R}$. However, the sequence itself being the set of those partial sums –

as a *set* – is not internal. Why? It cannot be described purely arithmetically, hence internally, but only externally. To characterize its elements the word "finite" is needed since it integrates only finite partial sums.

Sets of real numbers in $\mathbb{R}$ and real functions can be extended to internal sets and functions in $^*\mathbb{R}$ (cf. [226, Chapter 3]). Expressions about internal objects in $^*\mathbb{R}$ can generally be interpreted as expressions about real numbers – this is the *nonstandard method*. Just this has been done above. Even better: this procedure is a *part* of the nonstandard method. In fact, $\mathbb{R}$ is a subset of $^*\mathbb{R}$, and expressions in $^*\mathbb{R}$, to which just the "external expressions" belong, have a meaning beyond $\mathbb{R}$. Such external expressions about hyperreal numbers can then often be connected with the reals by building standard parts as we saw above and will see below.

6.4.4 On the justification of the naive infinitesimal calculus

Intervals $[a, b]$ are examples of internal sets in $^*\mathbb{R}$. In fact, $[a, b]$ is a set of ξ such that $a \leq \xi \leq b$, hence it is purely arithmetically defined. But how do hyperreal intervals arise from real intervals? How are real intervals $[a, b]$ "filled up" with hyperreal numbers to become hyperreal intervals $[a, b]$? Think about the construction of $^*\mathbb{R}$: Hyperreal numbers are represented by sequences (x_n) of real numbers, hence a hyperreal ξ in $[a, b]$ is represented by sequences whose members are real numbers x_n from the real interval $[a, b]$. All internal sets Γ arise in a more or less complicated way by such sequences of representatives [226, pp. 107 f.].

If f is a real function on the real interval $[a, b]$ then it can be extended to an internal hyperreal function f_h in the following way: If (x_n) is a sequence representing ξ in the hyperreal interval $[a, b]$ and $f(x_n) = y_n$ then we put $f_h(\xi) = \eta$ where η is represented by the sequence (y_n). Instead of f_h one can simply write f – and we have done so above. In such a way our permanent – and unnoticed – changes between real and hyperreal intervals and functions in the infinitesimal calculations above have been explained and justified.

Look at the method of partition of real intervals, say at the partition of $[a, b]$ into $\mu \gg 1$ infinitesimal intervals $[\xi_\kappa, \xi_{\kappa+1}]$ of the length $\delta \approx 0$ such that $\xi_0 = a$, $\xi_\kappa = a + \kappa \cdot \delta$ and $a + \mu \cdot \delta = b$. In this way we proceeded above. It looks at bit "non-arithmetically". In fact, there appear symbols like $\mu \gg 1$ and $\approx$ and expressions like "infinitesimal" and "real" are used. However, the set of ξ_κ as well as the set of hypernatural κ enumerating ξ_κ are internal. In fact, a, b, μ and δ are hyperreal numbers and the ξ_κ have been arithmetically determined. On the other hand the purely arithmetical situation has been considered and commented externally – just by using expressions like "real", "$\mu \gg 1$" etc.

Furthermore, in our proofs we have argued in such a way: "Choose the greatest κ in a (bounded) set of hypernatural numbers". Or: "Let $f(\xi_\lambda)$ be the greatest value of $f(\xi_\kappa)$" – hence in such a way as the set of μ-infinitely many $f(\xi_\kappa)$ were finite. But this is also *internal*, say "hyperfinite": There are μ-many $f(\xi_\kappa)$ and, since all is internal here, there exists the greatest value – similarly to the case of m-many values $f(\xi_k)$.

Let us formulate now the following theorem describing the situation in a general way.

Theorem 6.4.1. *Let $\mu \gg 1$ be a hypernatural number and let Γ be an internal set of μ values below a given bound M. Then there exists the greatest element of the values.*

Proof. The set Γ arises by a sequence of sets

$$G_n = \{r^n_{m_1}, r^n_{m_2}, \ldots, r^n_{m_n}\},$$

where $r^n_i \leq M$ and μ is represented by the sequence (m_n). If $\xi \in \Gamma$ then ξ is represented by a sequence (x_n) with $x_n \in G_n$. Every set G_n is finite and its elements are below M. In each of those sets there exists (at least one) greatest element r^n_k. A sequence of such greatest elements represents the greatest element of Γ. □

This is a typical and simple example of a transfer of real relations to the internal part of ${}^*\mathbb{R}$.

Let us summarize the methodic procedure: One uses the arithmetical properties of internal sets and functions in ${}^*\mathbb{R}$ as they hold in $\mathbb{R}$ or in $\mathbb{N}$. Simultaneously, one works externally with non-archimedean, special infinitesimal methods of ${}^*\mathbb{R}$ and it is commented and argued externally – from the point of view of ${}^*\mathbb{R}$.

The dual nature of ${}^*\mathbb{R}$ – internal model with everything that $\mathbb{R}$ has, external with its infinitesimal and infinite elements and methods – is significant, maybe unusual but effective. It is amazing that the methodic advantages of ${}^*\mathbb{R}$ are used so seldom and in teaching and learning rarely or not at all. What is the reason?

6.5 On the status of nonstandard numbers

Nonstandard numbers apparently are mathematically still underestimated. In particular their meaning for the understanding of fundamental concepts has not been (or is hesitantly) recognized. What are the reasons of such reserved perception of those "new" numbers? We have often indicated them and in Section 3.3.11 in a summarized way discussed them. Since those numbers now are explicitly given we want to turn back to such reasons once again and try to answer more precisely some questions sketched there.

Where does the reserve come from?

In mathematical logic many interesting results have been obtained – they are everywhere accepted and used. Only mathematics seems to be more or less unaffected. Despite incompleteness and uncertainty mathematics progresses "in a good temper".

One is far away from the warning formulated in 1921 by Herrmann Weyl whom we already quoted above:

> "In fact: every serious and honest reflection must lead to the understanding that drawbacks in the boarder domains of mathematics [i.e., set theory and logic – authors' remark] should be seen as alarming symptoms; they exhibit what the externally brilliant and smooth business in the center hides: the internal instability of the foundations on which the building of the kingdom is founded." ([366, p. 1].)

The mathematical business goes "smoothly", even "more brilliantly" than ever. It can be "in fact" understood that one has less interest to consider the foundations, for example the real numbers. The standard model works. So what are nonstandard models for? Why rethink moreover.

The reasons of the resistance and reservation lie in the history of the introduction of nonstandard numbers – disregarding their unfortunate name. By mathematics based on the concept of limit, infinitesimals have in the 19th century been declared as overcome and are treated today as "antique". Not the construction of nonstandard numbers but rather the logical proof of the existence of nonstandard models has been noticed in the 1960s. Nonstandard models do appear as logical products – coming from an outlying district of mathematics – and are considered to be irrelevant for mathematical practice and for the usual mathematical thinking as the results about incompleteness and truth. The analysis of the construction of nonstandard numbers indicates the "defect" – the usage of the Axiom of Choice. All this seems to relativize the mathematical perception of the usefulness of nonstandard numbers.

In fact, the objection of being not constructive due to the Axiom of Choice cannot be disregarded. However, it is in various aspects not substantial:

(1) It is known that there are many paradoxical consequences of the Axiom of Choice. However, for us this seems to be a phenomenon connected not only with choice. The reason is the fact that the actual infinite is used freely and without any limitations. In set theory one proceeds with infinite sets as with finite ones and it is something usual and unproblematic to choose elements from finite sets and to form new sets out of them. So it seems to be consistent – as a consequence of set-theoretical axioms – to transfer the choice to infinite sets of sets (cf. Section 4.4). Choosing is an elementary principle. The Axiom of Choice is nowadays widely accepted because it is used in various domains of mathematics.

(2) The treatment of classes in the construction proceeds largely on the base of the ring $\mathbb{R}^{\mathbb{N}}/V$. The construction starts with them and its elements can be concretely identified as classes in the construction of $\mathbb{R}$.

(3) The Axiom of Choice is in fact required in the *construction* of nonstandard models but not when they are used: the acceptance of nonstandard models does not lead to contradictions even when the Axiom of Choice is eliminated (cf. [350,

pp. 2 f.]). The concrete work in the nonstandard analysis hardly needs a reference to the construction of $^*\mathbb{R}$. This makes the argument of the non-constructiveness weaker – both for the purely mathematical as well as for the practical evaluation of nonstandard numbers. Mathematics can be done equally well with nonstandard numbers as with standard ones – and in a certain sense better [350, Preface and p. 3].

(4) In principle one can freely choose to work with standard or with nonstandard numbers. This is a similar situation as in the case of the Axiom of Choice. In fact, the acceptance of nonstandard models in the framework of ZF is to the same degree consistent as the rejection of them (cf. [12, p. 851]). And mathematics with them is *richer* than without them – analogously to the work with AC or without AC.

(5) Elementary calculus with infinitesimals is easier. We have seen this. However, – as we indicated – the construction of nonstandard numbers is not easier and the logical justification of the calculus is laborious (cf. for example [224, Chapter 2]). On the other hand the real numbers are accepted in teaching and practice rather pragmatically than axiomatically and the construction of them is seldom introduced (the difficulty of this construction is comparable with that of nonstandard numbers). Hence the confrontation of real and hyperreal numbers loses its meaning and significance. In fact, one can and will proceed with nonstandard numbers in a similar way.

Briefly: Problems connected with the constructions of real and hyperreal numbers are comparable. They rest in the extensive usage of the actual infinite without which neither the first nor the second construction is possible. The construction of the real numbers leads into the theoretical area, the construction of the hyperreal numbers continues this way. The application of the Axiom of Choice in the construction of the hyperreal numbers seems to us to be only a marginal adjunct that finally becomes invisible. In fact, the Axiom of Choice is not needed for the acceptance of models. From the point of view of calculating practice in the infinitesimal calculus whose advantages are apparent, the question concerning the Axiom of Choice is not relevant.

Mathematically seen the hyperreal nonstandard numbers are equally fundamental as the standard real numbers. The invention of infinitely small magnitudes in the 17th century, especially by Pascal and Leibniz, were so revolutionary and effective as the invention of transfinite sets by Cantor. In fact, both – infinitely large sets *and* infinitely small magnitudes, limit values *and* infinitely small numbers, real *and* hyperreal numbers – can exist together. Nonstandard numbers seem to be more difficult to be accepted because we are not accustomed to them. Their double character – internally as models of $\mathbb{R}$ and externally as non-models, namely as non-archimedean fields – is unusual but it is extremely effective. However, the hyperreal numbers are still in the deep shadow of the reals.

On the ontological status of hypernatural numbers

More fundamental is the *philosophical* question about the way of existence of nonstandard numbers. Let us first consider the nonstandard natural numbers.

We talked about *models* of natural numbers. This indicates that we are remote from the reality. Also the standard model $\mathbb{N}$ of natural numbers is called "model". However, it is distinguished by the fact that among all models it is just that which quasi represents the reality, hence is not only a "theoretical" one. We do not keep the secret that the problem of reality, on the ontological basis of standard natural numbers, is an old philosophical problem. The problem of the concept of number has been considered in a historical perspective in Chapter 2 and it was the main subject of Section 3.1. Generally, it is assumed that natural numbers are close to the reality.

It does not hold for nonstandard natural numbers ${}^*\mathbb{N}$. They are theoretical. This is stressed first of all by the purely *logical* character of the conclusion concerning their "existence". What does "existence" mean here? Logic is formal, one argues in it on the basis of the first-order predicate logic and it is stated that nonstandard models fulfill Peano axioms in the same way as the standard model. It is a *formal* existence – though mathematically significant, however of fully different type than the existence of the "real" standard natural numbers. Nonstandard natural numbers as elements of those formal models are first of all only formal-arithmetical phenomena.

From the set-theoretical point of view it is a bit different. The set-theoretical construction shows similarly the theoretical character of nonstandard natural numbers, but in another way. The construction of ${}^*\mathbb{N}$ without using actual infinity is not possible – it is *necessary* to grasp the infinite sequences that represent numbers. Actual infinity is theoretical. The construction however is not formal – in fact it produces nonstandard numbers as "tangible" set-theoretical objects with which one can calculate as with natural numbers.

Above we already gave the following visualization of natural nonstandard numbers ${}^*\mathbb{N}$:

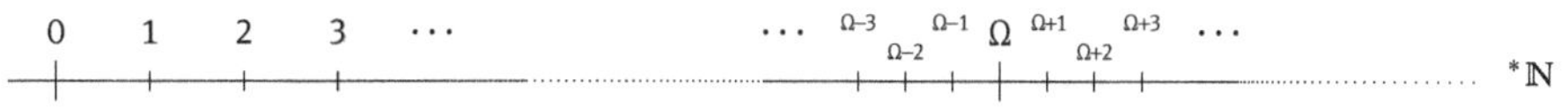

This visualization of natural nonstandard numbers is – though it sounds paradoxical – theoretical. In fact, this representation functions only indirectly through the abstract actually infinite set $\mathbb{N}$. One can think in an abstract *and simultaneously* intuitive way. On the one hand we can think in an *abstract* way about the actually infinite set of numbers $\mathbb{N}$ and about the line as actually infinite set of points. On the other hand we see how the numbers arise in an open process of counting and we have an *intuition* of a line that flows continuously as the time.

We have put the completed *abstract* actually infinite set $\mathbb{N}$ into the *open*, continuously-geometrical line and we see the nonstandard natural numbers as succeeding all natural numbers. This is a combination of the old Aristotle-geometrical and the new set-theoretical conception. Is it allowed? Yes, of course it is because the construction of nonstandard numbers and the logical proof of their existence entitle to do this – even more, they forced this. There is “only” the problem of intuition.

On the ontological status of hyperreal nonstandard numbers

In the case of real numbers the situation is different. Here the standard model whose elements are transfinite constructions has no reference to the reality as it was in the case of standard natural numbers. Thus it is not similarly an exceptional phenomenon among models. Its theoretical character is equal to the character of nonstandard models. Nonstandard models are archimedean or non-archimedean – according to the perspective. The real numbers however are *the* unique archimedean model. They provide the axioms. This makes them special.

Historically seen the real numbers arose from magnitudes – in an archimedean form. Its set-theoretical construction was the condition of the elimination of non-archimedean infinitesimals that were actively used. However, real numbers are constructs. They owe their “existence” to set theory which only pretendedly seems to be elementary and real in their sets. The actual infinity makes real numbers theoretical elements.

Axioms of the real numbers describing the theory of them links arithmetical and geometrical ideas. The archimedean axiom is borrowed from rational measured values approximating real numbers. The potentially non-archimedean feature of the linear continuum, being infinitesimal, that flourished from the 17th till the 19th century, has been repressed for a long time. The axioms of the real numbers are not as the Peano axioms that have been read from a quasi-real standard model. In fact, they are *theoretical* from the very beginning.

The formally logical conclusion on the formal existence of non-archimedean nonstandard models of axioms of real numbers appeared late in the history. It is worth noticing that the idea of a set-theoretical construction preceded them by few years. The set-theoretical and categorical characterization of the real numbers in the second-order logic that convinced mathematicians who were not logicians was perhaps a reason of the “delay” of nonstandard models; it reassured. In the first-order predicate logic there were some problems with the axiom of completeness – it is usually formulated in a set-theoretical manner (cf. Section 5.3.2). Since the first-order predicate logic turned out to be an appropriate background for mathematics, there appeared inevitably the problem of models of axioms for the real numbers.

The real numbers appear among models as other numbers do – not only when they are treated logically. However, nonstandard models generally have no or little *practical* meaning, yet they are not “less existing” than the standard model. They are treated as logical phenomena and are used rarely or not at all. This determines their apperception.

This apperception is *not* justified in the case of the nonstandard model of hyperreal numbers. The hyperreal numbers are as good as the reals, they can do everything, and even more than the reals can. They represent both the non-archimedean *and* the archimedean view of the continuum. Their relation to the axioms is the same as the relation of the standard model. Practically there is no distinction. We are of the opinion that just the tradition and the power of history are the reason for the fact that the one type of numbers is standard and the other nonstandard – or still nonstandard. It can be said that the hyperreal numbers are *the* non-archimedean *and* archimedean model.

The hyperreal numbers and the real numbers are closely connected as shown by their constructions and their structure. Both originate from Euclidean magnitudes which they have relieved and which they replace in different ways.

Real numbers are classes of various approximation sequences that approximate a considered magnitude (in the old sense). They *identify*, they *are* the identified approximation sequences. The identification eliminates the potential feature of being non-archimedean. They are the theoretical *quantitative* view of magnitudes that include in a transfinite way the practical approximations of magnitudes by measured values.

Hyperreal numbers *differentiate* the approximation sequences that approximate a magnitude. It can be said that they *are* the distinguished approximation sequences, "undercut" the feature of being archimedean and form infinitely small domains around the real numbers – "monads" as they are called. The hyperreal numbers are – theoretically as in the case of the reals – *qualitative* way to magnitudes. They undercut and exceed what is measurable and demonstrate what is immeasurable of the continuum.

Final remark

Logic shows – by indicating various models that fit into the continuum – that mathematics is theoretical when it puts the real numbers as the continuum. The set-theoretical construction of the hyperreal numbers adds new numbers to the number line. The philosophical opinion on the theoretical character holds both for standard and nonstandard numbers. In our opinion – though not substantially – it is possible to see a theoretical difference between real and hyperreal numbers – in fact the latter need the Axiom of Choice and the former not. Let us stress however that this is the case when one considers the construction of the hyperreal numbers and not when one works with them. They both – real and hyperreal numbers – belong to mathematics. Hyperreal numbers offer as a model of real numbers new instruments and an extended framework for mathematical work and practice.

7 Retrospection

The most beautiful and deepest experience
a man can have is the sense of the mysterious.
It is the underlying principle of religion
as well as of all serious endeavour in art and science. [1]

Albert Einstein

The real numbers (the concept of them has been announced for a long time in the history of mathematics and in the 19th century enforced in a revolutionary way), set theory and the axiomatic method made the enormous development of mathematics possible and cleared the way towards unexpected domains. What are the reasons for this rise? The renewed axiomatic method has canceled the old ontological commitments of mathematics, the infinite became the mathematical tool and the introduction of the domain $\mathbb{R}$ of real numbers decided partially age-old conceptual and methodic problems. The real numbers are the uncontested foundation of large parts of mathematics.

Should nowadays – after the decisions made to release mathematics from the philosophy and the general acceptance of $\mathbb{R}$ – the old methodological and philosophical questions still be asked? Has philosophy of mathematics still any meaning? We believe we have shown that the answer is clearly "yes". Decisions have been made but the questions were not answered. They remained and still remain. We are convinced that one should know the questions and the problems in order to be aware on which foundations we build.

Looking back at various important points of the previous text and summarizing some ideas we want to keep always an eye on the history of mathematics. One should be conscious of the fact that the foundation of today's mathematics, that what we today "believe", has been established only around one hundred years ago – after 2 400 years of "old" mathematics and against considerable resistance. How justified and how meaningful this resistance was, how virulent the old problems are cannot be adequately estimated without mathematical awareness of the historical development that *per se* is in a sense philosophical. It should not be forgotten that mathematics evolved from philosophy. Philosophical problems have not disappeared by the separation of mathematics from philosophy at the turn of the 19th and 20th centuries. As was said they have been decided – in particular by the factum $\mathbb{R}$ – only for the nowadays

1 Das Schönste und Tiefste, was der Mensch erleben kann, ist das Gefühl des Geheimnisvollen. Es liegt der Religion sowie allem tieferen Streben in Kunst und Wissenschaft zugrunde.

https://doi.org/10.1515/9783110468335-008

mathematics. New serious questions deeply concerning the basis of mathematics follow not least from old open questions.

We started this book by looking at the way from rational numbers to the real ones. Several old and new questions have been encountered there – conceptual, methodic, methodological and philosophical ones. Let us recall here the important fundamental questions and problems:

– The question about the concept of natural numbers.
– The problem of the construction of real numbers.
– The problem of the axiomatic method and of the evidence of axioms.
– The problem of the actual infinity.
– The classical problem of the continuum and of magnitudes.
– The question of the infinitely small.

Those questions were met in Chapter 2 when various positions from the history of mathematics and philosophy have been presented. The fundamental problems have been considered in Chapter 3 – in various cases we could refer to some of those positions. There we came to the conclusion that every of those problems disappears when $\mathbb{R}$ is introduced. Every fundamental problem will be decided but none would be solved. The decisions made determine the mathematics nowadays, and $\mathbb{R}$ is the *embodiment* of the decisions.

The new foundations of mathematics – set theory and logic – generate new questions and indicate new problems of mathematical and simultaneously of philosophical character. The meaning of those questions and problems goes far beyond mathematics. Let us mention some of them:

– The problem of the consistency of mathematics.
– The problem of completeness.
– The problem of mathematical truth.
– The new problem of the continuum and of the Axiom of Choice.

The point is the limits of mathematics and beyond that of the theoretical method. In Chapters 4 and 5 we have prepared those questions, indicated studies on them and explained possible answers.

7.1 Setting of real numbers

The algebraic axioms for $\mathbb{R}$ are axioms of an ordered field. They come from the domain of rational numbers. In the case of the completeness axiom the situation is different – it does not come from rational numbers but is transferred from geometry, i.e., from geometrical ideas concerning relations on the line and magnitudes, to the arithmetic. Hence there is no arithmetical obviousness in the case of the completeness axiom.

The *completeness axiom* is in fact an addition from the outside. It is an expression of an idea that appeared in the middle of the 19th century. Independently of each other and at the same time, some mathematicians – for example Dedekind, Weierstraß, Cantor, Heine – came to the idea of a complete domain of genuine *numbers*. There were reasons that forced this: one has calculated for centuries with irrational numbers without knowing how they should be understood. They were magnitudes or proportions of magnitudes. But what were magnitudes and what should intuitive magnitudes do in the domain of pure numbers? Irrational numbers should be genuine numbers. Here helped actually infinite sets – in Dedekind's cuts or in infinite sequences. Now it was known what irrational numbers are. This knowledge has been acquired – and bought – through the infinity that at that time was still uncertain and dark – in fact also nowadays mathematically seen it possesses still considerable transcendence. It was noticed at the end of Chapter 4 and we come back to it.

To be able to express it clearly we abbreviate and simplify the issue: By using axiomatically postulated or constructed real numbers one turned back to the geometry where the completeness came from, and eliminated there magnitudes as well as the intuitive continuum. At their position one put real numbers. Today lines are treated as copies of $\mathbb{R}$. The linear continuum that once was a medium of visualization of numbers became the number line. It has been dissolved into a set of single individual elements called points, numbers or sometimes still magnitudes. The medium has been replaced by that of which it was a medium.

If one wants to understand this revolution that took place one hundred years ago one should take into account this situation. What for millenniums was not thinkable, even "forbidden", has now been thought and done. So far mathematicians thought in a different way. From ancient times they quasi stood on the other side, on the geometrical fundament of the classical continuum and of magnitudes and from that position considered numbers. The change appeared to be too difficult for some of them. Today we consider it obvious to look from the fundament of real numbers at geometry that became a number space. It should be stressed that it was then not "only" the methodic modernization. The replacing of magnitudes and the continuum by putting $\mathbb{R}$ at their place was in fact a *revolution in thinking*. This revolution was so deep and complete that today we can hardly think in another way. For example the problem of incommensurability – if one wants to realize it – became thanks to irrational numbers a pseudo-problem and has only a nostalgic value. The same holds in the case of infinitesimals – it is hardly noticed that they mathematically came back.

7.2 Axiomatic method

Parallel to and maybe in the course of the separating of mathematical thinking from the geometrical "reality", the axiomatic method has been renewed. The renewed *axiomatic method* in mathematics is generally a procedure of avoiding the question about the nature of mathematical objects. The new axiomatic method is so to say the trade mark of modern mathematical presentation. The question “what?” everywhere became the question “how?”.

The *axiomatic method* possessed, one can say, still 120 years ago grip on reality. It came from Euclid who implemented ideas of Plato and Aristotle and who understood the primitive notions with regards to content, hence philosophically. Axioms were – according to the meaning of the appropriate Greek word – “right” requirements whose obviousness were founded on the reality. The new axiomatic method – since Hilbert’s *Grundlagen der Geometrie* and its formal conception of logic – “took leave” from reality. Geometrical intuition has been replaced by axiomatic strictness. Russell describes the new situation in the following way: “Mathematics can be defined as the subject in which we never know what we are talking about nor whether what we are saying is true.” [305, p. 84]. This is the price of taking leave.

The profit is the free interpretability of primitive notions and axioms, the insight into the theoretical interrelations, the clarity of the method. At the end there is something crucial: awareness of the limits of mathematics, i.e., of the axiomatic method and in principle of the theoretical method. They are stated in deep theorems of mathematical logic that we presented in Chapter 5. First of all it is the uncertainty whether mathematics is consistent, then the results on undecidability, incompleteness, non-axiomatizability, undefinability of the concept of truth, and the numerous nonstandard models that appear beside the “genuine” standard ones – last witnesses of the old reality on which the obviousness was formerly founded.

Particularly serious is the mathematical uncertainty of the security of mathematics that actually is proverbial. Since mathematics is in fact set theory – practically all mathematical notions can be reduced to set-theoretical notions – Gödel’s Second Incompleteness Theorem is meaningful and significant for the whole of mathematics: the consistency of mathematics cannot be proved by mathematical means. This makes that mathematics becomes a common science whose consistency should be confirmed by genuine applications or remains speculative. Mathematics is in principle a natural science or philosophy – however it is singular as a science not least by its results about its limits.

7.3 The concept of number

The axiomatic approach covers up the problem of the concept of number, the problem of natural as well as of real numbers. When real numbers are introduced in an axiomatic way then natural numbers simply become special real numbers.

Considering various constructions of real numbers one sees that in principle they go back to natural numbers. The basic question about this fundament on which real numbers are constructed is usually not asked. The natural numbers are treated here as being given by set-theoretical Peano axioms, as quasi real and genuine.

Sets allegedly say what numbers are. In fact, it can be seen on the set-theoretical base how complex and rich of aspects the concept of number is (cf. [20]). However, numbers are reduced to sets, and sets are mathematically only sets of sets. Hence one does not know what sets are and consequently what numbers are. The last attempt to found the *concept of number* in a mathematical way, namely logically – as an element of pure thought – has been done one hundred years ago.

Natural numbers cannot be justified in a purely logical way – as was noticed in Section 2.19. Theorems presented in Chapter 5 have shown that the concept of natural numbers cannot be grasped mathematically, hence in a theoretical-axiomatic way. This was indicated by the existence of nonstandard models as well as by Gödel's First Incompleteness Theorem and by theorems on decidability and on axiomatizability that hold not only for natural numbers but also for set theory and consequently for the whole of mathematics. Those results have meaning not only for mathematics but far beyond it – in particular individual sciences and in philosophy. They confirm what has been said at the end of Section 3.1 about the concept of number: there is and there will be *no* theoretical explanation of the concept of number. This is one of the philosophical meanings of those theorems of the foundations of mathematics.

Psychology and anthropology have been looking for a long time for the concept of number. There dominates the evolutionary standpoint presented by us in Section 2.24. We mentioned interesting empirical investigations (cf. Section 2.24.2) and noticed that the perspective from which numbers and their development are seen there is directed towards experience and consequently narrow. It seems that one does not know or does not take into account enough mathematics – which could give more clarity about the complexity and fundamentality of aspects of the concept of number. Empirical investigations just by being empirical cut off fundamental aspects of the concept of number.

Those researches are based on numerous presuppositions and suppositions concerning evolution and the development of children that in principle must remain hypothetical. Hence empirical statements are highly theoretical. This makes the results – which is sometimes forgotten – to be hypotheses that give however important and detailed insights into the development of numbers. An explanation of the concept of number, even of particular aspects, is not given.

So there remains the philosophical question about the nature of numbers. The varied philosophical approach complements necessarily other methods because they are incomplete or hypothetical. Mathematically the concept of number is a fundamental notion that appears in various forms in the base of mathematics (cf. [20]). Also from the empirical point of view numbers seem to belong to the fundament – via the phenomenon of time in the development and in the fundamental concept of acting (cf. Section 2.24.2). The reflection on the fundamental concepts independently of the domain and discipline is and remains the philosophical task.

It is the matter of important conceptions and backgrounds of the concept of number that appear in the philosophical reflection and that appear in the foreground in particular positions. We followed the development of the concept of number in the history of mathematics (cf. Chapter 2 and Section 3.1) and saw that there were and still are many various views. The complexity of those is restricted due to historical conditions. One finds no genuine answer to the question concerning numbers.

The concept of a natural number seems to be so small, however it lies at the bottom of thinking.

7.4 Infinity, Axiom of Choice and Continuum Hypothesis

The construction of real numbers provide an exact answer to the problem of the *actual infinity*. It says: it is no problem. The actual infinity belongs to the everyday and obvious repertoire of a mathematician – one cannot resign it. It is realized in real numbers. Only the *Axiom of Choice* (AC) – combined with the actual infinity – evokes by its paradoxical consequences some doubts. It is noticed that the Axiom of Choice is relatively consistent with other axioms of set-theoretical foundations of mathematics, and it is maybe ousted that the same holds for its negation. The Axiom of Choice is nowadays widely accepted in research practice even when its usage is sometimes for safety explicitly marked. It is – like the actual infinity – in many domains indispensable.

The acceptance of the Axiom of Choice is in fact "honest" and *consistent* if the actual infinity is accepted. If infinite sets are treated like finite sets then it should be also the case when the principle of choice is considered – choice is in the domain of finite sets something obvious and will be hardly seen as a special principle. However, there appears a very simple concrete problem of the actual infinite that in the case of other handling of the infinite is not immediately seen: in fact, it is something principally else when what is obvious in the finite case is extended into the *infinite*, in particular when one wants to choose elements not only from finite many sets, even "very many", but from infinitely many sets. Choosing as a process conflicts with the concept of a set that omits the process. An infinite closed process is something unthinkable. The *set* of chosen elements whose existence is required by the Axiom of Choice and which remains undetermined unlike other axiomatically required sets covers only the conflict. The range of problems connected with the Axiom of Choice have been considered in Chapter 4.

Also the *Continuum Hypothesis* (CH) has been discussed there. *Set theory* is the mathematical theory of the infinite. Where one is concerned with actually infinite sets there necessarily arises – similarly to the case of finite sets – the question about the number, cardinal number of their elements. The cardinal number of the set of reals is uncountable. And already here, at the first step into the infinite, in the transition from infinite countability of natural numbers to the uncountability of real numbers there arise problems. Is the cardinal number of the set of real numbers the next infinite cardinal number that follows the infinite cardinal number of the set of natural numbers? This is today's *continuum problem*. This new continuum problem is the result of the decision of the old continuum problem. It is the price that must be paid for accepting the decision to treat the continuum as a set.

The situation of the Continuum Hypothesis – i.e., of the statement that the cardinal number of real numbers *is* the next cardinal that follows the cardinal number of natural numbers – is similar to that of the Axiom of Choice. It can be either rejected or accepted. This was shown by Gödel's and Cohen's theorems. If it is rejected then one can choose one, two, three . . . , even "arbitrary many" transfinite cardinal numbers that lie between the cardinal number of $\mathbb{N}$ and the cardinal number of $\mathbb{R}$. This indicates how arbitrary and undetermined the situation beyond the finite is. Once again: the infinite was and still is obscure, even when one deals with it every day.

Both principles – AC and CH – are independent of other axioms of set theory that are regarded as unproblematic. One can speak about Cantorian set theory when both the Axiom of Choice and the Continuum Hypothesis are accepted, as well as about non-Cantorean set theory when both are rejected. This is similar to the situation in geometry where one has Euclidean and non-Euclidean geometries – however, the situation is not the same. In fact, there are two essentially different situations: when one has to choose – briefly speaking – whether two parallel lines intersect or not and when the fundament of the whole mathematics should be chosen. One expects a firm base and clear relations. However, set theory cannot provide us such a base, moreover it cannot do this in important points. When mathematics is set theory then AC and CH are outside mathematics.

The reason of this dilemma is the infinite which is treated as the finite and the continuum which is the base of our natural perception of space and which we disaggregate into elements and treat as a set. Just the infinite makes the choice a problem, and the disaggregation into elements forces the question about the cardinality of the "continuum". Today's continuum problem is a result of the transition from the classical continuum to the real numbers, from mathematics of magnitudes to set-theoretical mathematics. The nature of the infinity and of the continuum was and remains – this is indicated by results of the foundations of mathematics – a secret for mathematics.

7.5 Infinitely small and the continuum

To identify the continuum with $\mathbb{R}$ – which is today widely done – is in fact a far-reaching decision. The identification means that next to real numbers there is no place in the continuum. This was – at the time when $\mathbb{R}$ won mathematically recognition – a decision against the infinitely small, against the *infinitesimal*. Since $\mathbb{R}$ is so present, one has forgotten that this was in fact a matter of a *decision*. The infinitesimal has from the beginnings of mathematics inspired and accompanied the visions of mathematicians. In the infinitesimal calculus of Leibniz and in the later development of mathematics till the 19th century achieved the infinitely small its mathematical full bloom and brought rich fruits. Infinitesimals as well as the irrational numbers were at that time accepted and approved elements of mathematical thinking though it was in both cases to the same extent unknown what they actually are. Today infinitesimals are used in a formal and suggestive way as notations but simultaneously they are dismissed as *façon de parler* or their idea is declared as absurd. In the light of the fact that they have a long history and that they have been used by great mathematicians who thought infinitesimally this seems to be inappropriate. Cauchy for example used both: infinitesimals and limits.

The infinitesimal has been rehabilitated by nonstandard analysis founded in the 1960s. In the justification of it relatively strong algebraic and set-theoretical means are assumed or logical tools are needed. There was neither set theory nor logic when mathematical foundations of analysis were considered. No attempt to justify mathematically the method of infinitesimals has been at that time undertaken; older and perhaps more forward-pressing, maybe also more comprehensible, was the problem of irrational magnitudes that were approximated in processes of measuring. The concept of the limit value became the key notion. It was 150 years ago an essential stimulus for thinking in terms of the actual infinite and for the construction of the real numbers. Infinitesimals have been suppressed by limit values. Since the limit-values mathematics has been governing for a longer time, the chances of the infinitesimal mathematics in elementary teaching are small. The literature in analysis formulated in terms of limit-values mathematics is widely available. Today it determines the mathematical thinking.

However, the influence of nonstandard mathematics on methods, fundamental conceptions and notions cannot and should not be ignored. The nonstandard mathematics has rehabilitated the infinitely small and the approach of the old infinitesimal mathematics. In Chapter 6 we described this in detail. There is still place in the linear continuum. It is allowed to consider that $0.999\ldots$ is smaller than 1 and to treat the continuum as something different than simply a copy of the real numbers.

Just the influence on the concept of the continuum is spectacular. The real standard numbers are only a model, the standard model, among models. There are arbitrarily many nonstandard models. All those models, even "strongly uncountable" have their place in the linear continuum. The real numbers $\mathbb{R}$ are *not* the continuum. Even the conception of a point will be different. Every "real point" contains an infinitesimal continuum as neighborhood, in which there are again points – with neighborhoods – representing new numbers, nonstandard numbers. A point is only a snapshot. It was explicated in Section 3.3.3 on the concept of a point and in Section 3.3.9.

Behind those conceptions and insights that could not be omitted, there is mathematics which is called "nonstandard" – which is apparently too often treated as not being standard.

7.6 Applicability

Still another problem was always present in the philosophical reflection on mathematics. Though applications of mathematics were a permanent and universal factum, there arose the question why it can be applied in the "reality". However, there is certain difficulty, namely what is meant by reality. This is a problem that blows up the frames of philosophy of mathematics. In Section 3.4 naive notions like "real", "concrete", "actual" and "physical" have been used. It should be however noticed that the concept of the reality has been changed by modern physics. Mathematics plays here an important role. Both in microcosmic and in macrocosmic scopes realities seem to be visual only by and through mathematics.

The question about applicability arises again and newly in the case of modern – highly abstract, theoretical and formal – mathematics. Why is the nowadays pure infinite mathematics suitable for the finite "non-pure" world? How do mathematical theories jump over the abyss between formality and reality? How does one go from a shoreless world of sets into the real or concrete continuums of the physical world? Mathematics has reached such a degree of abstraction as never before. The clarity and visuality of concepts and the obviousness of axioms have been lost. The step towards the actual infinity and the world of sets opened for mathematics new and unimagined dimensions – and simultaneously led it into inconceivable and insecure areas. The old certainty can be restored neither by philosophy nor by mathematics themselves.

We have considered the problem of applicability from various philosophical positions and no convincing answer has been found (cf. Section 3.4). It is just the puzzling application itself from which mathematics retrieves the lost reality. Not beauty but application is the daily test of today's mathematics – it is also the everyday *justification* that can be given to it neither by philosophy nor by the foundations of mathematics. Application is the justification of mathematics on probation.

7.7 Theoretical limits

It has been remarked again and again that mathematics has definitely been separated from philosophy as a super-domain when it has transformed the foundations of mathematics to be a mathematical discipline. Clarifying the foundations of mathematics in a mathematical way has its repercussions. On the one hand philosophy remains to be an important discipline where the foundations are considered. On the other hand the mathematical clarifications of foundational problems have consequences and implications in philosophy and in the theory of science. We have mentioned such consequences in the case of the concept of number.

Only in mathematics one finds precise statements about theories. Mathematical logic achieved here something epoch-making. Its aim was to justify mathematics and to prove the infallibility of it. At the turn of the 19th and 20th centuries the atmosphere among logicians was optimistic, even euphoric with respect to those perspectives. The young Russell spoke in 1901 about "great triumphs" and "great hopes": "[...] pure thought may achieve, within our generation, such results as will place our time, in this respect, on a level with the greatest age of Greece" [305, p. 96]. Russell was in our opinion right after all. In fact, logic achieved something *greater* than to build only the sole fundament and to prove the absolute certainty of mathematical knowledge. It has determined with absolute certainty the borders of this knowledge. With the help of logic, mathematics "became conscious".

The philosophical meaning of logical statements about theories and arithmetic reaches in principle far beyond, when applied – with caution – also to natural sciences and philosophical theories as well as closed systems in general. Two fundamental presuppositions of those statements on borders of the theoretical method in general should be respected here: (1) Considered theories should be in principle axiomatically describable. This is a condition and simultaneously a requirement concerning theories formulated already by Hilbert (cf. Section 5.2). (2) The arithmetic of natural numbers should be explicitly or implicitly a part of the discussed theories.

If both these conditions are not satisfied then it is difficult to speak in fact about "theories". In general one should respect limits of the range of scientific statements. In the first case they do not come from firm theories but from not yet ready, diffuse and open experimental domains. This is the negative, alarming case in which, against loud tradition, it is forbidden to spread statements as sure or even conclusive. In the case of statements coming from sources of this type one has to do at best with hypotheses that should be marked as such. In the second case theories are within borders determined by mathematical logic. Then sobriety should be granted to certainty and the range of their statements. Statements of some theories that pretend to explain the world and simultaneously themselves should be treated with doubts.

7.8 Usage of computers

Various important philosophical problems concerning mathematics have only been mentioned in this book. For example the problem of *applying computers* in mathematics and of possible limits of such applications has been only briefly noticed in Section 2.25. Let us also now make only a short remark and indicate the basic problems. In fact, the applications of computers in mathematics, which are growing and becoming more and more comprehensive, generate new questions in philosophy of mathematics.

One can ask to what extent the usage of computers in pure theoretical mathematics is allowed. Can they be used in proofs or – being real calculators based on physical processes – can they be only experimentally allowed? If they were allowed as tools in proofs then mathematics would get a new status. It would lose its *a priori* or ideal character that is nowadays granted to it and would become *a posteriori*, experimental, empirical knowledge. There exist theorems, for example the Four-Color

Theorem, whose proofs are in fact based on the usage of computers. Do such theorems belong to mathematics? Or do they remain in fact unproven hypotheses that have been experimentally conformed?

7.9 What is the philosophy of mathematics and what does it provide?

Finally, before the general conclusion, let us try to consider what we did in fact undertake in this book. We asked the question what "philosophy of mathematics" actually is. We do not know the answer exactly – which partially is the consequence of the lack of knowledge concerning mathematics itself – and therefore we propose an answer by putting the question implicitly. Looking back we tried to describe the place, aspects, subjects, tasks and possibilities of the philosophy of mathematics.

One can speak about philosophy of mathematics as a philosophical discipline only since the 19th century. Before the 19th century philosophical statements about mathematics can be found in marginal notes of philosophers and in philosophical remarks of mathematicians who earlier were – more or less – also philosophers. Mathematics was still treated as being on the fringes of philosophy though it was becoming more and more separate. The real beginning of the philosophy of mathematics as a branch of philosophy were the attempts to justify the analysis – they were made by mathematicians being still philosophers. Mainly one should mention here B. Bolzano.

In the period of the turn from mathematics oriented at geometrical ideas and interpretations to set-theoretical and number-theoretical mathematics, discussions of philosophical problems of mathematics became more intensive among some mathematicians. At that time such conceptions as formalism, logicism and intuitionism have been formulated and developed (cf. Sections 2.19 to 2.22 and [132]). Their aim was the reduction of mathematics to various foundations: in logicism to logic, in formalism to formalization and "finitistic" mathematics, in intuitionism to *a priori* intuition of time perception in natural numbers. The fundamental questions have been formulated that also nowadays are relevant and unanswered – in spite of various mathematical results and decisions that set facts but do not solve problems. Decisions are often accepted by pragmatic reasons and being proposed by authoritative mathematicians dominating in mathematics and in the community of mathematicians.

The discussions have been continued in the new philosophy of mathematics, they have been extended and sophisticated or new positions have been developed. There arose also a tendency that rejected the concentration on those questions and attempts to reduce mathematics in one or another way to various foundations. Mathematics has been considered there from another point of view and emphasis is put on research practice, mathematicians and their community, its cultural and scientific environment and first of all on the development of mathematics and its preconditions. This tendency – oriented towards empirical and evolutionary investigations – provided hardly new contributions to *fundamental questions* which we considered and which determined the philosophy of mathematics in the first half of the 20th century.

Philosophy of mathematics begins in practice by looking carefully and in a reflexive way at mathematics and on what mathematicians are doing. During such observations and the reflection on them there arise questions which one tries to answer. This is the next important step: the outline and formulation of one's own views that can lead to answers.

However, philosophy of mathematics in the strict sense requires more. What is required is to reflect on one's own views, to ask further and deeper questions, to learn other conceptions, to presuppose other philosophical positions, to consider various answers as well as mathematical-philosophical approaches. Philosophy of mathematics requires also that the development of mathematics, of its concepts, methods and problems should be taken into account and that information from the history of mathematics proposed by philosophy should be recognized and considered. In spite of its autonomy and exceptional position mathematics still 140 years ago was rooted by many of its fundamental concepts just in philosophy. Through logic and set theory it turned itself to its foundations. However, classical philosophical, epistemological and ontological questions remained unanswered. New questions arose and are still arising in research practice in mathematics and are generated by approaches and results of the foundations of mathematics. Scientific foundations challenge philosophy *per se*. One of the tasks of the philosophy of mathematics is to consider attempts and methods of the foundations of mathematics from the point of view of philosophical conceptions, to interpret their results philosophically and to discuss their consequences.

Philosophy of mathematics makes it possible to overlook a wide environment of doing mathematics, to recognize origins of mathematics, to develop guidelines for considering the development, perspectives and aims of doing mathematics, to notice philosophical tendencies in mathematics as well as to find and to justify one's own points of view. Philosophy of mathematics provides those possibilities not only in the framework of philosophy itself, not only by taking into account mathematics and its applications, but also in the context of the general cultural, scientific, philosophical and historical framework.

Philosophy of mathematics lies between mathematics and philosophy and is connected with both of them. Nowadays it is practised – after the period of intensive discussions at the beginning of the 20th century – again more and more by mathematicians. This is the consequence of the enormous development of mathematics that can be hardly grasped by philosophers, as well as of the character of the foundations of mathematics that have been mathematized to a large extent. The mathematical discipline that primarily refers to the philosophy of mathematics are the foundations of mathematics. Those connections between them allow to formulate questions and problems as well as answers and solutions in a precise way.

Very important are also the connections between philosophy of mathematics and *history of mathematics*. It is obvious that mathematics as a science has been developed and changed. Structures, conceptions, methods and concepts have been changed and transformed. It is impossible to consider mathematics without taking into account the historical context. And vice versa: the history of mathematics is hardly possible without philosophical considerations. Those close relations between history and philosophy of mathematics can be characterized – paraphrasing Kant – in the following way: history of mathematics without philosophy is blind and philosophy of mathematics without history is empty.

One can distinguish in the philosophy of mathematics epistemological and ontological questions and problems. The most important ontological questions are: what are the objects of mathematics, what is their nature, how do they exist, what are the criteria of existence of mathematical objects? Further, one can ask about the basis of mathematics, about problems of the continuum and about the concept of a number, about the origin and development of mathematical concepts and last but not least about problems connected with the infinite in mathematics.

Among epistemological problems are the following: of what type is the mathematical knowledge, in particular what are in fact axioms and theorems, why are they, if they are, evident, are they of *a priori* or *a posteriori* character? Important is the problem of limits of mathematical knowledge and the question about methods of mathematics. Which methods are appropriate, what is the nature and what are the methods of proving in mathematics? There is the problem of systematizing mathematics, the problem of relations between mathematical theories and of their common basis. What are the criteria of accepting mathematical results? The problem of the dynamics of the development of mathematics appears as well as the question about the place of mathematics in culture and about the relations between mathematics and other sciences.

Ontological and epistemological aspects and problems are connected. Decisions and solutions in the ontology of mathematics have epistemological consequences and vice versa.

Finally, notice that the philosophy of mathematics that attempts to consider the phenomenon of mathematics as a science and to determine and discuss its foundations, describes on the one hand mathematics as it is done and developed and on the other hand should fix, reconsider and evaluate methodological norms. Hence philosophy of mathematics has both descriptive and normative character.

7.10 Evidence and transcendence

An important and fundamental problem of the philosophy of mathematics that has not been explicitly discussed so far in our book and that we want to formulate now concluding our considerations is the problem of evidence about which one speaks willingly and often (we sometimes did it as well). This question is important for the philosophy of mathematics especially in connection with axioms when they are not treated only as simple conventions or meaningless expressions which can be arbitrarily assumed. Even if axioms are treated nowadays in a strictly formal way, a concrete mathematician expects a certain meaning (with regard to content) that would give sense to doing mathematics.

In the case of evidence of meaningful statements it is the matter of accordance with the reality. On the one side there is the reality, on the other are thinking and thoughts expressed in statements. Often the word "intuition" is used in this context. Intuition is the fundament of logical *insight* as well as of insight based on experience. It is said that intuitive knowledge is a kind of direct perception in which the separation of object and subject, of object and thinking has been canceled. It seems to form the basis of mathematics through its axioms. It becomes clear that the question about evidence leads to transcendent parts of philosophy and that one touches here problems that seem to go far beyond philosophy.

The evidence of axioms which one assumes – in spite of an in principle formal approach – denies and overcomes in the mathematical (research) practice the widespread image of mathematics that is in fact formalistically stamped. Knowledge provided by such formalistically understood mathematics provides no evidence, it provides only models and has nothing to do with the reality – moreover this is not intended at all. Against that, mathematical results – in spite of all metamathematical restrictions concerning the range of mathematics – possess, thanks to methods that should withstand logical tests, unsurpassed certainty and evidence. The evidence of mathematical results is – provided mathematics is understood not only as a game or experiment – the heritage of the evidence of axioms that lie at the base of mathematical theories and it vitalizes mathematical theories.

We see with admiration the edifice of mathematics as a fascinating, certain and unique science consisting of many disciplines, countless subdisciplines and universal applications that nobody is able to overview. However, there is also the other side that reminds to be modest in spite of all mathematical enthusiasm – and generates new amazement.

The physicist Albert Einstein speaks in the motto to this chapter about the sense of the mysterious and its fascination. Also a mathematician stands before mysteries that he tries to investigate thoroughly and that in fact he cannot fathom. It is the matter of the foundations of his research – and again about the range of mathematics. Among the great fathomless mysteries are that of the space and time, of the continuum with which both mathematics as well as physics are confronted. We face the mystery of the concept of number and see the problems in connection with the reals. We face the mystery of evidence being (despite formalization) a source of all axioms and we stay helpless before the puzzle of the applicability of theories. There is the small mystery of a geometrical point that we cannot disclose but grasp only axiomatically. We speak about sets and elements that finally remain unexplained. There is the great mystery of the infinity that – though mathematics has attempted to penetrate – remains unattained. The fundamental elements of mathematical work remain transcendental.

There is also the mystery of mathematics itself, of its completeness, its truth, its consistency – mathematics learns about its indissolubility from logic. Mathematics is the domain full of beauty and truth – and nevertheless it stands before something deeply mysterious.

> "Who never had this experience seems for me to be, if not dead, then at least blind."[2]

With these words Albert Einstein concludes the sentence from the motto quoted at the beginning of this retrospection.

It was one of the aims of this book to ask questions and present unsolved and unsolvable problems that are connected with mathematics and are at the basis of it. We will not be able to uncover them – and we *know* with mathematical certainty that we cannot do it. This is in fact important and without a counterexample in other disciplines. We do not know what mathematics is; however, we have experienced and learned a lot about it and maybe look now at it in another way and through new eyes.

2 Wer dies nicht erlebt hat, erscheint mir, wenn nicht wie ein Toter, so doch wie ein Blinder.

Biographies

Aristotle (384–322 BC). Born in the city of Stagira, Chalkidice, on the northern periphery of Classical Greece. He was the son of the personal physician of the king of Macedon. At seventeen or eighteen years of age, he joined Plato's Academy in Athens and remained there until the age of 37 (ca. 347 BC) first as a pupil and later as a teacher and researcher. In 343–340 BC he was the teacher of Alexander the Great. In 335 BC he returned to Athens and established his own school known as the Lyceum. Many researchers were active in it. Aristotle organized the scientific life and founded the first European library. His school followed the patterns of Plato's Academy, however the scope of disciplines cultivated there was much broader, from natural sciences till humanities. Aristotle lead his school away from platonism towards a mental kind of empiricism. He himself not only studied almost every subject possible at the time, but made significant contributions to most of them. He was one of the greatest and most influential personalities of the history of science.

Bernays, Paul (1888–1977). Born in London he spent his childhood and youth in Berlin and studied mathematics, philosophy and theoretical physics in Berlin and Göttingen. He obtained his doctor's degree in 1912 in Göttingen – the subject of his doctoral dissertation belongs to the analytical number theory. In the same year he completed his habilitation in Zürich – his *Habilitationsschrift* was devoted to the theory of functions. In 1912–1917 he was appointed as *Privatdozent* at the university in Zürich. In 1917 he became the secretary and assistant of D. Hilbert in Göttingen and gave lectures there. On the base of the work on axiomatics of the propositional calculus of *Principia Mathematica* by A. N. Whitehead and B. Russell he was awarded *venia legendi*. In 1922 he became extraordinary professor without tenure. In 1933 he was dismissed from this post because of his Jewish ancestry. He moved with his family to Switzerland where he was granted 1939 the *venia legendi* at the ETH Zürich and 1945 appointed as professor. In collaboration with Hilbert he developed the formalism, wrote the monograph *Grundlagen der Mathematik* [179] and published several papers in the philosophy of mathematics [31].

Bolzano, Bernardus Placidus Johann Nepomuk (1781–1848). Born in Prague into a German-speaking family. He entered the University of Prague in 1796 and studied mathematics, philosophy and physics and obtained the doctor's degree in philosophy in 1804. Starting in 1800, he also began studying theology, becoming a Catholic priest in 1804. He was appointed as the new chair of philosophy of religion at Prague University in 1806. In his lectures he criticized the Austrian constitution and represented pacifistic and socialistic views. Therefore, in 1819 he was dismissed from the university by emperor Franz I. He devoted himself entirely to scientific work. The result was the book *Wissenschaftslehre* [41] in 1837, containing an extensive system of logic which

https://doi.org/10.1515/9783110468335-009

should be an introduction to the system of mathematics. The latter was planned to be presented in *Größenlehre* – the work on it was disrupted by his death in 1848. The importance of Bolzano's works has been rediscovered long after his death.

Boole, George (1815–1864). Born in Lincoln, Lincolnshire, England. He had a primary school education, and received lessons from his father, but had little further formal and academic teaching. He studied as autodidact works of great mathematicians and at the age of nineteen published a spectacular work about Newton. Since 1837 he worked for the newly established *Cambridge Mathematical Journal.* In 1849 he was appointed as the first professor of mathematics at Queen's College, Cork (now University College Cork (UCC)) in Ireland. In 1857 he was elected a Fellow of the Royal Society. He propagated as the first a purely formal approach to logic which he treated as the foundation of the pure mathematics, and developed the algebra of logic.

Brouwer, Luitzen Egbertus Jan (1881–1966). Born in Overschie (the Netherlands). In 1897–1904 he studied mathematics at the University of Amsterdam where in 1907 he was granted the doctor's degree. There in 1909 he was appointed as associate professor, in 1912 as extraordinary professor of mathematics and in 1913–1951 as ordinary professor. He worked in the field of topology, where he obtained significant results, as well as in the foundations of mathematics. He was the founder and main representative of the intuitionism and worked on the intuitionistic reconstruction of mathematics.

Cantor, Georg Ferdinand Ludwig Philip (1845–1918). Born in St. Petersburg. His father came from Denmark. In 1856 the family moved to Frankfurt/Main. In 1862 Cantor began his studies in Zürich and a year later moved to Berlin where he studied mathematics, physics and philosophy. There his teachers were among others Kummer, Weierstraß and Kronecker. The greatest influence on him had Weierstraß. In 1867 he obtained the doctor's degree in Berlin and in 1869 he completed his habilitation at the University of Halle where he accepted the position of *Privatdozent.* There in 1872 he was appointed as extraordinary and in 1879 as ordinary professor. The first works of Cantor were devoted to the number theory, the theory of trigonometric series and to the theory of functions. In 1874–1897 he published his famous works that established the completely new theory – set theory.

Chwistek, Leon (1884–1944). Born in Cracow. He studied philosophy and mathematics at the Jagiellonian University and as an unenrolled student at the Academy of Fine Arts in Cracow under the supervision of Józef Mehoffer. In 1906 he obtained his doctorate at the Jagiellonian University. From 1906, he taught mathematics in the Jan Sobieski Gymnasium (which he had completed) on and off for 20 years. In the years 1908–1909 he continued philosophical studies in Göttingen (where he listened to Hilbert's lectures). In 1910 he stayed in Vienna, being fascinated with Venetian Renaissance painting. In 1913–1914 he studied drawing in Paris. In 1914–1916 he fought in the First Brigade of the Polish Legions. From 1922 he lectured on mathematics to students in natural sciences at the Jagiellonian University. He presented his *Habilitationsschrift* in 1928 to this University, and in 1930 received the Chair of Mathematical Logic at the Mathematical-Natural Sciences Faculty at the Jan Kazimierz University of Lvov. In June 1941 he left Lvov together with the Russian Army. He settled in Tiflis (Tbilisi), where he taught mathematical analysis. From 1943, he lived in Moscow. He was active in the Association of Polish Patriots. He died in Moscow on August 21, 1944 (some sources give the date of August 27, 1944, in Barwicha near Moscow). In mathematical logic Chwistek is known first of all for his version of the theory of types. He was also active as a painter and an art historian.

Curry, Haskall Brooks (1900–1982). Born in Millis, Massachusetts, USA. He entered Harvard University in 1916 to study medicine but switched to mathematics before graduating in 1920. After two years of graduate work in electrical engineering at MIT, he returned to Harvard to study physics, earning an MA in 1924. In 1928 he moved to Göttingen where he could work with Heinrich Behmann and Paul Bernays. Curry was supervised by David Hilbert and worked closely with Bernays, receiving a Ph.D. in 1930 with a dissertation on combinatory logic. In 1929–1966 he taught at the Pennsylvania State College (since 1953 Pennsylvania State University). After retiring from Penn State in 1966, Curry accepted a position at the University of Amsterdam. In 1970, after finishing the second volume of his treatise on the combinatory logic, Curry retired from the University of Amsterdam and returned to State College, Pennsylvania.

Dedekind, Julius Wilhelm Richard (1831–1916). Born in Braunschweig, Germany. He studied in Göttingen where he was the last student of C. F. Gauß. In 1852 he obtained the doctor's degree and went to Berlin where in 1854 he was awarded the habilitation. After that he returned to Göttingen to teach as a *Privatdozent*. In 1858 he began teaching at the Polytechnic School in Zürich (now ETH Zürich). When the Collegium Carolinum was upgraded to a Polytechnic School in 1862 (since 1877 Technische Hochschule (Institute of Technology)) Dedekind returned to his birthplace Braunschweig, where he spent the rest of his life, teaching at the Institute. He retired in 1894. Dedekind was elected to the Academies of Berlin (1880) and Rome, and to the French Academy of Sciences (1900). He received honorary doctorates from the universities of Oslo, Zürich, and Braunschweig. Dedekind was one of the founders of the modern algebra, worked in the theory of groups and the theory of ideals – he introduced the concepts of an ideal, ring and field – as well as in the theory of numbers. Beside Weierstraß and Cantor he belonged to the main representatives of the new direction of research based on the works of A. Cauchy, C. F. Gauß and B. Bolzano – its aim was to provide solid foundations for mathematical concepts and to clarify them.

Descartes, René (1596–1650). Born in La Haye en Touraine (now Descartes, Indre-et-Loire), France. He was educated in the Jesuit College Royal Henry-Le-Grand at La

Flèche (1604–1612). In 1615–1616 he studied law and medicine in Poitiers and till 1618 mathematics in Paris. In 1618–1629 he spent time in the atmosphere of court life, traveled a lot and was even a soldier in the Thirty Years' War. In 1629 he settled in the Netherlands and devoted himself entirely to the scientific activity. In 1649 he was invited by Queen Christina of Sweden to her court to organize a new scientific academy and to tutor her in his ideas about love. The northern climate turned out to be too severe for him – he caught a cold which quickly turned into a serious respiratory infection. Descartes died in 1650. Descartes is treated as the founder of modern European philosophy. Traces of his ideas can be seen in the whole development of it. He was also a creative mathematician. He formulated the basic ideas of the analytic geometry connecting geometry and algebra. He introduced also many symbols into mathematics, some of them are used till today.

Euclid (ca. 365–ca. 300 BC). Sometimes called Euclid of Alexandria. Very few is known about his life. He lived in Alexandria at the time of Ptolemy I. He Studied in Athens at the Academy and inclined the whole life towards platonism. Only a part of his works survived. A special place among them has *Elements*. This work is a survey of results of Greek mathematicians from the previous three hundred years and provided a fundament for the further development of mathematics. It established also a paradigm of mathematics and scientific expositions in general that functioned till the 19th century.

Fraenkel, Adolf Halevi Abraham (1891–1965). Later: Halevi Abraham Fraenkel. Born in Munich, Germany into a merchant family. He studied mathematics at the University of Munich, University of Berlin, University of Marburg and University of Breslau. He obtained the doctor's degree in 1914 in Marburg (the supervisor was Kurt Hensel). In 1916 (being on furlough from the French front line) he completed his habilitation in Marburg, was promoted to a professor in 1922 and later on acceded to the chair of Kurt Hensel. After leaving Marburg in 1928, Fraenkel taught at the University of Kiel for a year. He then accepted a position at the Hebrew University of Jerusalem, which had been founded four years earlier, where he spent the rest of his career. He became the first Dean of the Faculty of Mathematics, and in 1938–1940 served as Rector of the University. Fraenkel worked in the area of algebra. However, he is known first of all by his contributions to set theory (problem of the Axiom of Choice), by his book *Einleitung in die Mengenlehre* (1919) and by improving Zermelo's set of axioms by the Replacement Axiom (1921). The completed system of axioms of set theory is today known as ZF set theory and became the standard one.

Frege, Friedrich Ludwig Gottlob (1848-1925). Born in Wismar, Mecklenburg-Schwerin (today part of Mecklenburg-Vorpommern). He studied mathematics in Jena and Göttingen. In 1873 he attained his doctorate in Göttingen and in 1874 completed the habilitation in Jena where he took the position of *Privatdozent*. In 1879 he was appointed as extraordinary and in 1896 as ordinary professor of the University in Jena. In 1918 he retired and moved to Bad Kleinen close to Wismar where he spent the rest of his life. He worked in the formal logic, the foundations of mathematics and in the logical semantics. He was one of the founders and one of the main representatives of the modern mathematical logic. The enormous significance of his works have not been recognized at the beginning. His *Begriffsschrift* (1879) is treated today as the beginning of mathematical logic.

Gauss (Gauß), Carl Friedrich (1777–1855). Born in Brunswick (Braunschweig), in the Duchy of Brunswick-Wolfenbüttel (now part of Lower Saxony, Germany), as the son of poor working-class parents. He was a mathematical prodigy. His abilities were discovered by a teacher in the primary school. Gauß's intellectual abilities attracted the attention of the Duke of Brunswick, who sent him to the Collegium Carolinum (now Braunschweig University of Technology), which he attended from 1792 to 1795, and to the University of Göttingen from 1795 to 1798 where he studied mathematics, physics and philology. At the age of seventeen he discovered the construction by compass and straightedge of the regular heptadecagon. In 1799, on the base of the dissertation providing the first complete proof of the fundamental theorem of algebra, he obtained his doctor's degree at the University in Helmstedt. In 1801 he published the famous *Disquisitiones arithmeticae* which had a significant influence on the development of number theory. In 1801 he predicted a position for Ceres (just about a year after its first sighting). In 1807 he was appointed as professor of astronomy and director of the astronomical observatory in Göttingen, a post he held for the remainder of his life. He rejected proposals of appointments from Berlin, Vienna, Paris and St. Petersburg. His contributions to mathematics, physics and astronomy were enormous. Many of his results have not been published. Many of them were discovered in his diaries (found in 1898). His works mark the beginning of the modern mathematics. He is called the Prince of Mathematics (*Mathematicorum Princeps* – as written on the medallion of the king of Hanover, 1856).

Gödel, Kurt Friedrich (1906–1978). Born in Brünn, Austria-Hungary (now Brno, Czech Republic) into an ethnic German family. In 1924 he began the studies of physics at the University of Vienna, a year later changed to mathematics. Since 1926 he took part in meetings of the Vienna Circle, however he never accepted fully its ideas. In 1930 he obtained the doctor's degree – his supervisor was Hans Hahn – and completed his habilitation in 1932. In 1933 he took the position of *Privatdozent.* In the years 1933–1934, 1935 and 1938–1939 Gödel worked in the USA in the Institute for Advanced Study (I. A. S.) in Princeton. In 1940 he emigrated to the USA and became an ordinary and in 1946 a permanent fellow of I. A. S. In 1953 he was appointed as professor there. Gödel worked first of all in the domain of mathematical logic and the foundations of mathematics as well as – especially in the last years of his life – in the theory of relativity.

His results belong to the most famous contributions to logic. He was interested also in philosophy, in particular in the philosophy of mathematics.

Heyting, Arend (1898–1980). Born in Amsterdam. In 1916–1922 he studied mathematics at the University of Amsterdam where he obtained the doctor's degree in 1925. Since 1923 he worked as a mathematician in Enschede. In 1936 he took the position of Privatdozent in Amsterdam, in 1937 of a lecturer and in 1948 was appointed as professor of mathematics and the philosophy of mathematics. In 1942 he became member of Koninklijke Nederlandse Akademie van Wetenschapen (Royal Netherlands Academy of Arts and Sciences). He died in Lugano, Switzerland. Heyting was a student of Luitzen Egbertus Jan Brouwer, continued his ideas of intuitionism and developed a system of intuitionistic logic.

Hilbert, David (1862–1943). Born in Königsberg (Province of Prussia, Kingdom of Prussia). In 1880–1884 he studied at the University of Königsberg – with the exception of the second semester spent at the University of Heidelberg. In 1885 he was granted the doctor's degree. In 1886 he completed his habilitation and took the position of *Privatdozent*. In 1892 Hilbert was appointed as professor in Königsberg and in 1895 as professor at the University in Göttingen where he worked till the end of his life. Hilbert was an exceptionally ingenious and creative mathematician. He was active in many domains of mathematics and obtained important results there, in particular in invariant theory, the foundations of geometry, the calculus of variations, integral equations, number theory and the foundations of mathematics. He was the founder and the main representative of the formalism in the philosophy of mathematics.

Husserl, Edmund (1859–1938). Born in Proßnitz, a town in the Margraviate of Moravia (then in the Austrian Empire, today called Prostějov in the Czech Republic). He studied mathematics, physics, philosophy and astronomy in Leipzig and Berlin by Karl Weierstraß and Leopold Kronecker. In 1881 he took part in lectures of the philosopher and psychologist Franz Brentano (1838–1917) in Vienna who influenced strongly his early works, his later career and his later philosophy. In 1883 Husserl obtained the doctor's degree – the doctoral dissertation was devoted to the calculus of variations. In 1886 he completed his habilitation at the University of Halle – the subject of his *Habilitationsschrift* entitled *Über den Begriff der Zahl* belonged to the area of psychology. In 1887 he took the position of *Privatdozent* at the University in Halle and, drawing on his prior studies in mathematics and philosophy, proposed a psychological context as the basis of mathematics. His first philosophical main work *Logische Untersuchungen* (Logical Investigations) (1900/1901) was written at the turn of the centuries. Husserl became to be a significant representative of antipsychologism. In 1901–1916 he was a professor at the University in Göttingen and in 1916 he moved to the University in Freiburg im Breisgau. He was professor there till his retiring in 1928. Husserl was the founder of the phenomenology which should found the philosophy as a science. He developed ideas in the logical semantics where he introduced the concept of a semantical category. His later main philosophical works are: *Ideen zu einer reinen Phänomenologie und phänomenologischen Philsopohie* (1913), *Cartesianische Meditationen und Pariser Vorträge* (1931), *Die Krisis der europäischen Wissenschaften und die transzendentale Phänemenologie* (1936).

Kant, Immanuel (1724–1804). Born in Königsberg, Prussia (since 1946 the city of Kaliningrad, Kaliningrad Oblast, Russia). In 1740–1755 he studied physics, theology and philosophy at the University in Königsberg. In 1755 Kant completed his habilitation and took the position of *Privatdozent*. Only in 1770 when his works were already widely known he was appointed as professor and became the chairman of the chair for logic and metaphysics. He taught till 1796. The whole life he was of a weak health. This was one of the reasons of never leaving the borders of Prussia (he even left the city of Königsberg very seldom) and of rejecting the proposals of appointments at universities in Erlangen (1769), Jena (1770) and Halle (1778) despite the fact that Königberg was then a provincial town without a particular cultural life, far away from the intellectual centers of Germany. The university in Königberg owes its fame to Kant.

Kronecker, Leopold (1823–1891). Born in Liegnitz (today Legnica, Poland). In 1841–1845 he studied mathematics at the universities in Breslau (today Wrocław, Poland), Bonn and Berlin where he later became a student and a friend of E. Kummer, his earlier teacher of mathematics in the gymnasium. After finishing his studies he went back to his hometown and focused on business. However, he continued to study mathematics as a hobby and corresponded with Kummer. Due to his business activity, Kronecker was financially comfortable, and thus he could return to Berlin in 1855 to pursue mathematics as a private scholar. He was elected a member of the Berlin Akademie der Wissenschaften in 1861. In 1868 he rejected the appointment at the University of Göttingen. In 1883 he was appointed as professor in Berlin as a successor of E. Kummer. He worked in algebra and number theory, in particular in the theory of quadratic forms, elliptic functions and the theory of groups. He was one of the representatives of the trend of "arithmetization" of analysis, rejected the attempts to found analysis on the base of set theory and battled Cantor's set theory.

Lakatos, Imre (1922–1974). Born Imre (Avrum) Lipschitz into a Jewish family in Debrecen (Hungary). Avoiding Nazi persecution of Jews he changed his name to Imre Molnár. His mother and grandmother were murdered in Auschwitz. He changed his surname once again to Lakatos (Locksmith) in honor of Géza Lakatos. Imre Lakatos studied mathematics, physics and philosophy in Debrecen and in 1948 was awarded the doctor's degree. He also studied at the Moscow State University under the supervision of Sofya Yanovskaya in 1949. After the Second World War he was an active communist and a high official at the ministry of education in Budapest. He was imprisoned on charges of revisionism from 1950 to 1953. After his release he returned to academic life, doing mathematical research and translating mathematical books into Hungarian. In 1956 he left Hungary and emigrated to England where he came under the influence of Karl Popper. He received a Ph.D. in philosophy in 1961 from the University of Cambridge. He became known as a specialist in the theory of science by his work *Proofs and Refutations: The Logic of Mathematical Discovery* [220] in 1963 and 1964 published as a series of four articles in *The British Journal for the Philosophy of Science* and later as a book.

Laugwitz, Detlef (1932–2000). Born in Breslau (today Wrocław, Poland). Since 1949 he studied mathematics, physics, and philosophy in Göttingen where he obtained the doctorate in 1954 on the base of his dissertation in differential geometry. Until 1956 he worked as a scholarship holder in the Mathematical Research Institute of Oberwolfach and Erlangen. In 1958 he became a lecturer at the Technical University of Munich, where he obtained his habilitation. In 1958 he moved to the Technical University of Darmstadt, where in 1962 he became a professor, and remained until his retirement. From 1976 to 1984 he was a visiting professor at Caltech. He became known by his contributions to the nonstandard analysis. In 1958 he published together with C. Schmieden a work devoted to the infinitesimal analysis [315] – still before A. Robinson [302] who is treated as the founder of the nonstandard analysis.

Leibniz, Gottfried Wilhelm (1646–1716). Born in Leipzig. At the age of fifteen he began to study law, mathematics and philosophy in Leipzig and Jena. In 1666, the University of Leipzig turned down Leibniz's doctoral application and refused to grant him a Doctorate in Law, most likely due to his relative youth. So he enrolled in the University of Altdorf where he earned in 1666 his license to practice law and his doctorate in law. He was offered an appointment at a university but rejected the proposal and did not hold an academic position his whole life. He entered the service of the Elector of Mainz. In 1672 he went in a diplomatic mission to Paris – he used the possibility to study the writings of Descartes and Pascal and to meet some scientists, in particular Christiaan Huygens, Nicolas Malebranche and Antoine Arnauld. In 1676 he was appointed by John Frederick of Brunswick as Privy Councillor (Hofrat) and scientific librarian (Bibliothekar) in Hanover. Leibniz was extremely versatile and active in many various domains, among others in mathematics (he developed independently of Isaac Newton the differential and integral calculus), natural sciences – in particular in mechanics – linguistics, logic, philosophy, theology (he was interested first of all in the problem of unifying various Christian denominations, hence in the ecumenic theology as well as in the theodicy) and in the law. He is counted as the last polymath in cultural history.

Leśniewski, Stanisław (1886–1939). Born in Sierpuchów, east of Moscow. After having graduated from a gymnasium in Irkutsk he studied in Germany (among other places in Munich; certain sources speak about his studies in Zurich, Heidelberg and Leipzig).

In 1912 he obtained his doctor's degree written under the supervision of Kazimierz Twardowski at the Jan Kazimierz University of Lvov. Then he stayed in France, Italy and in Saint Petersburg. In the years 1913–1915 he lived in Warsaw. He was a member of the Social Democracy of the Kingdom of Poland and Lithuania (SDKPiL) and it was Feliks Dzierżyński who accepted his candidature to the party. In 1915–1918 he taught mathematics in Polish secondary schools in Moscow and lectured in the so-called Higher Course. In Moscow he got to know Wacław Sierpiński, who introduced him to the group of Nikolai N. Luzin who dealt with set theory. In 1918, he returned to Warsaw where he was given the Chair of Philosophy of Mathematics at the University of Warsaw in 1919. He changed his social-political views, became conservative and anti-Semitic (but it should be stressed that he was not involved in any anti-Semitic disturbances of the 1930s). Together with Jan Łukasiewicz he co-founded the Warsaw School of Logic. He dealt with logic and the foundations of mathematics. Attempting to strengthen mathematics and eliminate antinomies, he created his own logical systems: propositional calculus (protothetic), calculus of names (ontology) and set theory in the collective sense (mereology). He died in Warsaw.

Lobachevsky, Nikolai Ivanovich (1792–1856). Born in or near the city of Nizhny Novgorod in the Russian Empire (now in Nizhny Novgorod Oblast, Russia). In 1807–1811 he studied at the University in Kazan. He was influenced there by Professor Johann Christian Martin Bartels, a former teacher and friend of German mathematician Carl Friedrich Gauß. He received a master's degree in physics and mathematics in 1811. In 1814, he became a lecturer at Kazan University, in 1816 he was promoted to associate professor, and in 1822, at the age of 30, he became a full professor teaching mathematics, physics and astronomy. In 1820–1821 and 1823–1825 he was dean of the Faculty of Mathematics and Physics, and in 1827–1846 he was the rector of Kazan University. In 1841 by the recommendation of Gauß he became a member of Göttinger Wissenschaftliche Gesellschaft. At the end of his life he was nearly blind and unable to walk. He died in poverty in 1856. His main achievement is the development (independently from János Bolyai) of a non-Euclidean geometry. His magnum opus *Geometriya* was completed in 1823 but was not published in its exact original form until 1909. Other of his works are: *New Foundations of Geometry* (1835–1838), *Geometrical Investigations on the Theory of Parallels* (1840) and *Pangeometry* (1855). In 1895 the Kazan Mathematical and Physical Society established the Lobachevsky Award for the excellent results in geometry, in particular in the non-Euclidean geometry. It was granted among others to D. Hilbert for his work *Grundlagen der Geometrie.*

Łukasiewicz, Jan (1879–1956). Born in Lvov. He studied philosophy under the guidance of Kazimierz Twardowski at the University of Lvov. He obtained his doctor's degree in 1902, and then he continued studies, mainly in Germany and Belgium. In 1906 he presented his *Habilitationsschrift* to the University of Lvov and became a docent of this University. In 1911 he was promoted to an extraordinary professor ("professor extraordinarius"). In 1915 he was invited to hold one of the chairs of the newly restored University of Warsaw. He was a professor of this university till 1944. In 1920 he was elected to the Chair of Philosophy at the Faculty of Mathematics and Natural Sciences. In the years 1918–1920, he was head of the Department of Higher Schools in the Ministry of Religious Denominations and Public Enlightenment, and in 1919 he directed this ministry in the cabinet of Ignacy Jan Paderewski. He was twice elected the rector of the University of Warsaw (1922-1923 and 1931-1932). During World War II he worked in the municipality and taught in clandestine courses. In 1944 with the help of Heinrich Scholz, a professor of the University of Münster, he left Poland. His destination was Switzerland. However, the bombardments of Germany forced him to change his plans. After staying in Münster, he went to Belgium. In 1946 he was appointed to the Chair of Logic at the Royal Irish Academy in Dublin, where he spent the last decade of his life. Łukasiewicz dealt with philosophy (especially in the first period of his scientific activities, i.e., between 1902 and 1918) and logic (mainly in the years 1918–1956). He was one of the first Polish scientists who dealt with mathematical logic in a professional way. He was also the first Polish lecturer on mathematical logic as a separate subject. With Stanisław Leśniewski he co-founded the Warsaw School of Logic.

Mill, John Stuart (1806–1873). Born in London. He was the son of James Mill, the Scottish philosopher (radical utilitarianism), historian and economist. Mill was educated by his father in the ancient languages Greek and Latin as well as in arithmetic, physics and astronomy – nearly isolated from the external world. At the age of fourteen he stayed a year in France. In particular in Montpellier he attended the winter courses on chemistry, zoology, logic of the Faculté des Sciences, as well as took a course in higher mathematics. His scientific works dealt with philosophy, logic and economy. He was also active in the political and administrative life. In 1823 he founded in London the Utilitarian Society. In 1823–1858 he worked at the British East India Company. Between 1865 and 1868 Mill served as Lord Rector of the University of St. Andrews. During the same period he was a Member of Parliament for City and Westminster (he became the first person in the history of Parliament to call for women to be given the right to vote). During his whole life he had no academic position and was in fact a private researcher. Mill is one of the most important representatives of the new empiricism; he developed first of all its methodology.

Mostowski, Andrzej (1913–1975). Born in Lvov on November 1, 1913. In the years 1923–1931, he attended the Stefan Batory Gymnasium in Warsaw. In 1931 he began mathematical studies at the University of Warsaw. There Alfred Tarski and Adolf Lindenbaum exerted the greatest influence on him. He completed his studies in 1936. In the academic year 1936/1937 he stayed in Vienna and in the year 1937/1938 in Zurich where he listened to the lectures by Kurt Gödel, Hermann Weyl and Wolfgang Pauli. After returning to Warsaw he defended his doctoral dissertation, written under the supervision of Kazimierz Kuratowski (Tarski, who was in fact Mostowski's supervisor, could not fulfil this function since he was not a professor then) in February 1939. Earlier in January 1939 he had begun working for the State Meteorological Institute in Warsaw. During the war he worked as an accountant. In the years 1942–1944, he was engaged in underground teaching at the University of Warsaw. He presented his habilitation thesis to the Jagiellonian University in 1945. From 1946 until his death, he worked at the University of Warsaw: from 1947 as an extraordinary professor and from 1951 as a full professor. He died in Vancouver (Canada). Mostowski dealt with mathematical logic and the foundations of mathematics, in particular set theory, recursion theory, model theory, logical calculi and proof theory.

Neumann, Johann von (1903–1957). Born in Budapest as Janoś von Neumann to a wealthy aristocratic family. After the emigration to the USA he was called John von Neumann. He was a prodigy – his mathematical talent has been discovered early. His first mathematical paper he published at the age of seventeen. In 1921–1923 he studied chemistry at the University of Berlin and at ETH in Zürich. In Zürich he attended lectures by Hermann Weyl and George Polya, and in 1923 he graduated from ETH. At the same time, von Neumann also entered Pázmâny Péter University in Budapest as a Ph.D. candidate in mathematics. He was granted the doctoral degree in 1926 – in his doctoral dissertation he proposed a new axiomatization of Cantor's set theory. He then went to the University of Göttingen to study mathematics under David Hilbert. His habilitation was completed in 1927 and he started his lectures as a *Privatdozent* at the University of Berlin in 1928. By the end of 1927, von Neumann had published twelve major papers in mathematics, and by the end of 1929, thirty-two papers, at a rate of nearly one major paper per month. In 1933 he was offered a lifetime professorship at the faculty of the Institute for Advanced Study and remained a mathematics professor there until his death. Since 1943 he served as a consultant primarily for the Navy, which in turn led to his involvement in the Manhattan Project. Von Neumann was an unconventional and extremely quick thinker. He worked in many mathematical domains, in particular in proof theory, set theory, where his contributions were very important and fundamental, game theory, which he founded as a mathematical discipline, functional analysis, ergodic theory, operator theory, measure theory, geometry (he founded the field of continuous geometry), lattice theory, mathematical economics, statistics, quantum mechanics (he was the first to establish a rigorous mathematical framework for it) and quantum logic as well as in the computer science. He died on cancer, which was probably the result of radioactive contamination during experiments in Los Alamos.

Nicholas of Cusa (1401–1464). Originally N. Krebs, Crypffs or Cryfftz, in Latin: Nicolaus Cusanus. Born in Kues (Latinized as Cusa) in southwestern Germany. He studied mathematics, astronomy and canonical law in Heidelberg and Padua (1418–1423) where he obtained the doctor's degree of canonical law in 1423. In 1425 he returned to Kues and studied theology (1425–1430) in Cologne. He was engaged in the church conflicts of that time, at the beginning a reformer on the side of the conciliarism but later switched to the pope's side. In 1448 he was nominated a cardinal and in 1450 a bishop of Brixen (today in south Tyrol). His philosophy and theology were influenced by Pythagoreanism and Neoplatonism as well as by nominalism, mysticism of J. Eckhart, Kabbalah and humanistic trends. He was one of the main representatives of the renaissance of pantheism. Nicholas of Cusa was one of the first German proponents of Renaissance humanism, made spiritual and political contributions in European history, and prepared the modern philosophy being the "first modern thinker". His works had a great influence on the further development of sciences.

Pascal, Blaise (1623–1662). Born in Clermont-Ferrand (Auvergne region, France). Educated by his father and private teachers. From his childhood he showed an amazing aptitude for mathematics and science. He was also literarily talented. He worked among others on mathematical and physical problems as well as in the philosophy. In 1646 he turned – after the first conversion – to religious problems. In 1654 he had an intense religious vision. His belief and religious commitment revitalized – for the next four years he was strongly connected with the Jansenist center in Port-Royal and led a life of an ascetic and mystic. However, he continued his philosophical, theological and mathematical investigations. In his correspondence with Fermat one finds the beginnings of the probability theory. The reading of one of Pascal's works during the visit to Paris encouraged in an essential way the young Leibniz to develop the differential calculus.

Peano, Giuseppe (1858–1932). Born in Spinetta near Turin in a rather poor environment. His mathematical abilities were discovered when he was in school. In 1876–1880 Peano studied mathematics at the University of Turin. He graduated in 1880 with high honors, after which the university employed him. In 1890 he was appointed as extraordinary and in 1895 as ordinary professor. In 1886–1901 he taught also at the Turin Military Academy. His scientific works belong to mathematical analysis, in particular to the differential and integral calculus, the foundations of mathematics as well as – at the end of his life – to comparative linguistic and attempts to develop an international language. Peano proposed an extensive symbolism for mathematics and logic. Many of the symbols proposed by him (eventually slightly modified) are today still used in logic and mathematics.

Peirce, Charles Sanders (1839–1914). Born in Cambridge, Massachusetts, USA as son of Benjamin Peirce, himself a professor of astronomy and mathematics at Harvard University and perhaps the first serious research mathematician in America. Educated by his father – who early discovered his abilities – in mathematics. The father even discussed with him in a critical way Kant's *Kritik der reinen Vernunft*. Peirce studied mathematics, astronomy, natural sciences and philosophy. His collaboration in geodetic tasks of his father led him in 1859 to the employment in various scientific capacities by the United States Coast Survey and its successor, the United States Coast and Geodetic Survey. This turned out to be very successful and productive and made him travel to Europe many times. In 1891, after the rejection of his work on experiments with the pendulum, Peirce resigned from the Coast Survey and never again held a regular employment. Giving – since 1864 – lectures in logic and the history of logic, Peirce developed his epistemology for which mathematics and the logic based on the theory of signs were important. In 1879 Peirce was appointed lecturer in logic at Johns Hopkins University in Baltimore from which he was dismissed in 1884 for not quite clear, probably social reasons – his difficult personality or the fact that his work was too far ahead of his time to be appreciated by the academic establishment of that day. He moved to Milford (Pennsylvania) where he – supported by some friends – worked as a private researcher doing some scientific and engineering consulting, and wrote much for meager pay, mainly encyclopedic dictionary entries and reviews. He died destitute in 1914. Peirce was an ingenious and difficult person. His scientific investigations were both theoretical as well as experimental. His philosophy was developed in the tension between a broad practical experience and high theoretical expertise. He is treated as the founder of the pragmatism. The modern science of signs, semiotics, has been founded by him. He left exceptionally extensive and scientifically deep works.

Plato (about 427–ca. 346 BC). Born into an aristocratic and influential family in Athens at a politically changeful period. He was a pupil of Socrates. After Socrates' execution Plato traveled to Italy, Sicily, Egypt and Cyrene. In 387 BC he established a new school in Athens, the famous Academy named after a holt in its neighbourhood. He contributed in an essential and seminal way to various scientific domains like metaphysics and epistemology, ethic, anthropology, political philosophy, cosmology, theory of art and philosophy of language. He was one of the most important and influential personalities in the intellectual history.

Poincaré, Henri (1854–1912). Born in the Cité Ducale neighborhood, Nancy, Meurthe-et-Moselle, into an influential family. He studied among others in École Normale Supérieure. In 1881 he was appointed as professor at the University of Paris (the Sorbonne) where he worked until his death. He was a versatile, important and influential mathematician, theoretical physicist, engineer and philosopher. His investigations indicated new ways in astronomy, geodesy, potential theory and quantum physics. He wrote also two novels (1879 and 1880) which unfortunately have been lost. In mathematics he contributed to topology – he was one of the founders of this discipline –, and to the theory of automorphic functions, differential equations and series. He provided the consistency proof of non-Euclidean geometry. In 1906 he established – almost at the same time as Einstein – the mathematical foundations of the special theory of relativity.

Proclus Diadochus (410–485). Proclus was one of the most important representatives and between 437 and 485 the leader of Athen's Neoplatonist school. He was an extremely versatile and productive scientific writer. He wrote more than fifty works some of which remained. He commented works of Aristotle and Plato as well as *Elements* by Euclid. In his *Commentary on the First Book of Euclid's 'Elements'* [285] he described and commented mathematics and the conceptions of his forerunners and developed an own Neoplatonist conception of mathematics.

Pythagoras (about 570–ca. 497 BC). Born on the Greek island of Samos, situated in the eastern Aegean. There is only uncertain information about his date of birth, and even less about his date of death. About 530 BC he founded in the Greek colony of Croton (today's Crotone, in Calabria) his own philosophical school in which he himself and later on his pupils – as Philolaos – taught mathematics, astronomy, philosophy and the theory of music as well as about religious, ethical and political subjects. The school of the Pythagoreans, pupils of Pythagoras, and his school existed after his death till the fourth century BC. Pythagoras himself left no writings.

Quine, Willard Van Orman (1908–2000). Born in Akron, Ohio, USA. In 1926–1930 he studied mathematics, philosophy and philology at Oberlin College and continued his studies till 1932 at Harvard University. He received the doctor's degree in philosophy from Harvard University in 1932. His thesis supervisor was Alfred North Whitehead. Afterwards he studied in Vienna, Prague and Warsaw where he had contacts with the Vienna Circle and the Warsaw School of Logic. There He met M. Schlick, O. Neurath, K. Gödel, R. Carnap, S. Leśniewski, J. Łukasiewicz and A. Tarski. Those contacts had a great influence on his future investigations. During World War II, Quine lectured on logic in Brazil, in Portuguese, and served in the United States Navy in a military intelligence role. He was professor at Harvard University since 1948. His main works belong to mathematical logic, set theory and the theory of science.

Robinson, Abraham (1918–1974). Born in Waldenburg (today: Wałbrzych in Poland) as Abraham Robinsohn. He studied from 1935 till 1939 at the Hebrew University in Jerusalem among others by Abraham Fraenkel, as well as at the Sorbonne in Paris. In the last moment he escaped from Paris when the Nazis invaded it in 1940. While in London, he joined the Free French Air Force. Since 1945 he was employed by the Royal Aircraft Establishment (RAe) in Farnborough. In 1949 he received the doctor's degree from the university in Jerusalem. After the war, Robinson worked in London, Toronto, and Jerusalem, but ended up at University of California, Los Angeles in 1962. He worked first of all in logic and model theory. In 1961 he constructed nonstandard models of the real numbers and provided in this way the mathematical background for the old infinitesimals of Leibniz. This provided also the fundament of the nonstandard analysis.

Russell, Bertrand (1872–1970). Born in Chapstow/Monmouthshire, into an influential and liberal family of the British aristocracy. In 1890–1894 he studied mathematics at Cambridge University. In 1908 he became a member of the Royal Society. In 1910–1916 Russell taught philosophy at Trinity College in Cambridge. In 1916 he was dismissed from Trinity College for his pacifist activities and in 1918 imprisoned for six months in Brixton prison. In 1919 he was reinstated but resigned in 1920. In 1920–1921 he taught philosophy in Peking. He returned to England in 1921. In 1938–1944 he lived in the USA where he taught logic and philosophy, since 1944 again at Cambridge University. He dealt with logic, mathematics and philosophy as well as ethics, education and social subjects. Beside his scientific activity he was engaged in political issues. In particular in 1969 he appealed to Secretary General U. Thant of the United Nations to support an international war crimes commission to investigate alleged torture and genocide by the United States in South Vietnam during the Vietnam War. In 1950 Russell got the Nobel Prize in Literature.

Socrates (469–399 BC). Born in Athens where he lived and worked his whole life. He developed the philosophical method of a dialogue, the so-called "Socratic dialogue". Nothing written by him remained. His special art of dialogue and his philosophy are conveyed to us only indirectly. Just his pupils – the most famous among them were Xenophon and Plato – wrote the Socratic dialogues. Accused of demoralizing the Athens youth by allegedly disregarding officially prescribed gods and the invention of new gods, he was sentenced to death. He refused to escape from prison and was executed. Socrates' activity marks the turning point in the Greek philosophy – from the cosmological philosophy of nature of his forerunners and the relativistic ethics of sophists to an anthropologically oriented ethics. All thinkers before him are called "pre-Socratic".

Tarski, Alfred (1901–1983). Born Alfred Tajtelbaum in Warsaw. In 1923 he changed his surname to "Tarski". In 1918 he began studying biology at the University of Warsaw. Influenced by Leśniewski he abandoned biology and enrolled in mathematics and philosophy. His teachers included Kotarbiński, Leśniewski, Łukasiewicz, Mazurkiewicz and Sierpiński. In 1924, he completed his doctorate under the supervision of Leśniewski; his *Habilitationsschrift* was entitled "O wyrazie pierwotnym logistyki" (On the Primitive Term of Logistic). He was habilitated a year later. In 1925–1939 he taught in the Stefan Żeromski Grammar School in Warsaw, and at the same time he held a temporary position of an associate professor at the University of Warsaw. In August 1939 he left for the USA to participate in the fifth International Congress for the Unity of Science. The outbreak of the war made him stay there. He lived in the USA till the end of his life (in 1946 the University of Warsaw wanted him to accept the post of associate professor but he refused the offer). He lectured at Harvard University (1939–1941), was a visiting professor at City College of New York (1940–1941) and a member of the Institute for Advanced Study in Princeton (1941–1942). He also lectured at the

University of California at Berkeley, where he was appointed a professor in 1946. There he created a strong research center for logic and the foundations of mathematics. He died in Berkeley, California. Tarski dealt with numerous domains of mathematics, in particular set theory, algebra, metamathematics and logic. He was one of the most outstanding logicians of the 20th century. One of his greatest achievements was the definition of the concept of truth. He initiated a new area of the foundations of mathematics – model theory. He conducted research on set theory and universal algebra. He launched the program of algebraization and topologization of mathematical logic. Tarski's works exerted a decisive influence on the development of mathematical logic and the foundations of mathematics. In 2000, the International Astronomical Union Commission named planetoid 13672, discovered in 1997, after Alfred Tarski.

Weyl, Hermann Klaus Hugo (1885–1955). Born in Elmshorn, a small town near Hamburg, in Germany. In 1904–1908 he studied mathematics – among others by Hilbert – as well as physics and philosophy in Göttingen and Munich. He received the doctor's degree in 1908 and the habilitation in 1910 in Göttingen where he then worked for three years as *Privatdozent*. In 1913–1930 he was professor at ETH in Zürich, and later, in 1930–1933, at the University of Göttingen as the successor of Hilbert. In 1933 he left Germany among others for political reasons – his wife Helene was Jewish – and went to the USA where he was appointed – on the initiative of Albert Einstein – as professor at the Institute for Advanced Study in Princeton. He worked in almost all domains of mathematics, especially in the theory of trigonometric and orthogonal series, period functions, complex functions, differential and integral equations, number theory, group theory and its applications in the differential geometry, in the theory of relativity and quantum mechanics. Weyl was engaged in the discussions on the foundations of mathematics and converted from the representative of mathematics based on set theory through the constructivistic position to the moderate formalism.

Whitehead, Alfred North (1861–1947). Born in Ramsgate, Kent, England, as a son of an anglican minister. In 1880 he began attending Trinity College, Cambridge, and studied mathematics. He earned his BA from Trinity in 1884. There he became fellow and assistant lecturer and in 1888 lecturer. In 1910 he left Cambridge and moved to London where he taught mathematics at University College. In 1914 he was appointed as professor at Imperial College of Science and Technology. In 1924 he went to the USA and became professor of philosophy at Harvard University. He worked in logic and mathematics, in physics, philosophy, in natural sciences and in the history of culture.

Wilder, Raymond Louis (1896–1982). Born in Palmer, Massachusetts, USA. In 1914 he entered the Brown University in Providence (Rhode Island) and studied there mathematics intending to become an actuary. He chose to do his Ph.D. at the University of Texas at Austin. There he discovered pure mathematics and topology. In 1923 he received the doctor's degree under the supervision of R. L. Moore. After a year as an instructor at

Texas, Wilder was appointed as assistant professor at the Ohio State University in 1924. In 1926 he joined the University of Michigan at Ann Arbor where he was appointed as professor in 1947. Retired from Michigan in 1967 Wilder became a research associate and occasional lecturer at the University of California at Santa Barbara. He was active in the American Mathematical Society. His main research domain was topology, but he later gradually acquired philosophical and anthropological interests propagating the culture-theoretical approach (cf. his *Mathematics as a Cultural System* [374] in 1981).

Wittgenstein, Ludwig Josef Johann (1889–1951). Born in Vienna. He studied at the Technische Hochschule in Charlottenburg in Berlin and at the Victoria University of Manchester. After graduating in 1908 his interests switched to philosophy and mathematics and he studied in 1911–1913 under Russell in Cambridge. Russell reckoned him as a genius and appreciated him as a partner in discussions. Wittgenstein read *Principia Mathematica*, worked on his own logical system and drafted his first philosophical work *Tractatus logico-philosophicus* on which he worked also in trenches during the First World War as an Austrian voluntary. *Tractatus* appeared only in 1921. This work strongly influenced the neopositivism. In 1920–1926 he worked (rather not successfully) as a primary school teacher near Vienna. He had contacts with the Vienna Circle and returned to his philosophical interests. In 1929 he was appointed as lecturer and in 1939 as professor in Cambridge. After the Second World War he resigned the professorship at Cambridge in 1947 to concentrate on his writing. Wittgenstein was the author of new approaches in the philosophy of logic and in the philosophy of language. Two small articles and *Tractatus* were his unique publications besides the posthumous *Philosophische Untersuchungen* [377] in 1953 and *Bemerkungen über die Grundlagen der Mathematik* [378] in 1956 in which he removed himself from his earlier ideas expressed in *Tractatus*.

Zermelo, Ernst Friedrich Ferdinand (1871–1953). Born in Berlin. In 1889 he started his study of mathematics, physics and philosophy at the universities of Berlin, Halle (Saale) and Freiburg. He finished his doctorate in 1894 at the University of Berlin, awarded for a dissertation on the calculus of variations. In 1897 he went to the University of Göttingen, then the leading center for mathematical research in the world, where he completed his habilitation thesis in 1899 – his *Habilitationsschrift* was devoted to hydrodynamics. In 1905 Zermelo was appointed as professor at the University of Göttingen. In 1910 he left Göttingen because he was appointed as the chair of mathematics at Zürich University, which he resigned in 1916 for health problems. He was appointed as an honorary chair at Freiburg im Breisgau in 1926. He was forced to resign this position in 1935 – he refused to begin his lectures by the Nazi's "Hitlergruß" and was denunciated by colleagues. At the end of World War II and at his request, Zermelo was reinstated in his honorary position in Freiburg. However, for his poor health he was unable to hold lectures. He worked on the calculus of variations and on the applications of the probability theory in the statistical physics. His mathematical investigations in set theory were seminal. In 1904 he formulated the Axiom of Choice and proved the famous well-ordering theorem stating that every set can be well-ordered. He was the author of the first system of axioms for set theory which – after being supplemented by the Replacement Axiom due to A. Fraenkel – is now the standard system.

Bibliography

[1] Aristoteles, *Physik*, Leipzig, 1829.
[2] Aristoteles, *Metaphysik*, Jena, 1907.
[3] Aristoteles, *Kleine Schriften zur Physik und Metaphysik*, Paderborn, 1957.
[4] Aristoteles, *Vom Himmel. Von der Seele. Von der Dichtkunst*, München, 1983.
[5] Aristoteles, *Prio Analytics. Translated by R. Smith*, Indianapolis, 1989.
[6] Aristoteles, *Topik*, Ditzingen, 2004.
[7] Aristoteles, *Physics*, Adealide, 2015.
[8] Aristoteles, *Posterior Analytics. Translated by G. R. G. Mure*, eBooks@Adelaide, Adelaide, 2016, https://ebooks.adelaide.edu.au/a/aristotle/a8poa/.
[9] Aristoteles *Metaphysics. Translated by W. D. Ross*, http://izt.ciens.ucv.ve/ecologia/Archivos/Filosofia-I/Aristotle.
[10] B. Artmann, *Der Zahlbegriff*, Göttingen, 1983.
[11] J. Barwise, *Handbook of Mathematical Logic*, Amsterdam, 1977.
[12] T. Bascelli, E. Bottazzi, F. Herzberg, V. Kanovei, K. U. Katz, M. G. Katz, T. Nowik, D. Sherry and S. Snider, Fermat, Leibniz, Euler and the Gang: The True History of the Concepts of Limit and Shadow, *Notices Amer. Math. Soc.* **61** (2014), no. 8, 848–864.
[13] T. Batóg, *Two Paradigms of Mathematics. A Study in the History and Philosophy of Mathematics*, Wydawnictwo Naukowe Uniwersytetu im. Adama Mickiewicza, Poznań, 1996.
[14] L. Bauer, Mathematik, Intuition, Formalisierung: Eine Untersuchung von Schülerinnen- und Schülervorstellungen zu $0,\overline{9}$, *J. Mathematikdidaktik* **32** (2011), no. 1, 79–102.
[15] O. Becker, Mathematische Existenz. Untersuchungen zur Logik und Ontologie mathematischer Phänomene, *Jahrbuch Philos. Phänemen. Forsch.* **8** (1927), 439–809.
[16] O. Becker, *Grundlagen der Mathematik in geschichtlicher Entwicklung*, Freiburg, 1954.
[17] O. Becker, *Zur Geschichte der griechischen Mathematik*, Darmstadt, 1965.
[18] T. Bedürftig, Archimedes, das Kontinuum und der Mathematikunterricht, in: *Konstruktionsprozesse und Mathematikunterricht – Festschrift für Bernd Zimmermann*, Hildesheim (2011), 33–58.
[19] T. Bedürftig, Über die Zahlengerade – historische und philosophische Anmerkungen, in: *Zeitläufte der Mathematik, Tagung zur Geschichte der Mathematik in Freising (1.6. bis 5.6. 2011), Reihe Algorithmus*, Augsburg (2012), 9–18.
[20] T. Bedürftig, *Zahlen und Zahlbegriff – Mathematisch-didaktische Studien*, Hildesheim, 2013.
[21] T. Bedürftig, Was ist ein Punkt? – Ein Streifzug durch die Geschichte, *Siegener Beiträge Gesch. Philos. Math.* **5** (2015), 1–21.
[22] T. Bedürftig, Über die Grundproblematik der Grenzwerte, *Math. Semesterber.* **65** (2018), no. 2, https://doi.org/10.1007/s00591-018-0220-0.
[23] T. Bedürftig and R. Murawski, *Zählen – Grundlage der elementaren Arithmetik*, Hildesheim, 2001.
[24] T. Bedürftig and R. Murawski, Alte und neue Ansichten über die Zahlen – aus der Geschichte des Zahlbegriffs, *Math. Semesterberichte* **51** (2004), 7–36; translation: From the history of the concept of number, in: Essays in the Philosophy and History of Logic and Mathematics, Amsterdam 2010, 95–122.
[25] E. Behrends, *Analysis I*, Braunschweig, 2003.
[26] E. Behrends, P. Gritzmann and G. M. Ziegler, *π und Co. – Kaleidoskop der Mathematik*, Berlin, 2008.
[27] E. T. Bell, *The Development of Mathematics*, 2nd ed., New York, 1945.

https://doi.org/10.1515/9783110468335-010

[28] P. Benacerraf and H. Putnam, *Philosophy of Mathematics: Selected Readings*, Englewood Cliffs, 1964; Selected readings, 2nd ed., Cambridge, 1983.

[29] K. Berka and L. Kreiser, *Logik-Texte, kommentierte Auswahl zur Geschichte der Logik*, Darmstadt, 1983.

[30] P. Bernays, A System of Axiomatic Set Theory, *J. Symb. Log.* **2** (1937), 65–77; 6 (1941), 1–17; 7 (1942), 65–89 u. 133–145; 8 (1943), 89–106; 13 (1948), 65–79; 19 (1954), 81-96; contained in: Sets and Classes, Amsterdam, 1976, 1–119.

[31] P. Bernays, *Abhandlungen zur Philosophie der Mathematik*, Darmstadt, 1976.

[32] E. W. Beth, H. J. Pos and H. J. A. Hollack, *Library of the Tenth International Congress in Philosophy, August 1948*, Amsterdam, 1949.

[33] A. Beutelspacher, *Albrecht Beutelspacher's Kleines Mathematikum – die 101 wichtigsten Fragen und Antworten zur Mathematik*, München, 2010.

[34] G. Biegel, K. Reich and T. Sonar, *Historische Aspekte im Mathematikunterricht an Schule und Universität*, Göttingen, 2008.

[35] H.-G. Bigalke, Zum Unendlichkeitsbegriff, *Praxis Math.* **12** (1968), 327–335.

[36] H.-G. Bigalke, Rekonstruktionen zur geschichtlichen Entwicklung des Begriffs der Inkommensurabilität, *J. Math. Didaktik* **4** (1983), 307–354.

[37] H. Boehme, Mathematik der Pythagoreer, in: *Wanderschaft in der Mathematik*, Augsburg (2006), 41–53.

[38] H. Boehme, Genetischer Aufbau der reellen Zahlen, in: *Historische Aspekte im Mathematikunterricht an Schule und Universität*, Göttingen (2008), 23–39.

[39] H. Boehme, Theaitetos – Proportionstheorie und Irrationale, in: *Eintauchen in die Mathematische Vergangenheit*, Augsburg (2011), 52–68.

[40] B. Bolzano, Rein analytischer Beweis des Lehrsatzes, daß zwischen je zwei Werten, die ein gegensätzliches Resultat gewähren, wenigstens eine reelle Wurzel der Gleichung liege, in: *Ostwald's Klassiker der exakten Wissenschaften Nr. 153*, Engelmann, Leipzg (1905), 1–36.

[41] B. Bolzano, *Wissenschaftslehre. Versuch einer ausführlichen und größtenteils neuen Darstellung der Logik mit steter Rücksicht auf deren bisherige Bearbeiter. Bd. I–IV*, Sulzbach, 1837.

[42] B. Bolzano, *Paradoxien des Unendlichen. Herausgegeben aus dem schriftlichen Nachlass von Fr. Přihonski*, Leipzig, 1851.

[43] B. Bolzano, *Paradoxes of the Infinite. Translated by D. A. Steele, Routledge and Kegan Paul*, London, 1950.

[44] G. Boole, *Mathematical Analysis of Logic, Being an Essay toward a Calculus of Deductive Reasoning (reprinted Oxford 1948, 1951)*, Cambridge, 1847.

[45] G. Boole, *An Investigation of the Laws of Thought on Which are Founded the Mathematical Theories of Logic and Probabilities (reprinted New York 1958)*, London, 1854.

[46] É. Borel, *Leçons sur la théorie des fonctions*, Paris, 1914.

[47] N. Bourbaki, *Élements d´ histoire des mathématiques*, Paris, 1969.

[48] J. Bouveresse, On the Meaning of the "Platonism" in the Expression "Mathematical Platonism", *Proc. Aristotelian Soc.* **105** (2005), 55–79.

[49] W. Breidert, *Das aristotelische Kontinuum in der Scholastik (2. verbesserte Auflage 1979)*, Münster, 1970.

[50] W. Breidert, Die nichteuklidische Geometrie bei Thomas Reid, *Sudhoffs Archiv Z. Wissenschaftsg.* **58** (1974), no. 3, 235–253.

[51] W. Breidert, Berkeleys Kritik an der Infinitesimalrechnung, *Stud. Leibnitiana Sonderh.* **14** (1984), 85–91.

[52] W. Breidert, *George Berkeley*, Basel, 1989.

[53] W. Breidert, Zur Effizienz der Mathematikgeschichte, in: *Mathematik im Fluß der Zeit*, Augsburg (2004), 220–236.

[54] L. E. J. Brouwer, *Intuitionisme en formalisme*, Groningen, 1912; also in Wiskundig Tijdschrift 9 (1912), 180–211, translation: Intuitionism and Formalism, Bull. Amer. Math. Soc. 20 (1913), 81–96.

[55] L. E. J. Brouwer, Consciousness, Philosophy and Mathematics, in: *Library of the Tenth International Congress in Philosophy. August 1948*, Amsterdam (1949), 1235–1249.

[56] J. R. Brown, *Philosophy of Mathematics - A Contemporary Introduction to the World of Proofs and Pictures*, 2nd ed., New York, 2008.

[57] M. Brunner, Didaktikrelevante Konzepte im Umfeld der Konzepte token und type, *J. Mathematikdid.* **34** (2013), no. 1, 53–72.

[58] M. Brunner, Diagrammatische Realität und Regelgebrauch, in: *Semiotische Perspektiven auf das Lernen von Mathematik*, Wiesbaden (2015), 9–32.

[59] W. Burkert, *Weisheit und Wissenschaft. Studien zu Pythagoras, Philolaos und Platon*, Nürnberg, 1962.

[60] H. J. Burscheid and H. Struve, *Mathematikdidaktik in Rekonstruktionen – Ein Beitrag zu ihrer Grundlegung*, Hildesheim, 2010.

[61] W. Büttemeyer, *Philosophie der Mathematik*, München, 2003.

[62] W. Büttemeyer, Philosophie in geometrischer Ordnung, in: *Philosophie in literarischen und ästhetischen Gestalten*, Oldenburg (2005), 15–33.

[63] G. Cantor, Über eine Eigenschaft des Inbegriffes aller reellen algebraischen Zahlen, *J. Reine Angew. Math.* **77** (1874), 258–262; in: Gesammelte Abhandlungen mathematischen und philosophischen Inhalts, Berlin (1932), 115–118.

[64] G. Cantor, Grundlagen einer allgemeinen Mannigfaltigkeitslehre, *Math. Ann.* **21** (1883), 545–591; in: Gesammelte Abhandlungen mathematischen und philosophischen Inhalts, Berlin (1932), 165–209.

[65] G. Cantor, Beiträge zur Begründung der transfiniten Mengenlehre, *Math. Ann.* **46** (1895), 481–512, 49 (1897), 207–246, in: Gesammelte Abhandlungen mathematischen und philosophischen Inhalts, Berlin (1932), 282–351.

[66] G. Cantor, *Gesammelte Abhandlungen mathematischen und philosophischen Inhalts*, Berlin, 1932.

[67] G. Cantor, *Contributions to the Founding of the Theory of Transfinite Numbers. Translated by Ph.E.B. Jourdain*, Dover Publications, New York, 1955.

[68] G. Cantor, *Briefe*, Berlin, 1991.

[69] G. Cantor, Foundations of a general theory of manifolds, in: *From Kant to Hilbert. A Source Book in the Foundations of Mathematics. Vol. 2*, Clarendon Press, Oxford (1996), 878–922.

[70] W. Capelle, *Die Vorsokratiker*, Stuttgart, 1968.

[71] R. Carnap, Die Antinomien und die Unvollständigkeit der Mathematik, *Monatsh. Math. Phys.* **41** (1934), 263–284.

[72] R. Carnap, Intellectual autobiography, in: *The Philosophy of Rudolf Carnap*, La Salle (1963), 3–84.

[73] E. Cassirer, *Philosophie der symbolischen Formen – Dritter Teil: Phänomenologie der Erkenntnis. Gesammelte Werke Bd. 13*, Hamburg, 2002.

[74] A. L. Cauchy, *Cours d'Analyse de l'École Polytechnique. Premier partie. Anaylse algébrique*, Paris, 1821.

[75] P. J. Cohen, Comments on the foundations of set theory, in: *Axiomatic Set Theory*, Proc. Sympos. Pure Math. 13, American Mathematical Society, Providence (1971), 9–16.

[76] R. Courant, David Hilbert zur Feier seines sechzigsten Geburtstages, *Naturw. Wochenschr. Fortschritte Naturw. Medizin Technik* **4** (1922), 65–101.

[77] R. Courant, *Vorlesungen über Differential- und Integralrechnung. 2 Bd.*, Berlin, 1927/1929.
[78] R. Courant and H. Robbins, *Was ist Mathematik*, Berlin, 2001.
[79] L. Couturat, *La logique de Leibniz*, Paris, 1901.
[80] H. B. Curry, *Outlines of a Formalist Philosophy of Mathematics*, Amsterdam, 1951.
[81] P. Damerow, Vorüberlegungen zu einer historischen Epistemologie der Zahlbegriffsentwicklung, in: *Der Prozeß der Geistesgeschichte, Studien zur ontogenetischen und historischen Entwicklung des Geistes*, Frankfurt am Main (1994), 248–322.
[82] P. Damerow, The material culture of calculation. A conceptual framework for an historical epistemology of the concept of number, Preprint 117, Max-Planck-Institut für Wissenschaftsgeschichte, Berlin, 1999.
[83] P. Damerow, Evolution of number systems, in: *International Encyclopedia of Social & Behavioral Sciences. Vol. 16*, Amsterdam (2001), 10753–10756.
[84] P. Damerow, R. K. Englund and H. J. Nissen, Die ersten Zahldarstellungen und die Entwicklung des Zahlbegriffs, *Spektrum Wiss.* **3** (1988), 46–55.
[85] R. Dedekind, *Stetigkeit und irrationale Zahlen*, Braunschweig, 1872.
[86] R. Dedekind, Continuity and irrational numbers, in: *Essays on the Theory of Numbers. Translated by W. W. Beman*, Dover Publications, New York (1963), 1–27.
[87] R. Dedekind, *Was sind und was sollen die Zahlen?*, Braunschweig, 1888.
[88] R. Dedekind, The nature and meaning of numbers, in: *Essays on the Theory of Numbers. Translated by W. W. Beman*, Dover, New York (1963), 44–115.
[89] R. Dedekind, *Gesammelte mathematische Werke*, Braunschweig, 1932.
[90] M. Dehn, Raum, Zeit, Zahl bei Aristoteles vom mathematischen Standpunkt aus, in: *Die Naturphilosophie des Aristoteles*, Darmstadt (1975), 199–218.
[91] O. Deiser, *Reelle Zahlen – Das klassische Kontinuum und die natürlichen Folgen*, Berlin, 2007/2008.
[92] O. Deiser, *Einführung in die Mengenlehre. 3. Auflage*, Berlin, 2010.
[93] H. de Monantheuil, *De puncto, primo geometriae principio, liber*, Leyden, 1600.
[94] R. Descartes, *Gespräch mit Burman (lateinisch-deutsch)*, Hamburg, 1982.
[95] R. Descartes, Discours de la méthode pour bien conduire sa raison et chercher la verité dans les sciences. Deutsche Übersetzung: Abhandlung über die Methode, richtig zu denken und die Wahrheit in der Wissenschaften zu suchen, in: *René Descartes philosophische Werke. Übersetzt, erläutert und mit einer Lebensbeschreibung des Descartes versehen von J. H. von Kirchmann*, L. Heimann, Berlin (1870), Abteilung 1, 19–83; English translation after http://www.gutenberg.org/files/59/59-h/59-h.htm.
[96] R. Descartes, *Regulae ad directionem ingenii. Regeln zur Ausrichtung der Erkenntniskraft (lateinisch-deutsch). Krit. regidiert, übersetzt und hrsg. von Heinrich Springmeyer, Lüder Gäbe und Hans Günter Zekl*, Hamburg, 1973.
[97] R. Descartes, *René Descartes philosophische Werke. Übersetzt, erläutert und mit einer Lebensbeschreibung des Descartes versehen von J. H. von Kirchmann*, L. Heimann, Berlin, 1870.
[98] W. Dörfler, Mathematische Objekte als Indizes in Diagrammen. Funktionen in der Analysis, in: *Sprache und Zeichen*, Hildesheim (2010), 25–48.
[99] G. Dux and U. Wenzel, *Der Prozeß der Geistesgeschichte, Studien zur ontogenetischen und historischen Entwicklung des Geistes*, Frankfurt am Main, 1994.
[100] K. Easwaran, Regularity and hyperreal credences, *Philos. Rev.* **123** (2014), no. 1, 1–41.
[101] H.-D. Ebbinghaus, H. Hermes, F. Hirzebruch, M. Koecher, K. Mainzer and A. Prestel und R. Remmert, *Zahlen*, Berlin, 1983.
[102] H.-D. Ebbinghaus, *Einführung in die Mengenlehre. 3. Auflage*, Mannheim, 1994.

[103] H.-D. Ebbinghaus, J. Flum and W. Thomas, *Einführung in die mathematische Logik. 5. Auflage*, Heidelberg, 2007.

[104] F. Engels, *Herrn Eugen Dühring's Umwälzung der Wissenschaft. Philosophie. Politische Ökonomie. Sozialismus*, Leipzig, 1878.; English translation: Herr Eugen Dühring's Revolution in Science, translated by Emile Burns, Stuttgart 1894.

[105] F. Engels, Dialektik der Natur, in: *Werke. Band 20*, Karl Dietz, Berlin (1962), 305–570.

[106] F. Engels, *Werke*, Karl Dietz, Berlin, 1962.

[107] M. Epple, Das Ende der Größenlehre: Grundlagen der Analysis 1860–1910, in: *Geschichte der Analysis*, Heidelberg (1999), 372–410.

[108] S. Ersch and G. Gruber, *Allgemeine Enzyklopädie der Wissenschaft und Künste, I. Section, Teil XC*, Leipzig, 1871.

[109] Euklid, *Die Elemente. Buch I–XIII. Hrsg. von Clemens Thaer*, Leipzig, 1933–1937; 6. unveränderter Nachdruck, Darmstadt 1975.

[110] Euklid, *The Elements: Books I-XIII – Complete and Unabridged. Translated by Sir Thomas Heath*, Barnes & Noble, New York, 2006.

[111] Euklid, *The Thirteen Books of Euclid's Elements Translation and Commentaries by Heath, Thomas L. in Three Volumes*, Dover, New York, 1956.

[112] W. Ewald, *From Kant to Hilbert: A Source Book in the Foundations of Mathematics. Vol. I, II*, Oxford, 1996.

[113] S. Feferman, Kurt Gödel: Conviction and Causation, Philosophia Naturalis 1984, a special issue. cf. [364]. Reprinted in [325].

[114] D. Fehling, *Materie und Weltbau in der Zeit der frühen Vorsokratiker. Wirklichkeit und Tradition*, Innsbruck, 1994.

[115] U. Felgner, *Mengenlehre*, Darmstadt, 1979.

[116] U. Felgner, Dichtung und Wahrheit – zur Geschichte der Antinomie des Lügners, in: *Mathematik – Logik – Philosophie, Ideen und ihre historischen Wechselwirkungen*, Frankfurt am Main (2012), 119–140.

[117] W. Felscher, *Naive Mengen und abstrakte Zahlen. Bd. I–III*, Zürich, 1978/79.

[118] H. Field, *Science Without Numbers*, Oxford, 1980.

[119] H. Fischer and S. Deschauer, *Zeitläufte der Mathematik. Tagung zur Geschichte der Mathematik in Freising (1.6. bis 5.6. 2011). Reihe Algorismus*, Augsburg, 2013.

[120] K. Flasch, *Was ist Zeit? Augustinus von Hippo. Das XI. Buch der Confessiones. Historisch-philosophische Studie (Text-Übersetzung-Kommentar)*, Frankfurt am Main, 2004.

[121] A. Fraenkel, Das Leben Georg Cantors, in: *Gesammelte Abhandlungen mathematischen und philosophischen Inhalts*, Berlin (1932), 452–483.

[122] G. Frege, *Begriffsschrift, eine der arithmetischen nachgebildete Formelsprache des reinen Denkens*, Halle, 1879.

[123] G. Frege, *Die Grundlagen der Arithmetik – Eine logisch-mathematische Untersuchung über den Begriff der Zahl*, Breslau, 1884.

[124] G. Frege, *Grundgesetze der Arithmetik – Begriffsschriftlich abgeleitet. Band 1 und 2*, Jena, 1893/1903.

[125] G. Frege, *Nachgelassene Schriften und wissenschaftlicher Briefwechsel. Hrsg. von H. Hermes, F. Kambartel und F. Kaulbach*, Hamburg, 1976.

[126] H. Friedrich, *Schü,lerinnen- und Schülervorstellungen vom Grenzwertbegriff beim Ableiten* Dissertation, Paderborn, 2001.

[127] T. Fritzlar, L. Haapasalo, F. Heinrich and H. Rehlich, *Konstruktionsprozesse und Mathematikunterricht – Festschrift für Bernd Zimmermann*, Hildesheim, 2011.

[128] G. Galilei, *Discorsi e dimostrazioni matematiche intorno a due nuove scienze attenenti alla meccanica e ai movimenti locali*, Leyden, 1638.

[129] C. F. Gauß, *Gesammelte Werke. Bd. I–XII*, Hildesheim, 1973.
[130] C. F. Gauß, *Gauß' Werke in Kurzfassung. Hrsg. von Karin Reich. Reihe Algorismus Heft 39*, Augsburg, 2002.
[131] C. F. Gauß – W. Bolyai Briefwechsel, in [316].
[132] A. George and D. J. Velleman, *Philosophies of Mathematics*, Malden, 2002.
[133] H. Gericke, *Mathematik in Antike und Orient*, Berlin, 1984.
[134] H. Gericke, *Mathematik im Abendland*, Berlin, 1990.
[135] K. Gödel, *Über die Vollständigkeit des Logikkalküls*, Ph.d. thesis, 1929; Nachdruck mit der englischen Übersetzung in Collected Works, New York (1986), 60–101.
[136] K. Gödel, Über formal unentscheidbare Sätze der "Principia Mathematica" und verwandter Systeme. I, *Monatsh. Math. Phys.* **38** (1931), 173–198.
[137] K. Gödel, *The Consistency of the Axiom of Choice and the Generalized Continuum Hypothesis*, Ann. of Math. Stud. 3, Princeton University Press, Princeton, 1940.
[138] K. Gödel, Russell's mathematical logic, in: *The Philosophy of Bertrand Russell*, New York (1944), 125–153.
[139] K. Gödel, What ist Cantor's continuum problem?, *Amer. Math. Monthly* **54** (1947), 515–525.
[140] K. Gödel, *Collected Works. Vol. I*, Oxford University Press, New York, 1986.
[141] K. Gödel, *Collected Works. Vol. III*, Oxford University Press, New York, 1995.
[142] K. Gödel, Is mathematics a syntax of language?, in: *Collected Works. Vol. III*, Oxford University Press, New York (1995), Version III 334–356, Version V 356–362.
[143] K. Gödel, Remarks before the Princeton Bicentennial Conference on Problems in Mathematics, in: *Collected Works. Vol. III*, Oxford University Press, New York (1995), 150–153.
[144] K. Gödel, Some basic theorems on the foundations of mathematics and their implications, in: *Collected Works. Vol. III*, Oxford University Press, New York (1995), 304–323.
[145] K. Gödel, The modern development of foundations of mathematics in the light of philosophy, in: *Collected Works. Vol. III*, Oxford University Press, New York (1995), 374–387.
[146] K. Gödel, The present situation in the foundations of mathematics, in: *Collected Works. Vol. III*, Oxford University Press, New York (1995), 45–53.
[147] K. Gödel, Undecidable Diophantine propositions, in: *Collected Works. Vol. III*, Oxford University Press, New York (1995), 164–175.
[148] N. D. Goodman, A world of individuals, in: *The Problem of Universals*, Notre Dame (1956), 198–210.
[149] N. D. Goodman, The knowing mathematician, *Synthese* **60** (1984), 21–38.
[150] N. D. Goodman, Mathematics as a natural science, *J. Symb. Log.* **55** (1990), 182–193.
[151] H. Griesel, Quantitative Messsysteme, ein Beitrag zu den Grundsatzfragen: Was ist quantitatives Messen? Wie hängen einerseits Messen und andererseits die Zahlen und Rechenoperationen zusammen?, NTAG im DIN, NA 152-01-56 AA N45.
[152] H. Griesel, Reform of the construction of the number system with reference to Gottlob Frege, *Math. Educ. ZDM* **39** (2007), no. 1–2, 31–38.
[153] L. Haaparanta, *Mind, Meaning and Mathematics*, Dordrecht, 1994.
[154] H. Hankel, *Theorie der complexen Zahlsysteme*, Leipzig, 1867.
[155] H. Hankel, Grenze, in: *Allgemeine Enzyklopädie der Wissenschaft und Künste. I. Section, Teil XC*, Leipzig (1871), 185–211.
[156] H. Hankel, Untersuchungen über die unendlich oft oszillierenden und unstetigen Funktionen, *Math. Ann.* **20** (1882), 63–112; in: Ostwald's Klassiker der exakten Wissenschaften Nr. 153, W. Engelmann, Leipzig (1905), 44–102.
[157] C. Hartshorn and A. W. Burks, *Collected Papers of Charles Sanders Peirce*, Cambridge, 1931–1958.

[158] J. L. Heiberg, *Archimedis opera omnia cum comentariis Eutocii. 2. Aufl. in 3 Bdn.*, Leipzig, 1910–1915; Nachdruck, Stuttgart, 1972..
[159] J. van Heijenoort, *From Frege to Gödel. A Source Book in Mathematical Logic. 1879–1931*, Cambridge, 1967.
[160] W. Hein and P. Ullrich, *Mathematik im Fluß der Zeit. Algorismus Bd. 44*, Augsburg, 2004.
[161] W. Heisenberg, *Der Teil und das Ganze – Gespräche im Umkreis der Atomphysik*, München, 1969.
[162] G. Hellman, *Mathematics without Numbers. Towards a Modal-Structured Interpretation*, Oxford, 1989.
[163] R. Hersh, Some proposals for reviving the philosophy of mathematics, *Adv. Math.* **31** (1979), 31–50.
[164] A. Heyting, Die intuitionistische Grundlegung der Mathematik, *Erkenntnis* **2** (1931), no. 1, 106–115.
[165] D. Hilbert, *Grundlagen der Geometrie*, Leipzig, 1899, 14. Auflage 1999; English translation: The Foundations of Geometry, authorized translation by E.J. Townsend, reprinted edition The Open Court Publishing House, La Salle, Illinois, 1950.
[166] D. Hilbert, Über den Zahlbegriff, *Jahresber. Dtsch. Math.-Ver.* **8** (1900), 180–184.
[167] D. Hilbert, Mathematische Probleme, *Arch. Math. Phys.* **1** (1901), 44–63, 213–237; Nachdruck in Gesammelte Abhandlungen, Berlin (1935), Bd. 3, 290–329.
[168] D. Hilbert, Logische Prinzipien des mathematischen Denkens, Vorlesungsnotizen von E. Hellinger, Mathematisches Institut, Georg-August-Universität Göttingen, 1905, unveröffentlichtes Manuskript.
[169] D. Hilbert, Über die Grundlagen der Logik und Arithmetik, in: *Verhandlungen des 3. Internationalen Mathematiker-Kongresses in Heidelberg*, Leipzig (1905), 174–185; English translation: On the foundations of logic and arithmetic, in: From Frege to Gödel. A Source Book in Mathematical Logic. 1879–1931, Cambridge (1967), 129–138.
[170] D. Hilbert, Prinzipien der Mathematik, Vorlesungsmitschrift von Paul Bernays, Mathematisches Institut, Georg-August-Universität Göttingen, 1917–18, unveröffentlichtes Manuskript.
[171] D. Hilbert, Axiomatisches Denken, *Math. Ann.* **78** (1918), 405–415; also in Gesammelte Abhandlungen. Bd. 3, Berlin (1935), 146–156.
[172] D. Hilbert, Neubegründung der Mathematik. Erste Mitteilung, *Abh. Math. Semin. Univ. Hambg.* **1** (1922), 157–177; in: Hilbertiana, fünf Aufsätze von David Hilbert, Wiss. Buchgesellschaft, Darmstadt, 1964.
[173] D. Hilbert, Über das Unendliche, *Math. Ann.* **95** (1925), 161–190; English translation: On the infinite, in: Philosophy of Mathematics: Selected Readings, Englewood Cliffs (1983), 134–151.
[174] D. Hilbert, Naturerkennen und Logik, *Naturwiss.* **18** (1930), 959–963; Nachdruck in: Gesammelte Abhandlungen. Bd. 3, Berlin (1935), 378–387.
[175] D. Hilbert, Probleme der Grundlegung der Mathematik, *Math. Ann.* **102** (1930), 1–9.
[176] D. Hilbert, Die Grundlegung der elementaren Zahlentheorie, *Math. Ann.* **104** (1931), 485–494; Nachdruck in: Gesammelte Abhandlungen. Bd. 3, Berlin (1935), 192–195.
[177] D. Hilbert, *Gesammelte Abhandlungen*, Berlin, 1935.
[178] D. Hilbert and W. Ackermann, *Grundzüge der theoretischen Logik*, Berlin, 1928.
[179] D. Hilbert and P. Bernays, *Grundlagen der Mathematik. Bd. I*, Berlin, 1934, Bd. II 1939.
[180] D. W. Hoffmann, *Grenzen der Mathematik – Eine Reise durch die Kerngebiete der mathematischen Logik*, Heidelberg, 2011.
[181] D. W. Hoffmann, *Die Gödelschen Unvollständigkeitssätze – Eine geführte Reise durch Kurt Gödels historischen Beweis*, Berlin, 2013.

[182] P. Howard and J. E. Rubin, *Consequences of the Axiom of Choice*, American Mathematical Society, Providence, 1998.
[183] K. Hülser, *Die Fragmente zur Dialektik der Stoiker. Bd. 3–4*, Stuttgart, 1987/88.
[184] D. Hume, *Enquiry Concerning Human Understanding*, London, 1777, deutsche Übersetzung: Eine Untersuchung über den menschlichen Verstand, hrsg. von R. Richter, Hamburg, 1964.
[185] E. Husserl, *Philosophie der Arithmetik. Psychologische und logische Untersuchungen*, Halle-Saale, 1891.
[186] E. Husserl, *Philosophie der Arithmetik. Mit ergänzenden Texten (1890–1901), Husserliana Band XII*, Springer, Dordrecht, 1970.
[187] E. Husserl, *Edmund Husserl Briefwechsel. Band VII: Wissenschaftlerkorrespondenz. Hrsg. von K. Schuhmann*, Springer, Dordrecht, 1994.
[188] E. Husserl, *Collected Works, Translated by D. Willard*, Dordrecht, 2003.
[189] E. Husserl, *Philosophy of Arithmetic. Psychological and Logical Investigations with Supplementary Texts from 1887–1901. Bd. 10 von Husserliana: Edmund Husserl - Collected Works*, Springer, Dordrecht, 2003.
[190] Husserl-Archiv Leuven, *Husserliana: Edmund Husserl Gesammelte Werke*, Den Haag, 1950, Berlin 2008.
[191] M. Hykšová and U. Reich, *Wanderschaft in der Mathematik*, Augsburg, 2006.
[192] M. Hykšová and U. Reich, *Eintauchen in die Mathematische Vergangenheit*, Augsburg, 2011.
[193] H. N. Jahnke, *Geschichte der Analysis*, Heidelberg, 1999.
[194] P. E. B. Jourdain, *Bolzano: Reelle Wurzeln, Hermann Hankel: Unstetige Funktionen. Ostwald's Klassiker der exakten Wissenschaften Nr. 153*, Leipzig, 1905.
[195] J. Jungius, Vom Prädikament der Relationen im allgemeinen, in: *Logik-Texte. Kommentierte Auswahl zur Geschichte der Logik*, Darmstadt (1983), 9–16.
[196] G. Kadunz, *Sprache und Zeichen*, Hildesheim, 2010.
[197] G. Kadunz, *Semiotische Perspektiven auf das Lernen von Mathematik*, Wiesbaden, 2015.
[198] A. Kanamori and M. Magidor, *The Evolution of Large Cardinals in Set Theory*, Lecture Notes in Math. 669, Springer, Berlin, 1978.
[199] I. Kant, *Critik der reinen Vernunft*, Riga, 1781; bzw. Kritik der reinen Vernunft, 2. Auflage, Riga 1787, Nachdruck des Felix Meiner Verlages, in der Ausgabe besorgt von Raymund Schmidt, 2. Auflage 1930, Nachdruck Hamburg 1956.
[200] I. Kant, *Prolegomena zu einer jeden künftigen Metaphysik, die als Wissenschaft wird auftreten können. Immanuel Kants Werke. Hg. v. Ernst Cassirer. Band III*, Berlin, 1913.
[201] I. Kant, *Critique of pure reason. Translated by P. Guyer and Allen W. Wood*, Cambridge University Press, Cambridge, 1998.
[202] I. Kant, *Prolegomena to any Future Metaphysics. Translated and Edited by G. Hatfield*, Cambridge University Press, Cambridge, 2004.
[203] I. Kant, *Prolegomena to Any Future Metaphysics. English Translation by James Fieser. Based on Paul Carus's 1902 translation*, http://philosophy.eserver.org/kant-prolegomena.txt.
[204] C. Kauferstein, *Transzendentalphilosophie der Mathematik*, Stuttgart, 2006.
[205] F. Kaufmann, *Das Unendliche in der Mathematik und seine Ausschaltung*, Leipzig, 1930.
[206] J. L. Kelley, *General Topology*, New York, 1955.
[207] K. L. Kettner and H. Putnam, *Das Denken und die Logik des Universums. Übersetzt und durch einen Anhang erweitert von H. Pape*, Frankfurt am Main, 2002.
[208] L. Kirby and J. Paris, Accessible independence results for Peano arithmetic, *Bull. Lond. Math. Soc.* **14** (1982), 285–293.
[209] F. Klein, *Elementarmathematik vom höheren Standpunkt aus. Bd. I und II*, Leipzig, 1908/1909.
[210] E. Kleinert, *Mathematik für Philosophen*, Leipzig, 2004.

[211] I. Kleiner, *Exkursions in the History of Mathematics*, New York, 2012.
[212] M. Kline, *Mathematical Thought From Ancient to Modern Times*, Oxford University Press, New York, 1972.
[213] C. Kloesel and H. Pape, *Charles S. Peirce. Semiotische Schriften. 3 Bände*, Frankfurt am Main, 2000.
[214] E. Köhler, Gödel und der Wiener Kreis, in: *Jour Fixe der Vernunft*, Wien (1991), 127–158; Hölder-Pichler-Tempsky. Auch in: E. Köhler, P. Weibel, M. Stöltzner, B. Buldt, C. Klein und W. Depauli-Schimanovich-Göttig (Hrsg.): Kurt Gödel. Wahrheit und Beweisbarkeit. Vol 1: Dokumente und historische Analysen, 83–108 (2002). Wien: öbt & hpt.
[215] A. Krazer, *Verhandlungen des dritten Internationalen Mathematiker-Kongresses in Heidelberg vom 8. bis 13. August 1904*, Leipzig, 1905.
[216] L. Kronecker, Über den Zahlbegriff, *J. Reine Angew. Math.* **101** (1887), 337–355.
[217] K. Kuhlemann, *Der Untergang von Mathemagica*, Heidelberg, 2015.
[218] K. Kuhlemann, Über die Technik der infiniten Vergrößerung und ihre mathematische Rechtfertigung, (2018), https://www.karlkuhlemann.net/start/forschung/.
[219] L. Kvasz, Sprache und Zeichen in der Algebra – ein Beitrag zur Theorie der Vergegenständlichung, *J. Math. Didaktik* **29** (2008), no. 2, 108–123.
[220] I. Lakatos, Proofs and refutations. The logic of mathematical discovery, *British J. Philos. Sci.* **14** (1963/64); als Buch veröffentlicht: Cambridge 1976; deutsche Übersetzung: Beweise und Widerlegungen. Die Logik mathematischer Entdeckungen. Hrsg. v. J. Worral und Elie Zahar, Braunschweig–Wiesbaden, 1979.
[221] I. Lakatos, A renaissance of empiricism in the recent philosophy of mathematics, in: *Problems in the Philosophy of Mathematics*, Amsterdam (1967), 199–202.
[222] I. Lakatos, *Problems in the Philosophy of Mathematics*, Amsterdam, 1967.
[223] G. Lakoff and R. E. Núñez, *Where Mathematics Comes from: How the Embodied Mind Brings Mathematics into Being*, New York, 2001.
[224] D. Landers and L. Rogge, *Nichtstandard Analysis*, Berlin, 1994.
[225] D. Laugwitz, *Infinitesimalkalkül – Eine elementare Einführung in die Nichtstandard-Analysis*, Mannheim, 1978.
[226] D. Laugwitz, *Zahlen und Kontinuum*, Mannheim, 1986.
[227] J. Legris, Universale Sprache und Grundlagen der Mathematik bei Ernst Schröder, in: *Mathematik – Logik – Philosophie, Ideen und ihre historischen Wechselwirkungen*, Frankfurt am Main (2012), 255–269.
[228] G. W. Leibniz, *Nouveaux essais sur l'entendement humain. 1704*; translation in: *Neue Abhandlungen über den menschlichen Verstand*, übers. und hrsg. von Ernst Cassirer, Hamburg. 1996.
[229] G. W. Leibniz, *Gesammelte Werke. Folge 3: Mathematik. Bd. 3–6. Hrsg. C. I. Gerhardt*, Halle, 1856–1860, Bd. 7 Halle 1863.
[230] G. W. Leibniz, *New Essays on Human Understanding. Translated by A.G. Langley*, Macmillan, London, 1896.
[231] G. W. Leibniz, *Die Philosophischen Schriften. Hrsg. C. J. Gerhardt, Nachdruck*, Hildesheim, 1960/1961.
[232] G. W. Leibniz, *Mathematische Schriften. Hrsg. C. J. Gerhardt, Nachdruck*, Hildesheim, 1971.
[233] G. Löffladt, *Mathematik – Logik – Philosophie, Ideen und ihre historischen Wechselwirkungen*, Frankfurt am Main, 2012.
[234] P. Lorenzen, Das Aktual-Unendliche in der Mathematik, *Philos. Natur.* **4** (1957), 4–11.
[235] J. Lützen, Grundlagen der Analysis im 19. Jahrhundert, in: *Geschichte der Analysis*, Heidelberg (1999), 191–244.
[236] P. Maddy, Perception and mathematical intuition, *Philos. Rev.* **89** (1980), 163–196.

[237] A. Marcja and C. Toffalori, *A Guide to Classical and Modern Model Theory*, Dordrecht, 2003.
[238] J. Mau, *Zum Problem des Infinitesimalen bei den antiken Atomisten*, Berlin, 1957.
[239] W. I. McLaughlin, Eine Lösung für Zenons Paradoxien, *Spektrum Wiss.* **1** (1995), 66–71.
[240] H. Meschkowski, *Mathematiker-Lexikon*, Mannheim, 1973.
[241] H. Meschkowski, Aus dem Briefwechsel Georg Cantors, *Arch. Hist. Ex. Sc.* **2** (1962–1966), 503–519.
[242] J. S. Mill, A System of Logic, Ratiocinative and Inductive. Being a Connected View of the Principles of Evidence and the Methods of Scientific Investigation. Deutsche Übersetzung: System der deductiven und inductiven Logik, in: *Gesammelte Werke. Bd. 2–4*, Leipzig (1872–1873).
[243] W. N. Molodschi, *Studien zu philosophischen Problemen der Mathematik*, Berlin, 1977.
[244] E. Moser Opitz, *Zählen – Zahlen – Zahlbegriff. 3. Auflage*, Bern, 2008.
[245] A. Mostowski, Widerspruchsfreiheit und Unabhängigkeit der Kontinuumshypothese, *Elemente Math.* **19** (1964), no. 6, 121–125.
[246] G. H. Müller, *Sets and Classes*, Amsterdam, 1976.
[247] R. Murawski, *Giuseppe Peano– Pioneer and Promoter of Symbolic Logic*, Komunikaty i Rozprawy Instytutu Matematyki Uniwersytetu im. Adama Mickiewicza, Poznań, 1985; also in: Essays in the Philosophy and History of Logic and Mathematics, Amsterdam (2010), 169–182.
[248] R. Murawski, Reverse Mathematik und ihre Bedeutung, *Math. Semesterber.* **40** (1993), 105–113.
[249] R. Murawski, Hilbert's program: Incompleteness theorems vs. partial realizations, in: *Philosophical Logic in Poland*, Kluwer Academic, Dordrecht (1994), 103–127.
[250] R. Murawski, *Recursive Functions and Metamathematics. Problems of Completeness and Decidability, Gödel's Theorems*, Dordrecht, 1999.
[251] R. Murawski, *Essays in the Philosophy and History of Logic and Mathematics*, Amsterdam, 2010.
[252] J. Mycielski and H. Steinhaus, A mathematical axiom contradicting the axiom of choice, *Bull. Acad. Polonaise Sci.* **10** (1962), 1–3.
[253] T. Nagel, *Mind and Cosmos: Why The Materialist Neo-Darwinian Coneception of Nature is Almost Certainly False*, Oxford, 2012.
[254] T. Nagel, *Geist und Kosmos – Warum die materialistische neodarwinistische Konzeption der Natur so gut wie sicher falsch ist*, Berlin, 2013.
[255] E. Nelson, A new aproach to nonstandard analysis, *Bull. Amer. Math. Soc.* **83** (1977), no. 6, 1165–1198.
[256] I. Newton, *Philosophiae Naturalis Principia Mathematica*, London, 1687.
[257] K. Oehler, *Charles Sanders Peirce*, München, 1993.
[258] F. Padberg, R. Danckwerts and M. Stein, *Zahlbereiche*, Heidelberg, 1995.
[259] H. Pape, *Charles S. Peirce – Zur Einführung*, Hamburg, 2004.
[260] J. Paris and L. Harrington, A mathematical incompleteness in peano arithmetic, in: *Handbook of Mathematical Logic*, Amsterdam (1977), 1133–1142.
[261] C. Parsons, Mathematical intuition, *Proc. Aristotelian Soc.* **80** (1980), 145–168; Abgedruckt in Mathematical Objects and Mathemtical Knowledge, Brookfield, 1995, 589–612.
[262] B. Pascal, *De l'esprit géométrique*, Paris, 1658.
[263] B. Pascal, *Pensées sur la religion et autres sujets* (1659), Paris, 1670; Deutsche Übersetzung: Gedanken über Religion und einige andere Themen. Hrsg. von Jean-Robert Armogathe. Übersetzt von Ulrich Kunzmann, Reclam Nr. 1623, Stuttgart 1997.

[264] B. Pascal, Betrachtungen über die Geometrie im allgemeinen – vom geometrischen Geist und von der Kunst zu überzeugen, in: *Kleine Schriften zur Religion und Philosophie*, Hamburg (2008).
[265] B. Pascal, *Kleine Schriften zur Religion und Philosophie. Latein/Deutsch. Hrsg. von Albert Raffelt*, Hamburg, 2008.
[266] G. Peano, *Arithmetices Principia Nova Methodo Exposita*, Torino, 1889.
[267] G. Peano, Démonstration de l'intégrabilité des equations differentielles ordinaires, *Math. Ann.* **37** (1890), 182–228.
[268] V. Peckhaus, Wozu Algebra der Logik? Ernst Schröders Suche nach einer universalen Theorie der Verknüpfungen, *Mod. Log.* **1** (1991), 174–205..
[269] C. S. Peirce, On the algebra of logic. A contribution to the philosophy of notation, *Amer. J. Math.* **1** (1884), 180–202.
[270] C. S. Peirce, *New Elements of Mathematics. Vols. I, II, III/1, III/2, IV. Ed. by Carolyn Eisele*, Mouton Publishers, Paris, 1976.
[271] J. Piaget, *Psychologie der Intelligenz*, Zürich, 1947.
[272] J. Piaget and A. Szeminska, *Die Entwicklung des Zahlbegriffs beim Kinde. Übersetzung der Originalausgabe "La genèse du nombre chez l'enfant" (Neuchâtel 1942) durch H. K. Weinert*, Stuttgart, 1965.
[273] J. Piaget, *Meine Theorie der geistigen Entwicklung*, Frankfurt am Main, 1983.
[274] Plato, *Meno. Translated by J. Holbo and B. Warring*, https://www.staff.ncl.ac.uk/joel.wallenberg/ContextsJoelGeoff/meno.pdf.
[275] Plato, *The Republic. Vol. II. Translated by P. Shorey*, London, 1932.
[276] Platon, *Werke. Hrsg. v. G. Eigler. 8 Bde.*, Darmstadt, 2005.
[277] H. Poincaré, *Science et l'hypothèse*, Paris, 1902; Deutsche Übersetzung: Wissenschaft und Hypothese, Berlin 1928.
[278] H. Poincaré, *Science and Hypothesis*, Walter Scott, London, 1905; https://www.gutenberg.org/files/37157/37157-pdf.pdf.
[279] H. Poincaré, *La valeur de la science*, Paris, 1905, Deutsche Übersetzung: Der Wert der Wissenschaft, Leipzig 1921.
[280] H. Poincaré, *Science et méthode*, Paris, 1908.
[281] H. Poincaré, *Dernières pensées*, Paris, 1913; Deutsche Übersetzung: Letzte Gedanken, Leipzig 1913.
[282] H. Poincaré, *Science and Method. Translated by Francis Maitland*, Thomas Nelson and Sons, London, 1914.
[283] H. Poincaré, *Wissenschaft und Methode*, Leipzig, 1914.
[284] A. Pringsheim, Grundlagen der allgemeinen Funktionenlehre, in: *Encyklopädie der mathematischen Wissenschaften mit Einschluss ihrer Anwendungen II A 1*, Leipzig, 1899, 1–53.
[285] Proklos Diadochos, *Kommentar zum ersten Buch von Euklids "Elementen". Aus dem Griechischen ins Deutsche übertragen und mit textkritischen Anmerkungen versehen von P. Leander Schönberger*, Halle(Saale), 1945.
[286] Proclus, *Commentary to the First Book of Euclid's Elements. Translated by Thomas Taylor*, London, 1788.
[287] H. Putnam, What Is Mathematical Truth?, in: *Mathematics, Matter and Method. Philosophical Papers. Vol. 1*, Cambridge (1975), 60–78.
[288] H. Putnam, *Mathematics, Matter and Method. Philosophical Papers. Vol. I*, Cambridge, 1975.
[289] W. V. O. Quine, New foundations of mathematicel logic, *Amer. Math. Monthly* **44** (1937), 70–80.

[290] W. V. O. Quine, *Mathematical Logic*, Cambridge, 1940, revisised edition 1951.
[291] W. V. O. Quine, On Carnaps views on ontology, *Philos. Stud.* **2** (1951).
[292] W. V. O. Quine, Two Dogmas of Empiricism, *Philos. Rev.* **60** (1951), no. 1, 20–43; also in: From a Logical Point of View, Cambridge 1953, 20–46.
[293] W. V. O. Quine, *From a Logical Point of View*, Cambridge, 1953.
[294] W. V. O. Quine, On what there is, in: *From a Logical Point of View*, Cambridge (1953), 1–19.
[295] K. Radbruch, *Bausteine einer Kulturgeschichte der Mathematik*, Leipzig, 2009.
[296] K. Radbruch, *Bausteine zu einer Kulturphilosophie der Mathematik*, Leipzig, 2009.
[297] K. Radbruch, Denken im Blick auf Leibniz: Das wissenschaftliche Werk von Christian Thiel, in: *Mathematik – Logik – Philosophie, Ideen und ihre historischen Wechselwirkungen*, Frankfurt am Main (2012), 23–36.
[298] C. Reid, *Hilbert*, Berlin, 1970.
[299] T. Reid, *An Inquiry into the Human Mind, on the Principles of Common Sense*, Edinburgh, 1764.
[300] M. D. Resnik, *Mathematical Objects and Mathemtical Knowledge*, Brookfield, 1995.
[301] M. D. Resnik, *Mathematics as Science of Patterns*, Oxford, 1997.
[302] A. Robinson, Non-standard analysis, *Indag. Math.* **23** (1961), 432–440.
[303] A. Robinson, *Non-standard Analysis*, Amsterdam, 1966.
[304] P. Ruben, *Philosophie und Mathematik*, Leipzig, 1979.
[305] B. Russell, Recent work in the philosophy of mathematics, *Internat. Monthly* **4** (1901), 83–101; Nachdruck als: Mathematics and the Metaphysicians, Mysticism and Logic and other Essays, London 1949.
[306] B. Russell, Mathematics and the metaphysicians, in: *Mysticism and Logic and other Essays. 8. Auflage*, London (1949), 74–96.
[307] B. Russell, *Mysticism and Logic and other Essays. 8. Auflage*, London, 1949.
[308] G. I. Ruzavin, *Die Natur der mathematischen Erkenntnis. Studien zur Methodologie der Mathematik*, Berlin, 1977.
[309] F. Schiller, *Friedrich Schiller – Schriften zur Philosophie und Kunst*, München, 1959.
[310] F. Schiller, Über naive und sentimentalische Dichtung, in: *Friedrich Schiller – Schriften zur Philosophie und Kunst*, München (1959), 150–221.
[311] F. Schiller, *Über die ästhetische Erziehung des Menschen in einer Reihe von Briefen. 23. Brief*, Stuttgart, 2004.
[312] P. A. Schilpp, *The Philosophy of Bertrand Russell*, New York, 1944.
[313] P. A. Schilpp, *The Philosophy of Rudolf Carnap*, La Salle, 1963.
[314] M. Schirm, *The Philosophiy of Mathematics Today (2. Auflage)*, New York, 2005.
[315] C. Schmieden and D. Laugwitz, Eine Erweiterung der Infinitesimalrechnung, *Math. Z.* **69** (1958), 1–39.
[316] , F. Schmidt and P. Stäckel, *Carl Friedrich Gauß – Wolfgang Bolyai Briefwechsel*, Leipzig, 1899; Carl Friedrich Gauß, Werke, Ergänzungsreihe Band II, Hildesheim–Zürich–New York 1987.
[317] I. Schneider, *Archimedes*, Darmstadt, 1979.
[318] E. Scholz, *Geschichte der Algebra – Eine Einführung*, Mannheim, 1990.
[319] H. Scholz, Leibniz und die mathematische Grundlagenforschung, *Jahresber. Dtsch. Math.-Ver.* (1943), 217–244.
[320] E. Schröder, *Vorlesungen über die Algebra der Logik. 3 Bände*, Leipzig, 1890–1895.
[321] R. Schulz, *Philosophie in literarischen und ästhetischen Gestalten*, Oldenburg, 2005.
[322] D. S. Scott, *Axiomatic Set Theory*, Proc. Sympos. Pure Math. 13 Part 1, American Mathematical Society, Providence, 1971.

[323] W. Sierpiński, Auswahlaxiom und Kontinuumshypothese, in: *Mengenlehre*, Darmstadt (1979), 143–158.
[324] G. A. Seek, *Die Naturphilosophie des Aristoteles*, Darmstadt, 1975.
[325] S. G. Shanker, *Gödel's Theorem in Focus*, London, 1988.
[326] S. Shapiro, Epistemic arithmetic and intuitionistic arithmetic, in: *Intensional Mathematics*, Amsterdam (1985).
[327] S. Shapiro, *Intensional Mathematics*, Amsterdam, 1985.
[328] S. Shapiro, *Philosophy of Mathematics. Structure and Ontology*, New York, 1997.
[329] S. G. Simpson, *Subsystems of Second Order Arithmetic*, Berlin, 1998.
[330] C. Smoryński, Hilbert's Programme, *CWI Qarterly* **1** (1988), 3–59.
[331] O. Spengler, *Der Untergang des Abendlandes. Bd. 1*, Wien, 1918, Bd. 2 München 1922, English translation: The Decline of the West, New York, vol. I – 1926, vol. II – 1928.
[332] P. Stekeler-Weithofer, *Formen der Anschauung – Eine Philosophie der Mathematik*, Berlin, 2008.
[333] A. Stephan and S. Walter, *Handbuch Kognitionswissenschaft*, Stuttgart, 2013.
[334] R. Strehl, *Zahlbereiche*, Hildesheim, 1996.
[335] A. Tarski, *Pojęcie prawdy w językach nauk dedukcyjnych*, Towarzystwo Naukowe Warszawskie, Warszawa, 1933.
[336] A. Tarski, *Über den Begriff der logischen Folgerung, Actes du Congrès International de Philosophie Scientifique. Bd. VII*, Paris, 1935.
[337] A. Tarski, Der Wahrheitsbegriff in den formalisierten Sprachen, *Stud. Philos. Comm. Soc. Philos. Polonorum* **1** (1936), 261–405.
[338] A. Tarski, *Einführung in die mathematische Logik und die Methodologie der Mathematik*, Wien, 1937; Zweite korrigierte Ausgabe: Einführung in die mathematische Logik, Göttingen 1966.
[339] R. Taschner, *Das Unendliche*, Berlin, 1995.
[340] C. Thiel, *Erkenntnistheoretische Grundlagen der Mathematik*, Hildesheim, 1982.
[341] C. Thiel, *Philosophie und Mathematik. Eine Einführung in ihre Wechselwirkungen und in die Philosophie der Mathematik*, Darmstadt, 1995.
[342] C. Thiel, Leibniz, Daedalus und der Faden der Ariadne, in: *Mathematik – Logik – Philosophie, Ideen und ihre historischen Wechselwirkungen*, Frankfurt am Main (2012), 41–58.
[343] R. Tieszen, Phenomenology and mathematical knowledge, *Synthese* **75** (1988), 373–403; abgedruckt in: Mathematical Objects and Mathemtical Knowledge, Brookfield (1995), 613–643.
[344] R. Tieszen, The philosophy of arithmetic: Frege and Husserl, in: *Mind, Meaning and Mathematics*, Dordrecht (1994), 85–112; Abgedruckt in: Phenomenology, Logic and the Philosophy of Mathematics, Cambridge (2005), 314–336.
[345] R. Tieszen, *Phenomenology, Logic and the Philosophy of Mathematics*, Cambridge, 2005.
[346] O. Toeplitz, Das Verhältnis von Mathematik und Ideenlehre bei Plato. Quellen und Studien zur Geschichte der Mathematik, Astronomie und Physik. Abteilung B: Studien. Band 1, in: *Zur Geschichte der griechischen Mathematik*, Darmstadt (1965), 1–33.
[347] A. Turing, Systems of logic based on ordinals, *Proc. Lond. Math. Soc. (2)* **45** (1939), 161–228.
[348] J. C. Tweddle, Weierstrass's construction of the irrational numbers, *Math. Semesterber.* **58** (2011), 47–58.
[349] B. L. van der Waerden, *Erwachende Wissenschaft*, Basel, 1956, 2. Auflage 1966.
[350] M. Väth, *Nonstandard Analysis*, Basel, 2007.
[351] N. von Cusanus, *Nicolae Cusae Cardinalis Opera*, Paris, 1514.
[352] N. von Kues, De docta ignorantia (1440), in: *Nicolae Cusae Cardinalis Opera*, Paris (1514).

[353] N. von Kues, De mathematica perfectione (1488), in: *Nicolae Cusae Cardinalis Opera*, Paris (1514).
[354] N. von Kues, Liber de mente (1448), in: *Nicolae Cusae Cardinalis Opera*, Paris (1514).
[355] J. von Neumann, Eine Axiomatisierung der Mengenlehre, *J. Reine Angew. Math.* **154** (1925), 219–240.
[356] J. von Neumann, Die Axiomatisierung der Mengenlehre, *Math. Z.* **27** (1928), 669–752.
[357] P. Vopěnka, *Mathematics in Alternative Set Theory*, Leipzig, 1979.
[358] P. Vopěnka and P. Hájek, *The Theory of Semisets*, Amsterdam, 1972.
[359] R. Wagner, *Gespräche mit Carl Friedrich Gauß in den letzten Monaten seines Lebens. Hrsg. von Heinrich Rubner (Nachrichten der Akademie der Wissenschaften Göttingen/Philologisch-Historische Klasse, 1975, 6)*, Göttingen, 1975.
[360] H. Wang, *From Mathematics to Philosophy*, London, 1974.
[361] H. Wang, Some facts about K. Gödel, *J. Symb. Log.* **46** (1981), 653–659.
[362] H. Wang, *Reflections on Kurt Gödel*, Cambridge, 1987.
[363] H. Wang, *A Logical Journey. From Gödel to Philosophy*, Cambridge, 1996.
[364] P. Weingartner, *Philosophy of Science — History of Science. A Selection of Contributed Papers of the 7th International Congress of Logic, Methodology and Philosophy of Science*, Salzburg, 1983, Meisenheim/Glan (Verlag Anton Hain).
[365] H. Weyl, *Das Kontinuum. Kritische Untersuchungen über die Grundlagen der Analysis*, Leipzig, 1918.
[366] H. Weyl, Über die neue Grundlagenkrise der Mathematik (Vorträge, gehalten im mathematischen Kolloquium Zürich), *Math. Z.* **10** (1921), 39–79; Nachdruck Darmstadt 1965.
[367] H. Weyl, *Philosophie der Mathematik und Naturwissenschaft*, München, 1928.
[368] H. Weyl, *Raum, Zeit, Materie. Vorlesungen über allgemeine Relativitätstheorie. 8. Auflage*, Wien, 1993.
[369] H. Weyl, *Gesammelte Abhandlungen. Bd. 1–4*, Heidelberg, 1968.
[370] A. N. Whitehead and B. Russell, *Principia Mathematica. Vol. I*, Cambridge, 1910, Vol. II 1912, Vol. III 1913.
[371] W. Wieland, Das Kontinuum in der Aristotelischen Physik, in: *Die Naturphilosophie des Aristoteles*, Darmstadt (1975), 261–300.
[372] E. P. Wigner, The unreasonable effectiveness of mathematics in the natural sciences, *Comm. Pure Appl. Math.* **13** (1960), 1–14.
[373] R. L. Wilder, *The Evolution of Mathematical Concepts. An Elementary Study*, New York, 1968.
[374] R. L. Wilder, *Mathematics as a Cultural System*, Oxford, 1981.
[375] T. Wilholt, *Zahl und Wirklichkeit – Eine philosophische Untersuchung über die Anwendbarkeit der Mathematik*, Paderborn, 2004.
[376] L. Wittgenstein, *Tractatus Logico-philosophicus*, New York, 1922.
[377] L. Wittgenstein, *Philosophical Investigations*, Oxford, 1953.
[378] L. Wittgenstein, *Remarks on the Foundations of Mathematics*, Oxford, 1956.
[379] H. Wunderling, *Analysis – als Infinitesimalrechnung*, Duden Schulbuchverlag, Berlin, 2007.
[380] H. Wußing, *Die Genesis des abstrakten Gruppenbegriffs*, Berlin, 1969.
[381] E. Zermelo, Beweis, daß jede Menge wohlgeordnet werden kann, *Math. Ann.* **59** (1904), 514–516.
[382] E. Zermelo, Neuer Beweis für die Möglichkeit einer Wohlordnung, *Math. Ann.* **65** (1908), 107–128.
[383] E. Zermelo, Untersuchungen über die Grundlagen der Mengenlehre I, *Math. Ann.* **65** (1908), 261–281.
[384] R. Ziegler, *Zahlen und ihre Struktur*, Würzburg, 2013.

[385] B. Zimmermann, *Heuristik als ein Element mathematischer Denk- und Lernprozesse. Fallstudien zur Stellung mathematischer Heuristik im Bild von Mathematik bei Lehrern und Schülern sowie in der Geschichte der Mathematik*, Habilitationsschrift, City Hamburg, 1991.
[386] J. Zink, *Kontinuum und Konstitution der Wirklichkeit*, Dissertation, Ludwig-Maximilians-Universität, München, 2004.

Index of names

Italic refers to places/pages where more detailed information on persons or their works is provided.

https://doi.org/10.1515/9783110468335-011

Index of symbols

Symbols are given in the order of their first appearance in the book. Italic indicates places where definitions or explanations of symbols are given.

https://doi.org/10.1515/9783110468335-012

Index of subjects

Italic refers to places where definitions, explanations or comments are provided.

https://doi.org/10.1515/9783110468335-013

Also of interest

Philosophie der Mathematik
Thomas Bedürftig, Roman Murawski, 2015
ISBN 978-3-11-033117-2, e-ISBN (PDF) 978-3-11-033118-9,
e-ISBN (EPUB) 978-3-11-038226-6

Relativistic World
Volume 1: Mechanics
Sergey Stepanov, 2018
ISBN 978-3-11-051587-9, e-ISBN (PDF) 978-3-11-051588-6,
e-ISBN (EPUB) 978-3-11-051600-5

Symmetry
Through the Eyes of Old Masters
Emil Makovicky, 2016
ISBN 978-3-11-041705-0, e-ISBN (PDF) 978-3-11-041714-2,
e-ISBN (EPUB) 978-3-11-041719-7

Metamathematics and the Philosophical Tradition
William Boos, 2018
ISBN 978-3-11-057221-6, e-ISBN (PDF) 978-3-11-057245-2,
e-ISBN (EPUB) 978-3-11-057239-1

Infinite Series in a History of Analysis
Stages up to the Verge of Summability
Hans-Heinrich Körle, 2015
ISBN 978-3-11-034372-4, e-ISBN (PDF) 978-3-11-035983-1,
e-ISBN (EPUB) 978-3-11-039916-5

www.ingramcontent.com/pod-product-compliance
Lightning Source LLC
Chambersburg PA
CBHW061153240926
51846CB00006B/998